The Most Successful Way of JEE Preparation

DPP

Daily Practice Problems

JEE Main & Advanced

Permutation-Combinations & Probability

MATHEMATICS Vol-2

AF529344

The Most Successful Way of JEE Preparation

DPP

Daily Practice Problems

JEE Main & Advanced

Permutation-Combinations & Probability

MATHEMATICS Vol-2

Amit M Agarwal

arihant
ARIHANT PRAKASHAN (SERIES), MEERUT

DPP
Daily Practice Problems

arihant

ARIHANT PRAKASHAN (SERIES), MEERUT

All Rights Reserved

© Author

No part of this publication may be re-produced, stored in a retrieval system or by any means, electronic mechanical, photocopying, recording, scanning, web or otherwise without the written permission of the publisher. Arihant has obtained all the information in this book from the sources believed to be reliable and true. However, Arihant or its editors or authors or illustrators don''t take any responsibility for the absolute accuracy of any information published, and the damages or loss suffered thereupon.

All disputes subject to Meerut (UP) jurisdiction only.

Administrative & Production Offices

Corporate Office 'Ramchhaya' 4577/15, Agarwal Road, Darya Ganj, New Delhi -110002
Tele: 011- 47630600, 43518550; Fax: 011- 23280316

Head Office Kalindi, TP Nagar, Meerut (UP) - 250002
Tel: 0121-2401479, 2512970, 4004199; Fax: 0121-2401648

Sales & Support Offices

Agra, Ahmedabad, Bengaluru, Bareilly, Chennai, Delhi, Guwahati, Hyderabad, Jaipur, Jhansi, Kolkata, Lucknow, Meerut, Nagpur & Pune

ISBN : 978-93-13193-50-0

Published by Arihant Publications (I) Ltd.

Production Team

Publishing Managers	Mahendra Singh Rawat & Keshav Mohan	*Inner Designer*	Mazher Chaudhary
Project Head	Sorabh Chaudhary	*Page Layouting*	Sunil Sharma, Hemant
Cover Designer	Shanu Mahad	*Proof Reading*	Himanshu

For further information about the products from Arihant, log on to www.arihantbooks.com or email to info@arihantbooks.com

/arihantpub /@arihantpub Arihant Publications /arihantpub

ABOUT DPP

Daily Practice Problems

Indian Institutes of Technology better known as IITs have always been the dream destination of each science group students who want to pursue their career as Engineers. These are the institutes which have got very high value not even in India but in the whole world. To get the admission in such institutes one need to face entrance Examinations; JEE Main & JEE Advanced, out of which JEE Main is considered as the preliminary exam & JEE Advanced is the final exam to get into the IITs. Approximately per year 12 lakh students enroll themselves for JEE Main and only the top 2.25 lakh in JEE Main merit can attempt for JEE Advanced, the fact clearly express the degree of toughness of JEE Main & JEE Advanced.

To have success in such an exam one needs to be very planned & systematic in its preparation. A very popular wise saying 'Rome was not built in a day' suggests that to achieve a big target like IITs we need to work continuously. For continuous work, one need to have proper planning that how much work he/she has to do on regular basis, i.e., on daily basis. Each student preparing for JEE has many books but unfortunately none of them gives the students a plan for daily studies, which is the most effective success key for JEE.

Keeping the daily practice needs of the students in mind we have come up with DPP Series (Daily Practice Problems) having approximately 26 books. The primary focus of this series is on achieving success through practice & proper planning. For that we have divided the whole syllabus of JEE Main & JEE Advanced; Physics, Chemistry & Mathematics, into different volumes (around 8 volumes of Mathematics & 9 volumes each of Physics & Chemistry), each volume has Daily Practice Problems Sheets having specific questions on various Topics of the individual chapter of the volume ensuring the complete Practice & Assessment of the topic.

DPP

Daily Practice Problems

In this way we have tried to give you an effective daily planning for JEE keeping in mind the daily hours that a student can devote to his practice, you are just required to practice one sheet of each subject on daily basis. This habit of daily practice will keep you away from exam stress and thoughts like; how much I have prepared/am I preparing properly or not/ do I need to study more on daily basis or what I am doing is ok, etc., which regularly tease your mind and can give you panic at any stage of your preparation.

Some special features that we have tried to incorporate in our Daily Practice Problems (DPP) series are discussed below

1. Micro Level coverage of all the topics of a Chapter/ Unit given in JEE Main & Advanced Syllabus.
2. Each DPP has various types of questions; Subjective Questions, Single/ Multiple Correct Options, Statements type, Comprehension based, Integer Value Answer type, Matching, etc., making you practice & ready for all the question formats of JEE.
3. New & innovative questions in each DPP challenging & improving your problem solving skills.
4. Along with Topical Coverage, Revisal DPPs for JEE Main & JEE Advanced with each Chapter, covering multidisciplinary questions proving comprehensive practice.
5. JEE Main & JEE Advanced Archive (collection of last 13 years' exams questions) with each chapter.
6. Complete Solutions for each DPP.

We hope that DPP series will prove to be very beneficial for the students preparing for JEE, by giving them daily practice in a systematic & planned manner. The suggestions/criticism from the students, teachers & parents about DPP series will be welcomed whole heartedly and we promise to update this series accordingly from time to time.

DPP

Daily Practice Problems

DAILY TASKS...

1 BINOMIAL THEOREM

2 PERMUTATION AND COMBINATION

DPP

Daily Practice Problems

3 PROBABILITY

1. Binomial Theorem

DPP

DAILY PRACTICE PROBLEMS

DPP-1 Binomial for Positive Index, General Term and Middle Term

Subjective Questions

Working Space

Direction *(Q. Nos. 1-3) These questions are subjective in nature, need to be solved completely on notebook.*

1 i. Find the 7th term of $\left(3x^2 - \frac{1}{3}\right)^{10}$.

ii. Find the term independent of x in the expansion $\left(2x^2 - \frac{3}{x^3}\right)^{25}$.

iii. Find the middle term in the expansion of

a. $\left(\frac{2x}{3} - \frac{3}{2x}\right)^6$ b. $\left(2x^2 - \frac{1}{x}\right)^7$

2 Find the coefficient of

i. x^7 in $\left(ax^2 + \frac{1}{bx}\right)^{11}$ ii. x^{-7} in $\left(ax - \frac{1}{bx^2}\right)^{11}$

3 Find the term independent of x in

i. $\left(2x - \frac{1}{\sqrt{x}}\right)^{15}$ ii. $\left(\frac{4x^2}{3} - \frac{3}{2x}\right)^9$

Only One Option Correct Type

Direction *(Q. Nos. 4-30) This section contains 27 multiple choice questions. Each question has four choices (a), (b), (c) and (d), out of which ONLY ONE option is correct.*

4 The term independent of x in $(1+x)^m \left(1+\frac{1}{x}\right)^n$, is

a. ${}^{m+n}C_m$ b. ${}^{m+n}C_{n+1}$
c. ${}^{m+n}C_{m-n}$ d. None of these

5 The $(p+2)$th term from the end in $\left(x - \frac{1}{x}\right)^{2n+1}$,is

a. $(-1)^p \cdot {}^{2n+1}C_{2n-p} \cdot x^{2p-2n+1}$ b. $(-1)^p \cdot {}^{2n+1}C_{2n-p} \cdot x^{2n-2p+1}$
c. $(-1)^p \cdot {}^{2n+1}C_p \cdot x^{2p-2n-1}$ d. None of these

6 The coefficient of x^{20} in the expansion of $(1+x^2)^{40} \cdot \left(x^2 + 2 + \frac{1}{x^2}\right)^{-5}$, is

a. ${}^{30}C_{10}$ b. ${}^{30}C_{25}$
c. 1   d. None of these

Working Space

7 The ratio of the coefficient of x^{10} in $(1-x^2)^{10}$ and the term independent of x in $\left(x-\frac{2}{x}\right)^{10}$, is

a. 1 : 16
b. 1 : 32
c. 1 : 64
d. None of these

8 The coefficient of x^3 in the expansion of $(1-x+x^2)^5$, is

a. 10
b. 8
c. –50
d. –30

9 If $(1+2x+3x^2)^{10} = a_0 + a_1x + a_2x^2 + \cdots + a_{20}x^{20}$, then a_1 equals to

a. 10
b. 20
c. 210
d. None of these

10 The coefficient of x^5 in the expansion of $(1+x)^{21} + (1+x)^{22} + \cdots + (1+x)^{30}$, is

a. ${}^{51}C_5$
b. ${}^{9}C_5$
c. ${}^{31}C_6 - {}^{21}C_6$
d. ${}^{30}C_5 + {}^{20}C_5$

11 The number of terms in the expansion of $\left(x^2+1+\frac{1}{x^2}\right)^n$, $n \in N$ is

a. $2n$
b. $n+1$
c. $(2n+1)$
d. $3n+1$

12 If in the expansion of $\left(\sqrt[3]{2}+\frac{1}{\sqrt[3]{3}}\right)^n$, the ratio of the seventh term from the beginning to the seventh term from end is equal to 1/6, then n is equal to

a. 3
b. 6
c. 9
d. None of these

13 The middle term in the expansion of $\left(x^2+\frac{1}{x^2}+2\right)^n$, is

a. $\frac{n!}{((n/2)!)^2}$
b. $\frac{(2n)!}{((n/2)!)^2}$
c. $\frac{1\cdot 3\cdot 5\cdots(2n-1)}{n!}\cdot 2^n$
d. $\frac{1\cdot 3\cdot 5\cdots(2n+1)}{n!}\cdot 2^n$

14 The number of terms which are free from radical signs in the expansion of $(y^{1/5}+x^{1/10})^{55}$, is

a. 5
b. 6
c. 7
d. None of the above

15 The coefficient of the term independent of x in the expansion of $\left(\frac{x-1}{x^{2/3}-x^{1/3}+1}-\frac{x-1}{x-x^{1/2}}\right)^{10}$, is

[JEE Main 2013]

a. 210
b. 105
c. 70
d. 112

16 If in the expansion of $(1+x)^m(1-x)^n$, the coefficients of x and x^2 and 3 and –6 respectively, then m is

a. 6
b. 9
c. 12
d. 24

Working Space

17 The value of $\sum_{r=0}^{n}(-1)^r \cdot {}^nC_r \dfrac{1+r\log_e 10}{(1+\log_e 10^n)^r}$, equals to

a. 1
b. −1
c. n
d. 0

18 If $\dfrac{1}{\sqrt{4x+1}}\left\{\left(\dfrac{1+\sqrt{4x+1}}{2}\right)^n - \left(\dfrac{1-\sqrt{4x+1}}{2}\right)^n\right\} = a_0 + a_1x + \cdots + a_5x^5$, then n is equal to

a. 11
b. 9
c. 10
d. None of these

19 The coefficient of x^{53} in $\sum_{r=0}^{100} {}^{100}C_r\,(x-3)^{100}\cdot 2^r$, is

a. ${}^{100}C_{51}$
b. ${}^{100}C_{52}$
c. $-{}^{100}C_{53}$
d. ${}^{100}C_{54}$

20 The coefficient of middle term in the expansion of $(1+x)^n$ when n is even, is

a. $\dfrac{1\cdot3\cdot5\cdots n}{2\cdot4\cdot6\cdots(n+1)}\cdot 2^n$
b. $\dfrac{1\cdot3\cdot5\cdots(n-1)}{2\cdot4\cdot6\cdots n}\cdot 2^n$
c. $\dfrac{2\cdot4\cdot6\cdots n}{1\cdot3\cdot5\cdots(n-1)}\cdot 2^n$
d. None of these

21 If the coefficients of pth, $(p+1)$th and $(p+2)$th terms in the expansion of $(1+x)^n$ are in AP, then

a. $n^2 - 2np + 4p^2 = 0$
b. $n^2 - n(4p+1) + 4p^2 - 2 = 0$
c. $n^2 - n(4p+1) + 4p^2 = 0$
d. None of the above

22 If 6th term in the expansion of $\left(\dfrac{1}{x^{8/3}} + x^2\log_{10} x\right)^8$ is 5600, then x is equal to

a. 8
b. 9
c. 10
d. None of these

23 If in the expansion of $(1+x)^n$, a, b and c are three consecutive coefficients, then n is equal to

a. $\dfrac{ac+ab+ac}{b^2+ac}$
b. $\dfrac{2ac+ab+bc}{b^2-ac}$
c. $\dfrac{ab+ac}{b^2-ac}$
d. None of these

24 In the expansion of $(5^{1/2} + 7^{1/8})^{1024}$, the number of integral terms is

a. 128
b. 129
c. 130
d. 131

25 The larger of $99^{50} + 100^{50}$ and 101^{50}, is

a. $99^{50} + 100^{50}$
b. Both are equal
c. 101^{50}
d. None of the above

Working Space

26 If n is positive integer, then the sum of
$$\left({}^nC_0 - {}^nC_1 \cdot \frac{(1+x)}{(1+nx)} + {}^nC_2 \cdot \frac{(1+2x)}{(1+nx)^2} - {}^nC_3 \frac{(1+3x)}{(1+nx)^3} + \cdots\right)$$ is equal to

a. 0

b. $2 \cdot \left(\frac{nx}{1+nx}\right)^n$

c. $\left(\frac{2nx}{1+nx}\right)^n$

d. None of these

27 If $(1+x-2x^2)^6 = a_0 + a_1x + a_2x^2 + \cdots + a_{12}x^{12}$, then value of $a_2 + a_4 + a_6 + \cdots + a_{12}$, is

a. 1024

b. 64

c. 32

d. 31

28 If $(1+x+x^2)^n = (a_0 + a_1x + a_2x^2 + \cdots + a_{2n}x^{2n})$, then $(a_0 + a_3 + a_6 + \cdots)$ equals to

a. 3^n

b. 3^{n-1}

c. 2^{n-1}

d. None of these

29 If $T_0, T_1, T_2, \cdots, T_n$ represents the terms in the expansion of $(x+a)^n$, then the value of $(T_0 - T_2 + T_4 - T_6 \cdots)^2 + (T_1 - T_3 + T_5 \cdots)^2$, is

a. $(x^2 - a^2)^n$

b. $(x^2 + a^2)^n$

c. $(a^2 - x^2)^n$

d. None of these

30 If $(3 + x^{2008} + x^{2009})^{2010} = a_0 + a_1x + a_2x^2 + \cdots + a_nx^n$, then value of
$$a_0 - \frac{a_1}{2} + \frac{a_2}{2} + a_3 - \frac{a_4}{2} - \frac{a_5}{2} + a_6 + \cdots,$$ is

a. 3^{2010}

b. 1

c. 2^{2010}

d. None of these

DPP-2 Divisibility, Remainder, Last Digits, Comparison of Numbers and $(A+\sqrt{B})^n$

Subjective Question

Working Space

Direction *(Q. No. 1) This question is subjective in nature, need to be solved completely on notebook.*

1 If n is a positive integer, show that

i. $9^n + 7$ is divisible by 8

ii. $3^{2n+1} + 2^{n+2}$ is divisible by 7

iii. $6^{2n} - 35n - 1$ is divisible by 1225

Only One Option Correct Type

Direction *(Q. Nos. 2-20) This section contains 19 multiple choice questions. Each question has four choices (a), (b), (c) and (d), out of which ONLY ONE option is correct.*

2 The remainder when 2^{100} is divided by 7, is

a. 1 b. 2
c. 3 d. 4

3 The remainder when 5^{99} is divided by 13, is

a. 10 b. 11
c. 12 d. 8

4 The remainder when 7^{103} is divided by 25, is

a. 15 b. 16
c. 17 d. 18

5 The last two digit of the number $(17)^{10}$, is

a. 49 b. 48
c. 47 d. 46

6 Which number is larger $(1.01)^{1000000}$ or 10000?

a. $(1.01)^{1000000}$ b. 10000
c. Cannot be discussed d. None of these

7 If $(7+4\sqrt{3})^n = \alpha + \beta$, where α is a positive integer and β is a proper fraction, then

a. $(1-\beta)(\alpha+\beta)=1$ b. $\beta(\alpha+\beta)=1$
c. $(\alpha+\beta)(2-\beta)=1$ d. $\beta(\alpha-\beta)=1$

8 If n is positive integer, then the integral part of $(7+4\sqrt{3})^n$, is

a. an odd integer b. an even integer
c. an composite integer d. None of these

Working Space

9 If n is the positive integer, then the integral part of $(7+5\sqrt{2})^{2n+1}$, is

a. an odd integer
b. an even integer
c. an composite integer
d. None of these

10 The last digit of $(3^P + 2)$ is, where $P = 3^{4n}$ and $n \in N$.

a. 1
b. 2
c. 4
d. 5

11 Let n be an odd number greater than 1. Then, number of zeros at the end of the sum $99^n + 1$, is

a. 1
b. 2
c. 3
d. 4

12 If $6^{83} + 8^{32}$ is divisible by 49, then the remainder is

a. 35
b. 5
c. 21
d. 0

13 If $(11)^{27} + (21)^{27}$ when divided by 16 leaves the remainder, is

a. 0
b. 1
c. 2
d. 14

14 The last three digits of the number $N = 7^{100} - 3^{100}$ are

a. 100
b. 300
c. 500
d. 000

15 The remainder, when $(15^{23} + 23^{23})$ is divided by 19, is

a. 4
b. 15
c. 0
d. 18

16 The last two digits of 3^{400} are

a. 01
b. 29
c. 43
d. 81

17 Let $(5+2\sqrt{6})^n = p + f$, where $n \in N$, $p \in N$ and $0 < f < 1$, then the value of $f^2 - f + pf - p$, is

a. a natural number
b. a negative number
c. a prime number
d. an irrational number

18 The last two digits of 21^{100}, is

a. 00
b. 01
c. 11
d. 21

19 The remainder, if $1 + 2 + 2^2 + \cdots + 2^{1999}$ is divided by 5, is

a. 0
b. 1
c. 2
d. 3

20 The remainder left out when $8^{2n} - (62)^{2n+1}$ is divided by 9, is **[AIEEE 2010]**

a. 7
b. 8
c. 0
d. 2

DPP-3 Properties of Binomial Coefficients Based on Summation

Subjective Questions

Working Space

Direction (Q. No. 1) *This question is subjective in nature, need to be solved completely on notebook.*

1 If $(1+x)^n = {}^nC_0 + {}^nC_1x + {}^nC_2x^2 + \cdots + {}^nC_nx^n$, then show that

i. ${}^nC_0 + {}^nC_1 + {}^nC_2 + \cdots + {}^nC_n = 2^n$

ii. ${}^nC_0 + {}^nC_2 + {}^nC_4 + \cdots = 2^{n-1}$

iii. ${}^nC_1 + {}^nC_3 + {}^nC_5 + \cdots = 2^{n-1}$

iv. ${}^nC_0 + 3\cdot{}^nC_1 + 3^2\cdot{}^nC_2 + \cdots + 3^n\cdot{}^nC_n = 4^n$

v. ${}^nC_0 - {}^nC_1 + {}^nC_2 - \cdots + (-1)^n\cdot{}^nC_n = 0$

vi. $\sum_{r=0}^{n}(-1)^r\cdot r\cdot{}^nC_r = 0$

2 If $(1+x)^n = \sum_{r=0}^{n}{}^nC_rx^r$, then prove the following series.

i. ${}^nC_1 + 2\cdot{}^nC_2 + 3\cdot{}^nC_3 + \cdots + n\cdot{}^nC_n = n\cdot 2^{n-1}$

ii. ${}^nC_0 + 3\cdot{}^nC_1 + 5\cdot{}^nC_2 + \cdots + (2n+1)\,{}^nC_n = (n+1)\cdot 2^n$

iii. $1^2\cdot{}^nC_1 + 2^2\cdot{}^nC_2 + 3^2\cdot{}^nC_3 + \cdots + n^2\cdot{}^nC_n = n(n+1)\cdot 2^{n-2}$

iv. ${}^nC_0 - 2\cdot{}^nC_1 + 3\cdot{}^nC_2 - \cdots + (-1)^n\cdot(n+1)\cdot{}^nC_n = 0$

v. ${}^nC_0 + 5\cdot{}^nC_1 + 9\cdot{}^nC_2 + \cdots + (4n+1)\,{}^nC_n = (2n+1)\cdot 2^n$

vi. $(1 + {}^nC_1 + {}^nC_2 + \cdots + {}^nC_n)^5 = 1 + {}^{5n}C_1 + {}^{5n}C_2 + \cdots + {}^{5n}C_{5n}$

vii. $1 - (a-1)\cdot{}^nC_1 + (a-2)\,{}^nC_2 - (a-3)\cdot{}^nC_3 + \cdots + (-1)^n\cdot(a-n)\cdot{}^nC_n = 0$

viii. ${}^nC_0 + \frac{{}^nC_1}{2} + \frac{{}^nC_2}{3} + \cdots + \frac{{}^nC_n}{n+1} = \frac{2^{n+1}-1}{n+1}$

ix. $2\cdot{}^nC_0 + 2^n\cdot\frac{{}^nC_1}{2} + 2^3\cdot\frac{{}^nC_2}{3} + \cdots + 2^{n+1}\cdot\frac{{}^nC_n}{n+1} = \frac{3^{n+1}-1}{n+1}$

x. $a\cdot{}^nC_0 + (a-b)\cdot{}^nC_1 + (a-2b)\,{}^nC_2 + \cdots + (a-nb)\,{}^nC_n = 2^{n-1}(a-nb)$

Only One Option Correct Type

Direction (Q. Nos. 3-11) *This section contains 9 multiple choice questions. Each question has four choices (a), (b), (c) and (d), out of which ONLY ONE option is correct.*

3 The sum of binomial coefficients of $\left(2x + \frac{1}{x}\right)^n$ is equal to 256. The constant term in the expansion is

a. 1120　　b. 2110
c. 1210　　d. None of these

Working Space

4 If $n \in N$ and n is even, $\frac{1}{1\cdot(n-1)!}+\frac{1}{3!(n-3)!}+\frac{1}{5!(n-5)!}+\cdots+\frac{1}{(n-1)!(1)!}$, is

a. 2^n
b. $\frac{2^{n-1}}{n!}$
c. $2^n \cdot n!$
d. None of these

5 The sum of all coefficients in $(2x^2-3x+1)^{11}$, is

a. 0
b. -1
c. 1
d. 2

6 The sum of the coefficients of all the even powers of x in the expansion of $(2x^2-3x+1)^{11}$, is

a. $2\cdot 6^{10}$
b. $3\cdot 6^{10}$
c. 6^{11}
d. None of these

7 If $(1+x-3x^2)^{2145}=a_0+a_1x+a_2x^2+\cdots$, then $a_0-a_1+a_2-a_3+\cdots$ ends with

a. 1
b. 3
c. 7
d. 9

8 If $(1+x)^n={}^nC_0+{}^nC_1x+{}^nC_2x^2+\cdots+{}^nC_nx^n$, then

$$\frac{{}^nC_1}{{}^nC_0}+\frac{2\cdot{}^nC_2}{{}^nC_1}+\frac{3\cdot{}^nC_3}{{}^nC_2}+\cdots+n\cdot\frac{{}^nC_n}{{}^nC_{n-1}},\text{ is}$$

a. $\frac{n(n-1)}{2}$
b. $\frac{n(n+2)}{2}$
c. $\frac{n(n+1)}{2}$
d. $\frac{(n-1)(n-2)}{2}$

9 The sum of the series $\sum_{r=0}^{10} {}^{20}C_r$, is

a. 2^{20}
b. 2^{19}
c. $2^{19}+\frac{1}{2}\,{}^{20}C_{10}$
d. $2^{19}-\frac{1}{2}\,{}^{20}C_{10}$

10 If ${}^nC_0, {}^nC_1, {}^nC_2, \cdots, {}^nC_n$ are binomial coefficients of different terms in the expansion of $(1+x)^n$, $(n \in N)$, then

$$\left(1+\frac{{}^nC_1}{{}^nC_0}\right)\left(1+\frac{{}^nC_2}{{}^nC_1}\right)\left(1+\frac{{}^nC_3}{{}^nC_2}\right)\cdots\left(1+\frac{{}^nC_n}{{}^nC_{n-1}}\right)$$

equals to

a. $\left(\frac{n+1}{n}\right)^n$
b. $\frac{((n+1)!)^n}{n!}$
c. $\frac{(n+1)^n}{n!}$
d. None of the above

11 If the sum of the coefficients in the expansion of $(\alpha^2x^2-2\alpha x+1)^{51}$ vanishes, then the value of α, is

a. 2
b. -1
c. 1
d. -2

Working Space

Statement Type

Direction (*Q. Nos. 12-15*) *This section is based on Statement I and Statement II. Select the correct answer from the codes given below.*

a. Both Statement I and Statement II are correct and Statement II is the correct explanation of Statement I
b. Both Statement I and Statement II are correct and Statement II is not the correct explanation of Statement I
c. Statement I is correct but Statement II is incorrect
d. Statement II is correct but Statement I is incorrect

12 Let $S_1 = \sum_{j=1}^{10} j(j-1)\cdot {}^{10}C_j$, $S_2 = \sum_{j=1}^{10} j\cdot {}^{10}C_j$ and $S_3 = \sum_{j=1}^{10} j^2\cdot {}^{10}C_j$ **[AIEEE 2010]**

Statement I $S_3 = 55 \times 2^9$.

Statement II $S_1 = 90 \times 2^8$, $S_2 = 10 \times 2^8$.

13 **Statement I** $\sum_{r=0}^{n} (r+1)\cdot {}^nC_r = (n+2)\cdot 2^{n-1}$.

Statement II $\sum_{r=0}^{r} (r+1)\cdot {}^nC_r \cdot x^r = (1+x)^n + nx\cdot(1+x)^{n-1}$. **[AIEEE 2008]**

14 If $(1+x)^n = \sum_{r=0}^{n} a_r x^r$ and $b_r = 1 + \dfrac{a_r}{a_{r-1}}$, then

Statement I $\prod_{r=1}^{n} b_r = \dfrac{(101)^{100}}{(100)!} \Rightarrow n = 100$.

Statement II $\sum_{r=0}^{n} \dfrac{1}{b_r} = 50 \Rightarrow n = 100$

15 **Statement I** If $q = \dfrac{1}{3}$ and $p + q = 1$, then $\sum_{r=0}^{15} r\cdot {}^{15}C_r p^r q^{15-r} = 5$.

Statement II If $p + q = 1$, $0 < p < 1$, then $\sum_{r=0}^{n} r\cdot {}^nC_r p^r q^{n-r} = np$.

Comprehension Type

Direction (*Q. Nos. 16-20*) *This section contains 2 paragraphs, each describing theory, experiments, data, etc. Five questions related to paragraphs have been given. Each question has only one correct answer among the four given options (a), (b), (c) and (d).*

Passage I

Integration plays a vital role in proving identities involving binomial coefficients. If we apply integration techniques, several elegent binomial identities can be proved.

For example, ${}^nC_0 - \dfrac{{}^nC_1}{2} + \dfrac{{}^nC_2}{3} + \cdots + (-1)^n\cdot\dfrac{{}^nC_n}{n+1} = \int_0^1 (1-x)^n\, dx = \dfrac{1}{n+1}$

16 The value of $\dfrac{{}^nC_0}{1} - \dfrac{{}^nC_1}{3} + \dfrac{{}^nC_2}{5} - \cdots + (-1)^n \dfrac{{}^nC_n}{2n+1}$, is

a. $\dfrac{n!}{(2n+1)!}$
b. $\dfrac{2^{2n}\cdot n!}{(2n+1)!}$
c. $\dfrac{2^{2n}\cdot (n!)^2}{(2n+1)!}$
d. $\dfrac{2^{2n}\cdot (2n+1)!}{n!}$

Working Space

17 The value of $\frac{^nC_0}{n(n+1)} - \frac{^nC_1}{(n+1)(n+2)} + \frac{^nC_2}{(n+2)(n+3)} - \cdots$ upto $(n+1)$ term, is

a. $\int_0^1 x^{n+1} \cdot (1-x)^{n+1} dx$ b. $\int_0^1 x^n (1-x)^{n+1} dx$

c. $\int_0^1 x^{n-1} \cdot (1-x)^{n+1} dx$ d. $\int_0^1 x^n (1-x)^n dx$

18 The value of

$$^nC_1 - \left(1+\frac{1}{2}\right) \cdot {^nC_2} + \left(1+\frac{1}{2}+\frac{1}{3}\right) {^nC_3} \cdots + (-1)^{n-1} \left(1+\frac{1}{2}+\cdots+\frac{1}{n}\right) \cdot {^nC_n}$$

is

a. 0 b. 1

c. n d. $\frac{1}{n}$

Passage II

If $(1+x)^n = a_0 + a_1x + a_2x^2 + \cdots + a_nx^n$, then

19 The value of $a_0 - a_2 + a_4 - a_6 + \cdots$, is equal to

a. $2^{n/2} \cdot \sin\frac{n\pi}{4}$ b. $2^{n/2} \cdot \cos\frac{n\pi}{4}$

c. $2^{n+1} \cdot \cos\left(\frac{n\pi}{3}\right)$ d. $2^{n/2} \cdot \sin\frac{n\pi}{3}$

20 The value of $a_1 - a_3 + a_5 - a_7 + \cdots$, is equal to

a. $2^{n/2} \cdot \sin\frac{n\pi}{4}$ b. $2^{n/2} \cdot \cos\frac{n\pi}{4}$

c. $2^{n/2} \cdot \cos\frac{n\pi}{3}$ d. $2^{n/2} \cdot \sin\frac{n\pi}{3}$

DPP-4 Product of Binomial Coefficients and Double Summation

Subjective Questions

Direction *(Q. Nos. 1-4) These questions are subjective in nature, need to be solved completely on notebook.*

Working Space

1 Prove that

i. ${}^nC_n + {}^{n+1}C_n + {}^{n+2}C_n + \cdots + {}^{n+k}C_n = {}^{n+k+1}C_{n+1}$

ii. ${}^mC_1 + {}^{m+1}C_2 + {}^{m+2}C_3 + \cdots + {}^{m+n-1}C_n = {}^nC_1 + {}^{n+1}C_2 + {}^{n+2}C_3 + \cdots + {}^{n+m-1}C_m$

2 If $(1+x)^n = {}^nC_0 + {}^nC_1x + {}^nC_2x^2 + \cdots + {}^nC_nx^n$, then prove that

i. ${}^nC_0 \cdot {}^nC_0 + {}^nC_1 \cdot {}^nC_1 + \cdots + {}^nC_n \cdot {}^nC_n = {}^{2n}C_n$

ii. ${}^nC_0 \cdot {}^nC_1 + {}^nC_1 \cdot {}^nC_2 + \cdots + {}^nC_{n-1} \cdot {}^nC_n = {}^{2n}C_{n-1}$

iii. ${}^nC_0 \cdot {}^nC_n + {}^nC_1 \cdot {}^nC_{n-1} + \cdots + {}^nC_n \cdot {}^nC_0 = {}^{2n}C_n$

iv. $\sum_{r=0}^{n} {}^nC_r \cdot {}^nC_{r+3} = {}^{2n}C_{n-3}$

v. ${}^mC_r \cdot {}^nC_0 + {}^mC_{r-1} \cdot {}^nC_1 + {}^mC_{r-2} \cdot {}^nC_2 + \cdots + {}^mC_0 \cdot {}^nC_r = {}^{m+n}C_r$

3 If $(1+x)^n = \sum_{r=0}^{n} {}^nC_r x^r$, then prove that

i. ${}^nC_0 \cdot {}^nC_0 - {}^nC_1 \cdot {}^nC_1 + {}^nC_2 \cdot {}^nC_2 - \cdots + (-1)^n \cdot {}^nC_n \cdot {}^nC_n = \begin{cases} 0, & \text{if } n \text{ is odd.} \\ (-1)^{n/2} \cdot {}^nC_{n/2}, & \text{if } n \text{ is even.} \end{cases}$

ii. ${}^{2n}C_0^2 - {}^{2n}C_1^2 + {}^{2n}C_2^2 - \cdots + (-1)^n \cdot {}^{2n}C_{2n}^2 = (-1)^n \cdot {}^{2n}C_n$

4 If $(1+2x+2x^2)^n = {}^{2n}C_0 + {}^{2n}C_1x + {}^{2n}C_2x^2 + \cdots + {}^{2n}C_{2n} \cdot x^{2n}$, then prove that

$${}^{2n}C_0 \cdot {}^{2n}C_{2n} - {}^{2n}C_1 \cdot {}^{2n}C_{2n-1} + {}^{2n}C_2 \cdot {}^{2n}C_{2n-2} - \cdots + {}^{2n}C_{2n} \cdot {}^{2n}C_0 = \begin{cases} 0, & \text{if } n \text{ is odd.} \\ 2^n \cdot {}^nC_{n/2}, & \text{if } n \text{ is even.} \end{cases}$$

Only One Option Correct Type

Direction *(Q. Nos. 5-14) This section contains 10 multiple choice questions. Each question has four choices (a), (b), (c) and (d), out of which ONLY ONE option is correct.*

5 The coefficient of x^{49} in the polynomial

$$\left(x - \frac{{}^{50}C_1}{{}^{50}C_0}\right)\left(x - 2^2 \cdot \frac{{}^{50}C_2}{{}^{50}C_1}\right)\left(x - 3^2 \cdot \frac{{}^{50}C_3}{{}^{50}C_2}\right)\cdots\left(x - 50^2 \cdot \frac{{}^{50}C_{50}}{{}^{50}C_{49}}\right), \text{ is}$$

a. -22100 b. -11220
c. 22100 d. 11220

Working Space

6 The sum of $\sum_{i=0}^{m} \binom{10}{i}\binom{20}{m-i}$ (where, $\binom{p}{q} = 0$ if $p < q$) is maximum, when m is **[IIT JEE 2000]**

a. 5
b. 10
c. 15
d. 20

7 The value of $\binom{30}{0}\binom{30}{10} - \binom{30}{1}\binom{30}{11} + \binom{30}{2}\binom{30}{12} - \dots + \binom{30}{20}\binom{30}{30}$ is, where $\binom{n}{r} = {}^nC_r$ **[IIT JEE 2005]**

a. $\binom{30}{10}$
b. $\binom{30}{15}$
c. $\binom{60}{30}$
d. $\binom{31}{10}$

8 For $r = 0, 1, 2, \dots, 10$, let A_r, B_r and C_r denote, respectively, the coefficient of x^r in the expansion of $(1+x)^{10}$, $(1+x)^{20}$ and $(1+x)^{30}$. Then, $\sum_{r=1}^{10} A_r(B_{10}B_r - C_{10}A_r)$ is equal to **[IIT JEE 2010]**

a. $B_{10} - C_{10}$
b. $A_{10}(B_{10}^2 - C_{10}A_{10})$
c. 0
d. $C_{10} - B_{10}$

9 If $(1+x)^n = {}^nC_0 + {}^nC_1x + {}^nC_2x^2 + \dots + {}^nC_nx^n$, then the value of $\sum_{r=0}^{n}\sum_{s=0}^{n} {}^nC_r \cdot {}^nC_s$ is equal to

a. 2^n
b. 2^{n+1}
c. 2^{2n}
d. 2^{3n}

10 If $(1+x)^n = {}^nC_0 + {}^nC_1x + {}^nC_2x^2 + \dots + {}^nC_nx^n$, then the value of $\sum_{0 \le r <} \sum_{s \le n} {}^nC_r \cdot {}^nC_s$ is equal to

a. $\frac{1}{2}[2^{2n} - {}^{2n}C_n]$
b. $\frac{1}{4}[2^{2n} - {}^{2n}C_n]$
c. $\frac{1}{2}[2^{2n} + {}^{2n}C_n]$
d. $\frac{1}{2}[2^{n} - {}^{2n}C_n]$

11 If $(1+x)^n = {}^nC_0 + {}^nC_1x + {}^nC_2x^2 + \dots + {}^nC_nx^n$, then the value of $\sum_{r=0}^{n}\sum_{s=0}^{n} ({}^nC_r + {}^nC_s)$ is equal to

a. $(n+1)2^{n+1}$
b. $(n-1)2^{n+1}$
c. $(n+1)2^n$
d. None of these

12 If $(1+x)^n = {}^nC_0 + {}^nC_1x + {}^nC_2x^2 + \dots + {}^nC_nx^n$, then the value of $\sum_{0 \le r <} \sum_{s \le n} ({}^nC_r + {}^nC_s)$ is equal to

a. 2^{2n}
b. 2^n
c. $(n+1)2^n$
d. $n \cdot 2^n$

13 The value of $\sum_{r=0}^{s}\sum_{s=1}^{n} {}^nC_s \cdot {}^sC_r$ is equal to

a. $3^n - 1$
b. $3^n + 1$
c. 3^n
d. None of these

Working Space

14 Let n be a natural number such that $(1+x+x^2)^n = a_0 + a_1x + a_2x^2 + \cdots + a_{2n} \cdot x^{2n}$, then $a_0 + a_1 + a_2 + \cdots + a_{n-1}$, is

a. $\frac{1}{2}(3^n + a_n)$
b. $\frac{1}{2}(3^{2n} - a_n)$
c. $\frac{1}{2}(3^n - a_n)$
d. None of these

One or More than One Options Correct Type

Direction (Q. Nos. 15-17) This section contains 3 multiple choice questions. Each question has four choices (a), (b), (c) and (d), out of which ONE or MORE THAN ONE are correct.

15 If $1+(1+x)+(1+x)^2+\cdots+(1+x)^n = \sum_{k=0}^{n} a_k x^k$, then

a. $\sum_{k=0}^{n} a_k = 2^{n+1} - 1$
b. $a_{n-2} = \frac{n(n+1)}{2}$
c. $a_p > a_{p-1}$ for $p < \frac{n}{2}$
d. $(a_9)^2 - (a_8)^2 = {}^{n+2}C_{10} \cdot ({}^{n+1}C_{10} - {}^{n+1}C_9)$

16 If $f(m) = \sum_{i=0}^{m} {}^{30}C_{30-i} \cdot {}^{20}C_{m-i}$, then

a. maximum value of $f(m) = {}^{50}C_{25}$
b. $\sum_{r=0}^{50} f(r) = 2^{50}$
c. $f(m)$ is divisible by 50
d. $\sum_{m=0}^{50} (f(m))^2 = {}^{100}C_{50}$

17 The value of ${}^{1000}C_{50} + {}^{999}C_{49} + {}^{998}C_{48} + \cdots + {}^{950}C_0$, is

a. ${}^{1001}C_{50}$
b. ${}^{1002}C_{51} - {}^{1001}C_{51}$
c. ${}^{1001}C_{951}$
d. ${}^{1002}C_{51} - {}^{1001}C_{950}$

Comprehension Type

Direction (Q. Nos. 18-20) This section contains a paragraph, describing theory, experiments, data, etc. Three questions related to the paragraph have been given. Each question has only one correct answer among the four given options (a), (b), (c) and (d).

Passage

If n is a positive integer and x is any real number, then $(1+x)^n = {}^nC_0 + {}^nC_1x + {}^nC_2x^2 + \cdots + {}^nC_nx^n$, where ${}^nC_0, {}^nC_1, {}^nC_2, \cdots, {}^nC_n$ are binomial coefficients, then

18 The value of $({}^nC_1 - 3\cdot{}^nC_3 + 5\cdot{}^nC_5 - \cdots)^2 + (2\cdot{}^nC_2 - 4\cdot{}^nC_4 + 6\cdot{}^nC_6 - \cdots)^2$, is

a. $n \cdot 2^{n-1}$
b. $n^2 \cdot 2^{n-1}$
c. $n(n-1) \cdot 2^{n-1}$
d. $n^2 \cdot 2^n$

19 $\frac{{}^nC_0}{n(n+1)} - \frac{{}^nC_1}{(n+1)(n+2)} + \frac{{}^nC_2}{(n+2)(n+3)} - \cdots$ upto $(n+1)$ terms is equal to

a. $\int_0^1 x^{n+1}(1-x)^{n+1}dx$
b. $\int_0^1 x^n(1-x)^{n+1}dx$
c. $\int_0^1 x^{n-1}\cdot(1-x)^{n+1}dx$
d. $\int_0^1 x^n(1-x)^{n-1}dx$

20 If $\sum_{r=0}^{n} \frac{1}{{}^nC_r} = a$, then $\sum_{0 \le i < j \le n} \sum \left(\frac{1}{{}^nC_i} + \frac{1}{{}^nC_j}\right)$ is equal to

a. na
b. n^2a
c. $\frac{na}{2}$
d. $\frac{n^2a}{2}$

DPP-5 Greatest Term, Binomial Theorem for any Index and Multinomial Theorem

Subjective Questions

Working Space

Direction (Q. Nos. 1-3) *These questions are subjective in nature, need to be solved completely on notebook.*

1 Find numerically greatest term in the expansion of

i. $(2+3x)^9$, when $x=\frac{3}{2}$

ii. $(3-5x)^{15}$, when $x=\frac{1}{5}$

2 Find the coefficient of

i. $x^2y^3z^4$ in the expansion of $(ax-by+cz)^9$.

ii. $a^2b^3c^4d$ in the expansion of $(a-b-c+d)^{10}$.

3 Find the coefficient of

i. x^4 in the expansion of $(1+x+x^2+x^3)^{11}$.

ii. x^4 in the expansion of $(2-x+3x^2)^6$.

Only One Option Correct Type

Direction (Q. Nos. 4-19) *This section contains 16 multiple choice questions. Each question has four choices (a), (b), (c) and (d), out of which ONLY ONE option is correct.*

4 The coefficient of $a^3b^4c^7$ in the expansion of $(ab+bc+ca)^8$, is

a. 280 b. 270

c. 260 d. 250

5 The number of terms in the expansion of $(a+b+c+d+e+f)^n$, is

a. ${}^{n+5}C_n$ b. ${}^{n+6}C_n$

c. ${}^{n+7}C_n$ d. ${}^{n+8}C_n$

6 The expansion of $\frac{1}{(4-3x)^{1/2}}$ by binomial theorem will be valid, if

a. $x<1$ b. $|x|<1$

c. $|x|<\frac{2}{\sqrt{3}}$ d. $|x|<\frac{4}{3}$

Working Space

7 $\dfrac{1}{(6-3x)^{1/3}}$ is equal to

a. $6^{1/3}\left(1+\dfrac{x}{6}+\dfrac{2x^2}{6^2}+\cdots\right)$
b. $6^{-1/3}\left(1+\dfrac{x}{6}+\dfrac{2x^2}{6^2}+\cdots\right)$
c. $6^{1/3}\left(1-\dfrac{x}{6}+\dfrac{2x^2}{6^2}+\cdots\right)$
d. $6^{-1/3}\left(1-\dfrac{x}{6}+\dfrac{2x^2}{6^2}-\cdots\right)$

8 The general term in the expansion of $(1-3x)^{-1/3}$, is

a. $\dfrac{1\cdot 4\cdot 7\cdots(3r-2)}{r!}x^r$
b. $\dfrac{1\cdot 4\cdot 7\cdots(3r-2)}{r!}\cdot 3^r\cdot x^r$
c. $\dfrac{1\cdot 4\cdot 7\cdots(3r-2)\cdot(-1)^r\cdot x^r}{r!}$
d. $\dfrac{1\cdot 4\cdot 7\cdots(3r-2)\cdot(-3)^r\cdot x^r}{r!}$

9 The coefficient of x^2 in $(1+3x)^{1/2}\,(1-2x)^{-1/3}$, is

a. $\dfrac{6}{13}$
b. $\dfrac{55}{72}$
c. $\dfrac{7}{19}$
d. $\dfrac{2}{9}$

10 The term independent of y in the binomial expansion of $\left(\dfrac{1}{2}\cdot y^{1/3}+y^{-1/5}\right)^8$, is

a. sixth
b. seventh
c. fifth
d. None of these

11 The square root of $1+2x+3x^2+4x^3+\cdots$, is

a. $(1-x)^{-2}$
b. $(1+x)^{-2}$
c. $(1+x)^{-1}$
d. $(1-x)^{-1}$

12 If $|x|<1$, then in the expansion of $(1+2x+3x^2+4x^3+\cdots)^{1/2}$, the coefficient of x^n, is

a. n
b. $n+1$
c. 1
d. -1

13 If $|x|<1$, then the coefficient of x^n in the expansion of $(1+x+x^2+x^3+x^4+\cdots)^2$, is

a. n
b. $n-1$
c. $n+2$
d. $n+1$

14 The $(r+1)$th term in the expansion of $(1-x)^{-4}$ will be

a. $\dfrac{x^r}{r!}$
b. $\dfrac{(r+1)(r+2)(r+3)}{6}x^r$
c. $\dfrac{(r+2)(r+3)}{2}x^r$
d. None of these

15 In the expansion of $\left(\dfrac{1+x}{1-x}\right)^2$, the coefficient of x^n will be

a. $4n$
b. $4n-3$
c. $4n+1$
d. None of these

Working Space

16 The value of $1+\frac{1}{4}+\frac{1\cdot 3}{4\cdot 8}+\frac{1\cdot 3\cdot 5}{4\cdot 8\cdot 12}+\cdots$, is

a. $\sqrt{2}$

b. $\frac{1}{\sqrt{2}}$

c. $\sqrt{3}$

d. $\frac{1}{\sqrt{3}}$

17 If $y=\frac{2}{5}+\frac{1\cdot 3}{2!}\left(\frac{2}{5}\right)^2+\frac{1\cdot 3\cdot 5}{3!}\left(\frac{2}{5}\right)^3+\cdots$, then the value of (y^2+2y-4), is

a. 1

b. -1

c. 0

d. None of the above

18 If the value of x is so small that x^2 and higher powers of x can be neglected, then $\frac{\sqrt{1+x}+\sqrt[3]{(1-x)^2}}{(1+x)+\sqrt{1+x}}$ is equal to

a. $1+\frac{5}{6}x$

b. $1-\frac{5}{6}x$

c. $1+\frac{2}{3}x$

d. $1-\frac{2}{3}x$

19 If the expansion in powers of x of the function $\frac{1}{(1-ax)(1-bx)}$ is $a_0+a_1x+a_2x^2+a_3x^3+\cdots$. Then, a_n is

a. $\frac{a^n-b^n}{a-b}$

b. $\frac{a^{n+1}-b^{n+1}}{b-a}$

c. $\frac{b^{n+1}-a^{n+1}}{b-a}$

d. $\frac{b^n-a^n}{a-b}$

One or More than One Options Correct Type

Direction (Q. Nos. 20-22) *This section contains 3 multiple choice questions. Each question has four choices (a), (b), (c) and (d), out of which ONE or MORE THAN ONE are correct.*

20 The 4th term in the expansion of $\left(2+\frac{3x}{8}\right)^{10}$ has the maximum numerical value, then the value of x can be, in the interval.

a. $\left[\frac{-64}{21},-2\right]$

b. $\left[2,\frac{64}{21}\right]$

c. $\left(-\infty,\frac{-64}{21}\right)$

d. $\left(\frac{64}{21},\infty\right)$

21 In the expansion of $(x+y+z)^{10}$ which of the following will be true

a. The expansion will contains 66 terms

b. The coefficient of $x^3y^2z^5$ will be $\frac{(10)!}{3!\,2!\,5!}$

c. There cannot be term of $x^3y^2z^4$

d. $\sum_{\alpha+\beta+\gamma=10}\frac{(10)!}{\alpha!\,\beta!\,\gamma!}=3^{10}$

Working Space

22 Which one of the following is true?

a. If $x^2 < 1$, the coefficient of x^n in $\dfrac{(1+2x)^n + (1+3x)^n}{(1-x)}$ is $3^n + 4^n$.

b. The value of $\sum_{r=0}^{n} (-1)^r \cdot {}^nC_r \left[\dfrac{1}{2^r} + \dfrac{3^r}{2^{2r}} + \dfrac{7^r}{2^{3r}} + \cdots \infty\right)$ is equal to $\left(\dfrac{1}{2^n - 1}\right)$.

c. The coefficient of x^2 in $(1+x+x^2)^{10}$ is ${}^{12}C_2$.

d. The coefficient of x^2 in $(1+x+x^2)^{10}$ is ${}^{11}C_2$.

Comprehension Type

Direction *(Q. Nos. 23-25) This section contains a paragraph, describing theory, experiments, data, etc. Three questions related to the paragraph have been given. Each question has only one correct answer among the four given options (a), (b), (c) and (d).*

Passage

For $n \in N$, consider $(1 + x + x^2)^n = \sum_{r=0}^{2n} a_r x^r$, then

23 $2(a_0 + a_1 + a_2 + \cdots + a_{n-1}) + a_n$, is

a. 2^{2n-1} b. 3^n

c. $\dfrac{3^n}{2}$ d. $\dfrac{3^n - 2}{2}$

24 If $a_0^2 - a_1^2 + a_2^2 - a_3^2 + \cdots + a_{2n}^2 = k \cdot a_n$ then k is equal to

a. 1 b. 2

c. $\dfrac{1}{3}$ d. 0

25 If $a = \sum_{k=0}^{[n/2]} {}^nC_{2k} \cdot {}^{2k}C_k$, where $[x]$ denotes greatest integer less than or equals to x, then $a_n - a$ is

a. 0 b. 1

c. −1 d. 2

SET - 1

Revisal Problems
for JEE Main

Only One Option Correct Type

Working Space

Direction (Q. Nos. 1-25) *This section contains 25 multiple choice questions. Each question has four choices (a), (b), (c) and (d), out of which ONLY ONE option is correct.*

1 $(2n+1)(2n+3)(2n+5)\cdots(4n-1)$ is equal to

a. $\dfrac{(4n)!}{2^n \cdot (2n)!(2n)!}$
b. $\dfrac{(4n)! \cdot n!}{2^n (2n)!(2n)!}$
c. $\dfrac{(4n)! \cdot n!}{(2n!)(2n)!}$
d. $\dfrac{(4n)! \cdot n!}{(2^n)(2n)!}$

2 If the 7th term in the binomial expansion of $\left(\dfrac{3}{\sqrt[3]{84}} + \sqrt{3}\log_e x\right)^9$, $x > 0$ is equal to 729, then x can be

a. e
b. $2e$
c. $\dfrac{e}{2}$
d. e^2

3 The sum of the rational terms in the binomial expansion of $(2^{1/2} + 3^{1/5})^{10}$, is

a. 25
b. 32
c. 9
d. 41

4 Set of value of r for which, ${}^{18}C_{r-2} + 2 \cdot {}^{18}C_{r-1} + {}^{18}C_r \geq {}^{20}C_{13}$ contains

a. 4
b. 5
c. 7
d. 10

5 Number of rational terms in the expansion of $(\sqrt{2} + \sqrt[4]{3})^{100}$, is

a. 25
b. 26
c. 27
d. 28

6 The remainder, when $(15^{23} + 23^{23})$ is divided by 19, is

a. 4
b. 15
c. 0
d. 18

7 Coefficient of α^t in the expansion of $(\alpha+p)^{m-1} + (\alpha+p)^{m-2}(\alpha+q) + (\alpha+p)^{m-3}(\alpha+q)^2 + \cdots + (\alpha+q)^{m-1}$, where $\alpha \neq -q$ and $p \neq q$, is

a. $\dfrac{{}^mC_t(p^t - q^t)}{p-q}$
b. $\dfrac{{}^mC_t(p^{m-t} - q^{m-t})}{p-q}$
c. $\dfrac{{}^mC_t(p^t + q^t)}{p-q}$
d. $\dfrac{{}^mC_t(p^{m-t} + q^{m-t})}{p-q}$

8 The term independent of x in the expansion of $\left(9x - \dfrac{1}{3\sqrt{x}}\right)^{18}$, $x > 0$, is α times the corresponding binomial coefficient, then α is

a. 3
b. $\dfrac{1}{3}$
c. $-\dfrac{1}{3}$
d. 1

9 The value of $\binom{99}{97} + \binom{98}{96} + \binom{97}{95} + \cdots + \binom{3}{1} + \binom{2}{0}$ is equal to

a. $\binom{99}{97}$
b. $\binom{100}{98}$
c. $\binom{99}{98}$
d. $\binom{100}{97}$

Working Space

10 If $\sum_{r=1}^{10} r(r-1)\cdot{}^{10}C_r = k\cdot 2^9$, then k is equal to

a. 10 b. 45
c. 90 d. 100

11 The largest real value of x such that $\sum_{k=0}^{4}\left(\frac{5^{4-k}}{(4-k)!}\right)\cdot\left(\frac{x^k}{k!}\right)=\frac{8}{3}$, is

a. $2\sqrt{2}-5$ b. $2\sqrt{2}+5$
c. $-2\sqrt{2}-5$ d. $-2\sqrt{2}+5$

12 The sum of the series ${}^{20}C_0 - {}^{20}C_1 + {}^{20}C_2 - {}^{20}C_3 + \cdots + {}^{20}C_{10}$, is

a. $\frac{1}{2}\cdot{}^{20}C_{10}$ b. 0
c. $-{}^{20}C_{10}$ d. ${}^{20}C_{10}$

13 In the binomial expansion of $(a-b)^n$, $n \ge 5$ the sum of 5th and 6th terms is zero, then $\frac{a}{b}$ equals

a. $\frac{6}{n-5}$ b. $\frac{n-5}{6}$ c. $\frac{n-4}{5}$ d. $\frac{5}{n-4}$

14 The coefficients of rth, $(r+1)$th and $(r+2)$th terms in the binomial expansion of $(1+y)^m$ are in AP, then m and r satisfy the equation

a. $m^2 - m(4r-1) + 4r^2 + 2 = 0$ b. $m^2 - m(4r+1) + 4r^2 - 2 = 0$
c. $m^2 - m(4r+1) + 4r^2 + 2 = 0$ d. $m^2 - m(4r-1) + 4r^2 - 2 = 0$

15 The coefficient of x^7 in the expansion of $(1-x-x^2+x^3)^6$, is

a. -144 b. 132 c. 144 d. -132

16 If n is a positive integer, then $(\sqrt{3}+1)^{2n} - (\sqrt{3}-1)^{2n}$, is

a. a rational number other than positive integers
b. an irrational number
c. an odd positive integer
d. an even positive integer

17 If ${}^{n-1}C_r = (k^2-3)\cdot{}^nC_{r+1}$, if $|k| \in$

a. $[-\sqrt{3}, \sqrt{3}]$ b. $(2, \infty)$
c. $(0, \infty)$ d. $(\sqrt{3}, 2]$

18 If $y = 3x + 6x^2 + 10x^3 + \cdots$. Then, the value of x in terms of y, is

a. $1-(1-y)^{-1/3}$ b. $1-(1+y)^{1/3}$
c. $1+(1-y)^{1/3}$ d. $1-(1+y)^{-1/3}$

19 If x is very small and $\frac{\left(1+\frac{3x}{4}\right)^{-4}\sqrt{16-3x}}{(8+x)^{2/3}} = P + Qx$, then

a. $P=1$ and $Q=\frac{305}{96}$ b. $P=1$ and $Q=\frac{-305}{96}$
c. $P=2$ and $Q=\frac{305}{48}$ d. $P=2$ and $Q=\frac{-305}{48}$

20 The coefficient of x^r in the expansion of $(1+3x+6x^2+10x^3+\cdots)^2$, is

a. $\frac{(r+1)(r+2)(r+3)}{5!}$ b. $\frac{(r+2)(r+3)(r+4)}{5!}$
c. $\frac{(r+1)(r+2)(r+3)(r+4)(r+5)}{5!}$ d. None of these

21 The sum of $1 + n\left(1-\frac{1}{x}\right) + \frac{n(n+1)}{2!}\left(1-\frac{1}{x}\right)^2 + \cdots\infty$, will be

a. x^n b. x^{-n} c. $\left(1-\frac{1}{x}\right)^n$ d. None of these

Working Space

22 The coefficient of x^n in the expansion of $(1-9x+20x^2)^{-1}$, is

a. $5^n - 4^n$
b. $5^{n+1} - 4^{n+1}$
c. $5^{n-1} - 4^{n-1}$
d. None of these

23 If x is nearly equal to 1, then the approximate value of $\frac{px^q - qx^p}{x^q - x^p}$, is

a. $\frac{p+q}{1-x}$
b. $\frac{1}{1+x}$
c. $\frac{1}{1-x}$
d. $\frac{p+q}{1+x}$

24 If x is positive, the first negative term in the expansion of $(1+x)^{27/5}$, is

a. 7th term
b. 5th term
c. 8th term
d. 6th term

25 If x is so small that x^3 and higher powers of x may be neglected, then $\frac{(1+x)^{3/2} - \left(1+\frac{x}{2}\right)^3}{(1-x)^{1/2}}$ may be approximated as

a. $\frac{x}{2} - \frac{3}{8}x^2$
b. $-\frac{3}{8}x^2$
c. $3x + \frac{3}{8}x^2$
d. $1 - \frac{3}{8}x^2$

Statement Type

Direction *(Q. Nos. 26-30) This section is based on Statement I and Statement II. Select the correct answer from the codes given below.*

a. Both Statement I and Statement II are correct and Statement II is the correct explanation of Statement I
b. Both Statement I and Statement II are correct and Statement II is not the correct explanation of Statement I
c. Statement I is correct but Statement II is incorrect
d. Statement II is correct but Statement I is incorrect

26 **Statement I** If n is an odd prime, then the integral part of $(\sqrt{5}+2)^n - 2^{n+1}$ is divisible by $20n$.

Statement II If n is a prime, then ${}^nC_1, {}^nC_2, \cdots, {}^nC_{n-1}$, must be divisible by n.

27 **Statement I** $11^{25} + 12^{25}$ when divided by 23 leaves the remainder zero.

Statement II $a^n + b^n$ is always divisible by $(a+b)$ for all $n \in N$.

28 **Statement I** The sum of the series, ${}^nC_0 \cdot {}^mC_r + {}^nC_1 \cdot {}^mC_{r-1} + {}^nC_2 \cdot {}^mC_{r-2} + \cdots + {}^nC_r \cdot {}^mC_0$, is equal to ${}^{n+m}C_r$.

Statement II Number of ways in which r children can be selected out of $(m+n)$ children consisting of n boys and m girls, if each selection may consist of any number of boys and girls is equal to ${}^{n+m}C_r$.

29 **Statement I** If $T_n = \sum_{r=1}^{n} \frac{1}{r}$ and $S_n = \sum_{r=1}^{n} {}^nC_r (-1)^{r-1}, \frac{1}{r}$. Then, $T_n = S_n$ for all $n \in N$.

Statement II $S_{n+1} - S_n = T_{n+1} - T_n$.

30 **Statement I** $\sum_{r=0}^{n} (-1)^r \cdot {}^nC_r (a-r)(b-r)(c-r) = 0$, where a, b and c are constants.

Statement II $\sum_{r=0}^{n} (-1)^r \cdot {}^nC_r = 0$, $\sum_{r=0}^{n} (-1)^r \cdot r \cdot {}^nC_r = 0$ and $\sum_{r=0}^{n} (-1)^r \cdot r^2 \cdot {}^nC_r = 0$.

In general $\sum_{r=0}^{n} (-1)^r \cdot r^p \cdot {}^nC_r = 0$ for each natural number P.

SET - 2

Revisal Problems for JEE Advanced

Only One Option Correct Type

Working Space

Direction *(Q. Nos. 1-11) This section contains 11 multiple choice questions. Each question has four choices (a), (b), (c) and (d), out of which ONLY ONE option is correct.*

1 If coefficient of x^k and x^{2k} $(k \in N)$ in the expansion of $1 + (1+x) + (1+x)^2 + \cdots + (1+x)^{16}$ are equal, then k equals

a. 3 b. 4
c. 5 d. 6

2 If p is the prime number and $p \neq 2$, then $[(2+\sqrt{5})^p] - (2)^{p+1}$ is always divisible by (where, $[x]$ denotes greatest integer function less than or equals to x)

a. $p+1$ b. p
c. $2p$ d. $4p$

3 The number 7^{1995}, when divided by 100 leaves the remainder

a. 43 b. 45
c. 50 d. None of these

4 The coefficient of x^9 in the polynomial given by, $(x+1)(x+2)\cdots(x+10) + (x+2)(x+3)\cdots(x+11) + \cdots + (x+11)(x+12)\cdots(x+20)$, is

a. 5511 b. 5151
c. 1515 d. 1155

5 In the expansion of $3(1+x)^n(1+y)^n(1+z)^n$, the sum of the coefficients of the terms of degree r, is

a. ${}^{n^3}C_r$ b. ${}^{n}C_{r^3}$
c. ${}^{3n}C_r$ d. $3 \cdot {}^{2n}C_r$

6 The value of $\sum_{r=0}^{n} (-1)^r \left(\frac{{}^nC_r}{{}^{r+3}C_r}\right)$, is

a. $\frac{n(n+1)}{2}$ b. $\frac{n+3}{3}$
c. $\frac{3}{n+3}$ d. $\frac{n+2}{2}$

7 The fractional part of $\frac{3^{2n}}{8}$, is equal to

a. $\frac{3}{8}$ b. $\frac{7}{8}$
c. $\frac{1}{8}$ d. None of these

8 If the unit digit of $13^n + 7^n - 3^n$, $n \in N$, is 7, then the value of n is of the form

a. $4k+1, k \in I$ b. $4k+2, k \in I$
c. $4k+3, k \in I$ d. $4k, k \in I$

9 The value of $S = \sum_{r=0}^{n-1} \frac{1}{(n-r)^2} \cdot \left(\frac{{}^nC_{r+1}}{{}^nC_r}\right)^2$ equals

a. $\frac{1}{n+1}$ b. $\frac{2}{n+1}\left(1+\frac{1}{2}+\frac{1}{3}+\cdots+\frac{1}{n}\right)$
c. $\frac{1}{n+1}\left(1+\frac{1}{2}+\cdots+\frac{1}{n}\right)$ d. $\frac{1}{1^2}+\frac{1}{2^2}+\frac{1}{3^2}+\cdots+\frac{1}{n^2}$

Working Space

10 $(x+m)^m - {}^mC_1(x+m-1)^m + {}^mC_2(x+m-2)^m - \cdots + (-1)^m \cdot x^m$ is equal to

a. $(m+1)!$
b. ${}^{m+1}P_m$
c. $m!$
d. ${}^mP_{m-2}$

11 The number $(2010)^n - 2(1010)^n - 4(510)^n + 5(10)^n$, $n \in N$ is always divisible by

a. 10^4
b. 10^3
c. 10^5
d. None of these

One or More than One Options Correct Type

Direction (Q. Nos. 12-16) *This section contains 5 multiple choice questions. Each question has four choices (a), (b), (c) and (d), out of which ONE or MORE THAN ONE are correct.*

12 If $(9+\sqrt{80})^n = I + f$, where I, n are integers and $0 < f < 1$, then

a. I is odd integer
b. I is an even integer
c. $(I+f)(1-f)=1$
d. $I - f = (9-\sqrt{80})^n$

13 If ${}^{100}C_6 + 4 \cdot {}^{100}C_7 + 6 \cdot {}^{100}C_8 + 4 \cdot {}^{100}C_9 + {}^{100}C_{10}$ has the value equal to ${}^xC_y$, then the value of $(x+y)$ can be

a. 114
b. 115
c. 198
d. 199

14 The value of the constant term in the binomial $\left(x^2 + \frac{1}{x^2} - 2\right)^{10}$ is also equal to

a. number of different dissimilar terms in $(x_1 + x_2 + \cdots + x_{10})^{10}$
b. $({}^{10}C_0)^2 + ({}^{10}C_1)^2 + \cdots + ({}^{10}C_{10})^2$
c. coefficient of x^{10} in $(1-x)^{20}$
d. number of linear arrangements of 20 things of which 10 alike of one kind and rest all are different, taken all at a times.

15 The coefficient of x^{50} in $\sum_{k=0}^{100} {}^{100}C_k (x-2)^{100-k} \cdot 3^k$ is also equal to

a. number of ways in which 50 identical books can be distributed in 100 students, if each student can get at most one book.
b. number of ways in which 100 different white balls and 50 identical red balls can be arranged in a circle, if no two red balls are together.
c. number of dissimilar terms in $(x_1 + x_2 + x_3 + \cdots + x_{50})^{51}$
d. $\frac{2 \cdot 6 \cdot 10 \cdot 14 \cdots 198}{(50)!}$

16 The value of ${}^{200}C_0 - {}^{200}C_1 + {}^{200}C_2 - {}^{200}C_3 + \cdots + {}^{200}C_{148}$ is equal to

a. ${}^{198}C_{148}$
b. ${}^{198}C_{51}$
c. ${}^{200}C_{148}$
d. ${}^{199}C_{148}$

Comprehension Type

Direction (Q. Nos. 17-26) *This section contains 3 paragraphs, each describing theory, experiments, data, etc. Ten questions related to the paragraphs have been given. Each question has only one correct answer among the four given options (a), (b), (c) and (d).*

Passage I

If $(1+px+x^2)^n = 1 + a_1x + a_2x^2 + \cdots + a_{2n}x^{2n}$, then

17 Which of the following is true for $1 < r < 2n$

a. $(np+pr)a_r = (r+1)a_{r+1} + (r-1)a_{r-1}$
b. $(np-pr)a_r = (r+1)a_{r+1} + (r-1-2n)a_{r-1}$
c. $(np-pr)a_r = (r+1)a_{r+1} + (r-1-n)a_{r-1}$
d. $(2np+pr)a_r = (r+1+n)a_{r+1} + (r+1-n)a_{r-1}$

18 The remainder obtained, when $a_1 + 5a_2 + 9a_3 + \cdots + (8n-3)a_{2n}$ is divided by $(p+2)$, is

a. 1
b. 2
c. 3
d. 0

19 The value of $a_1 + 3a_2 + 5a_3 + 7a_4 + \cdots + (4n-1)a_{2n}$, when $p = -3$ and $n \in$ even, is

a. n
b. $2n-1$
c. $2n-2$
d. $2n$

Working Space

Passage II

If $(1+x+x^2)^{100} = \sum_{r=0}^{200} a_r x^r$, then

20 Which of the following is true?

a. $a_{28} = a_{72}$
b. $a_{56} = a_{144}$
c. $a_{200} = a_{300}$
d. None of these

21 $a_0 + a_1 + a_2 + \cdots + a_{99}$, is equal to

a. $\dfrac{3^{99} - a_{99}}{2}$
b. $\dfrac{3^{101} - a_{99}}{2}$
c. $\dfrac{3^{100} - a_{100}}{2}$
d. None of these

22 $37a_{35}$, is equal to

a. $64a_{36} + 165a_{35}$
b. $64a_{35} + 148a_{36}$
c. $56a_{32} + 168a_{22}$
d. None of these

Passage III

Let n be a positive integer such that $(1+x)^n = {}^nC_0 + {}^nC_1 x + {}^nC_2 x^2 + \cdots + {}^nC_n x^n$, then

23 The value of $\sum_{0 \le r < s \le n} \sum (r+s)({}^nC_r + {}^nC_s)$, is

a. $n^2 \cdot 2^{n-1}$
b. $n(n-1)2^{n-2}$
c. $n^2 \cdot 2^{n-2}$
d. $n^2 \cdot 2^n$

24 The value of $\sum_{0 \le r < s \le n} \sum (r+s)({}^nC_r + {}^nC_s + {}^nC_r \cdot {}^nC_s)$, is

a. $n^2 \cdot 2^n - \frac{n}{2}[2^{2n} - {}^{2n}C_n]$
b. $n^2 \cdot 2^n + \frac{n}{2}[2^{2n} - {}^{2n}C_n]$
c. $n^2 \cdot 2^n - \frac{n}{2}[2^{2n} + {}^{2n}C_n]$
d. $n^2 \cdot 2^n + \frac{n}{2}[2^n - 1]$

25 The value of $\sum_{r=0}^{n} \sum_{s=0}^{n} \sum_{t=0}^{n} \sum_{u=0}^{n} (1)$, is

a. ${}^nC_4$
b. ${}^{n+1}C_4$
c. ${}^{4n+1}C_4$
d. $(n+1)^4$

26 The value of $\sum_{0 \le r < s < t < u \le n} \sum \sum \sum (2)$, is

a. $2 \cdot {}^nC_4$
b. $2 \cdot {}^{n+1}C_4$
c. $2 \cdot {}^{4n+1}C_4$
d. $2(n+1)^4$

One Integer Value Correct Type

***Direction** (Q. Nos. 27-31) This section contains 5 questions. When worked out will result in an integer from 0 to 9 (both inclusive).*

27 If $\sum_{k=0}^{100} \left(\dfrac{k}{k+1}\right) {}^{100}C_k = \dfrac{a(2^{100}) + b}{c}$, where a, b and $c \in N$, then find the number of digits in least value of $(a+b+c)$.

28 The value of $\lim_{n \to \infty} \sum_{r=0}^{n} \dfrac{{}^nC_r}{n^r (r+3)} = e - x$, then x is

29 $\lim_{x \to 0} \left(\sum_{r=0}^{n} (-1)^r \cdot {}^nC_r \cdot \left(\sum_{k=0}^{n-r} {}^{n-r}C_k \cdot x^k \cdot 2^k \right) \cdot (x^2 - x)^r \right)^{1/x} = e^{n\lambda}$, then λ is equal to

30 If ${}^nC_0 - {}^nC_1 + {}^nC_2 - {}^nC_3 + \cdots + (-1)^r\, {}^nC_r = 28$, then n is equal to

31 If $S = \sum_{k=1}^{\infty} \dfrac{1}{k^2}$ and $t = \sum_{k=1}^{\infty} \dfrac{(-1)^{k+1}}{k^2}$, then the value of $\dfrac{s}{t}$ is equal to

JEE Main & AIEEE Archive

(Compilation of Last 13 Years Questions)

1 If the coefficients of x^3 and x^4 in the expansion of $(1+ax+bx^2)(1-2x)^{18}$ in powers of x are both zero, then (a, b) is equal to **(JEE Main 2014)**

a. $\left(16, \frac{251}{3}\right)$ b. $\left(14, \frac{251}{3}\right)$

c. $\left(14, \frac{272}{3}\right)$ d. $\left(16, \frac{272}{3}\right)$

2 The term independent of x in the expansion of

$$\left(\frac{x+1}{x^{2/3}-x^{1/3}+1}-\frac{x-1}{x-x^{1/2}}\right)^{10} \text{ is}$$

a. 4 b. 120 **(JEE Main 2013)**

c. 210 d. 310

3 If n is a positive integer, then $(\sqrt{3}+1)^{2n}-(\sqrt{3}-1)^{2n}$ is **(AIEEE 2012)**

a. an irrational number

b. an odd positive integer

c. an even positive integer

d. a rational number other than positive integers

4 The coefficient of x^7 in the expansion of $(1-x-x^2+x^3)^6$ is **(AIEEE 2011)**

a. – 132 b. –144

c. 132 d. 144

5 Let $S_1=\sum_{j=1}^{10} j(j-1)\,{}^{10}C_j$,

$S_2=\sum_{j=1}^{10} j\,{}^{10}C_j$

and $S_3=\sum_{j=1}^{10} j^2\,{}^{10}C_j$.

Statement I $S_3=55\times 2^9$ **(AIEEE 2010)**

Statement II $S_1=90\times 2^8$

and $S_2=10\times 2^8$

a. Both Statement I and Statement II are correct and Statement II is the correct explanation of Statement I

b. Both Statement I and Statement II are correct and Statement II is not the correct explanation of Statement I

c. Statement I is correct but Statement II is incorrect

d. Statement II is correct but Statement I is incorrect

6 The remainder left out when $8^{2n}-(62)^{2n+1}$ is divided by 9, is **(AIEEE 2009)**

a. 0 b. 2

c. 7 d. 8

7 **Statement I** $\sum_{r=0}^{n}(r+1)\cdot{}^nC_r=(n+2)2^{n-1}$

Statement II $\sum_{r=0}^{n}(r+1)\,{}^nC_r\cdot x^r$

$=(1+x)^n+nx(1+x)^{n-1}$ **(AIEEE 2008)**

a. Both Statement I and Statement II are correct and Statement II is the correct explanation of Statement I

b. Both Statement I and Statement II are correct and Statement II is not the correct explanation of Statement I

c. Statement I is correct but Statement II is incorrect

d. Statement II is correct but Statement I is incorrect

8 In the binomial expansion of $(a-b)^n, n \geq 5$, the sum of 5th and 6th terms is zero, then $\frac{a}{b}$ is equal to **(AIEEE 2007)**

a. $\frac{5}{n-4}$ b. $\frac{6}{n-5}$

c. $\frac{n-5}{6}$ d. $\frac{n-4}{5}$

9 The sum of the series ${}^{20}C_0-{}^{20}C_1+{}^{20}C_2-{}^{20}C_3+\cdots+{}^{20}C_{10}$ is **(AIEEE 2007)**

a. $-\,{}^{20}C_{10}$ b. $\frac{1}{2}\,{}^{20}C_{10}$

c. 0 d. ${}^{20}C_{10}$

10 In the expansion, powers of x in the function $\frac{1}{(1-ax)(1-bx)}$ is $a_0+a_1x+a_2x^2+a_3x^3+\ldots$, then a_n is equal to **(AIEEE 2006)**

a. $\frac{a^n-b^n}{b-a}$ b. $\frac{a^{n+1}-b^{n+1}}{a-b}$

c. $\frac{b^{n+1}-a^{n+1}}{b-a}$ d. $\frac{b^n-a^n}{b-a}$

11 For natural numbers m and n, if $(1-y)^m(1+y)^n=1+a_1y+a_2y^2+\ldots$ and $a_1=a_2=10$, then (m, n) is **(AIEEE 2006)**

a. (35, 20) b. (45, 35)

c. (35, 45) d. (20, 45)

12 The value of ${}^{50}C_4+\sum_{r=1}^{6}{}^{56-r}C_3$ is **(AIEEE 2005)**

a. ${}^{56}C_4$ b. ${}^{56}C_3$

c. ${}^{55}C_3$ d. ${}^{55}C_4$

13 If the coefficients of rth, $(r+1)$th and $(r+2)$th terms in the binomial expansion of $(1+y)^m$ are in AP, then m and r satisfy the equation **(AIEEE 2005)**

a. $m^2 - m(4r-1) + 4r^2 + 2 = 0$
b. $m^2 - m(4r+1) + 4r^2 - 2 = 0$
c. $m^2 - m(4r+1) + 4r^2 + 2 = 0$
d. $m^2 - m(4r-1) + 4r^2 - 2 = 0$

14 If the coefficient of x^7 in $\left(ax^2 + \frac{1}{bx}\right)^{11}$ equal to the coefficient of x^{-7} in $\left(ax - \frac{1}{bx^2}\right)^{11}$, then a and b satisfy the relation **(AIEEE 2005)**

a. $ab = 1$ b. $\frac{a}{b} = 1$
c. $a + b = 1$ d. $a - b = 1$

15 If x is so small that x^3 and higher powers of x may be neglected, then $\frac{(1+x)^{3/2} - \left(1 + \frac{1}{2}x\right)^3}{(1-x)^{1/2}}$ may be approximated as **(AIEEE 2005)**

a. $\frac{x}{2} - \frac{3}{8}x^2$ b. $-\frac{3}{8}x^2$
c. $3x + \frac{3}{8}x^2$ d. $1 - \frac{3}{8}x^2$

16 The coefficient of the middle term in the binomial expansion in powers of x of $(1+\alpha x)^4$ and of $(1-\alpha x)^6$ is the same, if α is equal to **(AIEEE 2004)**

a. $-\frac{5}{3}$ b. $\frac{10}{3}$
c. $-\frac{3}{10}$ d. $\frac{3}{5}$

17 The coefficient of x^n in the expansion of $(1+x)(1-x)^n$ is **(AIEEE 2004)**

a. $(n-1)$
b. $(-1)^n(1-n)$
c. $(-1)^{n-1}(n-1)^2$
d. $(-1)^{n-1}n$

18 If $S_n = \sum_{r=0}^{n} \frac{1}{{}^nC_r}$ and $t_n = \sum_{r=0}^{n} \frac{r}{{}^nC_r}$, then $\frac{t_n}{S_n}$ is equal to **(AIEEE 2004)**

a. $\frac{n}{2}$ b. $\frac{n}{2} - 1$
c. $n - 1$ d. $\frac{2n-1}{2}$

19 The number of integral terms in the expansion of $(\sqrt{3} + \sqrt[8]{5})^{256}$ is **(AIEEE 2003)**

a. 32 b. 33
c. 34 d. 35

20 If x is positive, the first negative term in the expansion of $(1+x)^{27/5}$ is **(AIEEE 2003)**

a. 7th term
b. 5th term
c. 8th term
d. 6th term

21 The coefficient of x^5 in $(1 + 2x + 3x^2 + \dots)^{-3/2}$ is **(AIEEE 2002)**

a. 21 b. 25
c. 26 d. None of these

22 If $|x| < 1$, then the coefficient of x^n in the expansion of $(1 + x + x^2 + x^3 + \dots)^2$ is **(AIEEE 2002)**

a. n b. $n - 1$
c. $n + 2$ d. $n + 1$

JEE Advanced & IIT JEE Archive

(Compilation of Last 13 Years Questions)

1 Coefficient of x^{11} in the expansion of $(1+x^2)^4(1+x^3)^7(1+x^4)^{12}$ is **(2014 Adv., Single Option Correct)**

a. 1051
b. 1106
c. 1113
d. 1120

2 The coefficients of three consecutive terms of $(1+x)^{n+5}$ are in the ratio 5 : 10 : 14. Then, n is equal to **(2013 Adv., Integer Type)**

3 For $r = 0, 1, \ldots, 10$, let A_r, B_r and C_r denote, respectively, the coefficient of x^r in the expansions of $(1+x)^{10}$, $(1+x)^{20}$ and $(1+x)^{30}$. Then, $\sum_{r=1}^{10} A_r(B_{10}B_r - C_{10}A_r)$ is equal to **(2010, Single Option Correct)**

a. $B_{10} - C_{10}$
b. $A_{10}(B_{10}^2 - C_{10}A_{10})$
c. 0
d. $C_{10} - B_{10}$

4 $\binom{30}{0}\binom{30}{10} - \binom{30}{1}\binom{30}{11} + \cdots + \binom{30}{20}\binom{30}{30}$ is equal to **(2005, Single Option Correct)**

a. ${}^{30}C_{11}$
b. ${}^{60}C_{10}$
c. ${}^{30}C_{10}$
d. ${}^{65}C_{55}$

2. Permutation and Combination

DPP
DAILY PRACTICE PROBLEMS

DPP-1 Factorial and Exponent of Prime Value p in $(n!)$

Subjective Questions

Working Space

Direction *(Q. Nos. 1-8) These questions are subjective in nature, need to be solved completely on notebook.*

1 Prove that

i. $\frac{1}{4!}+\frac{1}{5!}+\frac{1}{6!}=\frac{37}{6!}$. ii. $\frac{1}{9!}+\frac{1}{10!}+\frac{1}{11!}+\frac{1}{12!}=\frac{122\times 13}{12!}$.

iii. $n(n-1)(n-2)\cdots(n-r+1)=\frac{n!}{(n-r)!}$ iv. $1\cdot 3\cdot 5\cdots(2n-1)(2n+1)\cdot 2^n=\frac{(2n+1)!}{n!}$

v. $2\cdot 6\cdot 10\cdot 14\cdots$ to n factors $=\frac{(2n)!}{n!}=(n+1)(n+2)\cdots(2n)$.

vi. $(100)!=2^{50}\cdot(50)!(1\cdot 3\cdot 5\cdot 7\cdots 99)$.

2 Solve for x

i. $(x)!=132(x-2)!$ ii. $(x+2)!=870(x)!$

iii. $\frac{(x+3)!(2x-1)!}{(2x+1)!(x+2)!}=\frac{7}{72}$ iv. $\frac{x!}{2!(x-2)!}:\frac{x!}{4!(x-4)!}=\frac{2}{1}$

3 Find exponent of 2 in (65).

4 Find exponent of 3 in (100).

5 Find the prime factorisation of 10!, i.e. if $10!=2^p\cdot 3^q\cdot 5^r\cdot 7^s$, then find p, q, r and s.

6 Find the exponent of 2 in $x=20\times 19\times 18\times 17\times\cdots\times 11$.

7 Find exponent of 12 in (50)!.

8 Find the number of zero's at the end of (60)!.

Only One Option Correct Type

Direction *(Q. Nos. 9-10) This section contains 2 multiple choice questions. Each question has four choices (a), (b), (c) and (d), out of which ONLY ONE option is correct.*

9 The number of prime numbers among the numbers $105!+2, 105!+3, 105!+4, \cdots, 105!+104, 105!+105$, is

a. 31 b. 32 c. 33 d. None of these

10 For some natural number N, the number of positive integral 'x' satisfying the equation, $1!+2!+3!+\cdots+(x!)=(N)^2$, is

a. one b. two c. infinite d. None of these

DPP-2 Fundamental Principle of Counting

Subjective Questions

Working Space

Direction (Q. Nos. 1-3) *These questions are subjective in nature, need to be solved completely on notebook.*

1 There are 6 roads between A and B and 4 roads between B and C.

i. In how many ways can one drive from A to C by way of B?
ii. In how many ways can one drive from A to C and back to A, passing through B on both trips?
iii. In how many ways can one drive the circular trip described in (ii) without using the same road more than once.

2 If repetitions are not permitted,

i. how many 3-digit numbers can be formed from the six digits 2, 3, 5, 6, 7 and 9.
ii. how many of these are less than 400.
iii. how many are even.
iv. how many are odd.
v. how many are multiples of 5.

3 i. Find the number of four letter word that can be formed from the letters of the word HISTORY. (each letter to be used at most once).
ii. How many of them contains only consonants.
iii. How many of them begin and end in a consonant.
iv. How many of them begin with a vowel.
v. How many contains the letters Y.
vi. How many begin with T and end in a vowel.
vii. How many begin with T and also contain S.
viii. How many contain both vowels.

Only One Option Correct Type

Direction (Q. Nos. 4-22) *This section contains* 19 *multiple choice questions. Each question has four choices (a), (b), (c) and (d), out of which ONLY ONE option is correct.*

4 3 different railway passes are allotted to 5 students, the number of ways this can be done is

a. 60 b. 20
c. 15 d. 10

5 A man has 3 jackets, 10 shirts and 5 pair of slacks. If an outfit consists of a jacket, a shirt, and a pair of slacks, the different outfits can the man make, is

a. 120 b. 150
c. 180 d. 210

Working Space

6 The number of ways in which 5 letters be mailed, if there are 3 different mailboxes available, if each letters can be mailed in any mailbox, is

a. 81 b. 243 c. 60 d. 125

7 The number of four-digit numbers are there which are divisible by 2, is

a. 9000 b. 4000
c. 4500 d. 1000

8 The number of natural numbers are their from 1 to 1000 which have none of their digits repeated, is

a. 738 b. 81
c. 648 d. 729

9 Number of natural numbers between 100 and 1000 such that atleast one of their digits is 7, is

a. 225 b. 243
c. 252 d. None of these

10 Number of odd numbers of five distinct digits can be formed with the digits 0, 1, 2, 3, 4, is

a. 24 b. 120
c. 48 d. 36

11 Number of ways in which 7 different colours in a rainbow can be arranged, if green is always in the middle, is

a. 5! b. 7!
c. 6! d. 8!

12 A letter lock consists of three rings each marked with 10 different letters. Then, the number of ways in which it is possible to make an unsuccessful attempts to open the lock, is

a. 1000 b. 999
c. 720 d. 10000

13 The number of 10-digit numbers that can be made with odd digits, so that no two consecutive digits are same, is

a. 5×4^9 b. 5^{10}
c. 5^9 d. 5×4^8

14 It is required to seat 5 men and 4 women in a row, so that the women occupy the even places, the number of such arrangements are

a. 2880 b. $5! \times 4$
c. $5 \times 4!$ d. None of these

15 The number of 5 lettered palindromes which can be formed using the letters from the English alphabets, is

a. 26^5 b. 26^4
c. 26^3 d. 26^2

16 The number of ways in which the letters of the word 'MIRACLE' can be arranged, if vowels always occupy the odd places, is

a. 144 b. 576
c. 7! d. 720

17 The number of words which can be formed using all the letters of word 'AKSHI', if each word begins with vowels or terminates in vowel, is

a. 96 b. 48
c. 84 d. 120

Working Space

18 The number of ways in which four passangers be accommodate in three railway carriages, if each carriage can accommodate any number of passangers, is

a. 24 b. 64
c. 81 d. None of these

19 The number of ways in which four persons can be accommodate in 3 different chairs, if each person can occupy only one chair, is

a. 24 b. 64
c. 81 d. None of these

20 How many natural number are there with the property that they can be expressed as the sum of the cubes of two natural numbers in two different ways

a. 9 b. 27
c. 64 d. infinitely many

21 If the letters of the word SACHIN are arranged in all possible ways and these words are written out as in dictionary, then the word SACHIN appears at serial number, is

a. 602 b. 603 c. 600 d. 601

22 The 120 permutations of MAHES are arranged in dictionary order, as if each were an ordinary five letter word, the last letter of the 86th word in the list, is

a. A b. H c. S d. E

Comprehension Type

Direction (Q. Nos. 23-27) *This section contains 2 paragraphs, each describing theory, experiments, data, etc. Five questions related to the paragraphs have been given. Each question has only one correct answer among the four given options (a), (b), (c) and (d).*

Passage I

Four prizes are distributed among 5 students, then

23 If no student get more than one prize, then the number of ways, is

a. 5^4 b. 120 c. 4^5 d. 620

24 If each student is eligible for all prizes, then the number of ways is

a. 625 b. 620 c. 1024 d. 1020

25 If no student gets all the prizes, then the number of ways, is

a. 625 b. 620 c. 1024 d. 1020

Passage II

A triangle is called an integer triangle, if all the sides are integers. If a, b and c are sides of an integer triangle, then we can assume $a \le b \le c$ (any other permutation will yield same triangle). Since, sum of two sides is greater than the third side therefore if c is fixed $(a + b)$ will vary from $(c + 1)$ to $2c$. The number of integer triangles can be formed by finding integer solution of

$$a + b = c + 1, a + b = c = 2, \cdots, a + b = 2c.$$

26 The number of integer isosceles or equilateral triangle none of whose side exceed 4 must be

a. 9 b. 10 c. 11 d. 12

27 The total number of integer isosceles or equilateral triangle none of whose sides exceed $2c$ must be

a. c^2 b. $2c^2$ c. $3c^2$ d. $\frac{3c^2}{2}$

DPP-3 Application of ^nP_r

Subjective Question

Working Space

Direction (Q. No. 1) *This question is subjective in nature, need to be solved completely on notebook.*

1 Find the numerical value of

i. $^{50}P_3$

ii. $^6P_4 + {}^8P_2$

Only One Option Correct Type

Direction (Q. Nos. 2-20) *This section contains 19 multiple choice questions. Each question has four choices (a), (b), (c) and (d), out of which ONLY ONE option is correct.*

2 If $^nP_8 = 12 \cdot {}^nP_6$, then the value of n, is

a. 8 b. 9
c. 10 d. 11

3 If $^{2n}P_4 = 396 \times {}^nP_2$, then the values of n, is

a. 4 b. 5
c. 6 d. 7

4 The value of $^{n-1}P_r + r \cdot {}^{n-1}P_{r-1}$, is

a. ^nP_r b. $^{n-1}P_r$
c. $^{n-1}P_{r+1}$ d. $^nP_{r+1}$

5 If $^{l+m}P_2 = 56$, and $^{l-m}P_2 = 12$, then $l + m$, is

a. 6 b. 8
c. 10 d. 12

6 The value of
$1 + 1 \cdot {}^1P_1 + 2 \cdot {}^2P_2 + 3 \cdot {}^3P_3 + \dots + n \cdot {}^nP_n$, is

a. $n!$ b. $(n-1)!$
c. $(n+1)!$ d. $(n+2)!$

7 The number of permutations of n different objects taken r at a time in which two specified objects always occur together, is

a. $\dfrac{(n-2)!}{(n-r)!}$ b. $\dfrac{(n-2)!}{(n-r)!} \cdot 2(r-1)$
c. $\dfrac{(n-2)!}{(n-r)!}(r-1)$ d. None of these

8 If m men and n women are to be seated in a row so that no two women sit together. If $m > n$, then the number of ways in which they can be seated, is

a. $\dfrac{m!(m+1)!}{(m-n+1)!}$ b. $\dfrac{(m+1)!}{(m-n+1)!}$
c. $m!n!$ d. None of these

Working Space

9 The number of ordered pairs (n, r), when $1 \le r \le n \le 8$ and such that $P(n, r)$ is a prime number, is

a. 4 b. 6
c. 7 d. 8

10 How many of the 900 three-digit numbers have atleast one even digit, is

a. 775 b. 875
c. 450 d. 750

11 The number of natural numbers from 1000 to 9999 (both inclusive) that do not have all 4 different digits, is

a. 4048 b. 4464
c. 4518 d. 4536

12 A new flag is to be designed with six vertical strips using some or all of the colours yellow, green, blue and red. Then, the number of ways this can be done such that no two adjacent strips have same colour, is

a. 12×81 b. 16×192
c. 20×25 d. 24×216

13 A 5-digit numbers divisible by 3 is to be formed using the numerals 0, 1, 2, 3, 4 and 5 without repetition the total number of ways this can be done, is

a. 3125 b. 600 c. 240 d. 216

14 Number of 9-digit numbers divisible by nine digits from 0 to 9, if each digit is used at most once is $k \cdot (8!)$, then k has the value equal to

a. 15 b. 17 c. 19 d. 21

15 Number of natural numbers less than 1000 and divisible by 5 can be formed with ten digits, each digit not occuring more than once in each number, is

a. 72 b. 64 c. 136 d. 154

16 The number of ways in which 5 different books can be distributed among 10 people, if each person can get atmost one book, is

a. 252 b. 10^5 c. 5^{10} d. ${}^{10}P_5$

17 The product of all odd positive integers less than 10000, is

a. $\frac{(10000)!}{(5000!)^2}$ b. $\frac{(10000)!}{2^{5000}}$
c. $\frac{(9999)!}{2^{5000}}$ d. $\frac{(10000)!}{2^{5000} \cdot (5000)!}$

18 A telegraph has x arms and each arm is capable of $(x-1)$ distinct positions, including the position of rest, the total number of signals that can be made, is

a. $(x-1)^x$ b. $x^{x-1}-1$
c. $(x-1)^x-1$ d. x^{x-1}

19 The number of n-digit numbers which consists of the digits 1 and 2 only if each digit is to be used atleast once, is equal to 512, then n is equal to

a. 7 b. 8 c. 9 d. 10

20 The tamer of wild animals has to bring one by one 5 lions and 4 tigers to the circus area. The number of ways this can be done, if no two tigers immediately follow each other, is

a. $5!\,{}^6P_4$ b. $5! \times 6!$
c. $5! \times 4!$ d. $6! \times 4!$

DPP-4 Application of ^nC_r

Only One Option Correct Type

Working Space

Direction *(Q. Nos. 1-17) This section contains 17 multiple choice questions. Each question has four choices (a), (b), (c) and (d), out of which ONLY ONE option is correct.*

1 If $^{10}P_r = 604800$ and $^{10}C_r = 120$, then r is equal to

a. 7 b. 5 c. 4 d. None of these

2 If $^nC_{12} = {}^nC_8$, then $^nC_{17}$ is equal to

a. 1040 b. 1140 c. 1240 d. None of these

3 The positive integer r, such that $^{15}C_{3r} = {}^{15}C_{r+2}$, is equal to

a. 3 b. 4 c. 5 d. None of these

4 If $^{18}C_r = {}^{18}C_{r+2}$, then $^rC_5$ is equal to

a. 23 b. 46 c. 56 d. None of these

5 $^{47}C_4 + \sum_{j=1}^{5} {}^{52-j}C_3$, is equal to

a. $^{52}C_4$ b. $^{51}C_4$ c. $^{51}C_3$ d. None of these

6 $^{15}C_8 + {}^{15}C_9 - {}^{15}C_6 - {}^{15}C_7$, is equal to

a. 8 b. 0 c. 6 d. None of these

7 The least positive integral value of x which satisfy the inequality $^{10}C_{x-1} > 2 \cdot {}^{10}C_x$, is

a. 7 b. 8 c. 9 d. 10

8 The value of
$({}^7C_0 + {}^7C_1) + ({}^7C_1 + {}^7C_2) + ({}^7C_2 + {}^7C_3) + \cdots + ({}^7C_6 + {}^7C_7)$, is

a. $2^8 - 2$ b. $2^8 - 1$ c. $2^8 + 1$ d. 2^8

9 The value of $\frac{^{20}C_n}{^{25}C_m}$, when both numerator and denominator have their greatest values, is

a. $\frac{143}{4025}$ b. $\frac{140}{4025}$

c. $\frac{141}{4025}$ d. None of these

10 If ^nC_r denotes the number of combinations of things r at a time, then the expression $^nC_{r+1} + 2 \cdot {}^nC_r + {}^nC_{r-1}$ equals **[AIEEE 2003]**

a. $^{n+2}C_r$ b. $^{n+2}C_{r+1}$

c. $^{n+1}C_r$ d. $^{n+1}C_{r+1}$

Working Space

11 The value of $^{50}C_4 + \sum_{r=1}^{6} {}^{56-r}C_r$, is

a. $^{56}C_4$
b. $^{54}C_3$
c. $^{55}C_3$
d. $^{55}C_4$

12 A student has to answer 10 out of 13 questions in an examination. The number of ways in which he can answer, if he must answer atleast 3 of the first five questions, is

a. 276
b. 267
c. 80
d. 1200

13 The number of different seven-digit numbers that can be written using only three digits 1, 2 and 3 under the condition that the digit 2 occurs exactly twice in each number, is

a. 672
b. 640
c. 512
d. None of these

14 Out of seven consonants and four vowels, the number of words of six letters formed by taking four consonants and two vowels, is

a. 210 b. 462 c. 151200 d. 332640

15 A bag contains 5 mangoes and 6 oranges. In how many ways can one make a selection, so as to take atleast one mango and one orange, is

a. 31×63
b. 2^{11}
c. 1024
d. None of these

16 The number of all three elements subsets of the set $\{a_1, a_2, \ldots, a_n\}$ which contains a_3, is

a. $^{n}C_3$
b. $^{n-1}C_3$
c. $^{n-1}C_2$
d. None of these

17 The number of positive integers satisfying the inequality $^{n+1}C_{n-2} - {}^{n+1}C_{n-1} \le 100$, is

a. 9
b. 8
c. 5
d. None of these

Comprehension Type

Direction (Q. Nos. 18-20) This section contain 1 paragraph, describing theory, experiments, data, etc. Three questions related to the paragraphs have been given. Each question has only one correct answer among the four given options (a), (b), (c) and (d).

Passage

In how many ways can 5 colours be selected out of 8 different colours including red, blue and green, when

18 Blue and green are always to be included, is

a. 10
b. 20
c. 21
d. 40

19 Red is always excluded, is

a. 10
b. 20
c. 21
d. 40

20 Red and blue are always included but green excluded, is

a. 10 b. 21 c. 31 d. 40

DPP-5 Restricted Arrangements

Subjective Questions

Working Space

Direction *(Q. Nos. 1-6) These questions are subjective in nature, need to be solved completely on notebook.*

1 Find the number of ways in which 5 boys and 5 girls be seated in a row, so that

i. no two girls may sit together

ii. all the girls sit together and all the boys sit together

iii. all the girls are never together

2 How many different words can be formed with the letters of the word MISSISSIPPI?

3 How many arrangements can be made with the letters of the word MATHEMATICS. In how many of them vowels are together?

4 i. How many different words can be formed with the letters of the word HARYANA?

ii. How many of these begin with *H* and end with *N*?

iii. In how many of these *H* and *N* are together?

5 How many words can be formed from the letters of the word DAUGHTER, so that

i. the vowels always come together.

ii. all vowels never come together.

6 In how many ways can 4 men and 4 women be seated at a round table, if

i. no restriction is imposed.

ii. each women is to be seated between two men.

iii. two particular women must sit together.

iv. two particular women must not sit together.

v. all women must sit together.

vi. there is exactly one person between two particular women.

Only One Option Correct Type

Direction *(Q. Nos. 7-13) This section contains 7 multiple choice questions. Each question has four choices (a), (b), (c) and (d), out of which ONLY ONE option is correct.*

7 The number of arrangements of the letters of the word BANANA in which the two *N*'s do not appear adjacently, is **[IIT JEE 2002]**

a. 40 b. 60 c. 80 d. 100

8 How many different nine-digit numbers can be formed from the number 223355888 by rearranging its digits, so that odd digits occupy even positions, is **[IIT JEE 2000]**

a. 16 b. 36 c. 60 d. 180

Working Space

9 How many necklaces can be made of 18 beads, 8 being blue, 5 red, 4 green and 1 yellow, is

a. $\frac{(17)!}{2\cdot 8!\, 5!\, 4!}$ b. $\frac{(17)!}{8!\, 5!\, 4!}$ c. $\frac{(17)!}{2\cdot 5!\, 4!}$ d. $\frac{(18)!}{2\cdot 5!\, 4!\, 8!}$

10 The numbers of ways in which four S come consecutively in the word MISSISSIPPI, is

a. 8! b. $\frac{8!}{4!}$ c. $\frac{8!}{4!\, 2!}$ d. None of these

11 In how many ways three different rings can be worn in four figures with atmost one ring in each figure, is

a. 4 b. 24 c. 81 d. 64

12 In how many ways five different rings can be worn in four figures with atleast one ring in each figure, is

a. 240 b. 6720 c. 480 d. None of these

13 The number of ways of putting five different rings on four fingers of left hand, is

a. 1240 b. 6720 c. 480 d. None of these

Comprehension Type

Direction *(Q. Nos. 14-20) This section contains 2 paragraphs, each describing theory, experiments, data, etc. Seven questions related to the paragraphs have been given. Each question has only one correct answer among the four given options (a), (b), (c) and (d).*

Passage I

There are r distinct objects and n distinct boxes such that $n > r$, then

14 The numbers of ways of putting r objects in n boxes, if each box is sufficient to contain r objects, is

a. n^r b. ${}^nC_r$ c. ${}^nP_r$ d. None of these

15 The number of ways of arranging r objects in n boxes so that each box contains only one object, is

a. n^r b. ${}^nC_r$ c. ${}^nP_r$ d. None of these

16 The number of ways of arranging r objects in n boxes if, a box may contain any number of objects, is

a. n^r b. ${}^nC_r$ c. ${}^nP_r$ d. ${}^{n+r-1}P_n$

Passage II

Five balls are to be placed in 3 boxes. Each can hold all the five balls. In how many ways can we place the balls so that no box remains empty, if

17 Balls and boxes are all different, is

a. 60 b. 30 c. 150 d. 120

18 Balls are identical but boxes are different, is

a. 3 b. 6 c. 9 d. None of these

19 Balls are different but boxes are identical, is

a. 20 b. 30 c. 50 d. 90

20 Balls as well as boxes are identical, is

a. 2 b. 3 c. 6 d. 9

DPP-6 Geometrical Application of ${}^nC_r$

Only One Option Correct Type

Working Space

Direction *(Q. Nos. 1-16) This section contains 16 multiple choice questions. Each question has four choices (a), (b), (c) and (d), out of which ONLY ONE option is correct.*

1 The number of diagonals that can be drawn by joining the vertices of an octagon, is

a. 20 b. 28
c. 48 d. None of these

2 The number of diagonals in a polygon of n sides, is

a. $\frac{n(n-1)}{2}$ b. $\frac{(n-1)(n-2)}{2}$
c. $\frac{n(n-3)}{2}$ d. None of these

3 The number of lines drawn through 6 points lying on a circle, is

a. 12 b. 15
c. 24 d. 30

4 The straight lines l_1, l_2, l_3 are parallel and lie in the same plane. A total number of m points are taken on l_1, n points on l_2, k points on l_3. The maximum number of triangles formed with vertices at these points, are

a. ${}^{m+n+k}C_3$ b. ${}^{m+n+k}C_3 - {}^mC_3 - {}^nC_3 - {}^kC_3$
c. ${}^mC_3 + {}^nC_3 + {}^kC_3$ d. None of these

5 Out of 18 points in a plane no three are in the same straight line except five points which are collinear. The number of straight lines that can be formed joining them, is

a. 143 b. 144
c. 153 d. None of these

6 The maximum number of points of intersections of 8 straight lines, is

a. 56 b. 28
c. 16 d. 8

7 A polygon has 44 diagonals. The number of its sides are

a. 13 b. 12
c. 11 d. 10

8 There are 10 points in a plane of which no three are collinear and 4 points are concyclic the number of different circles that can be drawn through atleast 3 of these points, is

a. 116 b. 120
c. 117 d. None of these

Working Space

9 The maximum number of points into which 4 circles and 4 straight lines intersect, is

a. 26 b. 50
c. 56 d. 72

10 The interior angles of a regular polygon measure 150° each the number of diagonals of the polygon, is

a. 35 b. 44
c. 54 d. 78

11 The number of parallelograms in given figure, is

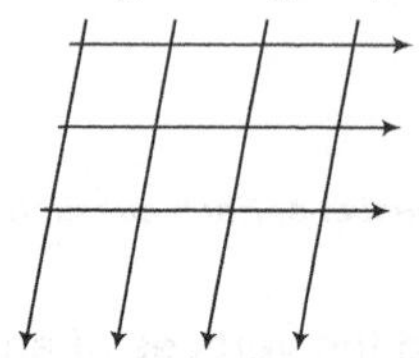

a. 6 b. 18
c. 12 d. 9

12 The number of squares in given figure, is

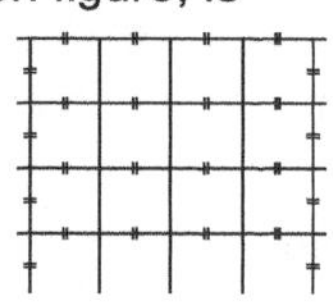

a. 8 b. 18
c. 20 d. 40

13 There are n concurrent lines and another line parallel to one of them. The number of different triangles that will be formed by the $(n+1)$ lines, is

a. $\frac{(n-1)n}{2}$ b. $\frac{(n-1)(n-2)}{2}$
c. $\frac{n(n+1)}{2}$ d. $\frac{(n+1)(n+2)}{2}$

14 18 points are indicated on the perimeter of a ΔABC (see figure). How many triangles are there with vertices at these points

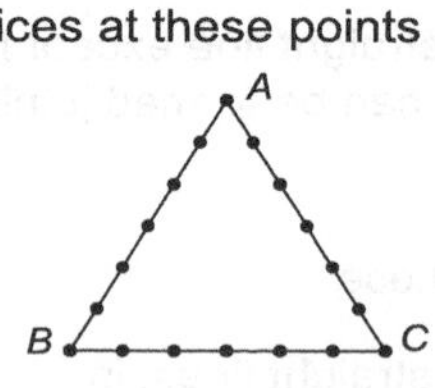

a. 331 b. 408
c. 710 d. 711

15 Let there be 9 fixed points on the circumference of a circle. Each of these points is joined to every one of the remaining 8 points by a straight line and the points are so positioned on the circumference that atmost 2 straight lines meet in any interior point of the circle. The number of such interior intersection points, is

a. 126
b. 351
c. 756
d. None of the above

Working Space

16 The 9 horizontal and 9 vertical lines on an 8×8 chessboard from 'r' rectangles and 's' squares. The ratio $\frac{s}{r}$ in its lowest term, is

a. $\frac{1}{6}$
b. $\frac{17}{108}$
c. $\frac{4}{27}$
d. None of these

Comprehension Type

***Direction** (Q. Nos. 17-20) This section contains 2 paragraphs, each describing theory, experiments, data, etc. Four questions related to the paragraphs have been given. Each question has only one correct answer among the four given options (a), (b), (c) and (d).*

Passage I

There are m points on a straight line AB and n points on the line AC none of them being the point A. Triangles are formed with these points as vertices, then

17 A is excluded, is

a. $m \cdot {}^nC_2 + n \cdot {}^mC_2$
b. ${}^{m+n}C_2$
c. ${}^{m+n}C_3$
d. ${}^{m+n+1}C_3$

18 A is included, is

a. $m \cdot {}^nC_2 + n \cdot {}^mC_2 + mn$
b. $m \cdot {}^nC_2 + n \cdot {}^mC_2$
c. ${}^{m+n+1}C_3$
d. ${}^{m+n}C_3$

Passage II

The triangle are constructed from the vertices of a convex n sided polygon are selected, then

19 Number of triangle's having one side common to polygon, is

a. $n(n-4)$
b. $(n-4)$
c. n
d. $\frac{n}{3}$

20 Number of triangles that have none of sides is common to polygon, is

a. ${}^nC_3 - n$
b. ${}^nC_3 - n - n(n-4)$
c. $n(n-4)$
d. None of these

DPP-7 Problems on Arrangements and Selection

Only One Option Correct Type

Working Space

Direction (Q. Nos. 1-10) *This section contains 10 multiple choice questions. Each question has four choices (a), (b), (c) and (d), out of which ONLY ONE option is correct.*

1 The number of three-digit numbers, whose middle digit is bigger than the extreme digits, is

a. 180
b. 240
c. 300
d. None of these

2 The number of three-digit numbers having only two consecutive digits identical, is

a. 153
b. 162
c. 180
d. 161

3 Number of 3-digits numbers in which the digit at hundreath's place is greater than the other two digit, is

a. 285
b. 281
c. 240
d. 204

4 If m denotes the number of 5-digit numbers if each successive digits are in their descending order of their magnitude and n is in their ascending order of their magnitude, then $(m-n)$ has the value

a. ${}^{10}C_4$
b. ${}^{9}C_5$
c. ${}^{10}C_3$
d. ${}^{9}C_3$

5 There are n persons and m monkey $(m > n)$. Number of ways in which each person may become the owner of one monkey, is

a. n^m
b. m^n
c. ${}^mP_n$
d. mn

6 Numbers of ways in which 7 green bottles and 8 blue bottles can be arranged in a row if exactly 1 pair of green bottle is side by side, is (assume all bottles to be alike except for the colour).

a. 84
b. 360
c. 504
d. 84

7 Number of three-digit numbers with atleast one 3 and atleast one 2, is

a. 58
b. 56
c. 54
d. 52

8 There are 10 red balls of different shades and 9 green balls of identical shades. Then, the number of ways to arrange them in a row so that two green balls are together, is

a. $(10!)^{11}P_9$
b. $(10!)^{11}C_9$
c. 10!
d. 10!9!

9 In a unique hockey series between India and Pakistan, they decide to play on till a team wins 5 matches. The number of ways in which the series can be won by India, if no match ends in a draw, is

a. 126
b. 252
c. 225
d. None of these

Working Space

10 The sum of all four-digit numbers that can be formed with the digits 0, 1, 2 and 3, is

a. 415968
b. $3!(6)\cdot\left(\frac{10^4-1}{10-1}\right)$
c. 15968
d. 455968

Comprehension Type

Direction (Q. Nos. 11-14) *This section contains* 1 *paragraph, each describing theory, experiments, data, etc. Four questions related to the paragraphs. Each question has only one correct answer among the four given options (a), (b), (c) and (d).*

Passage

Six married couple are sitting in a room. Then, the number of ways in which 4 people can be selected, so that

11 They do not form a couple, is

a. 240 b. 255 c. 480 d. 495

12 They form exactly one couple, is

a. 240 b. 255 c. 480 d. 495

13 They form atleast one couple, is

a. 240 b. 255 c. 480 d. 495

14 They form atmost one couple, is

a. 240 b. 255 c. 480 d. 495

One Integer Value Correct Type

Direction (Q. Nos. 15-20) *This section contains* 6 *questions. When worked out will result in an integer from* 0 *to* 9 *(both inclusive).*

15 Number of permutations of 1, 2, 3, 4, 5, 6, 7, 8 and 9 taken all at a time, such that the digit

1 appearing somewhere to the left of 2,

3 appearing to the left of 4 and

5 somewhere to the left of 6 is $k\cdot(7!)$, then k is

16 Number of different natural numbers which are smaller than two hundred million and using only the digits 1 or 2 is $k(2^8)-2$, then k is

17 5 Indian and 5 American couples meet at party and shake hand. If no wife shake hand with her own husband and no Indian wife shakes hands with a male, then the number of handshakes that take place in the party is $15\cdot k$, then

18 The number of permutations of the word 'AUROBIND' in which vowels appear in an alphabetical order is $\frac{8!}{k!}$, then k is

19 All the five digit numbers in which each successive digit exceeds its predecessor are arranged in the increasing order of their order of their magnitudes. The 97th number in the list does not contain the digit.

20 Define a 'Good word' as a sequence of letters that consists only of the letters A, B and C and in which A never immediately followed by B, B is never immediately followed by A. If the number of n letter good words is 384, then the value of n, is

DPP-8 Division of Objects into Groups of Equal and Unequal Groups, Number of Divisions and Sum of Divisions

Only One Option Correct Type

Direction *(Q. Nos. 1-23) This section contains 23 multiple choice questions. Each question has four choices (a), (b), (c) and (d), out of which ONLY ONE option is correct.*

Working Space

1 The number of ways in which 9 students can be equally distributed among 3 sections, is

a. $\frac{9!}{3(3!)^2}$ b. $\frac{9!}{(3!)^3}$

c. 9! d. None of these

2 The number of divisors of 9600 including 1 and 9600 are

a. 60 b. 58

c. 48 d. 46

3 The total number of ways of dividing 15 different things into groups of 8, 4 and 3 respectively, is

a. $\frac{(15!)}{8!\,4!\,3!\,3!}$ b. $\frac{(15)!}{8!\,4!\,3!}$

c. $\frac{(15)!}{8!\,4!\,2!}$ d. None of these

4 In how many ways can we put 12 different balls in three different boxes such that first box contains exactly 5 balls, is

a. $\frac{12!}{5!\,2^7}$ b. $\frac{12!}{5!\,7!}$

c. ${}^{12}C_5 \times 2^7$ d. None of these

5 The number of ways of distributing 50 identical things among 8 persons in which such a way that three of them get 8 things each, two of them get 7 things each and remaining 3 get 4 things each, is equal to

a. $\frac{8!}{3!\,2!}$ b. $\frac{8!}{(3!)^2 \cdot 2!}$

c. $\frac{50!\,8!}{(3!)^2 \cdot 2!}$ d. None of these

6 In how many ways can 10 balls be divided between two boys, one receiving two and other eight balls, is

a. 45 b. 75

c. 90 d. None of these

Working Space

7 The total number of different combinations of one or more letters which can be made from the letter of the word MISSISSIPPI, is

a. 150 b. 148
c. 149 d. None of these

8 The number of divisors of $a^p b^q c^r d^s$, where a, b, c, d are primes and $p, q, r, s, \in N$ excluding 1 and the number itself, is

a. $pqrs$ b. $(p+1)(q+1)(r+1)(s+1)-4$
c. $pqrs-2$ d. $(p+1)(q+1)(r+1)(s+1)-2$

9 The sum of the divisors of $2^5 \cdot 3^7 \cdot 5^3 \cdot 7^2$, is

a. $2^6 \cdot 3^8 \cdot 5^4 \cdot 7^3$ b. $2^6 \cdot 3^8 \cdot 5^4 \cdot 7^3 - 2 \cdot 3 \cdot 5 \cdot 7$
c. $2^6 \cdot 3^8 \cdot 5^4 \cdot 7^3 - 1$ d. None of these

10 How many divisors of 21600 are divisible by 10 but not by 15?

a. 10 b. 30 c. 40 d. None of these

11 The number of ways in which the number 27720 can be split into two factors which are coprimes, is

a. 15 b. 16 c. 25 d. 49

12 A box contains 2 white balls, 3 black balls and 4 red balls. In how many ways can three balls be drawn from the box if atleast one black ball is to be included in a draw (the balls of the same colour are difference).

a. 60 b. 64 c. 56 d. None of these

13 Number of ways in which 9 different toys be distributed among 4 children belonging to different age group in such a way that distribution among the 3 elder children is even and the youngest one is to receive one toy more, is

a. $\dfrac{(5!)^2}{8}$ b. $\dfrac{9!}{2}$
c. $\dfrac{9!}{3!(2!)^3}$ d. None of these

14 The number of ways in which 200 different things can be divided into groups of 100 pairs, is

a. $\dfrac{(200)!}{2^{100}}$ b. $\dfrac{(200)!}{(100)!\, 2^{100}}$
c. $(101)\cdot(102)\cdots(200)$ d. None of these

15 Seven different coins are to be divided amongst three persons. If no two of the persons receive the same number of coins but each receives atleast one coin and none is left over, then the number of ways in which the division may be made, is

a. 420 b. 630
c. 710 d. None of these

16 Number of different ways in which 8 different books can be distributed among 3 students, if each student receives atleast 2 books, is

a. 2940 b. 3940
c. 4940 d. 5940

17 In an election three districts are to be canvassed by 2, 3 and 5 men, respectively. If 10 men volunteer, then the number of ways they can be allotted to the different districts, is

a. $\dfrac{10!}{2!\, 3!\, 5!}$ b. $\dfrac{10!}{2!\, 5!}$ c. $\dfrac{10!}{(2!)^2 \cdot 5!}$ d. $\dfrac{10!}{(2!)^3 \cdot 3!\, 5!}$

Working Space

18 A shelf contains 20 different books of which 4 are in single volume and the others form sets of 8, 5 and 3 volumes, respectively. Number of ways in which the books, may be arranged on the shelf, if the volumes of each set are together and in their due order, is

a. $\frac{20!}{8!\,5!\,3!}$ b. $7!$ c. $8!$ d. $7 \cdot 8!$

19 There are n identical red balls and m identical green balls. The number of different linear arrangements consisting of "n red balls but not necessarily all the green balls" is ${}^xC_y$, then

a. $x = m + n, y = m$ b. $x = m + n + 1, y = m$
c. $x = m + n + 1, y = m + 1$ d. $x = m + n, y = n$

20 Number of ways in which 5A's and 6B's can be arranged in a row which reads the same backwards and forwards, is

a. 8 b. 9 c. 10 d. 11

21 A gentleman invites a party of $m + n (m \neq n)$ friends to a dinner and places m at one table T_1 and n at another table T_2, the table being round. If not all people shall have the same neighbour in any two arrangement, then the number of ways in which he can arrange the guests, is

a. $\frac{(m+n)!}{4mn}$ b. $\frac{1}{2} \cdot \frac{(m+n)!}{mn}$
c. $2 \cdot \frac{(m+n)!}{mn}$ d. None of these

22 Product of all even divisors of $N = 1000$, is

a. $32 \cdot 10^5$ b. $64 \cdot 2^{14}$
c. $64 \cdot 10^{18}$ d. $128 \cdot 10^6$

23 The sum of all the numbers formed from the digits 1, 3, 5, 7, 9 which are smaller than 10000 if repetition of digits is not allowed, is

a. $(28011)s$ b. $(28041)s$
c. $(28121)s$ d. $(29152)s$, where $s = 25$

Matching List Type

***Direction** (Q. No. 24) Choices for the correct combination of elements from Column I and Column II are given as options (a), (b), (c) and (d), out of which one is correct.*

24

	Column I		Column II
i.	Number of increasing permutations of m symbols are there from the n set numbers $\{a_1, a_2, \cdots, a_n\}$, where the order among the numbers is given by $a_1 < a_2 < a_3 < \cdots < a_{n-1} < a_n$, is	p.	n^m
ii.	There are m men and n monkey. Number of ways in which every monkey has a master, if a man can have any number of monkeys	q.	${}^mC_n$
iii.	Number of ways in which n red balls and $(m-1)$ green balls can be arranged in a line, so that no two red balls are together, is (balls of the same colour and alike)	r.	${}^nC_m$
iv.	Number of ways in which 'm' different toys can be distributed in 'n' children if every child may receive any number of toys, is	s.	m^n

Codes

	i	ii	iii	iv		i	ii	iii	iv
a.	p	q	r	s	b.	r	s	q	p
c.	s	p	r	q	d.	q	r	s	s

Working Space

Comprehension Type

Directions (Q. Nos. 25-33) This section contains 3 paragraphs, each describing theory, experiments, data, etc. Nine questions related to the paragraphs have been given. Each question has only one correct answer among the four given options (a), (b), (c) and (d).

Passage I

There are 4 oranges, 5 apples and 6 mangoes in a fruit basket. In how many ways can a person make a selection of fruits from among the fruits in the basket, if

25 All fruits of the same type are identical

a. 209 b. $2^{15}-1$ c. 2^{10} d. 210

26 All fruits of the same type are different

a. 209 b. $2^{15}-1$ c. 2^{10} d. 210

Passage II

If 52 cards are divided into 4 groups, then the number of ways 52 cards be

27 Distributed equally among four players in order

a. $\frac{(52)!}{(13!)^4}$ b. $\frac{(52)!}{(13!)^4 \cdot 4!}$ c. $\frac{52!}{20!\,15!\,10!\,7!}$ d. $\frac{(52)!}{(15!)^3 \cdot 3!\,7!}$

28 Divided into 4 groups of 13 cards each

a. $\frac{(52)!}{(13!)^4}$ b. $\frac{(52)!}{(13!)^4 \cdot 4!}$ c. $\frac{(52)!}{20!\,15!\,10!\,7!}$ d. $\frac{(52)!}{(15!)^3 \cdot 3!\,7!}$

29 Divided into four sets of 20, 15, 10, 7 each

a. $\frac{(52)!}{(13!)^4}$ b. $\frac{(52)!}{(13!)^4 \cdot 4!}$ c. $\frac{(52)!}{20!\,15!\,10!\,7!}$ d. $\frac{(52)!}{(15!)^3 \cdot 3!\,7!}$

30 Divided into four sets, three of them having 15 cards each and fourth having 7 cards

a. $\frac{(52)!}{(13!)^4}$ b. $\frac{(52)!}{(13!)^4 \cdot 4!}$ c. $\frac{(52)!}{20!\,15!\,10!\,7!}$ d. $\frac{(52)!}{(15!)^3 \cdot 3!\,7!}$

Passage III

Players $P_1, P_2, P_2, \cdots, P_{16}$ take part in a tennis tournament. Lower suffix player is better than any higher suffix player. These are to be divided into 4 groups each comprising of 4 players and the best from each group is selected for semifinals.

31 Number of ways in which 16 players can be divided into 4 equal groups, is

a. $\frac{35}{27}\prod_{r=1}^{8}(2r-1)$ b. $\frac{35}{24}\prod_{r=1}^{8}(2r-1)$

c. $\frac{35}{52}\prod_{r=1}^{8}(2r-1)$ d. $\frac{35}{6}\prod_{r=1}^{8}(2r-1)$

32 Number of ways in which they can be divided into 4 equal groups, if the players P_1, P_2, P_3, P_4 are in different groups, is

a. $\frac{(11)!}{36}$ b. $\frac{(11)!}{72}$ c. $\frac{(11)!}{108}$ d. $\frac{(11)!}{216}$

33 Number of ways in which these 16 players can be divided into four equal groups such that when the best player is selected from each group P_6 is one of them, is $(k)\cdot\frac{12!}{(4!)^3}$. The value of k, is

a. 36 b. 24 c. 18 d. 20

DPP-9 Derangement, Multinomial Selection and Arrangement from Identical Arrangement

Only One Option Correct Type

Working Space

Direction (Q. Nos. 1-23) *This section contains 23 multiple choice questions. Each question has four choices (a), (b), (c) and (d), out of which ONLY ONE option is correct.*

1 The number of ways can 10 letters be placed in 10 marked envelops, so that no letters is in the right envelope are

a. $10!\left\{1-\frac{1}{1!}+\frac{1}{2!}-\frac{1}{3!}+\cdots+\frac{1}{10!}\right\}$
b. $10!\left\{1+\frac{1}{1!}-\frac{1}{2!}-\frac{1}{3!}+\cdots+\frac{1}{10!}\right\}$
c. $10!\left\{1+\frac{1}{1!}+\frac{1}{2!}+\frac{1}{3!}+\cdots+\frac{1}{10!}\right\}$
d. $9!\left\{1+\frac{1}{1!}-\frac{1}{2!}+\frac{1}{3!}-\cdots-\frac{1}{10!}\right\}$

2 Ajay writes letters to his five friends and addresses the corresponding. The number of ways can the letters be placed in the envelops so that atleast two of them are in the wrong envelops are

a. 120
b. 125
c. 119
d. None of these

3 The total number of integral solutions for the equation $x+y+z+t=20$, where x, y, z and t are all ≥ -1, is

a. ${}^{27}C_3$
b. ${}^{20}C_4$
c. ${}^{23}C_3$
d. ${}^{27}C_4$

4 The number of integer solutions of $x+y+z=0$ with $x \geq -5$, $y \geq -5$, $z \geq -5$, is

a. 135
b. 136
c. 455
d. 105

5 The number of non-negative integral solutions of $x+y+z \leq n$, where $x, y, z \in N$, is

a. ${}^{n+3}C_3$
b. ${}^{n+4}C_3$
c. ${}^{n+5}C_5$
d. None of these

6 The number of integral solutions of the equation $3x+y+z=33$, where $x, y, z > 0$, is

a. 210
b. 120
c. 135
d. 520

7 The number of positive integral solution of $abc=30$, is

a. 30
b. 12
c. 27
d. 8

Working Space

8 Number of positive integral solutions satisfying the equation $(x_1 + x_2 + x_3)(y_1 + y_2) = 77$, is

a. 150 b. 270
c. 420 d. 1024

9 Number of positive integral solutions satisfying the equation $x_1 \cdot x_2 \cdot x_3 \cdot x_4 \cdot x_5 = 240$, is

a. 1875 b. 1600
c. 1250 d. None of these

10 The total number of integral solutions of $abc = 24$, is

a. 36 b. 90
c. 120 d. 30

11 If n objects are arranged in a row, then the number of ways of selecting three of these objects, so that no two of them are next to each other, is

a. ${}^{n-2}C_3$ b. ${}^{n-3}C_2$
c. ${}^{n-3}C_3$ d. None of these

12 The number of integers which lie between 1 and 10^6 and which have sum of digits equal to 12, is

a. 8550 b. 5382
c. 6062 d. 8055

13 The number of ways in which a score of 11 can be made from a throw by three persons, each throwing a single die once, is

a. 45 b. 18
c. 27 d. None of these

14 Two classrooms A and B having capacity of 25 and $(n - 25)$ seats, respectively. A_n denotes the number of possible seating arrangements of room 'A', when 'n' students are to be seated in these rooms, starting from room 'A' which is to be filled up full to its capacity. In $A_n - A_{n-1} = (25)!\,({}^{49}C_{25})$, then '$n$' is

a. 49 b. 50
c. 51 d. 52

15 The number of ways in which any four letters can be selected from the word CORGOO, is

a. 15 b. 11
c. 7 d. None of these

16 The number of different words that can be formed out of the letters of the word MORADABAD taken four at a time, is

a. 1500 b. 600
c. 620 d. 626

17 The number of words which can be formed by taking two same and two different letters from the letters of the word 'COMBINATION', is

a. 756 b. 1512
c. 252 d. None of these

18 The total number of ways of selecting five letters from the letters of the word 'INDEPENDENT', is

a. 72 b. 3320
c. 120 d. None of these

Working Space

19 If chocolates of a particular brand are all identical, then the number of ways in which we can choose 6 chocolates out of 8 different brands available in the market, is

a. ${}^{13}C_6$ b. ${}^{13}C_8$
c. 8^6 d. None of these

20 The number of ways in which $5X$'s can be placed in the squares of the figure so that no row remains empty, is

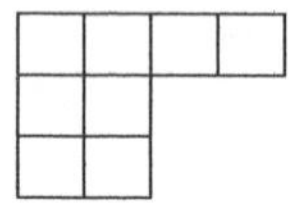

a. 97 b. 44
c. 100 d. 126

21 Number of ways in which *AAA BBB* can be placed in the squares of the figure as shown, so that, no two remains empty, is

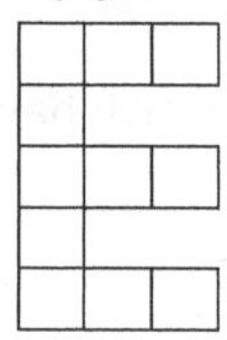

a. 2430 b. 2160
c. 1620 d. None of these

22 Six X have to be placed in the squares of figure below in such a way that each row contains atleast one X. In how many different ways can this be done

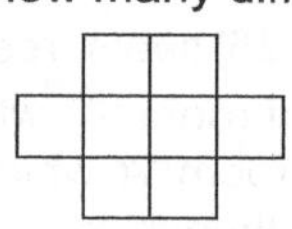

a. 28 b. 26
c. 24 d. 22

23 Between two junction stations A and B, there are 12 intermediate stations. The number of ways in which a train can be made to stop at 4 of these stations, so that no two of these halting stations are consecutive, is

a. ${}^{8}C_4$ b. ${}^{9}C_4$
c. ${}^{12}C_4 - 4$ d. None of these

Comprehension Type

Direction *(Q. Nos. 24-26) This section contains 1 paragraph, each describing theory, experiments, data, etc. Three questions related to the paragraphs have been given. Each question has only one correct answer among the four given options (a), (b), (c) and (d).*

Passage

Consider the word W = MISSISSIPPI.

24 If N denotes the number of different selections, of 5 letters from the word W = MISSISSIPI, then N belongs to the set

a. {15, 16, 17, 18, 19}
b. {20, 21, 22, 23, 24}
c. {25, 26, 27, 28, 29}
d. {30, 31, 32, 33, 34}

Working Space

25 Number of ways in which the letters of the word *W* can be arranged, if atleast one vowel is separated from rest of the vowel, is

a. $\dfrac{8!(161)}{(4)4!\,2!}$ b. $\dfrac{8!}{4!\,4!\,2!}$

c. $\dfrac{8!\cdot 161}{4!\cdot 2!}$ d. $\dfrac{8!}{4!\,2!}\cdot\dfrac{165}{4!}$

26 If the number of arrangements of the letters of the word *W*, if all the *S*'s and *P*'s are separated is $(K)\dfrac{10!}{4!\cdot 4!}$, then *K* equals

a. $\dfrac{6}{5}$ b. 1

c. $\dfrac{4}{3}$ d. $\dfrac{3}{2}$

One Integer Value Correct Type

***Direction** (Q. Nos. 27-31) This section contains 5 questions. When worked out will result in an integer from 0 to 9 (both inclusive).*

27 There are 2 identical white balls, 3 identical red balls and 4 green balls of different shades. The number of ways in which they can be arranged in a row so that atleast one ball is separated from the balls of same colour is $K(7! - 4!)$, then *k* is

28 There are four balls of different colours and four boxes of colours, same as those of the balls. The number of ways in which the balls, one each in a box could be placed such that a ball does not go to a box of its own colour, is

29 There are three piles of identical red, blue and green balls and each pile contains atleast 10 balls. The number of ways of selecting 10 balls, if twice as many red balls as green balls are to be selected, is

30 Let P_n denotes the number of ways of selecting 3 people out of '*n*' sitting in a row, if no two of them are consecutive and Q_n is the corresponding figure when they are in a circle. If $P_n - Q_n = 6$. Then, *n* is equal to 10^x, find value of *x*

31 There are 100 different books in a shelf. Number of ways in which 3 books can be selected so that no two of which are neighbours is ${}^{98}C_k$, then *k* is

SET - 1

Revisal Problems for JEE Main

Only One Option Correct Type

Working Space

Direction (Q. Nos. 1-34) *This section contains 34 multiple choice questions. Each question has four choices (a), (b), (c) and (d), out of which ONLY ONE option is correct.*

1 The number of signals that can be made with 3 flags each of different colours by hoisting 1 or 2 or 3 above the other, is

a. 3 b. 7
c. 15 d. 16

2 The number of numbers from 1000 to 9999 (both inclusive) that do not have all 4 different digits, is

a. 4048 b. 4464
c. 4518 d. 4536

3 How many words can be made with the letters of the word "GENIUS" if each word neither begins with G nor ends in S, is

a. 24 b. 240
c. 480 d. 504

4 5 boys and 3 girls are sitting in a row of 8 seats. Number of ways in which they can be seated, so that not all the girls sit side by side, is

a. 36000 b. 9080
c. 3960 d. 11600

5 The sum of all the numbers which can be formed by using 1, 3, 5, 7 all at a time and which have no digit repeated, is

a. $16 \times 4!$ b. $1111 \times 3!$
c. $16 \times 1111 \times 3!$ d. $16 \times 1111 \times 4!$

6 The number of permutations that can be formed by arranging all the letters of the word 'NINETEEN' in which no two E's occur together, is

a. $\frac{8!}{3! \cdot 3!}$ b. $\frac{5!}{3! \times {}^6C_2}$
c. $\frac{5!}{3!} \times {}^6C_3$ d. $\frac{8!}{5!} \times {}^6C_3$

7 The number of ways in which 8 different flowers can be strung to form a garland so that 4 particular flowers are never separated, is

a. $4! \cdot 4!$ b. $\frac{8!}{4!}$
c. 288 d. None of these

8 The number of ways in which 6 red roses and 3 white roses (all roses different) can form a garland so that all the white roses come together, is

a. 2170 b. 2165
c. 2160 d. 2155

9 The number of ways in which 4 boys and 4 girls can stand in a circle so that each boy and each girl is one after the other, is

a. $3! \cdot 4!$ b. $4! \cdot 4!$
c. $8!$ d. $7!$

10 The number of ways in which 5 beads, chosen from 8 different beads be threaded on to a ring, is

a. 672 b. 1344
c. 336 d. None of these

Working Space

11 12 guests at a dinner party are to be seated along a circular table. Supposing that the master and mistress of the house have fixed seats opposite one another, and that there are two specified guests who must always, be placed next to one another, the number of ways in which the company can be placed, is

a. $20\cdot(10)!$
b. $22\cdot(10!)$
c. $44\cdot(10!)$
d. None of these

12 Six persons *A*, *B*, *C*, *D*, *E* and *F* are to be seated at a circular table. The number of ways this can be done if *A* must have either *B* or *C* on his right and *B* must have either *C* or *D* on his right, is

a. 36
b. 12
c. 24
d. 18

13 The number of permutations which can be formed out of the word 'SERIES' taking three letters together, is

a. 120
b. 60
c. 42
d. None of these

14 The streets of a city are arranged like the lines of a chess board. There are *m* streets running North to South and *n* streets are running East to west. The number of ways in which a man can travel from NW to SE corner going to the shortest possible distance, is

a. $\sqrt{m^2+n^2}$
b. $\sqrt{(m-1)^2(n-1)^2}$
c. $\dfrac{(m+n)!}{m!\cdot n!}$
d. $\dfrac{(m+n-2)!}{(m-1)!\cdot(n-1)!}$

15 The number of ways in which a mixed double tennis game can be arranged from amongst 9 married couple if no husband and wife plays the game, is

a. 756
b. 3024
c. 1512
d. 6048

16 If all the letters of the word 'QUEUE' are arranged in all possible manner as they are in a dictionary, then the rank of the word 'QUEUE', is

a. 15
b. 16
c. 17
d. 18

17 In a conference 10 speakers are present. If S_1 wants to speak before S_2 and S_2 wants to speak after S_3, then the number of ways all the 10 speakers can give their speeches with the above restriction if the remaining seven speakers have no objection to speak at any number, is

a. ${}^{10}C_3$
b. ${}^{10}P_8$
c. ${}^{10}P_3$
d. $\dfrac{10!}{3}$

18 The number of ways selecting 8 books from a library which has 10 books each of Mathematics, Physics, Chemistry and English, if books of the same subjects are alike, is

a. ${}^{13}C_4$
b. ${}^{13}C_3$
c. ${}^{11}C_4$
d. ${}^{11}C_3$

19 Number of ways in which 2 Indians, 3 Americans, 3 Italians and 4 French men can be seated on a circle, if the people of the same nationality sit together, is

a. $2(4!)^2(3!)^2$
b. $2(3!)^3(4!)$
c. $2\cdot(3!)(4!)^3$
d. None of these

20 All possible three digits even numbers which can be formed with the condition that if 5 is one of the digit, then 7 is the next digit, is

a. 5
b. 325
c. 345
d. 365

21 The number of six-digit numbers that can be formed from the digits 1, 2, 3, 4, 5, 6 and 7 so that digits do not repeat and terminal digits are even, is

a. 144
b. 72
c. 288
d. 720

Working Space

22 There are 720 permutations of the digits 1, 2, 3, 4, 5, and 6. Suppose these permutations are arranged from smallest to largest numerical values, beginning from 123456 and ending with 654321. Then, the number falls on the 124th position, is

a. 213564 b. 321456
c. 231564 d. 251364

23 A question paper on mathematics consists of twelve questions divided into 3 parts. *A*, *B* and *C*, each containing four questions. In how many ways can an examines answer five questions, selecting atleast one from each part

a. 624 b. 208
c. 2304 d. None of these

24 In a certain algebraical exercise book. There are 4 examples on AP, 5 examples on *P* and *C* and 6 examples on Binomial theorem. Number of ways a teacher can select for his pupils atleast one but not more than 2 examples from each of these sets, is

a. 1350 b. 3150
c. 5130 d. 5310

25 The Kindergarten teacher has 25 kids in her class. She takes 5 of them at a time, to zoological garden as often as she can, without taking the same 5 kids more than once. Then, the number of visits, the teacher makes to the garden and also the number of the visits every kid makes, is

a. ${}^{24}C_5, {}^{25}C_5$ b. ${}^{24}C_4, {}^{25}C_4$
c. ${}^{25}C_5, {}^{24}C_4$ d. None of these

26 The number of 5-digit numbers such that the sum of digits is even, is

a. 50000 b. 45000
c. 60000 d. None of these

27 The number of ways in which 8 distinguishable apples can be distributed among 3 boys such that every boy should get atleast 1 apple and atmost 4 apples is $k \cdot {}^7P_3$, where *k* has the value equal to

a. 14 b. 66
c. 44 d. 22

28 A rack has 5 different pairs of shoes. The number of ways in which 4 shoes can be chosen from it so that there will be no complete pair, is

a. 1920 b. 200
c. 110 d. 80

29 *A* has 3 maps and *B* has 9 maps. All the 12 maps being distinct. Determine the number of ways in which they can exchange their maps, if each keeps his initial number of maps, is

a. 219 b. 220
c. 240 d. 270

30 Number of ways in which *n* distinct objects can be kept into two identical boxes, so that no boxes remains empty, is

a. 2^n b. $2^n - 2$
c. $2^{n-1} - 1$ d. None of these

31 Number of rectangles in the grid shown which are not squares is

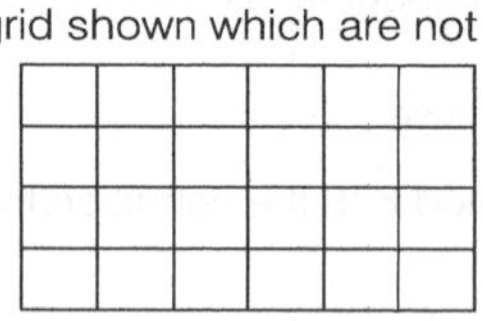

a. 160 b. 162
c. 170 d. 185

32 The number of combinations of 16 things, 8 of which are alike and the rest different, taken 8 at a time is

a. 2^6 b. 2^7
c. 2^8 d. $2^8 - 1$

Working Space

33 The number of different arrangements are possible with the factor of the term $a^2b^4c^5$ written at full length, is

a. 6930 b. 5930
c. 4930 d. 3930

34 The number of 4-digit numbers starting with 1 and having exactly two identical digits, is

a. 240 b. 270
c. 432 d. None of these

Statement Type

Direction (Q. Nos. 35-40) ***This section is based on Statement I and Statement II. Select the correct answer from the codes given below.***

a. Both Statement I and Statement II are correct and Statement II is the correct explanation of Statement I
b. Both Statement I and Statement II are correct and Statement II is not the correct explanation of Statement I
c. Statement I is correct but Statement II is incorrect
d. Statement II is correct but Statement I is incorrect

35 **Statement I** The number of ways of arranging the letters of the word ASSOCIATION such that the two S's come together and 2I's are not together is $\dfrac{8!}{2!\times 2!}\times {}^9P_2$.

Statement II If in the given n things and p things are alike of one kind and q things are alike of second kind, then the number of ways of arranging all the n things is $\dfrac{n!}{p!q!}$.

36 **Statement I** The number of selections of four letter words taken from the word PARALLEL must be 15.

Statement II If there are $l+m+n$ objects of which l are alike, m are alike and n are alike objects, then the number of possible arrangements of r objects.

$$= \text{Coeff. of } x^r \text{ in } r!\left\{\left(1+\frac{x}{1!}+\frac{x^2}{2!}+\cdots+\frac{x^l}{l!}\right)\left(1+\frac{x}{1!}+\cdots+\frac{x^m}{m!}\right)\left(1+\frac{x}{1!}+\cdots+\frac{x^n}{n!}\right)\right\}$$

37 **Statement I** The maximum number of points of intersection of 8 circle is 56.

Statement II The maximum number of points into which 4 circles and 4 straight lines intersect, is 50.

38 **Statement I** $\dfrac{(n+1)!}{(n-1)!}$ is divisible by 6 for some $n \in N$.

Statement II Product of three consecutive integers is divisible by 3!.

39 **Statement I** The maximum value of k such that $(50)^k$ divides (100)! by 2.

Statement II If p is any prime number, then power of p in $(n!)$ is equal to,

$\left[\dfrac{n}{p}\right]+\left[\dfrac{n}{p^2}\right]+\cdots+$ where, [·] denotes greatest integer functions.

40 **Statement I** If $f:\{a,b,c,d,e\}\rightarrow\{a,b,c,d,e\}$, f is onto and $f(x)\neq x$ for each $x\in\{a,b,c,d,e\}$ is equal to 44.

Statement II The number of dearrangement for n objects is $n!\left(1-\dfrac{1}{1!}+\dfrac{1}{2!}-\dfrac{1}{3!}\cdots+\dfrac{(-1)^n}{n!}\right)$.

SET - 2

Revisal Problems
for JEE Advanced

Only One Option Correct Type

Working Space

Direction *(Q. Nos. 1-20) This section contains 20 multiple choice questions. Each question has four choices (a), (b), (c) and (d), out of which ONLY ONE option is correct.*

1 The value of, $\sum\sum\sum_{0\le i<j<k<l\le n}\sum n$ is equal to

a. ${}^{n+1}C_4$ b. $n.{}^{n+1}C_4$ c. ${}^{n+1}C_3$ d. $n(n+1)$

2 Let $n_1 = x_1x_2x_3$ and $n_2 = y_1y_2y_3$ be two 3-digit numbers, then the pairs of n_1 and n_2 can be formed so that n_1 can be subtracted from n_2 without borrowing, is

a. $55\cdot(54)$ b. $45\cdot(55)$ c. $45\cdot(55)^2$ d. $55\cdot(45)^2$

3 Let $S_n = \sum_{r=1}^{n} r!; (n > 6)$. Then, $S_n - 7\left[\frac{S_n}{7}\right]$ (where, [·] denotes greatest integer function) is equal to

a. $\left[\frac{n}{7}\right]$ b. $n! - 7\left[\frac{n}{7}\right]$ c. 5 d. 3

4 A seven-digit numbers divisible by 9 is to be formed by using 7 out of numbers {1, 2, 3, 4, 5, 6, 7, 8, 9}. The number of ways in which they can be done, is

a. 7! b. 2(7)! c. 3(7)! d. 4(7)!

5 Among 9! permutations of the digit 1, 2, 3, ..., 9. Consider those arrangements which have the property that, if we take any five consecutive positions, the product of the digits in those positions is divisible by 7, the number of such arrangements is

a. 7! b. 3·(7!) c. 8! d. 4(7!)

6 The number of ways of arranging m numbers out of $1, 2, 3, \cdots, n$ so that maximum is $(n-2)$ and minimum is 2 (repetitions of numbers is allowed) such that maximum and minimum both occur exactly once, $(n > 5, m > 3)$ is

a. ${}^{n-3}C_{m-2}$ b. ${}^{m}C_2(n-3)^{m-2}$

c. $m(m-1)(n-5)^{m-2}$ d. ${}^{n}C_2 \cdot {}^{n}C_m$

7 The number of rational numbers lying in the interval (2014, 2015) all whose digits after the decimal points are non-zero and are in decreasing order, is

a. $\sum_{i=1}^{9} {}^9P_i$ b. $2^9 - 1$ c. $\sum_{i=1}^{10} {}^{10}P_i$ d. $2^{10} - 1$

8 The number of ways of dividing 15 men and 15 women into 15 couples, each containing of a man and a woman, is

a. 1240 b. 1840 c. 1820 d. 2005

9 'n' is selected from the set $\{1, 2, 3, \cdots, 100\}$ and the number $2^n + 3^n + 5^n$ is formed. Total number of ways of selecting 'n', so that the formed number is divisible by 4, is equal to

a. 50 b. 49 c. 48 d. None of these

10 In a shooting competition a man can score 5, 4, 3, 2 or 0 points for each shot. Then, the number of different ways in which he can score 30 in seven shots, is

a. 419 b. 418 c. 420 d. 421

11 The sum of the factors of which are odd and are the form $3m + 2$, where m is a natural number is

a. 8 b. 40

c. 45 d. 50

Working Space

12 The number of divisors of $n = 2^7 \cdot 3^5 \cdot 5^3$ which are of the form $4t + 1$, ($t \in N$), is

a. 12 b. 11 c. 10 d. 9

13 The number of ways of choosing triplets (x, y, z) such that $z \geq \max.\{x, y\}$ and $x, y, z, \in \{1, 2, 3, \cdots, (n+1)\}$, is

a. $n(n+1)(n+2)$ b. $n(n+1)(2n+1)$

c. $\frac{n(n-1)(2n-1)}{6}$ d. ${}^{n+1}C_3 + {}^{n+2}C_3$

14 The integers from 1 to 1000 are written in order around a circle, starting at 1, every fifteenth number is marked $\{1, 16, 31, \ldots\}$. This process is continued until a number is reached which has already been marked. Then, unmarked numbers are

a. 200 b. 400 c. 600 d. 800

15 The number of ways in which we can choose 3 squares on a chess board such that one of the squares has its two sides common to the other two squares, is

a. 290 b. 292 c. 294 d. 296

16 Let $f(n)$ denote the number of different ways the positive integer 'n' can be expressed as the sum of 1's and 2's for example $f(4) = 5$, since
$4 = 1+1+1+1, 1+1+2, 1+2+1, 2+1+1, 2+2$. (**Note** That order of 1's and 2's is important)
Then, $f(f(6))$ is

a. $f(6)$ b. $f(12)$ c. $f(13)$ d. None of these

17 A seven-digit numbers made up of all distinct digits 8, 7, 6, 4, 2, x and y is divisible by 3. Then, possible number of ordered pair (x, y) is

a. 4 b. 8 c. 2 d. None of these

18 A delegation of 5 students is to be formed from a group of 10 students. If three particular students want to remain together and two particular students do not want to remain together, then number of selections, is

a. 10 b. 20 c. 30 d. 40

19 Let $a, b, c \in I^+$ such that LCM of a and b is 1000, LCM of b and c is 2000 and LCM of c and a is 2000, then number of ordered triplet (a, b, c) is

a. 63 b. 90
c. 70 d. None of these

20 The number of ways to place 11 identical balls in three boxes, so that any two boxes together will contain more balls than the other one

a. 63 b. 21
c. 15 d. None of these

One or More than One Options Correct Type

Direction (Q. Nos. 21-23) *This section contains 3 multiple choice questions. Each question has four choices (a), (b), (c) and (d), out of which ONE or MORE THAN ONE are correct.*

21 Number of ways in which 3 numbers in AP can be selected from $1, 2, 3, \cdots, n$ is

a. $\left(\frac{n-1}{2}\right)^2$, if n is even b. $\frac{n(n-2)}{4}$, if n is odd

c. $\frac{(n-1)^2}{4}$, if n is odd d. $\frac{n(n-2)}{4}$, if n is even

22 The combinatorial coefficient ${}^{n-1}C_p$ denotes

a. The number of ways in which n things of which p are alike and rest different can be arranged in a circle
b. The number of ways in which p different things can be selected out of n different thing, if a particular thing is always excluded
c. Number of ways in which n alike balls can be distributed in p different boxes, so that no box remains empty and each box can hold any number of balls.
d. The number of ways in which $(n-2)$ white balls and p black balls can be arranged in a line. If black balls are separated, balls are all alike except for the colour.

Working Space

23 The number of ways in which five different books to be distributed among 3 persons, so that each person gets atleast one book, is equal to the number of ways in which

a. 5 persons are allotted 3 different residential flats, so that each person is allotted at most one flat and no two persons are allotted the same flat.
b. Number of parallelograms formed by one set of 6 parallel lines and other set of 5 parallel lines that goes in other direction.
c. 5 different toys are to be distributed among 3 children, so that each child get atleast one toy.
d. 3 mathematics professors are assigned five different lectures to be delivered, so that each professor gets atleast one lecturer.

Matching List Type

Direction *(Q. Nos. 24) Choices for the correct combination of elements from Column I and Column II are given as options (a), (b), (c) and (d), out of which one is correct.*

24

	Column I		Column II
i.	Four different movies are running in a town. Ten students go to watch these four movies. The number of ways in which every movie is watched by atleast one student, is (Assume each way differs only by number of students watching a movie).	p.	11
ii.	Consider 8 vertices of a regular octagon and its centre. If T denotes the number of triangles and S denotes the number of straight lines that can be formed with these 9 points, then the value of $(T - S)$, is	q.	36
iii.	In an examination, 5 children were found to have their mobiles in their pocket. The invigilator fixed them and took their mobiles in his possession. Towards the end of the test, invigilator randomly returned their mobiles. The number of ways in which atmost two children did not get their own mobiles, is	r.	52
iv.	The product of the digits of 3214 is 24. The number of 4 digit natural number such that the product of their digits is 12, is	s.	60
v.	The number of ways in which a mixed double tennis game can be arranged from amongst 5 married couple, if no husband and wife plays in the same game, is	t.	84

Codes

	i	ii	iii	iv	v		i	ii	iii	iv	v
a.	p	q	r	s	t	b.	q	r	s	t	p
c.	t	r	p	q	s	d.	r	s	t	p	q

Comprehension Type

Direction *(Q. Nos. 25-47) This section contains 8 paragraphs, each describing theory, experiments, data, etc. 23 questions related to the paragraphs. Each question has only one correct answer among the four given options (a), (b), (c) and (d).*

Passage I

Let a_n denotes the number of all n-digit positive integers formed by the digits 0, 1 or both such that no consecutive digit in them are 0. Let b_n = The number of such n-digit integers ending with digit 1 and C_n = the number of such n-digit integers ending with digit 0.

25 The value of b_6, is

a. 7 b. 8 c. 9 d. 11

26 Which of the following is correct?

a. $a_{17} = a_{16} + a_{15}$ b. $c_{17} \neq c_{16} + c_{15}$ c. $b_{17} \neq b_{16} + c_{16}$ d. $a_{17} = c_{17} + b_{16}$

Passage II

10-digit numbers are formed by using all the digits $0, 1, 2, \cdots, 9$ such that they are divisible by 11111. Then, answer the following questions.

27 The digit in tens place, in smallest of such number, is

a. 9 b. 8 c. 7 d. 6

Working Space

28 The digit in units place, in greatest of such numbers, is

a. 4 b. 3 c. 2 d. 1

29 The total number of such numbers, is

a. 6543 b. 5634 c. 3456 d. 4365

Passage III

A shopkeeper has unlimited red, green, white, yellow and black balls. All balls of same colour are identical. Then,

30 In how many ways a person can make a selection of 12 balls

a. ${}^{15}C_3$ b. ${}^{16}C_4$

c. ${}^{16}C_5$ d. None of these

31 In how many ways a person can select 12 balls, if each selection consists of balls of all colours

a. ${}^{12}C_3$ b. ${}^{11}C_4$

c. ${}^{13}C_4$ d. ${}^{15}C_4$

32 In how many ways a person can select 16 balls, if each selection consists of atleast two balls of each colour, is

a. ${}^{10}C_2$ b. ${}^{11}C_4$

c. ${}^{10}C_4$ d. None of these

Passage IV

Let $x_1x_2x_3x_4x_5x_6$ be a six-digit numbers, find the number of such numbers such that

33 $x_1 < x_2 < x_3 < x_4 < x_5 < x_6$, is

a. ${}^{9}C_6$ b. ${}^{10}C_3$ c. ${}^{10}C_6$ d. ${}^{10}C_5$

34 $x_1 < x_2 < x_3 = x_4 < x_5\ x_6$, is

a. ${}^{9}C_3$ b. ${}^{9}C_4$ c. ${}^{10}C_3$ d. ${}^{10}C_4$

35 $x_1 < x_2 < x_3 \le x_4 < x_5 < x_6$, is

a. ${}^{9}C_4$ b. ${}^{9}C_3$ c. ${}^{10}C_4$ d. ${}^{10}C_3$

Passage V

Five balls are to be placed in three boxes. Each can hold all the five balls. In how many different ways can we places the balls so that no box remains empty, if

36 Balls and boxes are all different

a. 50 b. 100 c. 125 d. 150

37 Balls are identical but boxes are different

a. 2 b. 6 c. 4 d. 8

38 Balls are different but boxes are identical

a. 25 b. 15 c. 10 d. 35

Passage VI

A triangle is called an integral triangle, if all its sides are integers. If a, b and c are sides of an integral triangle, then we can assume that $a \le b \le c$, because any other permutation of a, b and c will yield the same result. Now, answer the following questions.

39 The number of integral isosceles or equilateral triangle's, none of whose side exceeds 4, is

a. 9 b. 10 c. 11 d. 12

40 The number of integral isosceles or equilateral triangles, none of whose side exceeds $2p$, is

a. $4p^2 - 2p$ b. $2p^2 + 2p$

c. $3p^2$ d. $\frac{3p^2}{2} + 3p$

Working Space

41 If c is fixed and odd, then number of integral isosceles or equilateral triangle's whose sides are a, b, c $(a \le b \le c)$ must be

a. $\frac{2c-1}{2}$ b. $\frac{2c+1}{2}$ c. $\frac{3c+1}{2}$ d. $\frac{3c-1}{2}$

Passage VII

Consider the network of equally spaced parallel lines (6 horizontal and 9 vertical) shown in the figure. All small squares are of same size. A shortest from A to C is defined as the route consisting of 8 horizontal and 5 vertical steps.

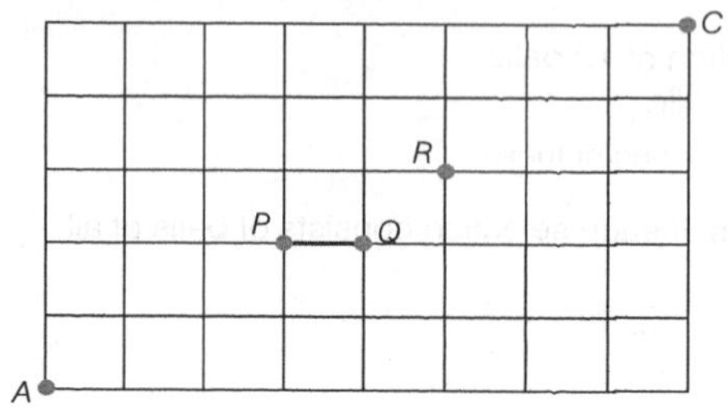

42 Number of shortest routes through P, is

a. 240 b. 216
c. 560 d. 640

43 The number of shortest routes through PQ, is

a. 324 b. 350
c. 512 d. 560

44 Number of shortest routes through P and R, is

a. 144 b. 240
c. 216 d. 324

Passage VIII

Define a function $\theta : N \to N$ as follows $\theta(1) = 1$, $\theta(r^n) = r^{n-1}(r-1)$, if r is prime and $n \in N$, $\theta(\ln) = \theta(l) \cdot \theta(n)$, if HCF $(l, n) = 1$.

45 $\theta(1672) =$

a. 720 b. 836
c. 832 d. 729

46 The number of natural numbers n such that $\theta(n)$ is odd, is

a. 1 b. 2
c. 3 d. infinite

47 If $\theta(61^n) = 2231260$, then n is equal to

a. 5 b. 61
c. 3 d. 10

One Integer Value Correct Type

Direction (Q. Nos. 48-50) ***This section contains 3 questions. When worked out will result in an integer from 0 to 9 (both inclusive).***

48 Let P_n denotes the number of ways in which three people can be selected out of 'n' people sitting in a row, if no two of them are consecutive. If, $P_{n+1} - P_n = 15$, then the value of n, is

49 Number of cyphers at the end of ${}^{2002}C_{1001}$, is

50 Given 11 points, in which 5 lie on one circle, other than these 5, no 4 lie on one circle. Then, the maximum number of circle that can be drawn, so that each contains atleast three of the given points, is $52 \cdot (k)$. Then, k is

JEE Main & AIEEE Archive

(Compilation of Last 13 Years Questions)

1 Let A and B be two sets containing 2 elements and 4 elements, respectively. The number of subsets of $A \times B$ having 3 or more elements is **(JEE Main 2013)**

a. 256 b. 220
c. 219 d. 211

2 Let T_n be the number of all possible triangles formed by joining vertices of an n-sided regular polygon. If $T_{n+1} - T_n = 10$, then the value of n is **(JEE Main 2013)**

a. 7 b. 5
c. 10 d. 8

3 Assuming the balls to be identical except for difference in colours, the number of ways in which one or more balls can be selected from 10 white, 9 green and 7 black balls is **(AIEEE 2012)**

a. 880 b. 629
c. 630 d. 879

4 **Statement I** The number of ways of distributing 10 identical balls in 4 distinct boxes such that no box is empty, is ${}^9C_3$.

Statement II The number of ways of choosing any 3 places from 9 different places is ${}^9C_3$.

a. Statement I is correct, Statement II is correct; Statement II is not a correct explanation of Statement I
b. Statement I is correct, Statement II is incorrect
c. Statement I is incorrect, Statement II is correct
d. Statement I is correct, Statement II is correct; Statement II is a correct explanation of Statement I **(AIEEE 2011)**

5 There are 10 points in a plane, out of these 6 are collinear. If N is the number of triangles formed by joining these points, then **(AIEEE 2011)**

a. $N > 190$
b. $N \le 100$
c. $100 < N \le 140$
d. $140 < N \le 190$

6 From 6 different novels and 3 different dictionaries, 4 novels and 1 dictionary are to be selected and arranged in a row on the shelf so that the dictionary is always in the middle. Then, the number of such arrangements is **(AIEEE 2009)**

a. atleast 500 but less than 750
b. atleast 750 but less than 1000
c. atleast 1000
d. less than 500

7 In a shop, there are five types of ice-creams available. A child buy six ice-creams.

Statement I The number of different ways the child can buy the six ice-creams is ${}^{10}C_5$. **(AIEEE 2008)**

Statement II The number of different ways the child can buy the six ice-creams is equal to the number of different ways of arranging 6 A's and 4 B's in a row.

a. Statement I is incorrect, Statement II is correct
b. Statement I is correct, Statement II is correct; Statement II is a correct explanation of Statement I
c. Statement I is correct, Statement II is correct; Statement II is not a correct explanation of Statement I
d. Statement I is correct, Statement II is incorrect

8 How many different words can be formed by jumbling the letters in the word 'MISSISSIPPI' in which no two S are adjacent? **(AIEEE 2008)**

a. $7 \cdot {}^6C_4 \cdot {}^8C_4$
b. $8 \cdot {}^6C_4 \cdot {}^7C_4$
c. $6 \cdot 7 \cdot {}^8C_4$
d. $6 \cdot 8 \cdot {}^7C_4$

9 The set $S = \{1, 2, 3, \ldots, 12\}$ is to be partitioned into three sets A, B and C of equal size.

Thus, $A \cup B \cup C = S$,

$A \cap B = B \cap C = A \cap C = \phi$ **(AIEEE 2007)**

The number of ways to partition S is

a. $12!/3!(4!)^3$
b. $12!/3!(3!)^4$
c. $12!/(4!)^3$
d. $12!/(3!)^4$

10 At an election, a voter may vote for any number of candidates not greater than the number to be elected. There are 10 candidates and 4 are to be elected. If a voter votes for atleast one candidate, then the number of ways in which he can vote, is **(AIEEE 2006)**

a. 6210 b. 385
c. 1110 d. 5040

11 If the letters of the word 'SACHIN' are arranged in all possible ways and these words are written out as in dictionary, then the word 'SACHIN' appears at serial number **(AIEEE 2005)**

a. 602 b. 603
c. 600 d. 601

12 The range of the function $f(x) = {}^{7-x}P_{x-3}$ is **(AIEEE 2004)**

a. {1, 2, 3}
b. {1, 2, 3, 4, 5, 6}
c. {1, 2, 3, 4}
d. {1, 2, 3, 4, 5}

13 How many ways are there to arrange the letters in the word 'GARDEN' with the vowels in alphabetical order? **(AIEEE 2004)**

a. 120
b. 240
c. 360
d. 480

14 The number of ways of distributing 8 identical balls in 3 distinct boxes, so that none of the boxes is empty, is **(AIEEE 2004)**

a. 5
b. 21
c. 3^8
d. ${}^8C_3$

15 A student is to answer 10 out of 13 questions in an examination such that he must choose atleast 4 from the first five questions. The number of choices available to him is **(AIEEE 2003)**

a. 140
b. 196
c. 280
d. 346

16 The number of ways in which 6 men and 5 women can dine at a round table, if no two women are to sit together, is given by **(AIEEE 2003)**

a. $6! \times 5!$
b. 30
c. $5! \times 4!$
d. $7! \times 5!$

17 If ${}^nC_r$ denotes the number of combinations of n things taken r at a time, then the expression ${}^nC_{r+1} + {}^nC_{r-1} + 2 \times {}^nC_r$ equal to **(AIEEE 2003)**

a. ${}^{n+2}C_r$
b. ${}^{n+2}C_{r+1}$
c. ${}^{n+1}C_r$
d. ${}^{n+1}C_{r+1}$

JEE Advanced & IIT JEE Archive

(Compilation of Last 10 Years Questions)

1 Let $n_1 < n_2 < n_3 < n_4 < n_5$ be positive integers such that $n_1 + n_2 + n_3 + n_4 + n_5 = 20$. The number of such distinct arrangements $(n_1, n_2, n_3, n_4, n_5)$ is **(2014 Adv., Integer Type)**

2 Let $n \geq 2$ be an integer. Take n distinct points on a circle and join each pair of points by a line segment. Colour the line segment joining every pair of adjacent points by blue and the rest by red. If the number of red and blue line segments are equal, then the value of n is **(2014 Adv., Integer Type)**

3 The total number of ways in which 5 balls of different colours can be distributed among 3 persons so that each person gets atleast one ball is **(2012, Single Option Correct Type)**

a. 75
b. 150
c. 210
d. 243

4 The number of seven digit integers, with sum of the digits equal to 10 and formed by using the digits 1, 2 and 3 only, is **(2009, Single Option Correct Type)**

a. 55
b. 66
c. 77
d. 88

5 Consider all possible permutations of the letters of the word ENDEANOEL. **(2008, Matching Type)**

	Column I		Column II
i.	The number of permutations containing the word ENDEA, is	p.	5!
ii.	The number of permutations in which the letter E occurs in the first and the last positions, is	q.	$2 \times 5!$
iii.	The number of permutations in which none of the letters D, L, N occurs in the last five positions, is	r.	$7 \times 5!$
iv.	The number of permutations in which the letters A, E, O occur only in odd positions, is	s.	$21 \times 5!$

Codes

	i	ii	iii	iv		i	ii	iii	iv
a.	p	q	r	s	b.	p	s	q	q
c.	q	r	s	p	d.	r	p	p	r

6 The letters of the word COCHIN are permuted and all the permutations are arranged in an alphabetical order as in an English dictionary. The number of words that appear before the word COCHIN is **(2007, Single Option Correct Type)**

a. 360 b. 192 c. 96 d. 48

7 If r, s, t are prime numbers and p, q are the positive integers such that LCM of p, q is $r^2s^4t^2$, then the number of ordered pairs (p, q) is **(2006, Single Option Correct Type)**

a. 252 b. 254 c. 225 d. 224

3. Probability

DPP

DAILY PRACTICE PROBLEMS

DPP-1 Equally Likely, Mutually Exclusive and Exhaustive Events, Classical Definition of Probability

Subjective Questions

Working Space

Direction (Q. Nos. 1-3) ***These questions are subjective in nature, need to be solved completely on notebook.***

1 Discuss whether the following are equally likely (EL) mutually exclusive (ME) and exhaustive events.

	Experiment	Events	*EL*	*ME*	Exhaustive
i.	Throwing of a die	A : Throwing an odd face. B : Throwing a composite.	...	...	...
ii.	A ball is drawn from an urn containing 2 white, 3 red and 4 green balls	E_1 : Getting a white ball. E_2 : Getting a red ball. E_3 : Getting a green ball.	...	...	...
iii.	Throwing a pair of dice	A : Throwing a doublet. B : Throwing a total of 10 or more.	...	...	...
iv.	From a well-shuffled pack of cards a card is drawn	E_1 : Getting a heart. E_2 : Getting a spade. E_3 : Getting a diamond. E_4 : Getting a club.	...	...	...
v.	From a well-shuffled pack of cards a card is drawn	A : Getting a heart. B : Getting a face card.	...	...	...

2 Two dice are thrown. The events A, B, C are as follow :

A. Getting an even number on the first die.
B. Getting an odd number on the first die.
C. Getting a sum of 2.

State whether the following statements are true or false.

i. A is a simple event.
ii. C is a simple event.
iii. A and B are mutually exclusive.
iv. A' and B' are mutually exclusive.
v. $A = B'$
vi. A and B are mutually exclusive.

3 Three coins are tossed together. Then, find the probability of getting

i. exactly two heads.
ii. atleast two heads.
iii. atleast one head and one tail.

Only One Option Correct Type

Working Space

Direction (Q. Nos. 4-26) *This section contains 23 multiple choice questions. Each question has four choices (a), (b), (c) and (d), out of which ONLY ONE option is correct.*

4 The probability that an ordinary or a non-leap year has 53 Sunday, is

a. 2/7 b. 1/7
c. 3/7 d. None of these

5 Two dice are thrown thrice. The probability that first throw shows 10, second 11 and third 12 will be

a. 1/216 b. 1/432
c. 1/7776 d. 1/648

6 A card is drawn at random from a pack of cards. The probability of this card being a red or a queen, is

a. 1/13 b. 1/26
c. 1/2 d. 7/13

7 If a party of n persons sit at a round table, then the odds against two specified individuals sitting next to each other are

a. $2:(n-3)$ b. $(n-3):2$
c. $(n-2):2$ d. $2:(n-2)$

8 Find the probability that the two-digit number formed by digits 1, 2, 3, 4, 5 is divisible by 4 (while repetition of digit is allowed).

a. 1/30 b. 1/20
c. 1/40 d. None of these

9 A bag contains 5 white, 7 black and 4 red balls. Three balls are drawn from the bag at random. The probability that all the three balls are white is

a. 3/16 b. 3/5
c. 1/60 d. 1/56

10 From a pack of 52 cards, one card is drawn at random, the probability that it is either a king or a queen in

a. 1/13 b. 2/13
c. 3/13 d. 4/13

11 Three persons A, B and C are to speak at a function along with the others. If they all speak in random order, the probability that A speaks before B and B speaks before C, is

a. 3/8 b. 1/6
c. 3/5 d. None of these

12 A bag contains 3 red and 3 white balls. Two balls are drawn one-by-one. The probability that they are of different colours, is

a. 3/10 b. 2/5
c. 3/5 d. None of these

13 A bag contains 4 red and 3 blue balls. Two balls are drawn twice. If the drawn balls are replaced before the second draw, then the probability that first two balls are red and second two balls are blue, will be

a. 2/49 b. 5/49
c. 1/35 d. 2/35

14 3 letters are placed in 3 envelopes randomly. The probability that only one letter goes in right envelope is

a. 1/2 b. 1/3 c. 1/6 d. 2/3

Working Space

15 There are 10 bulbs in a room, 4 of them are fused. The probability that the room will be lighted on pressing the switches of any three, is

a. 1/30 b. 29/30
c. 3/5 d. None of these

16 There are 4 red, 3 pink, 5 yellow and 8 white roses in a lawn. What is the probability that a blind man will touch a red or a white rose?

a. 1/5 b. 2/5
c. 3/5 d. None of these

17 Out of 40 consecutive natural numbers, two are chosen at random. Probability that sum of the numbers is odd, is

a. 14/29 b. 20/39
c. 1/2 d. None of these

18 Three integers are chosen at random from the first 20 integers. The probability that their product is even, is

a. 2/19 b. 3/29
c. 17/19 d. 4/19

19 A committee of five is to be chosen from a group of a 9 people. The probability that a certain married couple will either serve together or not at all, is

a. 1/2 b. 5/9
c. 4/9 d. 2/3

20 If a die is thrown twice, the probability of occurrence of 4 atleast once, is

a. 11/36 b. 7/12
c. 35/36 d. None of these

21 A cube with all six faces coloured is cut into 64 cubical blocks of the same size, which are throughly mixed. Then, the probability that the 2 randomly chosen blocks have 2 coloured faces each, is

a. 23/168 b. 51/168
c. 1/8 d. 1/16

22 6 married couples are standing in a room. If 4 people are chosen at random, then the chance that exactly one married couple is among the 4, is

a. 16/33 b. 8/33
c. 17/33 d. 24/33

23 Mr. A forgot to write down a very important phone number. All he remembers is that it started with 713 and that the next set of 4 digit involved are 1, 7 and 9 with one of these numbers appearing twice. He guesses a phone number and dials randomly. The odds in favour of dialing the correct telephone number, is

a. 1/35 b. 1/71
c. 1/23 d. 1/36

24 n different books ($n \geq 3$) are put at random in a shelf. Among these books there is a particular book A and a particular book B. The probability that there are exactly r books between A and B is

a. $\frac{2}{n(n-1)}$ b. $\frac{2(n-r-1)}{n(n-1)}$ c. $\frac{2(n-r-2)}{n(n-1)}$ d. $\frac{(n-r)}{n(n-1)}$

25 Consider a function $f(x)$ that has zeros 4 and 9. Given that, Mr. A randomly selects a number from the set $\{-10, -9, -8, \ldots, 8, 9, 10\}$, then the probability that Mr. A chooses a zero of $f(x^2)$ is

a. 4/21 b. 8/21 c. 11/21 d. 13/21

26 Mr. A lives at origin on the cartesian plane and has his office at (4, 5). His friend lives at (2, 3) on the same plane (as shown in figure). Mr. A can go his office travelling one block at a time either in the + y or + x direction. If all possible paths are equally likely, then the probability that Mr. A passed by his friends house, is

Working Space

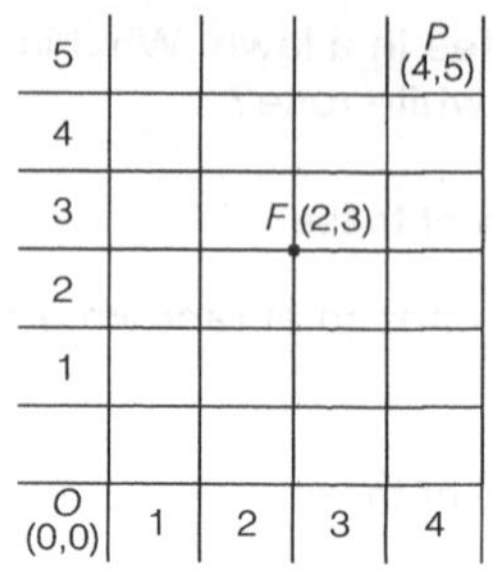

a. 1/2 b. 10/21 c. 1/4 d. 11/21

Comprehension Type

Direction (Q. Nos. 27-32) *This section contains 2 paragraphs, each describing theory, experiments, data, etc. Six questions related to the paragraphs have been given. Each question has only one correct answer among the four given options (a), (b), (c) and (d).*

Passage I

A is a set containing 10 elements. A subset P of A is chosen at random and set A is reconstructed by replacing the elements of P, another subset Q of A is now chosen at random. Then, the probability that, if

27 $P \cup Q = A$, is

a. $(1/2)^{10}$ b. $(2/3)^{10}$ c. $(3/4)^{10}$ d. $(2/5)^{10}$

28 $P \cup Q$ contains exactly 4 elements, is

a. $\frac{5805}{2^{19}}$ b. $\frac{8550}{2^{19}}$ c. $\frac{5058}{2^{19}}$ d. $\frac{8505}{2^{19}}$

29 P and Q have no common elements, is

a. $(2/3)^{10}$ b. $(3/4)^{10}$ c. $(4/5)^{10}$ d. $(5/6)^{10}$

Passage II

In an objective paper, there are two sections of 10 questions each. For Section I, each question has 5 options and only one option is correct and for Section II each question has 4 options with one or more than one correct and marks in each section is awarded only if he ticks all correct answers. Marks for each question in Section I is 1 and in Section II is 3. (there is no negative marking)

30 If a candidate attempts only two questions, one from Section I and one from Section II. The probability that be scores in both questions, is

a. 74/75 b. 1/25 c. 1/15 d. 1/75

31 If a candidate in total attempts 4 questions, then the probability of scoring 10 marks, is

a. $\frac{1}{5}\left(\frac{1}{15}\right)^3$ b. $\frac{4}{5}\left(\frac{1}{15}\right)^3$ c. $\frac{1}{5}\left(\frac{14}{15}\right)^3$ d. None of these

32 The probability of getting a score less than 40 in this paper is

a. $(1/75)^{10}$ b. $1-(1/75)^{10}$ c. $(74/75)^{10}$ d. $(1/5)^{10}$

Working Space

Statement Type

Direction (Q. Nos. 33-34) *This section is based on Statement I and Statement II. Select the correct answer from the codes given below.*

a. Both Statement I and Statement II are correct and Statement II is the correct explanation of Statement I
b. Both Statement I and Statement II are correct and Statement II is not the correct explanation of Statement I
c. Statement I is correct but Statement II is incorrect
d. Statement II is correct but Statement I is incorrect

33 Four numbers are chosen at random (without replacement from the set (1, 2, 3, ..., 20).

Statement I The probability that the chosen numbers when arranged in same order will form an AP is $\frac{1}{85}$.

Statement II In the four chosen numbers form an AP, the set of all possible values of common differences is $\{\pm 1, \pm 2, \pm 3, \pm 4, \pm 5\}$. **[AIEEE 2010]**

34 Consider the system of equations $ax + by = 0$, $cx + dy = 0$, where $a, b, c, d \in \{0, 1\}$.

Statement I The probability that the system of equations has a unique solution is $\frac{3}{8}$.

Statement II The probability that the system of equations has a solution is 1. **[IIT JEE 2008]**

One Integer Value Correct Type

Direction (Q. Nos. 35-40) *This section contains 6 questions. When worked out will result in an integer from 0 to 9 (both inclusive).*

35 A quadratic equation is chosen from the set of all the quadratic equations, which are unchanged by squaring their roots. The chance that the chosen equation has equal roots is $\frac{m}{n}$, then $(n - m)$ is

36 The probability that a positive two-digit numbers selected at random has its ten's digit atleast three more than its unit's digit is $m : n$. Then, $(n - 3m)$ is

37 A 5-digit number is formed by using the digits 0, 1, 2, 3, 4 and 5 without repetition. The probability that the number is divisible by 6 is m%, then number of digits in m is

38 Nine numbers 1, 2, 3, ..., 9 are arranged in a rectangular array of matrix of order 3, so that each number occurs exactly once. Then, the probability that the sum of the numbers in atleast one horizontal row is greater than 21, is $m : n$, then $(m + n)$ is

39 I have 3 normal dice, one red, one blue, one green and I roll all three simultaneously. Let P be the probability that the sum of the numbers on the green die. If P is the written in lowest term as $\frac{a}{b}$, then numbers of digit in $(a + b)$ is

40 If two of the 64 squares are chosen at random on a chess board, the probability that they have a side in common is, $\frac{m}{n}$ is lowest form, then numbers of digit in $(m + n)$ is

DPP-2 Addition and Subtraction Theorems of Probability

Subjective Questions

Working Space

Direction (Q. Nos. 1-3) *These questions are subjective in nature, need to be solved completely on notebook.*

1 If A and B are two events such that $P(A \cup B) = 3/4$, $P(A \cap B) = 1/4$ and $P(A^C) = 2/3$. Then, find

i. $P(A)$

ii. $P(B)$

iii. $P(A \cap \overline{B})$

iv. $P(\overline{A} \cap B)$

2 Draw Venn diagram of

i. $(A^C \cap B^C) \cup (A \cap B)$

ii. $B^C \cup (A^C \cap B)$

3 In a class of 125 students, 70 passed in English, 55 in Mathematics and 30 in both. Then, find the probability that a student selected at random from the class has passed in

i. atleast one subject.

ii. only one subject.

Only One Option Correct Type

Direction (Q. Nos. 4-17) *This section contains 14 multiple choice questions. Each question has four choices (a), (b), (c) and (d), out of which ONLY ONE option is correct.*

4 If A and B are two mutually exclusive events, then which of the following is true?

a. $P(A) \le P(\overline{B})$

b. $P(\overline{A} \cap \overline{B}) = P(\overline{A}) + P(B)$

c. $P(\overline{A} \cup \overline{B}) = 0$

d. $P(\overline{A} \cap B) = P(A)$

5 Three numbers are chosen at random without replacement from 1, 2, 3, ..., 10. The probability that the minimum of the chosen number is 4 or their maximum is 8, is

a. 11/40

b. 3/10

c. 1/40

d. None of these

6 A bag contains 6 white, 5 black and 4 red balls. Then, find the probability of getting either a white or a black ball in a single draw.

a. 11/15

b. 12/20

c. 1/2

d. 11/20

Working Space

7 Let two events be A and B. If odds against A are as 2 : 1 and those in favour of $A \cup B$ are as 3 : 1, then the range of $P(B)$ is

a. $\frac{5}{12} \le P(B) \le \frac{3}{4}$
b. $\frac{5}{12} \le P(B) \le 1$
c. $0 \le P(B) \le 1$
d. None of these

8 A die is thrown. Let A be the event that the number obtained is greater than 3 and B be the event that the number obtained is less than 5. Then, $P(A \cup B)$ is **[AIEEE 2008]**

a. 3/5
b. 0
c. 1
d. 2/5

9 Three distinct numbers are selected from first 100 natural numbers, then the probability that all the three numbers are divisible by 2 and 3 is **[IIT JEE 2004]**

a. 4/25
b. 4/35
c. 4/55
d. 4/1155

10 From a pack of 52 cards, face cards and tens are removed and kept a side, then a card is drawn at random from the remaining cards. If

A : The events that the card drawn is an ace.
H : The events that the card drawn is a heart.
S : The events that the card drawn is a spade.

Then, which of the following holds?

a. $9P(A) = 4P(H)$
b. $P(S) = 4P(A \cap H)$
c. $3P(H) = P(A \cup S)$
d. None of these

11 3 students A, B and C are in a swimming race. A and B have the same probability of winning and each is twice as likely to win as C. Then, the probability that B or C wins, is (Assume no two reach the winning point simultaneously)

a. 3/5
b. 2/5
c. 1/5
d. None of these

12 If E and F are events with $P(E) \le P(F)$ and $P(E \cap F) > 0$, then

a. occurrence of $E \Rightarrow$ occurrence of F
b. occurrence of $F \Rightarrow$ occurrence of E
c. non-occurrence of $E \Rightarrow$ non-occurrence of F
d. None of the above implications holds

13 In an examination, one hundred candidates took paper in Physics and Chemistry. Twenty five candidates failed in Physics only. Twenty candidates failed in Chemistry only. Fifteen failed in both Physics and Chemistry. A candidate is selected at random. The probability that he failed either in Physics or in Chemistry but not in both, is

a. 9/20
b. 3/5
c. 2/5
d. 11/20

14 The probability that atleast one of the events A and B occurs is 0.6. If A and B occur simultaneously with probability 0.2, then $P(\overline{A}) + P(\overline{B})$ is

a. 1.4
b. 1.2
c. 1.3
d. 1.1

15 If $\frac{1+3p}{3}, \frac{1-p}{4}, \frac{1-2p}{2}$ are probabilities of three mutually exclusive events, then

a. $-1/3 \le p \le 1/2$
b. $1/2 \le p \le 2/3$
c. $1/6 \le p \le 1/2$
d. None of these

Working Space

16 If A, B and C are three events such that $P(B) = 3/4$, $P(A \cap B \cap C') = 1/3$ and $P(A' \cap B \cap C') = 1/3$, then $P(B \cap C)$ is equal to

a. 1/12 b. 1/6
c. 1/15 d. 1/9

17 A die is rolled, so that the probability of face is proportional to i, where $i = 1, 2, \ldots, 6$. The probability of an even number occurring when the die is rolled, is

a. 4/7 b. 5/7
c. 6/7 d. None of these

Comprehension Type

Direction (Q. Nos. 18-20) *This section contains a paragraph, each describing theory, experiments, data, etc. Three questions related to the paragraph have been given. Each question has only one correct answer among the four given options (a), (b), (c) and (d).*

A is power set of a given set X containing n elements and $P, Q, R \in A$ (may or may not be distinct).

18 If P, Q, R are disjoint sets, the probability that $P \cup Q \cup R = X$, is

a. $(4/7)^{n-1}$ b. $(3/4)^n$
c. $(3/7)^n$ d. $(5/7)^n$

19 If $P \cap Q \cap R$ contains just 1 element, then the probability that P also contain just 1 element, is

a. $(4/7)^{n-1}$ b. $(3/4)^n$
c. $(3/7)^n$ d. $(5/7)^n$

20 If $P \cap Q \cap R$ contain no elements, then the probability that $Q = R$, is

a. $(4/7)^{n-1}$ b. $(3/4)^n$
c. $(3/7)^n$ d. $(5/7)^n$

DPP-3 Conditional Probability and Multiplication Theorem on Probability

Only One Option Correct Type

Working Space

Direction *(Q. Nos. 1-15) This section contains 15 multiple choice questions. Each question has four choices (a), (b), (c) and (d), out of which ONLY ONE option is correct.*

1 If $4P(A) = 6P(B) = 10P(A \cap B) = 1$, then $P(B/A)$ is equal to

a. 2/5 b. 3/5
c. 7/10 d. 19/60

2 If A and B are two events such that $P(A) \neq 0$ and $P(B) \neq 1$, then $P(\overline{A}/\overline{B})$ is equal to

a. $1 - P(A/B)$ b. $1 - P(\overline{A}/B)$
c. $\dfrac{1 - P(A \cup B)}{P(\overline{B})}$ d. $\dfrac{P(\overline{A})}{P(\overline{B})}$

3 If A and B are two events such that $P(A \cup B) = P(A \cap B)$, then the true relation is

a. $P(A) + B(B) = 0$ b. $P(A) + P(B) = P(A) \cdot P(B/A)$
c. $P(A) + P(B) = 2P(A) \cdot P(B/A)$ d. None of there

4 A coin is tossed three times in succession. If E is the event that there are atleast two heads and F is the event in which first throw is a head, then $P(E/F)$ is equal to

a. 3/4 b. 3/8
c. 1/2 d. 1/8

5 Two dice are thrown. Then, the probability that the numbers appeared have a sum of 8, if it known that the second die always exhibits 4, is

a. 1/2 b. 1/3
c. 1/4 d. 1/6

6 A bag contains 3 red, 6 white and 7 black balls. Two balls are drawn one-by-one, then the probability that first ball is white and second ball is blue when first drawn ball is not replaced in the bag, is

a. 7/40 b. 9/40
c. 11/40 d. 13/40

7 A bag contains 4 red and 4 blue balls. Four balls are drawn one-by-one from the bag, then the probability that the drawn balls are in alternate colour, is

a. $\dfrac{1}{35}$ b. $\dfrac{7}{31}$
c. $\dfrac{6}{35}$ d. $\dfrac{9}{31}$

8 If two events A and B are such that $P(\overline{A}) = 0.3$, $P(B) = 0.4$ and $P(A\overline{B}) = 0.5$. Then, $P[B/(A \cup \overline{B})]$ is

a. 1/2 b. 1/3 c. 1/4 d. 1/5

Working Space

9 Three cards are drawn successively, without replacement from a pack of 52 cards. Then, the probability that the drawn cards are face cards of same suits, is

a. $\frac{1}{5525}$ b. $\frac{11}{5525}$

c. $\frac{111}{5525}$ d. $\frac{1111}{5525}$

10 A bag contains 2 black, 4 white and 3 red balls. One balls is drawn at random from the bag and kept a side. From the remaining balls another ball is drawn and kept aside the first. This process is repeated till all the balls are drawn. Then, probability that the balls are in sequence of 2 black, 4 white and 3 red is

a. $\frac{1}{1260}$ b. $\frac{1}{7560}$

c. $\frac{1}{210}$ d. None of these

11 A pair of numbers is picked up randomly (without replacement) from the set {1, 2, 3, 5, 7, 11, 12, 13, 17, 19}. The probability that the number 11 was picked, given that the sum of the numbers was even, is nearly

a. 0.1 b. 0.125

c. 0.24 d. 0.18

12 For a biased die, the probability for the different faces to turn up are given below :

Faces	1	2	3	4	5	6
Probabilities	0.12	0.32	0.21	0.15	0.05	0.17

The die is tossed and you are told that either face one or face two has turned up. Then, the probability that it is face one, is

a. 1/6 b. 1/10

c. 5/49 d. 5/21

13 The probability that an automobile will be stolen and found within one week is 0.0006. The probability that an automobile will be stolen is 0.0015. The probability that a stolen automobile will be found in one week, is

a. 0.3 b. 0.4

c. 0.5 d. 0.6

14 The probability that a person will get an electric contract is 2/5 and the probability that he will not get plumbing contract is 4/7. If the probability of getting atleast one contract is 2/3. Then, the probability that he will get both, is

a. 17/105 b. 3/7

c. 4/7 d. None of these

15 There are 6 red balls and 6 green balls in a bag. Five balls are drawn out at random and placed in a red box. The remaining seven balls are put in a green box. If the probability that the number of red balls in the green box plus the number of red balls is not a prime number, is $\frac{p}{q}$, where p and q are relatively prime. Then, the value of $(p+q)$ is

a. 35 b. 36

c. 37 d. 38

DPP-4 Independent Events

Only One Option Correct Type

Working Space

Direction *(Q. Nos. 1-28) This section contains 28 multiple choice questions. Each question has four choices (a), (b), (c) and (d), out of which ONLY ONE option is correct.*

1 Let A and B be two events. If $P(A) = 0.4$, $P(B) = p$ and $P(A \cup B) = 0.7$. Then, the value of p for which A and B are independent, is

a. 1/3 b. 1/4 c. 1/2 d. 1/5

2 A licence plate is 3 capital letters (of English alphabets) followed by 3 digits. If all possible licence plate are equally likely, the probability that a plate has either a letter palindrome or a digit palindrome (or both), is

a. 7/52 b. 9/65 c. 8/65 d. None of these

3 Events A and C are independent. If the probabilities relating A, B and C are $P(A) = 1/5$, $P(B) = 1/6$, $P(A \cap C) = 1/20$ and $P(B \cup C) = 3/8$, then

a. events B and C are independent
b. events B and C are mutually exclusive
c. events B and C are neither independent nor mutually exclusive
d. events A and C are equiprobable

4 For two events A and B, if $P(A) = P(A/B) = 1/4$ and $P(B/A) = 1/2$, then

a. A and B are independent b. $P(\overline{A}/B) = 3/4$
c. $P(\overline{B}/\overline{A}) = 1/2$ d. All of these

5 The probabilities of happening of three independent events are p_1, p_2 and p_3, then the probability of happening atleast one of them, is

a. $p_1 + p_2 + p_3$ b. $p_1 p_2 p_3$
c. $(1-p_1)(1-p_2)(1-p_3)$ d. $1-(1-p_1)(1-p_2)(1-p_3)$

6 Let A and B be two events such that $P(\overline{A \cup B}) = 1/6$, $P(A \cap B) = 1/4$ and $P(\overline{A}) = 1/4$, where $\overline{A}$ stands for complement of A. Then, events A and B are **[AIEEE 2005]**

a. mutually exclusive and independent
b. independent but not equally likely
c. equally likely but not independent
d. equally likely and mutually exclusive

7 Three houses are available in a locality. Three persons apply for the houses. Each applies for one house without consulting others. The probability that all three apply for the same house is **[AIEEE 2005]**

a. 7/9 b. 8/9 c. 1/9 d. 2/9

8 If C and D are two events such that $C \subset D$ and $P(D) \neq 0$, then the correct statement among the following is **[AIEEE 2011]**

a. $P(C/D) < P(C)$ b. $P(C/D) = \dfrac{P(D)}{P(C)}$
c. $P(C/D) = P(C)$ d. $P(C/D) \geq P(C)$

Working Space

9 Let A, B and C be pairwise independent events with $P(C) > 0$ and $P(A \cap B \cap C) = 0$. Then, $P(\overline{A} \cap \overline{B}/C)$ is **[IIT JEE 2007, AIEEE 2011]**

a. $P(\overline{A}) - P(B)$ b. $P(A) - P(\overline{B})$
c. $P(\overline{A}) + P(\overline{B})$ d. $P(\overline{A}) - P(\overline{B})$

10 An experiment has 10 equally likely outcomes. Let A and B be two non-empty events of the experiment. If A consists of 4 outcomes, the number of outcomes that B must have, so that A and B are independent, is **[IIT JEE 2008]**

a. 2, 4 or 8 b. 3, 6 or 9 c. 4 or 8 d. 5 or 10

11 A card is drawn and replaced in an ordinary pack of 52 playing cards. Minimum number of times must a card be drawn, so that there is atleast an even chance of drawing a heart, is

a. 2 b. 3 c. 4 d. All of these

12 Whenever horses a, b and c race together, their respective probabilities of winning the race are 0.3, 0.5 and 0.2, respectively. If they race three times the probability that 'the same horse wins all the races' and the probability that a, b and c each wins one race, are respectively (assume no dead heat), is

a. 4/25, 9/50 b. 16/100, 3/100
c. 12/50, 15/50 d. 10/50, 8/50

13 Two cubes have their faces painted either red or blue. The first cube has five red faces and one blue face. When the two cubes are rolled simultaneously, the probability that the two top faces show the same colour is (1/2). Number of red faces on the second cube, is

a. 1 b. 2 c. 3 d. 4

14 Four persons independently solve a certain problem correctly with probabilities 1/2, 3/4, 1/4, 1/8. Then, the probability that the problem is solved correctly by atleast one of them, is **[IIT JEE 2013]**

a. $\frac{235}{256}$ b. $\frac{21}{256}$ c. $\frac{3}{256}$ d. $\frac{253}{256}$

15 A committee of three persons is to be randomly selected from a group of three men and two women and the chair person will be randomly selected from the committee. The probability that the committee will have exactly two women and one man and that the chair person will be a women, is/are

a. 1/5 b. 8/15 c. 2/3 d. 3/10

16 A fever medicine cures 75% persons. Three persons suffering from fever used it. The probability that all the three persons will be benefited, is

a. 27/64 b. 27/32 c. 1/64 d. None of these

17 A man can kill a bird once in three shots on this assumption he fires three shots. The probability that the bird is killed is

a. 8/27 b. 19/27 c. 7/27 d. 11/27

18 The odds against solving a problem by A are 4 : 3 and in favour of B are 7 : 5, then probability that only one will solve the problem, is

a. 16/21 b. 5/21
c. 43/84 d. 41/84

19 A man alternately tosses a coin and throws a die beginning with the coin. The probability that he gets a head in the coin before he gets 5 or 6 in the die, is

a. 3/4 b. 1/2
c. 1/3 d. None of these

20 A pair of fair dice is rolled together till a sum of either 5 or 7 is obtained. Then, the probability that 5 comes before 7, is

a. 1/5 b. 2/5 c. 4/5 d. None of these

21 One bag contains 3 white balls, 2 black balls and another contains 2 white balls, 3 black balls. A ball is drawn from the second bag and placed in the first, then a ball is drawn from the first bag and placed in the second. When the pair of the operations is repeated, then probability that the first bag will contains 5 white ball, is

a. 1/25 b. 1/125 c. 1/225 d. 2/15

22 The probability of India winning a test match against West Indies is 1/2. Assuming independence from match-to-match, the probability that in a 5 match series India's second win occurs at the third test, is

a. 2/3 b. 1/2 c. 1/4 d. 1/8

23 An unbaised die maked with 1, 2, 2, 3, 3, 3 is rolled 3 times. The probability of getting a total score of 4 or 6, is

a. 16/216 b. 50/216 c. 60/216 d. None of these

24 A biased coin with probability P, $0 < P < 1$, of heads is tossed until a head appears for the first time. If the probability that the number of tosses required is even is 2/5, then the value of P is

a. 1/4 b. 1/6 c. 1/3 d. 1/2

25 Two numbers a and b are selected from the set of natural numbers, then the probability that $a^2 + b^2$ is divisible by 5, is

a. 9/25 b. 7/18 c. 11/36 d. 17/81

26 A butterfly randomly lands on one of the six squares of T-shaped figure shown and then randomly moves to an adjacent square. The probability that the butterfly ends up on the R square, is

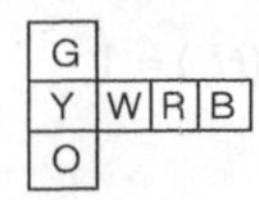

a. 1/4 b. 1/3 c. 2/3 d. 1/6

27 South African cricket captain lost the toss of a coin 13 times out of 14. The chance of this happening was

a. $7/2^{13}$ b. $1/2^{13}$ c. $13/2^{14}$ d. $13/2^{13}$

28 If all the mappings that can be defined from the set $A:\{1, 2, 3, 4\} \to B:\{5, 6, 7, 8, 9\}$ a mapping is randomly selected. The change that the selected mapping is strictly monotonic, is

a. 1/125 b. 2/125 c. 5/4096 d. 5/2048

One or More than One Options Correct Type

***Direction** (Q. Nos. 29-30) This section contains 2 multiple choice questions. Each question has four choices (a), (b), (c) and (d), out of which ONE or MORE THAN ONE are correct.*

29 Let X and Y be two events such that $P(X/Y) = 1/2$, $P(Y/X) = 1/3$ and $P(X \cap Y) = 1/6$. Which of the following is/are correct? **[IIT JEE 2012]**

a. $P(X \cup Y) = 2/3$ b. X and Y are independent

c. X and Y are not independent d. $P(\overline{X} \cap Y) = 1/3$

30 Let E and F be two independent events. The probability that exactly one of them occurs is 11/25 and the probability of none of them occurring is 2/25. If $P(T)$ denotes the probability of occurrence of the events T, then **[IIT JEE 2011]**

a. $P(E) = 4/5, P(F) = 3/5$ b. $P(E) = 1/5, P(F) = 2/5$

c. $P(E) = 2/5, P(F) = 1/5$ d. $P(E) = 3/5, P(F) = 4/5$

Working Space

Statement Type

Direction (Q. Nos. 31-33) *This section is based on Statement I and Statement II. Select the correct answer from the codes given below.*

a. Both Statement I and Statement II are correct and Statement II is the correct explanation of Statement I
b. Both Statement I and Statement II are correct and Statement II is not the correct explanation of Statement I
c. Statement I is correct but Statement II is incorrect
d. Statement II is correct but Statement I is incorrect

31 Four children A, B, C and D have 1, 3, 5 and 7 identical unbiased dice respectively and roll them with the condition that one who obtains an even score, wins. They keep playing till some one or the other wins.

Statement I All the four children are equally likely to win provided they roll their dice simultaneously.

Statement II The child A is most probable to win the game, if they roll dice in order A, B, C and D, respectively.

32 In one day test match between India and Australia the umpire continues tossing a fair coin until the two consecutive throws either HT or TT are obtained for the first time. If it is HT, then India wins and if it is TT, then Australia wins.

Statement I Both India and Australia have equal probabilities of winning the toss.

Statement II If a coin is tossed twice, then event HT or TT is equiprobable.

33 Let $H_1, H_2, \ldots, H_n$ be mutually exclusive and exhaustive events with $P(H_i) > 0, i = 1, 2, \ldots, n$. Let E be any other event with $0 < P(E) < 1$.

Statement I $P(H_i/E) > P(E/H_i) \cdot P(H_i)$, for $i = 1, 2, \ldots, n$.

Statement II $\sum_{i=1}^{n} P(H_i) = 1$

[IIT JEE 2007]

Comprehension Type

Direction (Q. Nos. 34-36) *This section contain a paragraph, each describing theory, experiments, data, etc. Three questions related to the paragraph have been given. Each question has only one correct answer among the four given options (a), (b), (c) and (d).*

The commander of an batalian is punishing two of his soldiers X and Y. He arranged a duel between them. The rules of the duel are that they are to pick up their guns and shoot at each other simultaneously. If one or both are hit, then the duel is over. If both shot miss, then they repeat the process. Suppose that the results of the shots are independent and that each shot of 'X' will hit 'Y' with the probability 2/5 and each shot 'Y' will hit 'X' with probability 1/5.

34 The probability that X is not hit, is

a. 3/25
b. 7/25
c. 5/13
d. 8/13

35 The probability that both the duelists are hit, is

a. 2/13
b. 5/13
c. 8/13
d. None of these

36 The probability that the duel ends after the nth round of shots, is

a. $\left(\frac{12}{5}\right)^{n-1} \cdot \frac{13}{5^{n+1}}$
b. $\left(\frac{12}{13}\right)^{n-1} \cdot \frac{13}{5^{2n}}$
c. Independent of n
d. None of these

One Integer Value Correct Type

Working Space

Direction *(Q. Nos. 37-45) This section contains 9 questions. When worked out will result in an integer from 0 to 9 (both inclusive).*

37 In a throwing 3 dice, the probability that atleast 2 of the three numbers obtained are same is $\frac{m}{n}$, then $(n - m)$ is

38 There are 4 defective items in a lot consisting of 10 items. From this lot, we select 5 items at random. The probability that there will be 2 defective items among them is $\frac{m}{n}$, then $(3m - n)$ is

39 There are ten prizes, five A's, three B's and two C's, placed in identical sealed envelopes for the top ten contestants in a Mathematics contest. The prizes are awarded by allowing winners to select an envelope at random from those remaining. When the 8th contestant goes to select the prize, the probability that the remaining three prizes are one A, one B and one C, is $\frac{m}{n}$, then $|m - n|$ is

40 A box contains 7 tickets, numbered from 1 to 7 inclusive. If 3 tickets are drawn from the box without replacement, one at a time, the probability that they are alternatively either odd-even-odd or even-odd-even is $\frac{m}{n}$, then $|m - n|$ is

41 Two red counters, 3 green counters and 4 blue counters are placed in a row in random order. The probability that no 2 blue counters are adjacent is $\frac{m}{n}$, then $|8m - n|$ is

42 A fair die is thrown 3 times. The chance that the sum of three numbers appearing on the die is less than 11, is equal to $\frac{m}{n}$, then $(m + n)$ is

43 The natural number X is randomly selected from the set of first 100 natural numbers. Then, the probability that it satisfies the inequality $x + \frac{100}{x} > 50$ is $\frac{m}{n}$, then $|m - n|$ is

44 Let a red die, a blue die and a white die are rolled once, the dice being fair. The outcomes on the red, blue, green and a white die denote the numbers a, b, c and d respectively. Let E denotes the events that absolute value of $(a - 1)(b - 2)(c - 3)(d - 6) = 1$. If $P(E) = \frac{1}{abc}$, then $(a + b - c)$ is

45 5 different marbles are placed in 5 different boxes randomly. Then, the probability that exactly two boxes remain empty is $\frac{m}{n}$, then $(2m - n)$ is

DPP-5 Theorem Law of Total Probability

Only One Option Correct Type

Working Space

Direction *(Q. Nos. 1-15) This section contains 15 multiple choice questions. Each question has four choices (a), (b), (c) and (d), out of which ONLY ONE option is correct.*

1 A purse contains 4 copper and 3 silver coins another purse contains 6 copper and 2 silver coins. One coin is drawn from any one of these two purses. The probability that it is a copper coin, is

a. $\frac{4}{7}$ b. $\frac{3}{4}$ c. $\frac{2}{7}$ d. $\frac{37}{56}$

2 Three groups A, B and C are contesting for positions on the Board of Directors of a company. The probabilities of their winning are 0.5, 0.3 and 0.2, respectively. If the group A wins, the probability of introducing a new product is 0.7 and the corresponding probabilities for groups B and C are 0.6 and 0.5, respectively. Then, the probability that the new product will be introduced, is

a. 0.63 b. 0.35 c. 0.18 d. 0.10

3 The probability of a number n showing in a throw of a die marked 1 to 6 is proportional to n. Then, the probability of the number 3 showing in a throw, is

a. $\frac{1}{2}$ b. $\frac{1}{6}$ c. $\frac{1}{7}$ d. $\frac{1}{21}$

4 There are n bags such that ith bag $(1 \le i \le n)$ contains i black and 2 white balls. Two balls are drawn from a randomly selected bag out of given n bags. Then, the probability that the both drawn balls are white, is

a. $\frac{1}{n}$ b. $\frac{1}{n+1}$ c. $\frac{1}{n+2}$ d. $\frac{n}{n+2}$

5 A child throws 2 fair dice. If the number showing are equal, he adds them together to get his final score. On the other hand, if the numbers showing are equal, he throws 2 more dice and adds all 4 numbers showing to get his final score. The probability his final score is 6, is

a. $\frac{145}{1296}$ b. $\frac{146}{1296}$ c. $\frac{147}{1296}$ d. $\frac{148}{1296}$

6 A bag contains 3 red and 3 green balls and a person draws out 3 at random. Then, he drops 3 blue balls into the bag and again draws out 3 at random. The chance that the 3 later balls being all of different colours, is

a. 15% b. 20% c. 27% d. 40%

7 When a missile is fired from a ship, the probability that it is intercepted is 1/3. The probability that the missile hits the target, given that it is not intercepted is 3/4. If three missiles are fired independently from the ship, the probability that all three hits the target, is

a. $\frac{1}{12}$ b. $\frac{1}{8}$ c. $\frac{3}{8}$ d. $\frac{3}{4}$

Working Space

8 An unbiased die with the numbers 1, 2, 3, 4, 6 and 8 on its six faces is rolled. After this roll, if an odd number appears on the top face, all odd numbers on the die are doubled. If an even number appears on the top face, all the even numbers are halved. If the given die changes in this way, then the probability that the face 2 will appear on the second roll, is

a. $\frac{2}{18}$ b. $\frac{3}{18}$ c. $\frac{2}{9}$ d. $\frac{5}{18}$

9 *A* is one of the 6 horses entered for a race and is to be ridden by one of two jockeys *B* or *C*. It is 2 to 1 that *B* rides *A*, in which case all the horses are equally likely to win, if *C* rides *A*, his chance is trebled, then the probability *A* wins the race, is

a. $\frac{13}{18}$ b. $\frac{5}{18}$ c. $\frac{7}{18}$ d. None of these

10 Let *A* consists of 3 good and 2 defective articles. Let *B* consists of 4 good and 1 defective article. A new to *C* is formed by taking 3 articles from *A* and 2 from *B*. The probability that an article chosen at random form *C* is defective, is

a. $\frac{1}{3}$ b. $\frac{2}{5}$ c. $\frac{8}{25}$ d. None of these

11 A purse contains 2 six sided dice. One is a normal fair die, while the other has 2 ones, 2 threes and 2 fives. A die is picked up and rolled. Because of some secret magnetic attraction of the unfair die, there is 75% chance of picking the unfair die and 25% chance of picking a fair die. The die is rolled and shows up the face 3. The probability that a fair die was picked up, is

a. $\frac{1}{7}$ b. $\frac{1}{4}$ c. $\frac{1}{6}$ d. $\frac{1}{24}$

12 A real estate man has eight master keys to open several new houses. Only one master key will open a given house. If 40% of these houses are usually unlocked, the probability that the real estate man can get into a specific house, if he selects three master keys at random, is

a. $\frac{3}{8}$ b. $\frac{5}{8}$ c. $\frac{7}{8}$ d. $\frac{1}{8}$

13 A bowl has 6 red marbles and 3 green marbles. The probability that a blind folded person will draw a red marble on the second draw from the bowl without replacing the marble from the first draw, is

a. $\frac{2}{3}$ b. $\frac{1}{3}$ c. $\frac{1}{4}$ d. $\frac{3}{4}$

14 The probability that a radar will detect an object in on cycle is *p*. The probability that the object will be detected in *n* cycles, is

a. $1-p^n$ b. p^n c. $p(1-p)^{n-1}$ d. $1-(1-p)^n$

15 There are three main political parties namely 1, 2 and 3. If in the adjoining table P_{ij}, $(i, j = 1, 2, 3, \cdots)$ denotes the probability that party *j* wins the general elections contested when party *i* is in the power. What is the probability that the party 2 will be in power after the next two elections, given that the party 1 is in power?

$P_{11} = 0.7$	$P_{12} = 0.2$	$P_{13} = 0.1$
$P_{21} = 0.5$	$P_{22} = 0.3$	$P_{23} = 0.2$
$P_{31} = 0.3$	$P_{32} = 0.4$	$P_{33} = 0.3$

a. 0.27 b. 0.24 c. 0.14 d. 0.06

DPP-6 Baye's Theorem

Only One Option Correct Type

Working Space

Direction (Q. Nos. 1-11) *This section contains 11 multiple choice questions. Each question has four choices (a), (b), (c) and (d), out of which ONLY ONE option is correct.*

1 A bag A contains 2 white balls, 3 red balls and bag B contains 4 white balls, 5 red balls. One ball is drawn at random from a randomly chosen bag and is found to be red. The probability that it was drawn from bag B, was

a. $\frac{5}{14}$ b. $\frac{5}{16}$ c. $\frac{5}{18}$ d. $\frac{25}{52}$

2 A box contains 10 mangoes out of which 4 are rotten. Two mangoes are taken out together. If one of them is found to be good, the probability that the other is also good, is

a. $\frac{5}{13}$ b. $\frac{8}{15}$ c. $\frac{5}{18}$ d. $\frac{2}{3}$

3 Box I contains 5 red and 2 blue balls, while box II contains 2 red and 6 blue balls. A fair coin is tossed. If it turns up head, a ball is drawn from box I, else a ball is drawn from box II. The probability ball drawn is from box I, if it is blue, is

a. $\frac{27}{56}$ b. $\frac{8}{29}$ c. $\frac{21}{29}$ d. $\frac{29}{56}$

4 A letter is known to have come either from 'KRISHNAGIRI' or 'DHARMAPURI'. On the post mark only the two consecutive letters 'RI' are visible. Then, the chance that it came from 'KRISHNAGIRI' is

a. $\frac{3}{5}$ b. $\frac{2}{3}$ c. $\frac{9}{14}$ d. $\frac{5}{14}$

5 A purse contains 100 coins of unknown value, a coin drawn at random is found to be a rupee. The chance that it is the only rupee in the purse, is (assume all numbers of rupee coins in the purse to be equally likely)

a. $\frac{1}{5050}$ b. $\frac{2}{5151}$ c. $\frac{1}{4950}$ d. $\frac{2}{4950}$

6 On a Saturday night, 20% of all drivers in the USA are under the influence of alcohol. The probability that a driver under the influence of alcohol will have an accident is 0.001. The probability that a sober driver will have an accident is 0.0001. If a car on a Saturday night smashed into a tree, then the probability that the driver was under the influence of alcohol, is

a. $\frac{3}{7}$ b. $\frac{4}{7}$ c. $\frac{5}{7}$ d. $\frac{6}{7}$

7 A box has four dice in it. Three of them are fair dice but the fourth one has the number five on all of its faces. A die is chosen at random from the box and is rolled three times and shows up the face five on all the three occassions. The chance that the die chosen was a rigged die, is

a. $\frac{216}{217}$ b. $\frac{215}{219}$ c. $\frac{216}{219}$ d. None of these

Working Space

8 A and B are two inaccurate arithmeticians whose chance of solving a given question correctly are 1/8 and 1/12, respectively. They solve a problem and obtained the same result. If it is 1000 to 1 against their making the same mistake, then chance that the result is correct, is

a. $\frac{13}{14}$ b. $\frac{1}{14}$ c. $\frac{1}{7}$ d. $\frac{6}{7}$

9 In a certain factory, machines A, B and C produce bolts of their production, machines A, B and C produce 2%, 1% and 3% defective bolts, respectively. Machine A produces 35% of the total output of bolts, machine B produces 25% and machine C produces 40%. A bolt is chosen at random from the factory's production and is found to be defective. The probability it was produced on machine C, is

a. $\frac{6}{11}$ b. $\frac{23}{45}$ c. $\frac{24}{43}$ d. $\frac{3}{11}$

10 Two buses A and B are scheduled to arrive at a town central bus station at noon. The probability that bus A will be late, is 1/5. The probability that bus B will be late, is 7/25. The probability that the bus B is late given that bus A is late, is 9/10, then the probability bus A is late given that bus B is late, is

a. $\frac{12}{18}$ b. $\frac{7}{10}$ c. $\frac{18}{28}$ d. $\frac{2}{28}$

11 A person goes to office either by car, scooter, bus or train, the probability of which being 1/7, 3/7, 2/7 and 1/7, respectively. Probability that he reaches office late, if he takes car, scooter, bus or train are 2/9, 1/9, 4/9 and 1/9, respectively. Given that, he reached office in time, then the probability that he travelled by car, is **[IIT JEE 2005]**

a. $\frac{1}{5}$ b. $\frac{1}{6}$ c. $\frac{1}{7}$ d. $\frac{1}{8}$

Comprehension Type

Direction *(Q. Nos. 12-27) This section contains 6 paragraphs, each describing theory, experiments, data, etc. Sixteen questions related to the paragraphs have been given. Each question has only one correct answer among the four given options (a), (b), (c) and (d).*

Passage I

There are n urns, each containing $(n+1)$ balls such that ith urn contains i white balls and $(n+1-i)$ red balls. Let u_i be the event of selecting ith urn, $i = 1, 2, 3, \ldots, n$ and w denotes the event of getting a white ball. **[IIT JEE 2006]**

12 If $P(u_i) \propto i$, where $i = 1, 2, 3, \ldots, n$, then $\lim_{n \to \infty} P(w)$ is equal to

a. 1 b. $\frac{2}{3}$ c. $\frac{3}{4}$ d. $\frac{1}{4}$

13 If $P(u_i) = c$, where c is a constant, then $P(u_n/w)$ is equal to

a. $\frac{2}{n+1}$ b. $\frac{1}{n+1}$ c. $\frac{n}{n+1}$ d. $\frac{1}{2}$

14 If n is even and E denotes the event of choosing even numbered urn $P(u_i) = \frac{1}{n}$, then the value of $P(w/E)$ is

a. $\frac{n+2}{2n+1}$ b. $\frac{n+2}{2(n+1)}$

c. $\frac{n}{n+1}$ d. $\frac{1}{n+1}$

Working Space

Passage II

A fair die is tossed repeatedly until a six is obtained. Let X denotes the number of tosses required, then **[IIT JEE 2009]**

15 The probability that $X = 3$, equals

a. $\frac{25}{216}$ b. $\frac{25}{36}$ c. $\frac{5}{36}$ d. $\frac{125}{216}$

16 The probability that $X \geq 3$, equals

a. $\frac{125}{216}$ b. $\frac{25}{36}$ c. $\frac{5}{36}$ d. $\frac{25}{216}$

17 The conditional probability that $X \geq 6$, given $X > 3$, equals

a. $\frac{125}{216}$ b. $\frac{25}{216}$ c. $\frac{5}{36}$ d. $\frac{25}{36}$

Passage III

Let U_1 and U_2 be two urns such that U_1 contains 3 white and 2 red balls and U_2 contains only 1 white ball. A fair coin is tossed. If head appears, then 1 ball is drawn at random from U_1 and put into U_2. However, if tail appears, then 2 balls are drawn at random from U_1 and put into U_2. Now, 1 ball is drawn at random from U_2. **[IIT JEE 2011]**

18 The probability of the drawn ball from U_2 being white, is

a. $\frac{13}{30}$ b. $\frac{23}{30}$ c. $\frac{19}{30}$ d. $\frac{11}{30}$

19 Given that, the drawn ball from U_2 is white, the probability that head appeared on the coin, is

a. $\frac{17}{23}$ b. $\frac{11}{23}$ c. $\frac{15}{23}$ d. $\frac{12}{23}$

Passage IV

A box B_1 contains 1 white ball, 3 red balls and 2 black balls. Another box B_2 contains 2 white balls, 3 red balls and 4 black balls. A third box B_3 contains 3 white balls, 4 red balls and 4 black balls. **[IIT JEE 2013]**

20 If 2 balls are drawn (without replacement) from a randomly selected box and one of the balls is white and the other ball is red, then the probability that these 2 balls are drawn from box B_2, is

a. $\frac{116}{181}$ b. $\frac{126}{181}$ c. $\frac{65}{181}$ d. $\frac{55}{181}$

21 If 1 ball is drawn from each of the boxes B_1, B_2 and B_3, then the probability that all 3 drawn balls are of same colour, is

a. $\frac{82}{648}$ b. $\frac{90}{648}$ c. $\frac{558}{648}$ d. $\frac{566}{648}$

Passage V

A JEE aspirant estimates that she will be successful with an 80% chance, if she studies 10 h per day, with a 60% chance, if she studies 7 h per day and with a 40% chance, if she studies 4 h per day. She further believes that she will study 10 h, 7 h and 4 h per day with probabilities 0.1, 0.2 and 0.7, respectively.

22 The chance she will be successful, is

a. 0.28 b. 0.38 c. 0.48 d. 0.58

Working Space

23 Given that, she is successful, the chance she studied for 4 h, is

a. $\frac{6}{12}$ b. $\frac{7}{12}$ c. $\frac{8}{12}$ d. $\frac{9}{12}$

24 Given that, she does not achieve success, the chance she studied for 4 h, is

a. $\frac{18}{26}$ b. $\frac{19}{26}$

c. $\frac{20}{26}$ d. $\frac{21}{26}$

Passage VI

Urn I contains 5 red balls and 1 blue ball and urn II contains 2 red balls and 4 blue balls.

A fair die is tossed. If it results in an even number, balls are repeatedly with drawn one at a time with replacement from urn I. If it is an odd number, then balls are repeatedly with drawn one at a time with replacement from urn II. Given that, the first two draws both have resulted in a blue ball.

25 Conditional probability that the first two draws have resulted in blue balls given urn II is used, is

a. $\frac{1}{2}$ b. $\frac{4}{9}$

c. $\frac{1}{3}$ d. None of these

26 If the probability that the urn I is being used is p and q is the corresponding figure for urn II, then

a. $q = 16p$ b. $q = 4p$ c. $q = 2p$ d. $q = 3p$

27 The probability of getting a red ball in the third drawn, is

a. $\frac{1}{3}$ b. $\frac{1}{2}$ c. $\frac{37}{102}$ d. $\frac{41}{102}$

One Integer Value Correct Type

***Direction** (Q. Nos. 28-30) This section contains 3 questions. When worked out will result in an integer from 0 to 9 (both inclusive).*

28 In a factory, which manufactures bolts, machines A, B and C manufacture respectively 25%, 35% and 40% of the bolts of their outputs 5%, 4% and 2% are respectively defective bolts. A bolt is drawn at random from the product and is found to be defective. The probability that it is manufactured by the machine B is $\frac{m}{n}$. Then, $\frac{|2m - n|}{13}$ is equal to

29 A bag A contains 2 white balls, 3 red balls and a bag B contains 4 white balls, 5 red balls. One ball is drawn at random from one of the bags and it is found to be red. The probability that it was drawn from bag B, is $\frac{m}{n}$. Then, $|2m - n|$ is equal to

30 In a test, an examinee either guesses or copies or knows the answer to a multiple choice question with four choices. The probability that he makes a guess is 1/3 and the probability that he copies the answer is 1/6. The probability that his answer is correct given that he copied it, is 1/8. The probability that he knew the answer to the question given that he correctly answered it is m/n. Then, $|m - n|$ is equal to

DPP-7 Binomial Distribution and Probability Distribution

Subjective Question

Direction (*Q. No.* 1) *This question is subjective in nature, need to be solved completely on notebook.*

Working Space

1 If a fair coin is tossed 10 times, then find the probability of getting

i. exactly six heads

ii. atleast six heads

iii. atmost six heads

Only One Option Correct Type

Direction (*Q. Nos.* 2-23) *This section contains 22 multiple choice questions. Each question has four choices (a), (b), (c) and (d), out of which ONLY ONE option is correct.*

2 India and Pakistan play a 5 match test series of hockey, then the probability that India wins atleast three matches is

a. 1/2　　b. 3/5
c. 4/5　　d. 5/16

3 A coin is tossed 7 times. Each time a man call head. The probability that he wins the toss on more than three occasions is

a. 1/4　　b. 5/8
c. 1/2　　d. None of these

4 A man take a step forward with probability 0.4 and backward with probability 0.6. The probability that at the end of eleven steps he is one step away from the starting point, is

a. 0.35　　b. 0.36
c. 0.37　　d. 0.38

5 8 coins are tossed simultaneously. The probability of getting atleast 6 heads, is

a. 57/64　　b. 229/256
c. 7/64　　d. 37/256

6 The probability that a man can hit a target is 3/4. He tries 5 times. The probability that he will hit the target atleast three times is

a. 291/364　　b. 371/464
c. 471/502　　d. 459/519

7 If two coins are tossed 5 times, then the probability of getting 5 heads and 5 tails is

a. 63/256　　b. 1/1024
c. 2/205　　d. 9/64

Working Space

8 If X is a random variable with distribution given below

X	0	1	2	3
$P(X = x)$	k	$3k$	$3k$	k

Then, the value of k and its variance are

a. 1/8, 22/27 b. 1/8, 23/27
c. 1/8, 24/27 d. 1/8, 3/4

9. A person can kill a bird with probability 3/4. He tries 5 times. What is the probability that he may not kill the bird?

a. 243/1024 b. 781/1024
c. 1/1024 d. 1023/1024

10 The probability that a student is not a swimmer is 1/5. The probability that out of 5 students, 4 are swimmers, is

a. ${}^5C_4\,(4/5)^4 \cdot 1/5$ b. $(4/5)^4 \cdot 1/5$
c. ${}^5C_1 \cdot (4/5)^4 \cdot 1/5 \cdot {}^5C_4$ d. None of these

11 If the mean and variance of a binomial variate X are 2 and 1 respectively, then the probability that X takes a value greater than or equal to 1, is

a. 2/3 b. 4/5
c. 7/8 d. 15/16

12 A dice is thrown 5 times, then the probability that an even number will come up exactly 3 times, is

a. 5/16 b. 1/2
c. 3/16 d. 3/2

13 A fair coin is tossed n times. If the probability that head occurs 6 times is equal to the probability that head occurs 8 times, then n is equal to

a. 15 b. 14
c. 12 d. 7

14 An experiment succeeds twice as often as it fails. The probability that in 4 trials there will be atleast three successes, is

a. 4/27 b. 8/27
c. 16/27 d. 24/27

15 If a die is thrown 7 times, then the probability of obtaining 5 exactly 4 times is

a. $(1/6)^3\,(5/6)^4$ b. ${}^7C_4\,(1/6)^3\,(5/6)^4$
c. $(1/6)^4\,(5/6)^3$ d. ${}^7C_4\,(1/6)^4\,(5/6)^3$

16 Two coins are tossed five times, the probability that an odd number of heads are obtain, is

a. $(1/2)^5$ b. 3/5
c. 2/5 d. None of these

17 A man can hit at bird with probability 1/4. How many times must he fires so that the probability of his hitting the target atleast once is greater than 2/3, is

a. 2 b. 3
c. 4 d. 8

18 If a coin is tossed n times, then probability that the head comes odd times, is

a. 1/2 b. $1/2^n$
c. $1/2^{n-1}$ d. None of these

Working Space

19 A random variable X has the probability distribution

X	1	2	3	4	5	6	7	8
$P(X)$	0.15	0.23	0.12	0.10	0.20	0.20	0.07	0.05

For the events $E = \{x \text{ is a prime number}\}$ and $F = \{x < 4\}$, the probability $P(E \cup F)$ is **[AIEEE 2004]**

a. 0.35 b. 0.77 c. 0.87 d. 0.50

20 A pair of fair dice is thrown independently three times. The probability of getting a score of exactly 9 twice, is

a. 8/729 b. 8/243 c. 1/729 d. 8/9

21 Consider 5 independent Bernoulli's trials each with probability of success is p. If the probability of atleast one failure is greater than or equal to 31/32, then p lies in the interval, is **[AIEEE 2011]**

a. [3/4, 11/12]
b. [0, 1/2]
c. [11/12, 1]
d. [1/2, 3/4]

22 In a binomial distribution $B(n, p = 1/4)$, if the probability of atleast one success is greater than or equal to 9/10, then n is greater than **[AIEEE 2009]**

a. $\dfrac{9}{\log_{10} 4 - \log_{10} 3}$
b. $\dfrac{4}{\log_{10} 4 - \log_{10} 3}$
c. $\dfrac{1}{\log_{10} 4 - \log_{10} 3}$
d. $\dfrac{1}{\log_{10} 4 + \log_{10} 3}$

23 A six faced die is thrown until 1 comes, the probability that 1 comes in even number of trials, is **[IIT JEE 2005]**

a. 5/11 b. 5/6 c. 6/11 d. 1/6

Matching Type

Direction (Q. Nos. 24-25) Choices for the correct combination of elements from Column I and Column II are given as options (a), (b), (c) and (d), out of which one is correct.

24

	Column I		Column II
i.	The probability of a bomb hitting a bridge is 1/2. Two direct hits are needed to destroy it. The least number of bombs required, so that the probability of the bridge being destroyed is greater than 0.9, is	p.	4
ii.	A bag contains 2 red, 3 white and 5 black balls. A ball is drawn, its colour is noted and replaced. Minimum number of times, a ball must be drawn so that the probability of getting a red ball for the first time is atleast even, is	q.	5
iii.	A hunter knows that a deer is hidden in one of two near bushes. The probability of its being hidden in bush I being 4/5. The hunter having a rifle containing 10 bullets decides to fire them at bush I or II. It is known that each shot may hit one of the two bushes, independently of the other with probability 1/2. Number of bullets must he fire on bush I to hit the animal with maximum probability is (assume that the bullet hitting the bush also hit the animal)	r.	6
		s.	7

Codes

	i	ii	iii
a.	p	q	r
b.	s	p	r
c.	q	r	s
d.	q	r	s

Working Space

25

	Column I		Column II
i.	In a knock out tournament, 2^n equally skilled players ; $S_1, S_2, \dots, S_n$, are participating. In each round, players are divided in pair at random and winner from each pair moves in the next round. If S_2 reaches the semifinal, then the probability that S_1 wins the tournament is 1/20. The value of n equals	p.	3
ii.	In a multiple choice question, there are four alternative answers of which one or more than one is correct. A candidate will get marks on the question only if he ticks all the correct answers. The candidate ticks the answer at random. If the probability of the candidate getting marks on the question is to be greater than or equal to 1/3, then the least number of chances he should be allowed, is	q.	4
iii.	All the face cards from a pack of 52 playing cards are removed. From the remaining pack, half of the cards are randomly removed without looking at them and then randomly drawn two cards simultaneously from the remaining. If the probability that, two cards drawn are both aces is $\frac{p \cdot ({}^{38}C_{20})}{{}^{40}C_{20} \cdot {}^{20}C_2}$, then the value of p is	r.	5
iv.	An unbiased normal coin is tossed n times. Let E_1 : Event that both heads and tails are present in n tosses. E_2 : Event that the coin shows up heads at most once. The value of n for which E_1 and E_2 are independent is	s.	6

Codes

	i	ii	iii	iv		i	ii	iii	iv
a.	p	q	r	s	b.	q	r	s	p
c.	q	r	s	r	d.	p	q	s	r

Comprehension Type

Direction *(Q. Nos. 26-31) This section contains 2 paragraphs, each describing theory, experiments, data, etc. Six questions related to the paragraphs have been given. Each question has only one correct answer among the four given options (a), (b), (c) and (d).*

Passage I

A chess match between two grandmasters X and Y is won by whoever first wins a total of two games. X's chances of winning, drawing, losing any particular game are a, b and c, respectively. The games are independent and $a + b + c = 1$. Then,

26 The probability that X wins the match after $(n + 1)$ games $(n \geq 1)$, is

a. $na^2 \cdot n^{n-1}$

b. $a^2 [nb^{n-1} + n(n-1) b^{n-2} c]$

c. $na^2 bc^{n-1}$

d. $(n-1) a^2 b^{n-2} (b + nc)$

27 The probability that Y wins the match in the 4th game is

a. $3bc^2 (b + 2a)$ b. $bc^2 (3b + a)$

c. $2ac^2 (b + c)$ d. $abc (2a + b)$

28 The probability that X wins the match, is

a. $\frac{a^3 + a^2 c}{(a+c)^3}$ b. $\frac{a^3 + 3a^2 c}{(a+c)^3}$

c. $\frac{a^3}{(a+c)^3}$ d. $\frac{a(a^2 + 3c)}{(1-b)^3}$

Working Space

Passage II

A coin shows head with probability p, $0 < p < 1$. x is a random variable which denotes the number of tosses before the appearance of the first head. Thus $TTTH \Rightarrow x = 3$ and $H \Rightarrow x = 0$. If $P(x = n)$ denotes the probability of the first head appearing at the $(n + 1)$th trial.

29 An expression for $P(x = 5)$, is

a. $(1-p)^5 \cdot p$ b. $(1-p)^4 \cdot p$
c. $(1-p)^4$ d. $(1-p)^5$

30 An expression for $P(3 \le x \le 6)$, is

a. $(1-p)^3 - (1-p)^6$ b. $(1-p)^2 - (1-p)^7$
c. $(1-p)^2 - (1-p)^6$ d. $(1-p)^3 - (1-p)^7$

31 An expression $P(x \ge 4)$, is

a. $(1-p)^4$ b. $(1-p)^2$
c. $1-(1-p)^3$ d. None of these

One Integer Value Correct Type

Direction *(Q. Nos. 32-35) This section contains 4 questions. When worked out will result in an integer from 0 to 9 (both inclusive).*

32 A coin that comes up with probability $p > 0$ and tails with probability $1-p > 0$ independently meach flip is flipped eight times. Suppose the probability of three heads and five tails is equal to 1/25 of the probability of five heads and three tails. Let $p = \frac{m}{n}$, where m and n are relatively prime positive integers. The value of $(m + n - 5)$ equals

33 The probability of man hitting a target on fire is 1/5. Then, the minimum number of fire he must follow in order to make his chance of hitting the target more than 3/4, is equal to

34 Let x denotes the number of times heads occur in n tosses of a fair coin. If $P(x = 4)$, $P(x = 5)$ and $P(x = 6)$ are in an AP. Then, least value of n is

35 A man has 3 coins A, B and C out of which A is fair. The probability of head appearing when tossed are 2/3 and 1/3 for B and C, respectively. One of the coins chosen at random, is tossed three times giving 2 heads and one tail. If the probability that the chosen coin was A is $k/25$, then k is

DPP-8 Probability Through Statistical (Stochatic) Tree Diagram

Subjective Questions

Working Space

Direction (Q. Nos. 1-3) *These questions are subjective in nature, need to be solved completely on notebook.*

1 A bag initially contains 1 red ball and 2 blue balls. A trial consists of selecting a ball at random, noting its colour and replacing it together with an additional ball of the same colour. Given that three trials are made, draw a tree diagram illustrating the various probabilities. Hence, then find the probability that

i. at least one blue ball is drawn.
ii. exactly one blue ball is drawn.
iii. Given that all three balls drawn are of the same colour find the probability that they are all red.

2 Three coins are tossed. Two of them are fair and one is biased so that a head is three times as likely as a tail. Then, find the probability of getting two heads and a tail.

3 In three multiple choice of 3 questions, there are five alternative answers given to the first two questions each and four alternative answers given to the last question. If a candidate guesses answer at random, what is the probability that he will get

i. exactly one right answer.
ii. at least one right answer.

Only One Option Correct Type

Direction (Q. Nos. 4-8) *This section contains 5 multiple choice questions. This question has four choices (a), (b), (c) and (d), out of which ONLY ONE option is correct.*

4 Suppose families always have one, two or three children, with probabilities 1/4, 1/2 and 1/4, respectively. Assume everyone eventually gets married and has children, the probability of a couple having exactly four grand children, is

a. 27/128 b. 37/128
c. 25/128 d. 20/128

5 Miss *C* has either tea or coffee at morning break. If she has tea one morning, the probability she has tea the next morning is 0.4. If she has coffee one morning, the probability she has tea on the following Wednesday morning is

a. 0.46 b. 0.49
c. 0.51 d. 0.61

Working Space

6 Sixteen players $S_1, S_2, \ldots, S_{16}$ play in a tournament. They are divided into eight pairs at random. From each pair, a winner is decided on the basis of a game played between the two players of the pair. Assume that all the players are of equal strength. The probability that 'exactly one of the two players S_1 and S_2 is among the eight winners', is

a. 4/15 b. 7/15
c. 8/15 d. 9/15

7 The number a is randomly selected from the set {0, 1, 2, 3, ..., 98, 99}. The number b is selected from the same set. Probability that the number $3^a + 7^b$ has a digit equal to 8 at the unit place, is

a. 1/16 b. 2/16
c. 4/16 d. 3/16

8 A signal which can be green or red with probability 4/5 and 1/5 respectively, is received by station A and then transmitted to station B. The probability of each station receiving the signal correctly is (3/4). If the signal received at station B is green, then the probability that the original signal was green, is **[IIT JEE 2010]**

a. 3/5 b. 6/7 c. 20/23 d. 9/23

One or More than One Options Correct Type

Direction (Q. No. 9) This section contains 1 multiple choice question. Each question has four choices (a), (b), (c) and (d), out of which ONE or MORE THAN ONE are correct.

9 A boy has a collection of blue and green marbles. The number of blue marbles to the sets {2, 3, 4, ..., 13}. If two marbles are chosen simultaneously and at random from his collection, then the probability that they have different colours is 1/2. Possible number of blue marbles, is

a. 2 b. 3 c. 6 d. 10

Matching Type

Direction (Q. No. 10) Choices for the correct combination of elements from Column I and Column II are given as options (a), (b), (c) and (d), out of which one is correct.

10

	Column I		Column II
i.	Two different numbers are taken from the set {0, 1, 2, 3, 4, ..., 10}. The probability that their sum and positive difference are both multiples of 4, is $\frac{x}{55}$, then x equals	p	4
ii.	There are two red, two blue, two white and certain number (greater than 0) of green socks in a drawer. If two socks are taken at random from the drawer without replacement, the probability that they are of the same colour is 1/5, then the number of green socks are	q	6
		r	8
iii.	A drawer contains red socks and blue socks, atmost 17 in all. It so happens that when two socks are selected randomly without replacement, there is a probability of exactly 1/2 that both red or both are blue. The largest possible number of red socks in the drawer that is consistent with this data, is	s	10

Codes

	i	ii	iii		i	ii	iii
a.	q	p	r	b.	p	p	s
c.	p	r	q	d.	r	s	p

Comprehension Type

Working Space

***Direction** (Q. Nos. 11-16) This section contains 2 paragraphs, each describing theory, experiments, data, etc. Six questions related to the paragraphs have been given. Each question has only one correct answer among the four given options (a), (b), (c) and (d).*

Passage I

A bag contains 6 balls of 3 different colours namely white, green and red, atleast one ball of each different colour. Assume all possible probability distributions are equally likely. Then,

11 The probability that the bag contains 2 balls of each colour, is

a. 1/3 b. 1/5
c. 1/10 d. 1/4

12 Three balls are picked up at random from the bag and found to be one of each different colour. The probability that the bag contained 4 red balls, is

a. 1/14 b. 2/14
c. 3/14 d. 4/14

13 Three balls are picked up at random from the bag and are found to be one of each different colour. The probability that the bag containeds equal number of white and green balls, is

a. 4/14 b. 3/14
c. 5/14 d. 6/14

Passage II

Two fair dice are rolled. Let $P(A_i) > 0$ denotes the probability of the event that the sum of the number appearing on the faces of the dice is divisible by i. Then,

14 Which one of the following events is most probable?

a. A_3 b. A_4
c. A_5 d. A_6

15 For which one of the following pairs (i, j) are the events A_i and A_j are independent?

a. (3, 4) b. (4, 6)
c. (2, 3) d. (4, 2)

16 Number of all possible ordered pairs (i, j) for which the events A_i and A_j are independent is

a. 6 b. 12
c. 13 d. 25

SET - 1

Revisal Problems for JEE Main

Only One Option Correct Type

Working Space

Direction (*Q. Nos.* 1-40) *This section contains 40 multiple choice questions. Each question has four choices (a), (b), (c) and (d), out of which ONLY ONE option is correct.*

1 The probability that the 13th day of a month, selected at random, will be Friday, is

a. 1/7 b. 1/12
c. 1/84 d. None of these

2 There are four letters and four envelops of corresponding address. If letters are put in these envelops at random, then the probability that all letters are not in correct envelops, is

a. 1/24 b. 1
c. 23/24 d. 9/24

3 For two events A and B, if $P(A) = P(A/B) = 1/4$ and $P(B/A) = 1/2$, then

a. A and B are independent b. $P(A'/B) = 3/4$
c. $P(B'/A') = 1/2$ d. All of these

4 The probability that a leap year selected at random contains either 53 Sunday or 53 Monday is,

a. 2/7 b. 4/7
c. 3/7 d. 1/7

5 A piece of equipment will function only when all the three components A, B and C are working, the probability of A failing during one year is 0.15, that of B is 0.05 and that of C is 0.10. The probability that the equipment will fail before the end of the year, is

a. 0.72675 b. 0.27325
c. 1 d. 0.95

6 The probability that in a year of the 22nd century chosen at random, there will be 53 Sunday, is

a. 3/28 b. 2/28
c. 7/28 d. 5/28

7 Dialing a telephone number, an old man forgets the last two digits remembering only that these are different dialled at random. The probability that the number is dialled correctly is

a. 1/45 b. 1/90
c. 1/100 d. None of these

8 Five coins whose faces are marked 2, 3 are tossed. The chance of obtaining a total of 12, is

a. 1/32 b. 1/16
c. 3/16 d. 5/16

9 Two persons A and B take turns in throwing a pair of dice. The first person to throw 9 from both dice will be awarded the prize. If A throws first, then the probability that B wins the game, is

a. 9/17 b. 8/17 c. 8/9 d. 1/9

10 A die is thrown twice, when the sum of their digits is 6, then the probability that atleast one of them will show 4, is

a. 1/6 b. 2/5 c. 3/5 d. 1/2

11 India plays two matches each with West Indies and Australia. In any match, the probability of India getting point 0,1 and 2 are 0.45, 0.05 and 0.50 respectively. Assuming that the outcomes are independents, the probability of India getting atleast 7 points, is

a. 0.8750 b. 0.0875 c. 0.0625 d. 0.0250

Working Space

12 A bag contains $(n + 1)$ coins. If it is known that one of these coins shows heads on both sides, whereas the other coins are fair. One coin is selected at random and tossed. If the probability that toss results in heads is 7/12, then the value of n is

a. 3 b. 4
c. 5 d. None of these

13 The sum of two positive numbers is 100. The probability that their product is greater than 1000, is

a. 7/9 b. 7/10
c. 2/5 d. None of these

14 Let A and B be two independent events. The probability that both A and B occur together is 1/6 and the probability that neither of them occurs is 1/3. The probability of occurrence of A, is

a. 0 and 1 b. 1/2 or 1/3 c. 1/2 or 1/4 d. 1/3 or 1/4

15 A problem of Mathematics is given to three students A, B and C whose chances of solving it are 1/2, 1/3 and 1/4 respectively, then probability that the problem will be solved, is

a. 1/2 b. 1/4 c. 3/4 d. 2/3

16 The mean and the variance of a binomial distribution are 4 and 2, respectively. Then, the probability of 2 successes, is

a. 128/256 b. 219/256 c. 37/256 d. 28/256

17 One ticket is selected at random from 50 tickets numbered 00, 01, 02, ..., 49. Then, the probability that the sum of the digits on the selected ticket is 8, given that the product of these digits is zero equals

a. 1/14 b. 1/7 c. 5/14 d. 1/50

18 An urn contains nine balls of which three are red, four are blue and two are green. Three balls are drawn at random without replacement from the urn. The probability that the three balls have different colours, is

a. 1/3 b. 2/7 c. 1/21 d. 2/23

19 If the letters of the word 'ASSASSIN' are written down at random in a row, the probability that no two S's occur together, is

a. 1/5 b. 1/7 c. 1/35 d. None of these

20 There are 4 machines and it is known that exactly 2 of them are faulty. They are tested, one by one, in a random order till both the faulty machines are identified. Then, the probability that only 2 tests are needed, is

a. 1/3 b. 1/6 c. 1/2 d. 1/4

21 The probability that a student passes in Mathematics, Physics and Chemistry are m, p and c respectively. Of these subjects, the student has a 75% chance of passing in atleast one, 50% chance of passing in atleast two and 40% chance of passing in exactly two. Which of the following relations are true?

a. $pmc = 1/4$ b. $pmc = 1/10$ c. $pmc = 1/18$ d. $pmc = 1/6$

22 Let E and F be two independent events. The probability that exactly one of them occurs is $\frac{11}{25}$ and the probability of none of them occurs is $\frac{2}{25}$. If $P(T)$ denotes the probability of occurrence of the event T, then

a. $P(E) = 4/5, P(F) = 3/5$ b. $P(E) = 1/15, P(F) = 2/5$
c. $P(E) = 2/5, P(F) = 1/5$ d. None of these

23 Fifteen coupons are numbered 1,2, ..., 15, respectively. Seven coupons are selected at random one at a time with replacement. The probability that the largest number appearing on a selected coupon is 9, is

a. $(9/16)^6$ b. $(8/15)^7$
c. $(3/5)^7$ d. None of these

24 Three of the six vertices of a regular hexagon are chosen at random. The probability that the triangle with these three vertices is equilateral, equals

a. 1/2 b. 1/5 c. 1/10 d. 1/20

Working Space

25 For the three events A, B and C, P (exactly one of the events A or B occurs) = P (exactly one of the events B or C occurs) = P (exactly one of the events C or A occurs) = p and P (all the three events occurs simultaneously) = p^2, with $0 < p < 1/2$, then the probability of atleast one of the three events A, B and C occurring, is

a. $\frac{3p + 2p^2}{2}$ b. $\frac{p + 3p^2}{4}$ c. $\frac{p + 3p^2}{2}$ d. $\frac{3p + 2p^2}{4}$

26 7 white balls and 3 black balls are randomly placed in a row. The probability that no two black balls are placed adjacently, equals

a. 1/2 b. 7/15 c. 2/15 d. 1/3

27 If the integers m and n are chosen at random between 1 and 100, then the probability that a number of the form $7^m + 7^n$ is divisible by 5, equals

a. 1/4 b. 1/7 c. 1/8 d. 1/49

28 One Indian and four American men and their wives are to be seated randomly around a circular table. Then, the conditional probability that the Indian man is seated adjacent to his wife given that each American man is seated adjacent to his wife, is

a. 1/2 b. 1/3 c. 2/5 d. 1/5

29 A multiple choice examination has 5 questions. Each question has three alternative answers of which exactly one is correct. The probability that a student will get 4 or more correct answers just by guessing is

a. $17/3^5$ b. $13/3^5$ c. $11/3^5$ d. $10/3^5$

30 Four fair dice D_1, D_2, D_3 and D_4, each having six faces numbered 1, 2, 3, 4, 5 and 6 are rolled simultaneously. The probability that D_4 shows a number appearing on one of D_1, D_2 and D_3, is

a. 91/216 b. 108/216 c. 125/216 d. 127/216

31 Two dice are rolled simultaneously. The probability that the sum of the two numbers on the top faces will be atleast 10, is

a. 1/6 b. 1/12 c. 1/18 d. None of these

32 $2n$ boys are randomly divided into two subgroups containing n boys each. The probability that the two tallest boys are in different group, is

a. $\frac{n}{2n-1}$ b. $\frac{n-1}{2n-1}$
c. $\frac{2n-1}{4n^2}$ d. None of these

33 The chance that 13 cards combination from a pack of 52 playing cards is dealt to a player in a game of bridge, in which 9 cards are of the same suit, is

a. $\frac{4 \cdot {}^{13}C_9 \cdot {}^{39}C_4}{{}^{52}C_{13}}$ b. $\frac{4! \cdot {}^{13}C_9 \cdot {}^{39}C_4}{{}^{52}C_{13}}$
c. $\frac{{}^{13}C_9 \cdot {}^{39}C_4}{{}^{52}C_{13}}$ d. None of these

34 In throwing a pair of dice, the events 'coming up of 6 on first die' and 'a total of 7 on both the die' are

a. mutually exclusive b. exhaustive events
c. independent d. dependent

35 Out of 13 applicants for a job, there are 5 women and 8 men. It is desired to select 2 persons for the job. The probability that atleast one of the selected persons will be a woman, is

a. 25/39 b. 14/39 c. 5/13 d. 10/13

36 A pair of dice is thrown. If total numbers turned up on both dice is 8, then the probability that the number up on second die is 5' is

a. 5/36 b. 1/6 c. 1/5 d. 2/5

37 In a series of 3 independent trials, the probability of exactly 2 success is 12 times as large as the probability of 3 successes. The probability of a success in each trial is

a. 1/5 b. 2/5 c. 3/5 d. 4/5

Working Space

38 A bag contains 2 white and 4 black balls. A ball is drawn 5 times, each being replaced before another is drawn. The probability that atleast 4 of the balls are white, is

a. 4/81   b. 10/243
c. 11/243   d. None of these

39 A examination consists of 8 questions in each of which one of the 5 alternatives is the correct one. On the assumption that a candidate who has done no preparatory work choses for each question any one of the five alternatives with equal probability that he gets more than one correct answer, is equal to

a. $(0.8)^8$ b. $3(0.8)^8$ c. $1-(0.8)^8$ d. $1-3(0.8)^8$

40 Letters of the word 'VECTOR' are arranged in the grid shown below :

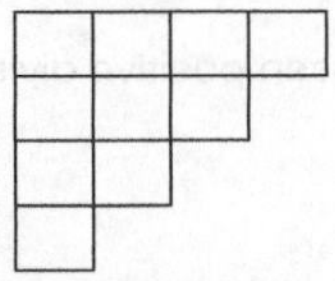

Then, the chance that no row remains empty, is

a. 7/15   b. 8/15   c. 6/15   d. 9/15

Statement Type

Direction (Q. Nos. 41-45) ***This section is based on Statement I and Statement II. Select the correct answer from the codes given below.***

a. Both Statement I and Statement II are correct and Statement II is the correct explanation of Statement I
b. Both Statement I and Statement II are correct and Statement II is not the correct explanation of Statement I
c. Statement I is correct but Statement II is incorrect
d. Statement II is correct but Statement I is incorrect

41 **Statement I** If p is chosen at random in the closed interval [0, 5], then the probability that the equation $x^2 + px + 1/4(p+2) = 0$ has real root is 3/5.

Statement II If discriminant ≥ 0, then roots of the quadratic equation are always real.

42 Four numbers are chosen at random (without replacement) from the set $\{1, 2, 3, \ldots, 20\}$.

Statement I The probability that the chosen numbers when arranged in some order will form an AP is 1/85.

Statement II If the four chosen numbers form an AP, then the set of all possible values of common difference is $\{\pm 1, \pm 2, \pm 3, \pm 4, \pm 5\}$.

43 **Statement I** Since, sample space of the experiment 'A coin is tossed, if it turns up head, a die is thrown' is $\{(H, 1)\ (H, 2)\ (H, 3)\ (H, 4)\ (H, 5)\ (H, 6), T\}$. Then , the probability of the event $\{(H, 1), (H, 2), (H, 5)\}$ is 3/7.

Statement II If all the sample points in the sample space of an experiment are pair wise mutually exclusive, equally likely and exhaustive, then probability of an event E is defined as

$$P(E) = \frac{\text{Number of sample points favourable to the event } E}{\text{Total number of sample points in the sample space}}$$

44 **Statement I** If A and B are two independent events such that $P(A) \neq 0$, $P(B) \neq 0$, then A and B cannot be mutually exclusive.

Statement II For independent events A and B, we have $P(A/B) = P(A)$, which is not so for mutually exclusive events.

45 **Statement I** If $\frac{1+4p}{4}, \frac{1-p}{4}, \frac{1-2p}{4}$ are probabilities of three pairwise mutually exclusive events, then the possible values of P belongs to the set $\left[-\frac{1}{4}, \frac{1}{2}\right]$.

Statement II If three events are pairwise mutually exclusive and exhaustive, then the sum of there probability is equal to 1.

SET - 2

Revisal Problems
for JEE Advanced

Only One Option Correct Type

Working Space

Direction *(Q. Nos. 1-12) This section contains 12 multiple choice questions. Each question has four choices (a), (b), (c) and (d), out of which ONLY ONE option is correct.*

1 If m/n (in lowest term) be the probability that a randomly chosen positive divisor of 10^{99} is an integral multiple of 10^{88}, then the value of $(m+n)$ is

a. 634   b. 638
c. 616   d. None of these

2 A determinant is chosen at random from the set of all determinants of order 2 with elements 0 or 1 only. The probability that the determinant chosen has the value non-negative, is

a. 3/16   b. 6/16
c. 10/16   d. 13/16

3 An urn contains 10 balls coloured either black or red. When selecting two balls from the urn at random, the probability that a ball of each colour is selected is 8/15. Assuming that the urn contains more black balls than red balls, then the probability that atleast one black ball is selected, when selecting two balls, is

a. 18/45   b. 30/45
c. 39/45   d. 41/45

4 A fair die is tossed repeatedly. Mr. A wins if it is 1 or 2 on two consecutive tosses and Mr. B wins, if it is 3, 4, 5 or 6 on two consecutive tosses. The probability that *A* wins, if the die is tossed indefinitely, is

a. 1/3   b. 5/21
c. 1/4   d. 2/5

5 Mr. A makes a bet with Mr. B that in a single throw with two dice he will throw a total of seven before B throws four. Each of them has a pair of dice and they throw simultaneously until one of them wins, equal throws being disregarded. Probability that B wins, is

a. 1/3   b. 4/11
c. 5/16   d. 5/17

6 If a, b and c are three numbers (not necessarily different) chosen randomly and with replacement from the set $\{1, 2, 3, 4, 5\}$, the probability that $(ab+c)$ is even, is

a. 35/125   b. 59/125
c. 64/125   d. 75/125

7 Mr. A and Mr. B each have a bag that contains one ball of each of the colours blue, green, orange, red and voilet. '*A*' randomly select one ball from his bag and puts it into B's bag. '*B*' randomly select one ball from his bag and puts it into *A*'s bag. The probability that after this process the contents of the two bags are the same, is

a. 1/6   b. 1/5
c. 1/3   d. 1/2

8 Two boys *A* and *B* find the jumble of n ropes lying on the floor. Each takes hold of one loose and randomly. If the probability that they are both holding the same rope is $\frac{1}{101}$. Then, the number of ropes is equal to

a. 101   b. 100   c. 51   d. 50

9 A letter is taken at random from the letters of the word 'STATISTICS' and another letter is taken at random from the letters of the word 'ASSISTANT'. The probability that they are same, is

a. 1/45   b. 13/90   c. 19/90   d. 5/18

Working Space

10 A fair coin is tossed 7 times, then probability that atleast 3 consecutive heads occurs is equal to

a. 1/8 b. 3/8 c. 5/8 d. 3/7

11 Let $S = \{a \in N, a \leq 1000\}$. If equation $[\tan^2 2x] - \tan 2x - a = 0$ has real roots, then probability of getting element 'a' from the set(s), which satisfy the equation

a. $p = 25/999$ b. $p = 125/999$ c. $p = 31/1000$ d. $p = 27/1000$

12 Three numbers are drawn from the set $\{1, 2, 3, \ldots, n\}$ with replacement. The probability that their sum is $2n$, is

a. $1/2n$ b. $1/2$ c. $\frac{(n-1)(n+4)}{2n^3}$ d. $\frac{(n+1)(n+2)}{n^3}$

One or More than One Options Correct Type

***Direction** (Q. Nos. 13-15) This section contains 3 multiple choice questions. Each question has four choices (a), (b), (c) and (d), out of which ONE or MORE THAN ONE are correct.*

13 Two whole numbers are randomly selected and multiplied. Consider two events E_1 and E_2 defined as

E_1 : Their product is divisible by 5.

E_2 : Units place in their product is 5.

Which of the following statements(s) is/are correct?

a. E_1 is twice as likely to occur as E_2 b. E_1 and E_2 are disjoint
c. $P(E_2/E_1) = 1/4$ d. $P(E_1/E_2) = 1$

14 If A and B are two events such that $P(B) \neq 1$, B^C denotes the event complementary to B, then

a. $P(A/B^C) = \frac{P(A) - P(A \cap B)}{1 - P(B)}$

b. $P(A \cap B) \geq P(A) + P(B) - 1$

c. $P(A) >< P(A/B)$ according as $P(A/B^C) >< P(A)$

d. $P(A/B^C) + P(A^C/B^C) = 1$

15 Which of the following statements(s) is/are correct?

a. 3 coins are tossed once. Two of them atleast must land the same way. No matter whether they land heads or tails, the third coin is equally likely to land either the same way or oppositely. So, the chance that all the three coins land the same way is 1/2.

b. Let $0 < P(B) < 1$ and $P(A/B) = P(A/B^C)$, then A and B are independent.

c. Suppose an urn contains 'w' white and 'b' black balls and a ball is drawn from it and is replaced along with 'd' additional balls of the same colour. Now, a second ball is drawn from it. The probability that the second drawn ball is white is independent of the value of 'd'.

d. A, B, C simultaneously satisfy $P(ABC) = P(A) \cdot P(B) \cdot P(C)$ and $P(AB\overline{C}) = P(A) \cdot P(B) \cdot P(\overline{C})$ and $P(A\overline{B}C) = P(A) \cdot P(\overline{B}) \cdot P(C)$ and $P(\overline{A}BC) = P(\overline{A}) \cdot P(B) \cdot P(C)$. Then, A, B, C are independent.

Comprehension Type

***Direction** (Q. Nos. 16-35) This section contains 7 paragraphs, each describing theory, experiments, data, etc. Twenty questions related to the paragraphs have been given. Each question has only one correct answer among the four given options (a), (b), (c) and (d).*

Passage I

From a lot containing 3 defective and 6 good items, 4 items are chosen at random, which are to be checked for their qualities. As a result of this inspection, first 2 of the 4 items chosen found to be both good.

16 The probability that the 4 chosen items for inspection where all good, is

a. 1/7 b. 2/7 c. 3/7 d. 4/7

17 If the checking for quantity continues, then probability that the third inspected item will also be good, is

a. 1/7 b. 2/7 c. 3/7 d. 4/7

Working Space

Passage II

A box contains b red balls, $2b$ white balls and $3b$ blue balls, where b is a positive integer. 3 balls are selected at random from the box.

18 If balls are drawn without replacement and A denotes the event that 'No two of the selected balls have the same colour', then

a. there is no value of b for which $P(A) = 0.3$
b. there is exactly one value of b for which $P(A) = 0.3$ and this value is less than 10
c. there is exactly one value of b for which $P(A) = 0.3$ and they value is greater than 10
d. there is more than one value of b for which $P(A) = 0.3$

19 If balls are drawn without replacement and B denotes the event that 'No two of the 3 drawn balls are blue', then

a. $P(B) = 1/3$, if $b = 1$
b. $P(B) = 2/3$, if $b = 2$
c. $P(B) = 1/4$, if $b = 4$
d. $P(B) = 1/2$ for all value of b

20 If $P(A) = 0.3$, then the value of $P(A/B)$ is

a. 3/5 b. 3/10
c. 1/2 d. 2/3

Passage III

There are four boxes A_1, A_2, A_3 and A_4. Box A_i has i balls and on each ball a number is printed, the number are from 1 to i. A box is selected randomly, the probability of selection of box A_i is $\left(\frac{i}{10}\right)$ and then a ball is drawn. Let E_i represents the event that a ball with number 'i' is drawn.

21 The value of $\sum_{i=1}^{4} P(E_i \cap A_i)$ is

a. 1/5 b. 1/10
c. 2/5 d. 1/4

22 $P(A_3/E_2)$ is equal to

a. 1/4 b. 1/3
c. 2/3 d. 1/2

23 One ball is drawn from each box randomly. The probability that the sum of four numbers appearing on the balls is less than 5, is

a. 1/6 b. 1/12
c. 1/15 d. 1/24

Passage IV

A and B play a game in which they alternately call out positive integer less than or equal to n, according to the following rules :

A goes first and always calls out an odd number, B always out an even number and each player must call out a number, which is greater than the previous number (except for A's first turn). The game ends when one player cannot call out a number.

24 For $n = 4$, number of games possible, is

a. 3 b. 4
c. 6 d. None of these

25 For $n = 8$, number of games possible, is

a. 20 b. 21
c. 4 d. None of these

26 For $n = 5$, what is the probability that A wins?

a. 1 b. 0
c. 1/2 d. None of these

Working Space

Passage V

An urn contains 4 white and 9 black balls. r balls are drawn with replacement. Let $P(r)$ be the probability that no two white balls appear in succession.

27 The value of $P(4)$ must be

a. $\frac{81}{(169)^2}$ b. $\frac{81 \times 273}{(169)^2}$

c. $1-\left(\frac{4}{13}\right)^4$ d. None of these

28 The recursion relation of $P_{(r)}$ must be

a. $P_{(r)} = P_{(r-2)} \cdot \left(\frac{9}{13}\right)\left(\frac{4}{13}\right) + P_{(r-1)}\left(\frac{9}{13}\right)$

b. $P_{(r)} = P_{(r-2)} \cdot \left(\frac{9}{13}\right)^2 + P_{(r+1)}\left(\frac{9}{13}\right)$

c. $P_{(r)} = P_{(r-2)} \cdot \left(\frac{4}{13}\right)^2 + P_{(r-1)} \cdot \left(\frac{9}{13}\right)$

d. None of the above

29 $P_{(r)}$ must be equal to

a. $\frac{16 \cdot (12^r) - (-3)^r}{12 \cdot (13)^r}$ b. $\frac{16 \cdot (12^r) - (-3)^r}{9^r \cdot (13)^r}$

c. $\frac{16 \cdot (12^r) - (-3)^r}{15 \cdot (13)^r}$ d. None of these

Passage VI

Let S be the set of students $\{a_1, a_2, \ldots, a_n\}$. The students plan to visit two institutes next month. Any student can accompany or with draw from the visit. So, on both occassions any subset of S can go.

30 Probability that there is no common student in two visits, is

a. $1/4^n$ b. $(3/4)^n$

c. $1/2^n$ d. $3/4^n$

31 Probability that each student either goes in first visit or in second or in both, is

a. $1/4^n$ b. $1/2^n$

c. $(3/4)^n$ d. 1

32 Probability that same number of students go in both visits, is

a. $\frac{{}^{2n}C_n}{8^n}$ b. $1/4^n$

c. $(3/4)^n$ d. $\frac{{}^{2n}C_n}{4^n}$

Passage VII

A box contains nine slipe bearing the numbers $-4, -3, -2, -1, 0, 1, 2, 3, 4$ (one number for one slip).

33 A slip is chosen at random, its number is noted and replaced in the box. If the experiment is repeated 5 times, then the probability that exactly 3 numbers appear in the noted list, is

a. $\frac{{}^9C_3 \times 90}{9^5}$ b. $\frac{{}^9C_3 \times 243}{9^5}$

c. $\frac{{}^9C_3 \times 150}{9^5}$ d. $\frac{{}^9C_3 \times 240}{9^5}$

34 If a slip is chosen at random from the box and the experiment is repeated nine times (with replacement) to form 3×3 matrix. The chosen number are arranged rowwise, then the probability that the matrix, so formed is skew-symmetric, is

a. $(1/9)^8$ b. $(1/9)^6$

c. $(1/9)^3$ d. $(1/9)^4$

35 A slip is chosen at random from the box, its number is noted and replaced in the box. If the experiment is repeated 5 times, then the probability that the sum of five number is zero, is

a. $\frac{3491}{9^5}$ b. $\frac{3951}{9^5}$

c. $\frac{3923}{9^5}$ d. None of these

Working Space

One Integer Value Correct Type

Direction (Q. Nos. 36-40) This section contains 5 questions. When worked out will result in an integer from 0 to 9 (both inclusive).

36 A positive integer that reads the same backwards and forwards is called a palindrome. A number greater than 10 and less than 10^6 is chosen at random. If it is known that it is a five-digit number and the probability that it is a palindrome is p, then the value of 300 p, is

37 A drawer contains several pair of socks. Not wanting to count Mr. A asked miss X 'how many pairs of socks are there in the drawer'? Miss X, not wanting to give answer replied, 'well, each sock has exactly one matching sock and the probability that two socks drawn from the drawer from a matching pair is (1/15)'. Then, the answer to Mr. A's question, is

38 If the papers of 4 students can be checked by one of the 7 teachers. If the probability that all the 4 papers are checked by exactly 2 teachers is A, then the value of 49 A must be

39 A function is selected from all function of set $S = \{1, 2, 3, \ldots, n\}$ to itself. If the probability that it is one-one is 3/32, then n is equal to

40 A bag contains n identical red balls, $2n$ identical black balls and $3n$ identical white balls. If probability of drawing n balls of same colour is greater than or equal to 1/6, then the minimum number of red balls in the bag is equal to

JEE Main & AIEEE Archive

(Compilation of Last 13 Years Questions)

1 Let A and B be two events such that $P(\overline{A \cup B}) = \frac{1}{6}, P(A \cap B) = \frac{1}{4}$ and $P(\overline{A}) = \frac{1}{4}$, where $\overline{A}$ stands for the complement of the event A. Then, the events A and B are **(JEE Main 2014, AIEEE 2005)**

a. independent but not equally likely
b. independent and equally likely
c. mutually exclusive and independent
d. equally likely but not independent

2 A multiple choice examination has 5 questions. Each question has three alternative answers of which exactly one is correct. The probability that a student will get 4 or more correct answers just by guessing is **(JEE Main 2013)**

a. $\frac{17}{3^5}$ b. $\frac{13}{3^5}$
c. $\frac{11}{3^5}$ d. $\frac{10}{3^5}$

3 Three numbers are chosen at random without replacement from numbers $\{1, 2, 3, ..., 8\}$. The probability that their minimum is 3, given that their maximum is 6, is **(AIEEE 2012)**

a. $\frac{3}{8}$ b. $\frac{1}{5}$
c. $\frac{1}{4}$ d. $\frac{2}{5}$

4 Consider 5 independent Bernoulli's trials each with probability of success p. If the probability of atleast one failure is greater than or equal to $\frac{31}{32}$, then p lies in the interval **(AIEEE 2011)**

a. $\left(\frac{3}{4}, \frac{11}{12}\right]$ b. $\left[0, \frac{1}{2}\right]$
c. $\left(\frac{11}{12}, 1\right]$ d. $\left(\frac{1}{2}, \frac{3}{4}\right]$

5 If C and D are two events such that $C \subset D$ and $P(D) \neq 0$, then the correct statement among the following is **(AIEEE 2011)**

a. $P(C/D) \geq P(C)$
b. $P(C/D) < P(C)$
c. $P(C/D) = \frac{P(D)}{P(C)}$
d. $P(C/D) = P(C)$

6 Let A, B and C be pairwise independent events with $P(C) > 0$ and $P(A \cap B \cap C) = 0$. Then, $P(A^C \cap B^C / C)$ is equal to **(AIEEE 2011)**

a. $P(A^C) - P(B)$
b. $P(A) - P(B^C)$
c. $P(A^C) + P(B^C)$
d. $P(A^C) - P(B^C)$

7 An urn contains nine balls of which three are red, four are blue and two are green. Three balls are drawn at random without replacement from the urn. The probability that the three balls have different colours, is **(AIEEE 2010)**

a. $\frac{1}{3}$ b. $\frac{2}{7}$
c. $\frac{1}{21}$ d. $\frac{2}{23}$

8 Four numbers are chosen at random (without replacement) from the set $\{1, 2, 3, ..., 20\}$.

Statement I The probability that the chosen numbers when arranged in some order will form an AP, is $\frac{1}{85}$.

Statement II If the four chosen numbers form an AP, then the set of all possible values of common difference is $\{\pm 1, \pm 2, \pm 3, \pm 4, \pm 5\}$. **(AIEEE 2010)**

a. Both Statement I and Statement II are correct and Statement II is the correct explanation of Statement I
b. Both Statement I and Statement II are correct and Statement II is not the correct explanation of Statement I
c. Statement I is correct but Statement II is incorrect
d. Statement II is correct but Statement I is incorrect

9 In a Binomial distribution $B\left(n, p = \frac{1}{4}\right)$, if the probability of atleast one success is greater than or equal to $\frac{9}{10}$, then n is greater than **(AIEEE 2009)**

a. $\frac{1}{\log_{10} 4 - \log_{10} 3}$ b. $\frac{1}{\log_{10} 4 + \log_{10} 3}$
c. $\frac{9}{\log_{10} 4 - \log_{10} 3}$ d. $\frac{4}{\log_{10} 4 - \log_{10} 3}$

10 One ticket is selected at random from 50 tickets numbered 00, 01, 02, ..., 49. Then, the probability that the sum of the digits on the selected ticket is 8, given that the product of these digits is zero, is equal to **(AIEEE 2009)**

a. $\frac{1}{14}$ b. $\frac{1}{7}$
c. $\frac{5}{14}$ d. $\frac{1}{50}$

11 It is given that, the events A and B are such that $P(A) = \frac{1}{4}, P(A/B) = \frac{1}{2}$ and $P(B/A) = \frac{2}{3}$. Then, $P(B)$ is equal to **(AIEEE 2008)**

a. $\frac{1}{2}$ b. $\frac{1}{6}$
c. $\frac{1}{3}$ d. $\frac{2}{3}$

JEE Archive

12 A die is thrown. Let A be the event that the number obtained is greater than 3. Let B be the event that the number obtained is less than 5. Then, $P(A \cup B)$ is **(AIEEE 2008)**

a. $\frac{2}{5}$ b. $\frac{3}{5}$
c. 0 d. 1

13 A pair of fair dice is thrown independently three times. The probability of getting a score of exactly 9 twice, is **(AIEEE 2007)**

a. 1/729 b. 8/9
c. 8/729 d. 8/243

14 Two aeroplanes I and II bomb a target in succession. The probabilities of I and II scoring a hit correctly are 0.3 and 0.2, respectively. The second plane will bomb only if the first misses the target. The probability that the target is hit by the second plane, is **(AIEEE 2007)**

a. 0.06 b. 0.14
c. 0.32 d. 0.7

15 At a telephone enquiry system, the number of phone calls regarding relevant enquiry follow. Poisson distribution with an average of 5 phone calls during 10 min time intervals. The probability that there is atmost one phone call during a 10 min time period, is **(AIEEE 2006)**

a. $\frac{5}{6}$ b. $\frac{6}{55}$
c. $\frac{6}{e^5}$ d. $\frac{6}{5^e}$

16 A fair die is tossed eight times. The probability that a third six is observed on the eight throw, is **(AIEEE 2002)**

a. $\frac{{}^7C_2 \times 5^5}{6^7}$

b. $\frac{{}^7C_2 \times 5^5}{6^8}$

c. $\frac{{}^7C_2 \times 5^5}{6^6}$

d. None of the above

17 Three houses are available in a locality. Three persons apply for the houses. Each applies for one house without consulting others. The probability that all the three apply for the same house, is **(AIEEE 2005)**

a. $\frac{7}{9}$ b. $\frac{8}{9}$
c. $\frac{1}{9}$ d. $\frac{2}{9}$

18 A random variable X has Poisson distribution with mean 2. Then, $P(X > 1.5)$ equal to **(AIEEE 2005)**

a. $\frac{3}{e^2}$ b. $1-\frac{3}{e^2}$
c. 0 d. $\frac{2}{e^2}$

19 The probability that A speaks truth is 4/5 while this probability for B is 3/4. The probability that they contradict each other when asked to speak on a fact, is **(AIEEE 2004)**

a. $\frac{3}{20}$ b. $\frac{1}{5}$
c. $\frac{7}{20}$ d. $\frac{4}{5}$

20 A random variable X has the probability distribution.

X	1	2	3	4	5	6	7	8
$P(X)$	0.15	0.23	0.12	0.10	0.20	0.08	0.07	0.05

For the events, $E = \{X \text{ is a prime number}\}$ and $F = \{X < 4\}$, the probability $P(E \cup F)$ is **(AIEEE 2004)**

a. 0.87 b. 0.77
c. 0.35 d. 0.50

21 The mean and the variance of a Binomial distribution are 4 and 2, respectively. Then, the probability of 2 successes is **(AIEEE 2004)**

a. $\frac{37}{256}$ b. $\frac{219}{256}$
c. $\frac{128}{256}$ d. $\frac{28}{256}$

22 Five horses are in a race. Mr. A selects two of the horses at random and bets on them. The probability that Mr. A selected the winning horse, is **(AIEEE 2003)**

a. $\frac{4}{5}$ b. $\frac{3}{5}$
c. $\frac{1}{5}$ d. $\frac{2}{5}$

23 Events A, B and C are mutually exclusive events such that $P(A) = \frac{3x+1}{3}$, $P(B) = \frac{1-x}{4}$ and $P(C) = \frac{1-2x}{2}$. The set of possible values of x are in the interval **(AIEEE 2003)**

a. $\left[\frac{1}{3}, \frac{1}{2}\right]$

b. $\left[\frac{1}{3}, \frac{2}{3}\right]$

c. $\left[\frac{1}{3}, \frac{13}{3}\right]$

d. [0, 1]

24 The mean and variance of a random variable X having a Binomial distribution are 4 and 2 respectively, then $P(X = 1)$ is **(AIEEE 2003)**

a. $\frac{1}{32}$ b. $\frac{1}{16}$
c. $\frac{1}{8}$ d. $\frac{1}{4}$

25 A problem in Mathematics is given to three students A, B, C and their respective probability of solving the problem is $\frac{1}{2}, \frac{1}{3}$ and $\frac{1}{4}$. Probability that the problem is solved, is **(AIEEE 2002)**

a. 3/4 b. 1/2
c. 2/3 d. 1/3

26 A and B play a game, where each is asked to select a number from 1 to 25. If the two numbers match, both of them win a prize. The probability that they will not win a prize in a single trial, is **(AIEEE 2002)**

a. $\frac{1}{25}$
b. $\frac{24}{25}$
c. $\frac{2}{25}$
d. None of the above

27 If A and B are two mutually exclusive events, then **(AIEEE 2002)**

a. $P(A) < P(\overline{B})$
b. $P(A) > P(\overline{B})$
c. $P(A) < P(B)$
d. None of the above

28 The probability of India winning a test match against the West Indies is 1/2 assuming independence from match-to-match. The probability that in a match series India's second win occurs at the third test, is **(AIEEE 2002)**

a. $\frac{1}{8}$ b. $\frac{1}{4}$
c. $\frac{1}{2}$ d. $\frac{2}{3}$

29 A biased coin with probability p, $0 < p < 1$, of heads is tossed until a head appears for the first time. If the probability that the number of tosses required is even, is 2/5, then p equal to **(AIEEE 2002)**

a. $\frac{1}{3}$ b. $\frac{2}{3}$
c. $\frac{2}{5}$ d. $\frac{3}{5}$

JEE Advanced & IITJEE Archive

(Compilation of Last 10 Years Questions)

1 Three boys and two girls stand in a queue. The probability that the number of boys ahead of every girl is atleast one more than the number of girls ahead of her, is **(2014 Adv., Single Option Correct Type)**

a. 1/2 b. 1/3
c. 2/3 d. 3/4

Passage for Q. Nos. (2-3)

Box I contains three cards bearing numbers 1, 2, 3 ; box II contains five cards bearing numbers 1, 2, 3, 4, 5; and box III contains seven cards bearing numbers 1, 2, 3, 4, 5, 6, 7. A card is drawn from each of the boxes. Let x_i be the number on the card drawn from the i, th box $i = 1, 2, 3$. **(2014 Adv., Comprehension Type)**

2 The probability that $x_1 + x_2 + x_3$ is odd, is

a. $\frac{29}{105}$ b. $\frac{53}{105}$
c. $\frac{57}{105}$ d. $\frac{1}{2}$

3 The probability that x_1, x_2, x_3 are in an arithmetic progression, is

a. $\frac{9}{105}$ b. $\frac{10}{105}$
c. $\frac{11}{105}$ d. $\frac{7}{105}$

4 Four persons independently solve a certain problem correctly with probabilities $\frac{1}{2}, \frac{3}{4}, \frac{1}{4}, \frac{1}{8}$. Then, the probability that the problem is solved correctly by atleast one of them is **(2013 Adv., Single Option Correct Type)**

a. $\frac{235}{256}$ b. $\frac{21}{256}$
c. $\frac{3}{256}$ d. $\frac{253}{256}$

5 Four fair dice D_1, D_2, D_3 and D_4, each having six faces numbered 1, 2, 3, 4, 5 and 6 are rolled simultaneously. The probability that D_4 shows a number appearing on one of D_1, D_2 and D_3, is **(2012, Single Option Correct Type)**

a. $\frac{91}{216}$
b. $\frac{108}{216}$
c. $\frac{125}{216}$
d. $\frac{127}{216}$

6 A ship is fitted with three engines E_1, E_2 and E_3. The engines function independently of each other with respective probabilities 1/2, 1/4 and 1/4. For the ship to be operational atleast two of its engines must function. Let X denote the event that the ship is operational and let X_1, X_2, X_3 dentoes, respectively the events that the engines E_1, E_2 and E_3 are functioning. Which of the following is/are correct? **(2012, One or More than One Options Correct Type)**

a. $P(X_1^c / X) = 3/16$

b. P [exactly two engines of the ship are functioning X] $= \frac{7}{8}$

c. $P(X / X_2) = \frac{5}{16}$

d. $P(X / X_1) = \frac{7}{16}$

7 Let X and Y be two events such that $P(X / Y) = \frac{1}{2}, P(Y / X) = \frac{1}{3}$ and $P(X \cap Y) \frac{1}{6}$. Which of the following is/are correct? **(2012, One or More than One Options Correct Type)**

a. $P(X \cup Y) = 2/3$
b. X and Y are independent
c. X and Y are not independent
d. $P(X^c \cap Y) = 1/3$

8 Let E and F be two independent events. The probability that exactly one of them occurs is $\frac{11}{25}$ and the probability of none of them occurring is $\frac{2}{25}$. If $P(T)$ denotes the probability of occurrence of the event T, then **(2011, One or More than One Options Correct Type)**

a. $P(E) = \frac{4}{5}, P(F) = \frac{3}{5}$ b. $P(E) = \frac{1}{5}, P(F) = \frac{2}{5}$
c. $P(E) = \frac{2}{5}, P(F) = \frac{1}{5}$ d. $P(E) = \frac{3}{5}, P(F) = \frac{4}{5}$

Passage for Q. Nos. (9-10)

Let U_1 and U_2 be two urns such that U_1 contains 3 white and 2 red balls and U_2 contains only 1 white ball. A fair coin is tossed. If head appears then 1 ball is drawn at random from U_1 and put into U_2. However, if tail appears then 2 balls are drawn at random from U_1 and put into U_2. Now, 1 ball is drawn at random from U_2. **(2011, Comprehension Type)**

9 The probability of the drawn ball from U_2 being white, is

a. $\frac{13}{30}$ b. $\frac{23}{30}$
c. $\frac{19}{30}$ d. $\frac{11}{30}$

10 Given that, the drawn ball from U_2 is white, then probability that head appeared on the coin, is

a. $\frac{17}{23}$ b. $\frac{11}{23}$

c. $\frac{15}{23}$ d. $\frac{12}{23}$

11 A signal, which can be green or red with probability $\frac{4}{5}$ and $\frac{1}{5}$ respectively, is received by station A and then transmitted to station B. The probability of each station receiving the signal correctly is $\frac{3}{4}$. If the signal received at station B is green, then the probability that the original signal green is **(2010, Single Option Correct Type)**

a. $\frac{3}{5}$ b. $\frac{6}{7}$

c. $\frac{20}{23}$ d. $\frac{9}{20}$

12 Let ω be a complex cube root of unity with $\omega \neq 1$. A fair die is thrown three times. If r_1, r_2 and r_3 are the numbers obtained on the die, then the probability that $\omega^{r_1} + \omega^{r_2} + \omega^{r_3} = 0$, is **(2010, Single Option Correct Type)**

a. 1/18 b. 1/9

c. 2/9 d. 1/36

Passage for Q. Nos. (13-15)

A fair die is tossed repeatedly until a six is obtained. Let X denote the number of tosses required. **(2009, Comprehension Type)**

13 The probability that $X = 3$, equals

a. $\frac{25}{216}$ b. $\frac{25}{36}$

c. $\frac{5}{36}$ d. $\frac{125}{216}$

14 The probability that, $X \geq 3$, equals

a. $\frac{125}{216}$

b. $\frac{25}{36}$

c. $\frac{5}{36}$

d. $\frac{25}{216}$

15 The conditional probability that $X \geq 6$, given $X > 3$, equals

a. $\frac{125}{216}$

b. $\frac{25}{216}$

c. $\frac{5}{36}$

d. $\frac{25}{36}$

16 Consider the system of equations

$$ax + by = 0$$
$$cx + dy = 0,$$

where $a, b, c, d \in \{0, 1\}$.

Statement I The probability that the system of equations has a unique solution, is 3/8.

Statement II The probability that the system of equations has a solution, is 1. **(2008, Statement Type)**

17 An experiment has 10 equally likely outcomes. Let A and B be two non-empty events of the experiment. If A consists of 4 outcomes, the number of outcomes that B must have so that A and B are independent, is **(2008, Single Option Correct Type)**

a. 2, 4 or 8

b. 3, 6 or 9

c. 4 or 8

d. 5 or 10

18 Let E^c denotes the complement of an event E. Let E, F, G be pairwise independent events with $P(G) > 0$ and $P(E \cap F \cap G) = 0$. Then, $P(E^c \cap F^c \mid G)$ equals **(2007, Single Option Correct Type)**

a. $P(E^c) + P(F^c)$

b. $P(E^c) - P(F^c)$

c. $P(E^c) - P(F)$

d. $P(E) - P(F^c)$

19 One Indian and four American men and their wives are to be seated randomly around a circular table. Then, the conditional probability that the Indian man is seated adjacent to his wife given that each American man is seated adjacent to his wife, is **(2007, Single Option Correct Type)**

a. $\frac{1}{2}$

b. $\frac{1}{3}$

c. $\frac{2}{5}$

d. $\frac{1}{5}$

20 Let $H_1, H_2, \ldots, H_n$ be mutually exclusive events with $P(H_i) > 0$, $i = 1, 2, \ldots, n$. Let E be any other event with $0 < P(E) < 1$.

Statement I $P(H_i/E) > P(E/H_i) \cdot P(H_i)$ for $i = 1, 2, \ldots, n$.

Statement II $\sum_{i=1}^{n} P(H_i) = 1$ **(2007, Statement Type)**

a. Both Statement I and Statement II are correct and Statement II is the correct explanation of Statement I

b. Both Statement I and Statement II are correct and Statement II is not the correct explanation of Statement I

c. Statement I is correct but Statement II is incorrect

d. Statement II is correct but Statement I is incorrect

Passage for Q. Nos. (21-23)

There are n urns each containing $(n+1)$ balls such that the ith urn contains 'i' white balls and $(n+1-i)$ red balls. Let u_i be the event of selecting ith urn, $i = 1, 2, 3, \ldots, n$ and W denotes the event of getting a white balls.

(2006, Comprehension Type)

21 If $P(u_i) \propto i$, where $i = 1, 2, 3, \ldots, n$, then $\lim_{n \to \infty} P(W)$ is equal to

a. 1
b. $\frac{2}{3}$
c. $\frac{1}{4}$
d. $\frac{3}{4}$

22 If $P(u_i) = c$, where c is a constant, then $P(u_n/W)$ is equal to

a. $\frac{2}{n+1}$
b. $\frac{1}{n+1}$
c. $\frac{n}{n+1}$
d. $\frac{1}{2}$

23 If n is even and E denotes the event of choosing even numbered urn $\left(P(u_i) = \frac{1}{n}\right)$, then the value of $P(W/E)$ is

a. $\frac{n+2}{2n+1}$
b. $\frac{n+2}{2(n+1)}$
c. $\frac{n}{n+1}$
d. $\frac{1}{n+1}$

24 A fair die is rolled. The probability that the first time 1 occurs at the even throw, is

(2005, Single Option Correct Type)

a. 1/6
b. 5/11
c. 6/11
d. 5/36

25 Of the three independent events E_1, E_2 and E_3, the probability that only E_1 occurs is α, only E_2 occurs is β and only E_3 occurs is γ. Let the probability p that none of events E_1, E_2 or E_3 occurs satisfy the equations $(\alpha - 2\beta)$, $p = \alpha\beta$ and $(\beta - 3\gamma)\, p = 2\beta\gamma$. All the given probabilities are assumed to lie in the interval (0, 1).

Then, $\frac{\text{probability of occurrence of } E_1}{\text{probability of occurrence of } E_3}$ is equal to

(2005, Integer Type)

Answers with Explanations

Answers

1. Binomial Theorem

DPP-1 Binomial for Positive Index, General Term and Middle Term

1. (i) $\frac{70}{3}x^8$ (ii) $\frac{25!}{10!15!}(2)^{15}(3)^{10}$ (iii) (a) -20, (b) $280\,x^2$ **2.** (i) ${}^{11}C_5 \cdot a^6/b^5$ (ii) ${}^{11}C_6 \cdot a^5/b^6$

3. (i) 96096 (ii) 2268 **4.** *a* **5.** *a* **6.** *b* **7.** *b* **8.** *d* **9.** *b* **10.** *c* **11.** *c*
12. *c* **13.** *c* **14.** *b* **15.** *a* **16.** *c* **17.** *d* **18.** *a* **19.** *c* **20.** *b* **21.** *b*
22. *c* **23.** *b* **24.** *b* **25.** *c* **26.** *a* **27.** *d* **28.** *b* **29.** *b* **30.** *c*

DPP-2 Divisibility, Remainder, Last Digits Comparision of Numbers and $(A+\sqrt{B})^n$

2. *b* **3.** *d* **4.** *d* **5.** *a* **6.** *a* **7.** *a* **8.** *a* **9.** *b* **10.** *d* **11.** *b*
12. *c* **13.** *a* **14.** *d* **15.** *c* **16.** *a* **17.** *b* **18.** *b* **19.** *a* **20.** *d*

DPP-3 Properties of Binomial Coefficient Based on Summation

3. *a* **4.** *b* **5.** *a* **6.** *b* **7.** *b* **8.** *c* **9.** *c* **10.** *c* **11.** *c* **12.** *c*
13. *a* **14.** *b* **15.** *d* **16.** *c* **17.** *c* **18.** *d* **19.** *b* **20.** *a*

DPP-4 Product of Binomial Coefficients and Double Summation

5. *a* **6.** *c* **7.** *a* **8.** *d* **9.** *c* **10.** *a* **11.** *a* **12.** *d* **13.** *a* **14.** *c*
15. *a,b,c,d* **16.** *a,b,d* **17.** *a,b,c,d* **18.** *b* **19.** *c* **20.** *a*

DPP-5 Greatest Term, Binomial Theorem for Any Index and Multinomial Theorem

1. (i) $\frac{7\cdot 3^{13}}{2}$, (ii) 455×3^{12} **2.** (i) $-1260\,a^2b^3c^4$, (ii) -12600 **3.** (i) 990, (ii) 3660 **4.** *a* **5.** *a*

6. *d* **7.** *b* **8.** *a* **9.** *b* **10.** *a* **11.** *d* **12.** *c* **13.** *d* **14.** *b* **15.** *a*
16. *a* **17.** *c* **18.** *b* **19.** *c* **20.** *a,b* **21.** *a,b,c,d* **22.** *a,b,d* **23.** *b* **24.** *a*
25. *b*

Revisal Problems (JEE Main)

1. *b* **2.** *a* **3.** *d* **4.** *c* **5.** *b* **6.** *c* **7.** *b* **8.** *d* **9.** *d* **10.** *b*
11. *a* **12.** *a* **13.** *c* **14.** *b* **15.** *a* **16.** *b* **17.** *d* **18.** *d* **19.** *b* **20.** *c*
21. *a* **22.** *b* **23.** *c* **24.** *c* **25.** *b* **26.** *a* **27.** *c* **28.** *a* **29.** *b* **30.** *a*

Revisal Problems (JEE Advanced)

1. *c* **2.** *c* **3.** *a* **4.** *d* **5.** *c* **6.** *c* **7.** *c* **8.** *a* **9.** *d* **10.** *c*
11. *b* **12.** *a,c,d* **13.** *a,c* **14.** *b,c* **15.** *a,d* **16.** *a* **17.** *b* **18.** *c* **19.** *d* **20.** *b*
21. *c* **22.** *a* **23.** *d* **24.** *b* **25.** *d* **26.** *b* **27.** 3 **28.** 2 **29.** 3 **30.** 9
31. 2

JEE Main & AIEEE Archive

1. *d* **2.** *c* **3.** *a* **4.** *b* **5.** *d* **6.** *b* **7.** *b* **8.** *d* **9.** *b* **10.** *b*
11. *c* **12.** *a* **13.** *b* **14.** *a* **15.** *b* **16.** *c* **17.** *b* **18.** *a* **19.** *b* **20.** *c*
21. *d* **22.** *d*

JEE Advanced & IIT JEE Archive

1. *c* **2.** 6 **3.** *d* **4.** *c*

2. Permutation and Combination

DPP-1 Factorial and Exponent of Prime Value p in $(n!)$

2. (i) $x=12$, (ii) $x=28$, (iii) $x=4$, (iv) $x=5$ **3.** 63 **4.** 48 **5.** $p=8, q=4, r=2, s=1$

6. 10 **7.** 23 **8.** 14 **9.** *d* **10.** *c*

DPP-2 Fundamental Principle of Counting

1. (i) 24 ways, (ii) 576 ways, (iii) 360 ways,

2. (i) 120, (ii) 40, (iii) 40, (iv) 80, (v) 20

3. (i) 840, (ii) 120, (iii) 400, (iv) 240, (v) 480, (vi) 40, (vii) 60, (viii) 240

4. *a*	**5.** *b*	**6.** *b*	**7.** *c*	**8.** *a*	**9.** *c*	**10.** *d*	**11.** *c*	**12.** *b*	**13.** *a*
14. *a*	**15.** *c*	**16.** *b*	**17.** *c*	**18.** *c*	**19.** *a*	**20.** *d*	**21.** *d*	**22.** *d*	**23.** *b*
24. *a*	**25.** *b*	**26.** *d*	**27.** *c*						

DPP-3 Application of ^nP_r

1. (i) 117600 (ii) 416	**2.** *c*	**3.** *c*	**4.** *a*	**5.** *b*	**6.** *c*	**7.** *b*	**8.** *a*	**9.** *a*
10. *a*	**11.** *b*	**12.** *a*	**13.** *d*	**14.** *b*	**15.** *d*	**16.** *d*	**17.** *d*	**18.** *c*
19. *c*	**20.** *a*							

DPP-4 Application of ^nC_r

1. *a*	**2.** *b*	**3.** *d*	**4.** *c*	**5.** *a*	**6.** *b*	**7.** *b*	**8.** *a*	**9.** *a*	**10.** *b*
11. *a*	**12.** *a*	**13.** *a*	**14.** *c*	**15.** *a*	**16.** *c*	**17.** *b*	**18.** *b*	**19.** *c*	**20.** *a*

DPP-5 Restricted Arrangements

1. (i) $5!\times 6!$ ways, (ii) $2!(5!)^2$ ways, (iii) $10!-6!5!$ ways

2. $\dfrac{11!}{4!4!2!}$ words **3.** 4989600, 120960

4. (i) 840 words, (ii) 20 words, (iii) 240 words **5.** (i) 4320 words, (ii) 36000 words

6. (i) 5040 ways, (ii) 144 ways, (iii) 1440 ways, (iv) 3600 ways, (v) 576 ways, (vi) 1440 ways

7. *a*	**8.** *c*	**9.** *a*	**10.** *c*	**11.** *b*	**12.** *c*	**13.** *b*	**14.** *a*	**15.** *c*	**16.** *d*
17. *c*	**18.** *b*	**19.** *c*	**20.** *a*						

DPP-6 Geometrical Application of ^nC_r

1. *a*	**2.** *c*	**3.** *b*	**4.** *b*	**5.** *b*	**6.** *b*	**7.** *c*	**8.** *c*	**9.** *b*	**10.** *c*
11. *b*	**12.** *c*	**13.** *b*	**14.** *d*	**15.** *a*	**16.** *b*	**17.** *a*	**18.** *a*	**19.** *a*	**20.** *b*

DPP-7 Problems on Arrangements and Selection

1. *b*	**2.** *b*	**3.** *a*	**4.** *b*	**5.** *c*	**6.** *c*	**7.** *d*	**8.** *b*	**9.** *a*	**10.** *a*
11. *a*	**12.** *a*	**13.** *b*	**14.** *c*	**15.** 9	**16.** 3	**17.** 9	**18.** 4	**19.** 5	**20.** 8

DPP-8 Division of Objects into Groups of Equal and Unequal Groups, Number of Divisions and Sum of Divisions

1. *b*	**2.** *c*	**3.** *b*	**4.** *c*	**5.** *b*	**6.** *c*	**7.** *c*	**8.** *d*	**9.** *d*	**10.** *a*
11. *b*	**12.** *b*	**13.** *c*	**14.** *b*	**15.** *b*	**16.** *a*	**17.** *a*	**18.** *c*	**19.** *b*	**20.** *c*
21. *a*	**22.** *c*	**23.** *b*	**24.** *b*	**25.** *a*	**26.** *b*	**27.** *a*	**28.** *b*	**29.** *c*	**30.** *d*
31. *a*	**32.** *c*	**33.** *d*							

DPP-9 Derangement, Multinomial Selection and Arrangement from Identical Arrangement

1. *a* 2. *c* 3. *a* 4. *b* 5. *d* 6. *a* 7. *c* 8. *c* 9. *d* 10. *c*
11. *a* 12. *c* 13. *c* 14. *b* 15. *c* 16. *d* 17. *a* 18. *a* 19. *a* 20. *b*
21. *c* 22. *b* 23. *b* 24. *c* 25. *a* 26. *b* 27. 6 28. 9 29. 4 30. 1
31. 3

Revisal Problems (JEE Main)

1. *c* 2. *b* 3. *d* 4. *a* 5. *c* 6. *c* 7. *c* 8. *c* 9. *a* 10. *a*
11. *a* 12. *d* 13. *c* 14. *d* 15. *c* 16. *c* 17. *d* 18. *d* 19. *b* 20. *d*
21. *d* 22. *a* 23. *a* 24. *b* 25. *c* 26. *b* 27. *d* 28. *d* 29. *a* 30. *c*
31. *a* 32. *c* 33. *a* 34. *c* 35. *d* 36. *d* 37. *b* 38. *b* 39. *d* 40. *a*

Revisal Problems (JEE Advanced)

1. *b* 2. *c* 3. *c* 4. *d* 5. *c* 6. *c* 7. *b* 8. *a* 9. *b* 10. *c*
11. *b* 12. *b* 13. *d* 14. *d* 15. *b* 16. *c* 17. *b* 18. *b* 19. *c* 20. *c*
21. *c,d* 22. *b,d* 23. *b,c,d* 24. *c* 25. *b* 26. *a* 27. *d* 28. *a* 29. *c* 30. *b*
31. *b* 32. *c* 33. *a* 34. *b* 35. *c* 36. *d* 37. *b* 38. *a* 39. *d* 40. *c*
41. *d* 42. *c* 43. *b* 44. *b* 45. *a* 46. *b* 47. *c* 48. 8 49. 1 50. 3

JEE Main & AIEEE Archive

1. *c* 2. *b* 3. *d* 4. *a* 5. *b* 6. *c* 7. *a* 8. *a* 9. *c* 10. *b*
11. *d* 12. *a* 13. *c* 14. *b* 15. *b* 16. *a* 17. *b*

JEE Advanced & IIT JEE Archive

1. 7 2. 5 3. *b* 4. *c* 5. *b* 6. *c* 7. *c*

3. Probability

DPP-1 Equally Likely, Mutually Exclusive and Exhaustive Events and Classical Definition of Probability

2. (i) False, (ii) True, (iii) True, (iv) True, (v) True, (vi) True 3. (i) 3/8, (ii) 1/2, (iii) 3/4 4. *a*
5. *c* 6. *d* 7. *b* 8. *d* 9. *d* 10. *b* 11. *b* 12. *c* 13. *a* 14. *a*
15. *b* 16. *c* 17. *b* 18. *c* 19. *c* 20. *a* 21. *a* 22. *a* 23. *a* 24. *b*
25. *a* 26. *b* 27. *c* 28. *d* 29. *b* 30. *d* 31. *a* 32. *b* 33. *c* 34. *b*
35. 1 36. 3 37. 2 38. 8 39. 2 40. 2

DPP-2 Addition and Subtraction Theorems of Probability

1. (i) 1/3, (ii) 2/3, (iii) 1/12, (iv) 5/12 3. (i) 19/25, (ii) 13/125 4. *a* 5. *a* 6. *a*
7. *a* 8. *c* 9. *d* 10. *a* 11. *a* 12. *d* 13. *a* 14. *b* 15. *a* 16. *a*
17. *a* 18. *c* 19. *d* 20. *c*

DPP-3 Conditional Probability and Multiplication Theorem on Probability

1. *a* 2. *c* 3. *c* 4. *a* 5. *d* 6. *a* 7. *c* 8. *c* 9. *a* 10. *a*
11. *c* 12. *d* 13. *b* 14. *a* 15. *c*

DPP-4 Independent Events

1. *c*	2. *a*	3. *a*	4. *d*	5. *d*	6. *b*	7. *c*	8. *d*	9. *a*	10. *d*
11. *b*	12. *a*	13. *c*	14. *a*	15. *a*	16. *a*	17. *b*	18. *c*	19. *a*	20. *b*
21. *c*	22. *c*	23. *b*	24. *c*	25. *a*	26. *a*	27. *a*	28. *b*	29. *a,b*	30. *a,d*
31. *b*	32. *d*	33. *d*	34. *d*	35. *a*	36. *a*	37. 5	38. 9	39. 3	40. 5
41. 2	42. 3	43. 9	44. 1	45. 1					

DPP-5 Theorem of Total Probability

1. *d*	2. *a*	3. *c*	4. *c*	5. *d*	6. *c*	7. *b*	8. *c*	9. *b*	10. *c*
11. *a*	12. *b*	13. *a*	14. *d*	15. *b*					

DPP-6 Baye's Theorem

1. *d*	2. *a*	3. *b*	4. *c*	5. *a*	6. *c*	7. *c*	8. *a*	9. *c*	10. *c*
11. *c*	12. *b*	13. *a*	14. *b*	15. *a*	16. *b*	17. *d*	18. *b*	19. *d*	20. *d*
21. *a*	22. *c*	23. *b*	24. *d*	25. *b*	26. *a*	27. *c*	28. 1	29. 2	30. 5

DPP-7 Binomial Distribution and Probability Distribution

1. (i) 105/512, (ii) 193/512, (iii) 53/64 2. *a* 3. *c* 4. *c* 5. *d* 6. *d* 7. *a*

8. *d*	9. *c*	10. *a*	11. *d*	12. *a*	13. *b*	14. *c*	15. *d*	16. *d*	17. *c*
18. *a*	19. *b*	20. *b*	21. *b*	22. *c*	23. *a*	24. *b*	25. *b*	26. *b*	27. *a*
28. *b*	29. *a*	30. *d*	31. *a*	32. 6	33. 7	34. 7	15. 9		

DPP-8 Probability Through Statistical (Stochatic) Tree Diagram

1. (i) 9/10, (ii) 1/5, (iii) 1/5 2. 7/16 3. (i) 2/5, (ii) 13/25 4. *a* 5. *b* 6. *c* 7. *d*

8. *c*	9. *b,c,d*	10. *b*	11. *c*	12. *a*	13. *b*	14. *a*	15. *c*	16. *d*

Revisal Problems (JEE Main)

1. *c*	2. *c*	3. *d*	4. *c*	5. *b*	6. *d*	7. *b*	8. *d*	9. *b*	10. *b*
11. *b*	12. *c*	13. *a*	14. *b*	15. *c*	16. *d*	17. *a*	18. *b*	19. *d*	20. *a*
21. *b*	22. *a*	23. *d*	24. *c*	25. *a*	26. *b*	27. *a*	28. *c*	29. *c*	30. *a*
31. *a*	32. *a*	33. *a*	34. *c*	35. *a*	36. *c*	37. *a*	38. *c*	39. *d*	40. *a*
41. *a*	42. *c*	43. *d*	44. *a*	45. *b*					

Revisal Problems (JEE Advanced)

1. *a*	2. *d*	3. *c*	4. *b*	5. *a*	6. *b*	7. *c*	8. *c*	9. *c*	10. *b*
11. *c*	12. *c*	13. *c,d*	14. *a,b,c,d*		15. *b,c,d*	16. *b*	17. *d*	18. *b*	19. *d*
20. *a*	21. *c*	22. *b*	23. *d*	24. *a*	25. *b*	26. *a*	27. *b*	28. *b*	29. *c*
30. *b*	31. *c*	32. *d*	33. *c*	34. *b*	35. *b*	36. 3	37. 8	38. 6	39. 4
40. 5									

JEE Main & AIEEE Archive

1. *a*	2. *c*	3. *b*	4. *b*	5. *a*	6. *a*	7. *b*	8. *c*	9. *a*	10. *a*
11. *c*	12. *d*	13. *d*	14. *c*	15. *a*	16. *b*	17. *c*	18. *b*	19. *c*	20. *b*
21. *d*	22. *d*	23. *a*	24. *a*	25. *a*	26. *b*	27. *d*	28. *b*	29. *a*	

JEE Advanced & IIT JEE Archive

1. *a*	2. *b*	3. *c*	4. *a*	5. *a*	6. *b,d*	7. *a,b*	8. *a,d*	9. *b*	10. *d*
11. *c*	12. *c*	13. *a*	14. *b*	15. *d*	16. *b*	17. *d*	18. *c*	19. *c*	20. *d*
21. *b*	22. *a*	23. *b*	24. *b*	25. 6					

1. Binomial Theorem

DPP-1 Binomial for Positive Index, General Term and Middle Term

1. (i) Given, $\left(3x^2 - \frac{1}{3}\right)^{10}$

To find 7th term in the expansion of $\left(3x^2 - \frac{1}{3}\right)^{10}$.

We know that, the general term in the expansion of $(x + a)^n$ is

$$T_{r+1} = {}^nC_r \cdot x^{n-r} \cdot a^r$$

Here, $n = 10$, $x = 3x^2$, $a = -\frac{1}{3}$

$$T_{6+1} = {}^{10}C_6 (3x^2)^4 \left(-\frac{1}{3}\right)^6$$

$$= {}^{10}C_6 \frac{3^4}{3^6} \cdot x^8 = {}^{10}C_6 \frac{1}{3^2} x^8$$

$$= \frac{10!}{6!4!} \times \frac{1}{9} \times x^8 = \frac{210}{9} x^8 = \frac{70}{3} x^8$$

(ii) Given, $\left(2x^2 - \frac{3}{x^3}\right)^{25}$

To find Term independent of x in the expansion of $\left(2x^2 - \frac{3}{x^3}\right)^{25}$.

Let $(r + 1)$th term be independent of x.

$$\therefore \quad T_{r+1} = {}^{25}C_r (2x^2)^{25-r} \left(-\frac{3}{x^3}\right)^r$$

$$= (-1)^r \, {}^{25}C_r (2)^{25-r} (3)^r (x)^{50-5r} \quad \ldots \text{(i)}$$

For the term independent of x, put $50 - 5r = 0$

$\Rightarrow \quad r = 10$

Put $r = 10$ in Eq. (i), we get

$$T_{11} = {}^{25}C_{10} (2)^{15} (3)^{10}$$

$$= \frac{25!}{10!15!} (2)^{15} (3)^{10}$$

(iii) (a) Given, $\left(\frac{2x}{3} - \frac{3}{2x}\right)^6$

Here, $n = 6$ (even)

$\therefore$ Only one middle term is possible and $\left(\frac{n}{2} + 1\right)$th term will be middle term.

$$\frac{n}{2} + 1 = \frac{6}{2} + 1 = 4\text{th term}$$

$$T_4 = T_{3+1} = {}^6C_3 \left(\frac{2x}{3}\right)^3 \left(-\frac{3}{2x}\right)^3$$

$$= -{}^6C_3 \left(\frac{2}{3}\right)^3 \left(\frac{3}{2}\right)^3 = -\frac{6!}{3!3!} = -20$$

(b) Given, $\left(2x^2 - \frac{1}{x}\right)^7$

Here, $n = 7$ (odd)

$\therefore$ There are two middle terms.

First middle term $= \left(\frac{n+1}{2}\right)$th $= \left(\frac{7+1}{2}\right)$th $= \left(\frac{8}{2}\right)$th

= 4th term

Second middle term $= \left(\frac{n+3}{2}\right)$th $= \left(\frac{7+3}{2}\right)$th

= 5th term

$$\therefore \quad T_4 = T_{3+1} = {}^7C_3 (2x^2)^4 \cdot \left(-\frac{1}{x}\right)^3$$

$$= -{}^7C_3 (2)^4 x^5 = -\frac{7!}{3!4!} 16 x^5$$

$$= -35 \times 16 x^5 = -560 x^5$$

and $\quad T_5 = T_{4+1} = {}^7C_4 (2x^2)^3 \left(-\frac{1}{x}\right)^4$

$$= {}^7C_4 (2)^3 x^2 = \frac{7!}{4!3!} (2)^3 x^2$$

$$= 35 \times 8 \times x^2 = 280 x^2$$

2. (i) Given, $\left(ax^2 + \frac{1}{bx}\right)^{11}$

To find Coefficient of x^7 in $\left(ax^2 + \frac{1}{bx}\right)^{11}$.

Let $(r + 1)$th term contains x^7.

$$\therefore \quad T_{r+1} = {}^{11}C_r (ax^2)^{11-r} \left(\frac{1}{bx}\right)^r$$

$$T_{r+1} = {}^{11}C_r \frac{(a)^{11-r}}{b^r} x^{22-3r} \quad \ldots \text{(i)}$$

To get x^7, put $22 - 3r = 7$

$\Rightarrow \quad 3r = 15$

$\Rightarrow \quad r = 5$

Put $r = 5$ in Eq. (i), we get

$$T_6 = {}^{11}C_5 \frac{a^6}{b^5} x^7$$

$\therefore$ Coefficient of x^7 is ${}^{11}C_5 \frac{a^6}{b^5}$.

(ii) Given, $\left(ax - \frac{1}{bx^2}\right)^{11}$

To find Coefficient of x^{-7} in $\left(ax - \frac{1}{bx^2}\right)^{11}$.

Let $(r + 1)$th term contains x^{-7}

$$\therefore \quad T_{r+1} = {}^{11}C_r (ax)^{11-r} \left(-\frac{1}{bx^2}\right)^r$$

$$T_{r+1} = (-1)^r \, {}^{11}C_r \frac{a^{11-r}}{b^r} x^{11-3r} \quad \ldots \text{(i)}$$

To get x^{-7}, put $11 - 3r = -7$

$\Rightarrow \quad -3r = -7 - 11$

$\Rightarrow \quad -3r = -18$

$\therefore \quad r = 6$

Put $r = 6$ in Eq. (i), we get

$$T_7 = {}^{11}C_6 \frac{a^5}{b^6} x^{-7}$$

$\therefore$ Coefficient of x^{-7} is ${}^{11}C_6 \frac{a^5}{b^6}$.

3. (i) Given, $\left(2x - \frac{1}{\sqrt{x}}\right)^{15}$

To find Term independent of x.

Let $(r + 1)$th term be independent of x.

$$\therefore \quad T_{r+1} = {}^{15}C_r\,(2x)^{15-r}\left(-\frac{1}{\sqrt{x}}\right)^r$$

$$= (-1)^r\,{}^{15}C_r\,(2)^{15-r}\,x^{15-r-r/2}$$

$$T_{r+1} = {}^{15}C_r\,(2)^{15-r}\,x^{\frac{30-3r}{2}} \quad \ldots\text{(i)}$$

To get term independent of x, put $\frac{30-3r}{2} = 0$

$\Rightarrow \quad 30 - 3r = 0 \Rightarrow r = 10$

Put $r = 10$ in Eq. (i), we get

$$T_{11} = {}^{15}C_{10}\,(2)^5 = \frac{15!}{10!5!} \times (2)^5$$

$$= 3003 \times 32 = 96096$$

(ii) Given, $\left(\frac{4x^2}{3} - \frac{3}{2x}\right)^9$

To find The term independent of x.

Let $(r + 1)$th be independent of x.

$$\therefore \quad T_{r+1} = {}^9C_r\left(\frac{4x^2}{3}\right)^{9-r}\left(-\frac{3}{2x}\right)^r$$

$$= (-1)^r\,{}^9C_r\left(\frac{4}{3}\right)^{9-r}\left(\frac{3}{2}\right)^r x^{18-2r-r}$$

$$T_{r+1} = (-1)^r\,{}^9C_r\left(\frac{4}{3}\right)^{9-r}\left(\frac{3}{2}\right)^r x^{18-3r} \quad \ldots\text{(i)}$$

To get term independent of x, put $18 - 3r = 0$

$\Rightarrow \quad r = 6$

Put $r = 6$ in Eq. (i), we get

$$T_7 = {}^9C_6\left(\frac{4}{3}\right)^3\left(\frac{3}{2}\right)^6 = \frac{9!}{6!3!} \times (3)^3$$

$$= 84 \times 27 = 2268$$

4. (a) Given, $(1 + x)^m \cdot \left(1 + \frac{1}{x}\right)^n$

To find Term independent of x,

Coefficient of x^o in $\left\{(1 + x)^m \cdot \left(\frac{x+1}{x}\right)^n\right\}$

$\Rightarrow$ Coefficient of x^o in $\left\{\frac{(1+x)^{m+n}}{x^n}\right\}$

$\Rightarrow$ Coefficient of x^n in $(1 + x)^{m+n}$

$\Rightarrow$ Coefficient of x^n in $\{{}^{m+n}C_r\,x^r\}$

$\Rightarrow$ ${}^{m+n}C_n$, put $r = n$.

$\therefore$ Term independent of $x = {}^{m+n}C_n$ or ${}^{m+n}C_m$.

5. (a) Given, $\left(x - \frac{1}{x}\right)^{2n+1}$

To find $(p + 2)$th term from the end of the expansion of $\left(x - \frac{1}{x}\right)^{2n+2}$.

Now, $(p + 2)$th term from the end $= (2n + 1 - p - 2 + 2)$ term from the beginning.

$(p + 2)$th term from the end $= (2n - p + 1)$th term from the beginning

[$\because$ In the binomial expansion of $(x + a)^n$, the rth term from the end is $(n - r + 2)$th term from the beginning]

$$\therefore \quad T_{(2n-p+1)} = {}^{2n+1}C_{(2n-p)}(x)^{(2n+1)-(2n-p)}\left(-\frac{1}{x}\right)^{2n-p}$$

$$= (-1)^{2n-p}\,.\,{}^{2n+1}C_{2n-p}\cdot x^{p+1}\cdot\frac{1}{x^{2n-p}}$$

$$= (-1)^{2n-p}\,.\,{}^{2n+1}C_{2n-p}\cdot x^{1+p-2n+p}$$

$$= (-1)^{2n-p}\,.\,{}^{2n+1}C_{2n-p}\cdot x^{1+2p-2n}$$

$$\text{or} \quad T_{(2n-p+1)} = (-1)^p\,.\,{}^{2n+1}C_{2n-p}\cdot x^{2p-2n+1}$$

$$\left[\because (-1)^{2n-p} = (-1)^{2n}\cdot(-1)^{-p} = 1\cdot\frac{1}{(-1)^p} = (-1)^p\right]$$

6. (b) Given, $(1 + x^2)^{40}\cdot\left(x^2 + 2 + \frac{1}{x^2}\right)^{-5}$

$$\Rightarrow \quad (1 + x^2)^{40}\left\{\left(x + \frac{1}{x}\right)^2\right\}^{-5}$$

$$\Rightarrow \quad (1 + x^2)^{40}\left(x + \frac{1}{x}\right)^{-10} = x^{10}\,(1 + x^2)^{40}\,(1 + x^2)^{-10}$$

$$= x^{10}\,(1 + x^2)^{30}$$

To find Coefficient of x^{20} in the expansions of $(1 + x^2)^{20}\left(x + \frac{1}{x}\right)^{-10}$ i.e. $x^{10}\,(1 + x^2)^{30}$

$$\text{Now, } (1 + x^2)^{30} = {}^{30}C_0 + {}^{30}C_1(x^2) + {}^{30}C_2(x^2)^2 + {}^{30}C_3(x^2)^3 + {}^{30}C_4(x^2)^4 + \cdots + {}^{30}C_{30}(x^2)^{30} \quad \ldots\text{(i)}$$

$$x^{10}\,(1 + x^2)^{30} = x^{10} + {}^{30}C_1x^{12} + {}^{30}C_2x^{14} + {}^{30}C_3x^{16} + {}^{30}C_4x^{18} + {}^{30}C_5x^{20} + \cdots + x^{70}$$

$\therefore$ Coefficient of x^{20} is ${}^{30}C_5$ or ${}^{30}C_{25}$. $\quad [\because {}^nC_r = {}^nC_{n-r}]$

7. (b) Given, $(1 - x^2)^{10}$ and $\left(x - \frac{2}{x}\right)^{10}$

To find Ratio of coefficient of x^{10} in $(1 - x^2)^{10}$ and the term independent of x in $\left(x - \frac{2}{x}\right)^{10}$.

Now, let $(r + 1)$th term in $(1 - x^2)^{10}$ contains x^{10}.

$$\therefore \quad T_{r+1} = {}^{10}C_r\,(-x^2)^r$$

$$= (-1)^r\,{}^{10}C_r\,x^{2r}$$

To get x^{10}, put $2r = 10$

$\Rightarrow \quad r = 5$

$$\therefore \quad T_6 = (-1)^5\,{}^{10}C_5\,x^{10}$$

Thus, coefficient of x^{10} is $-\,{}^{10}C_5$.

Again, let term be independent of x in $\left(x - \frac{2}{x}\right)^{10}$.

$$\therefore \quad T_{r+1} = {}^{10}C_r(x)^{10-r}\left(-\frac{2}{x}\right)^r$$

$$= (-1)^r\,{}^{10}C_r\,(2)^r\,(x)^{10-2r}$$

To get term independent of x, put $10 - 2r = 0$

$\Rightarrow \quad r = 5$

Thus, the term independent of x in $\left(x - \frac{2}{x}\right)^{10}$ is $-{}^{10}C_5(2)^5$.

$$\therefore \quad \text{Required ratio} = \frac{-\,{}^{10}C_5}{-\,{}^{10}C_5(2)^5}$$

$$= \frac{1}{(2)^5} = \frac{1}{32}$$

$$= 1 : 32$$

8. (d) The general term in the expansion of $(1 - x + x^2)^5$ is

$\frac{5!}{r!\,s!\,t!}\cdot(1)^r(-x)^s(x^2)^t$, where $r + s + t = 5$

$= \frac{5!}{r!\,s!\,t!}(1)^r(-1)^s(x)^{s+2t}$, where $r + s + t = 5$

For the coefficient of x^3, we must have

$$s + 2t = 3 \quad \ldots\text{(i)}$$

But $r + s + t = 5$ ...(ii)

$\Rightarrow r + 3 - 2t + t = 5$ [using Eq. (i)]

$\Rightarrow r - t = 2$

$\Rightarrow r = 2 + t$ ...(iii)

where, $0 \le r, s, t \le 6$

Now, following table gives the possible values of r, s and t.

t	r	s
0	2	3
1	3	1

Hence, two terms containing x^3.

$\therefore$ Coefficient of $x^3 = -\frac{5!}{2!\,3!\,0!} - \frac{5!}{3!\,1!\,1!}$

$= -5!\left[\frac{1}{12} + \frac{1}{6}\right] = -5! \times \left[\frac{1+2}{12}\right]$

$= -5! \times \frac{3}{12}$

$= -5! \times \frac{1}{4}$

$= -\frac{120}{4}$

$= -30$

9. (b) Given,

$(1 + 2x + 3x^2)^{10} = a_0 + a_1x + a_2x^2 + a_3x^3 + \cdots + a_{20}x^{20}$ ...(i)

To find a_1 in the expansions.

Now, differentiating Eq. (i) w.r.t x, we get

$10(1 + 2x + 3x^2)^9(2 + 6x)$

$= a_1 + 2a_2x + 3a_3x^2 + \cdots + 20a_{20}x^{19}$ ...(ii)

Put $x = 0$ in Eq. (ii), we get

$10 \times 2 = a_1 \Rightarrow a_1 = 20$

10. (c) Given, $(1 + x)^{21} + (1 + x)^{22} + \cdots + (1 + x)^{30}$

To find Coefficient of x^5 in $(1 + x)^{21} + (1 + x)^{22} + (1 + x)^{23} + \cdots + (1 + x)^{30}$

Now, $(1 + x)^{21} + (1 + x)^{22} + (1 + x)^{23} + \cdots + (1 + x)^{30}$

$\Rightarrow (1 + x)^{21}(1 + (1 + x) + (1 + x)^2 + \cdots + (1 + x)^9)$

$\Rightarrow (1 + x)^{21}\left[\frac{1[(1 + x)^{10} - 1]}{1 + x - 1}\right]$

$\left[\because 1 + r + r^2 + \cdots + r^{n-1} = \frac{1\cdot(r^n - 1)}{r - 1}, r > 1\right]$

$\Rightarrow (1 + x)^{21}\left(\frac{(1 + x)^{10} - 1}{x}\right) = \frac{(1 + x)^{31}}{x} - \frac{(1 + x)^{21}}{x}$

We know that, coefficient of x^r in the expansion of $(1 + x)^n$ is ${}^nC_r$.

$\therefore$ Coefficient of x^6 in $(1 + x)^{31}$ is ${}^{31}C_6$.

$\Rightarrow$ Coefficient of x^5 in $\frac{(1 + x)^{31}}{x}$ is ${}^{31}C_6$

and coefficient of x^6 in $(1 + x)^{21}$ is ${}^{21}C_6$.

$\Rightarrow$ Coefficient of x^5 in $\frac{(1 + x)^{21}}{x}$ is ${}^{21}C_6$.

Now, coefficient of x^5 in $\frac{(1 + x)^{31}}{x} - \frac{(1 + x)^{21}}{x} = {}^{31}C_6 - {}^{21}C_6$

11. (c) Given, $\left(x^2 + 1 + \frac{1}{x^2}\right)^n, n \in N$

To find Number of terms in the expansion of $\left(x^2 + 1 + \frac{1}{x^2}\right)^n$.

We have, $\left(x^2 + 1 + \frac{1}{x^2}\right)^n = \frac{1}{x^{2n}}(1 + x^2 + x^4)^n$

Put $t = x^2$

$\Rightarrow \left(x^2 + 1 + \frac{1}{x^2}\right)^n = \frac{1}{t^n}(1 + t + t^2)^n$

Clearly, $(1 + t + t^2)^n$, when expanded in a polynomial of degree $2n$.

Hence, there are $(2n + 1)$ terms.

12. (c) Given, $\left(\sqrt[3]{2} + \frac{1}{\sqrt[3]{3}}\right)^n$

To find n in $\left(\sqrt[3]{2} + \frac{1}{\sqrt[3]{3}}\right)^n$

Now, $T_{6+1} = {}^nC_6\cdot\{(2)^{1/3}\}^{n-6}\cdot\left\{\frac{1}{(3)^{1/3}}\right\}^6$

$T_7 = {}^nC_6(2)^{\frac{n-6}{3}}\cdot\frac{1}{(3)^2}$ ...(i)

and seventh term from the end $= (n - 7 + 2)$th term from beginning $= (n - 5)$th term from beginning.

$\therefore T_{n-5} = T_{(n-6+1)} = {}^nC_{n-6}\{(2)^{1/3}\}^6\left\{\frac{1}{(3)^{1/3}}\right\}^{n-6}$

$= {}^nC_{n-6}(2)^2\frac{1}{(3)^{\frac{n-6}{3}}}$ ...(ii)

On dividing Eq. (i) by Eq. (ii), we get

$\therefore$ Required ratio $= \dfrac{{}^nC_6\dfrac{(2)^{\frac{n-6}{3}}}{(3)^2}}{{}^nC_{n-6}\dfrac{(2)^2}{(3)^{\frac{n-6}{3}}}} = \frac{(2)^{\frac{n-6}{3}}}{(3)^2} \times \frac{(3)^{\frac{n-6}{3}}}{(2)^2}$

$= (2)^{\frac{n-6}{3}-2} \times (3)^{\frac{n-6}{3}-2}$

$= (2 \times 3)^{\frac{n-6}{3}-2} = (6)^{\frac{n-12}{3}}$

It is given that the ratio of seventh term from beginning and from end is $\frac{1}{6}$.

$\Rightarrow (6)^{\frac{n-12}{3}} = \frac{1}{6}$

$\Rightarrow (6)^{\frac{n-12}{3}} = (6)^{-1}$

$\Rightarrow \frac{n - 12}{3} = -1$

$\Rightarrow n - 12 = -3$

$\therefore n = -3 + 12 = 9$

13. (c) Given, $\left(x^2 + \frac{1}{x^2} + 2\right)^n$

$\Rightarrow \left\{\left(x + \frac{1}{x}\right)^2\right\}^n \Rightarrow \left(x + \frac{1}{x}\right)^{2n}$

Here, $2n$ is even.

Therefore, total terms $(2n + 1)$ is odd.

Thus, only one middle term exist.

and $\left(\frac{2n}{2} + 1\right)$ i.e. $(n + 1)$th term will be middle term.

$\therefore\ T_{n+1} = {}^{2n}C_n (x)^{2n-n}\left(\frac{1}{x}\right)^n = {}^{2n}C_n\, x^{2n-2n} = {}^{2n}C_n = \frac{2n!}{n!n!}$

$= \frac{(2n)(2n-1)(2n-2)(2n-3)(2n-4)\cdots 1}{n!n!}$

$= \frac{\{(2n-1)(2n-3)(2n-5)\cdots 1\} \times \{2n\cdot 2(n-1)\cdot 2(n-2)\cdots 2\}}{n!n!}$

$= \frac{\{(2n-1)(2n-3)(2n-5)\cdots 1\} \times \{2^n\cdot n(n-1)(n-2)\cdots 1\}}{n!\,n!}$

$= \frac{\{(2n-1)(2n-3)(2n-5)\cdots 1\}\, 2^n\, n!}{n!\,n!}$

$[\because n! = n(n-1)(n-2)\cdots 1]$

$= \frac{1\cdot 3\cdot 5\cdots(2n-1)\,2^n}{n!}$

14. (b) Given, $(y^{1/5} + x^{1/10})^{55}$

To find Number of terms free from radical sign.

Now, let $(r+1)$th terms is independent of radical sign.

$\therefore\quad T_{r+1} = {}^{55}C_r \{(y)^{1/5}\}^{55-r} (x^{1/10})^r$

$T_{r+1} = {}^{55}C_r\, y^{11-\frac{r}{5}}\, x^{\frac{r}{10}}$ ...(i)

Clearly, T_{r+1} will be independent of radical sign, if $r/5$ and $r/10$ are integers, where $0 \le r \le 55$.

$\therefore\ r = 0, 10, 20, 30, 40, 50.$

Hence, there are 6 terms in the expansion of the given expression, which are independent of radical sign.

15. (a) Given, $\left(\frac{x+1}{x^{2/3} - x^{1/3} + 1} - \frac{x-1}{x - x^{1/2}}\right)^{10}$

To find Coefficient of the term independent of x in

$$\left(\frac{x+1}{x^{2/3} - x^{1/3} + 1} - \frac{x-1}{x - x^{1/2}}\right)^{10}$$

We have,

$\frac{x+1}{x^{2/3} - x^{1/3} + 1} - \frac{x-1}{x - x^{1/2}} = \frac{(x^{1/3})^3 + 1^3}{x^{2/3} - x^{1/3} + 1} - \frac{(x^{1/2})^2 - (1)^2}{x^{1/2}(x^{1/2} - 1)}$

$= \frac{(x^{1/3}+1)(x^{2/3} - x^{1/3} + 1)}{(x^{2/3} - x^{1/3} + 1)} - \frac{(x^{1/2}+1)(x^{1/2}-1)}{x^{1/2}(x^{1/2}-1)}$

$[\because a^3 + b^3 = (a+b)(a^2 - ab + b^2),\ a^2 - b^2 = (a+b)(a-b)]$

$= (x^{1/3} + 1) - \frac{x^{1/2}+1}{x^{1/2}} = x^{1/3} + 1 - 1 - \frac{1}{x^{1/2}}$

$= x^{1/3} - x^{-1/2}$

$\therefore\ \left[\left(\frac{x+1}{x^{2/3} - x^{1/3} + 1}\right) - \left(\frac{x-1}{x - x^{1/2}}\right)\right]^{10} = (x^{1/3} - x^{-1/2})^{10}$

Let $(r+1)$th term be independent of x.

$\therefore\quad T_{r+1} = {}^{10}C_r (x^{1/3})^{10-r} (-x^{-1/2})^r$

$= (-1)^r\, {}^{10}C_r\, x^{\frac{10-r}{3}} \cdot x^{-r/2}$

$= (-1)^r\, {}^{10}C_r\, x^{\frac{20-5r}{6}}$

For this term to be independent of x.

Put $\frac{20-5r}{6} = 0 \Rightarrow 20 - 5r = 0$

$\Rightarrow\quad r = 4$

So, required coefficient $= {}^{10}C_4 (-1)^4 = {}^{10}C_4$

$= \frac{10!}{4!6!} = 210$

16. (c) Given, in the expansion of $(1+x)^m (1-x)^n$, coefficient of x and x^2 are 3 and -6, respectively.

To find m in $(1+x)^m (1-x)^n$

We have,

$(1+x)^m (1-x)^n = ({}^mC_0 + {}^mC_1 x + {}^mC_2 x^2 + \cdots + {}^mC_m x^m)$

$\times ({}^nC_0 - {}^nC_1 x + {}^nC_2 x^2 + \cdots (-1)^n\, {}^nC_n x^n)$

$= {}^mC_0{}^nC_0 - ({}^mC_0{}^nC_1 - {}^nC_0{}^mC_1)x$

$+ ({}^mC_0{}^nC_2 + {}^nC_0{}^mC_2 - {}^mC_1{}^nC_1)x^2 + \cdots$

$\therefore$ Coefficient of $x = 3$

$\Rightarrow\quad -({}^mC_0{}^nC_1 - {}^nC_0{}^mC_1) = 3$ ...(i)

Coefficient of $x^2 = -6$

$\Rightarrow\quad ({}^mC_0{}^nC_2 + {}^nC_0{}^mC_2 - {}^mC_1{}^nC_1) = -6$ ...(ii)

From Eq. (i), we have

$-(1\cdot n - 1\cdot m) = 3$

$\Rightarrow\quad m - n = 3$ ...(iii)

From Eq. (ii), we have

$1\cdot\frac{n(n-1)}{2!} + 1\cdot\frac{m(m-1)}{2!} - mn = -6$

$\Rightarrow\quad n(n-1) + m(m-1) - 2mn = -12$

$\Rightarrow\quad n^2 - n + m^2 - m - 2mn = -12$

$\Rightarrow\quad (m-n)^2 - (m+n) = -12$

$\Rightarrow\quad (3)^2 - (m+n) = -12$ [using Eq. (iii)]

$\Rightarrow\quad 9 - (m+n) = -12$

$\Rightarrow\quad m + n = 21$ ...(iv)

On solving Eqs. (iii) and (iv), we get

$m = 12,\ n = 9$

17. (d) Given, $\sum_{r=0}^{n} (-1)^r\, {}^nC_r \frac{1 + r\log_e 10}{(1 + \log 10^n)^r}$

Let $\log_e 10 = x$

$\Rightarrow\quad \sum_{r=0}^{n} (-1)^r\, {}^nC_r \frac{1+rx}{(1+nx)^r}$

$\Rightarrow \sum_{r=0}^{n} (-1)^r\, {}^nC_r \left(\frac{1}{1+nx}\right)^r + \sum_{r=0}^{n} (-1)^r \frac{n}{r}\, {}^{n-1}C_{r-1} \frac{rx}{(1+nx)^r}$

$\left[\because {}^nC_r = \frac{n}{r}\, {}^{n-1}C_{r-1}\right]$

$\Rightarrow\quad \sum_{r=0}^{n} (-1)^r\, {}^nC_r \left(\frac{1}{1+nx}\right)^r - \frac{nx}{1+nx}$

$\sum_{r=1}^{n} (-1)^{r-1}\, {}^{n-1}C_{r-1} \left(\frac{1}{1+nx}\right)^{r-1}$

$\Rightarrow \left(1 - \frac{1}{1+nx}\right)^n - \frac{nx}{1+nx}\left(1 - \frac{1}{1+nx}\right)^{n-1}$

$= \left(\frac{nx}{1+nx}\right)^n - \left(\frac{nx}{1+nx}\right)^n = 0$

18. (a) Given,

$\frac{1}{\sqrt{4x+1}}\left\{\left(\frac{1+\sqrt{4x+1}}{2}\right)^n - \left(\frac{1-\sqrt{4x+1}}{2}\right)^n\right\}$

$= a_0 + a_1 x + \cdots + a_5 x^5$ ...(i)

To find n in the given expansion.

We have,

$$\frac{1}{\sqrt{4x+1}}\left\{\left(\frac{1+\sqrt{4x+1}}{2}\right)^n-\left(\frac{1-\sqrt{4x+1}}{2}\right)^n\right\}$$

$$=\frac{1}{2^n\sqrt{4x+1}}2\{{}^nC_1\sqrt{4x+1}+{}^nC_3(\sqrt{4x+1})^3+\cdots\}$$

$$[\because (1+x)^n-(1-x)^n=2[{}^nC_1x+{}^nC_3x^3+\cdots]$$

$$=\frac{1}{2^{n-1}}\{{}^nC_1+{}^nC_3(\sqrt{4x+1})^2+{}^nC_5(\sqrt{4x+1})^4+\cdots\}$$

Since, RHS of Eq. (i) contains x^5. So, this expression should go upto $(\sqrt{4x+1})^{10}$ and hence, $n=11$.

19. (c) Given, $\sum_{r=0}^{100}{}^{100}C_r\,(x-3)^{100-r}\cdot 2^r=[(x-3)+2]^{100}$

$$[\because \sum_{r=0}^{n}{}^nC_r x^{n-r}a^r=(x+a)^n]$$

$$=(x-1)^{100}=(1-x)^{100}$$

Let $(r+1)$th contains x^{53}

$\therefore$ $T_{r+1}=(-1)^r\,{}^{100}C_r\,x^r$

Here, $r=53$

$T_{54}=-{}^{100}C_{53}\,x^{53}$

$\therefore$ Required coefficient $=-{}^{100}C_{53}$.

20. (b) Given, $(1+x)^n$, where n is even.

To find Coefficient of middle term.

Here, x is even, so only one middle term exist and $\left(\frac{n}{2}+1\right)$th term will be middle term.

$T_{(n/2+1)}={}^nC_{n/2}\,x^{n/2}$

Required coefficient $={}^nC_{n/2}$

Let $n=2m$ (even) $={}^{2m}C_m=\frac{2m!}{m!m!}$

$$=\frac{2m\,(2m-1)\,(2m-2)\cdots 1}{m!\,m!}$$

$$=\frac{2^m m!\,[1\cdot 3\cdot 5\cdots(2m-1)]}{m!m!}$$

$$=\frac{2^m\,[1\cdot 3\cdot 5\cdots(2m-1)]}{m!}$$

$$=\frac{2^m\,[1\cdot 3\cdot 5\cdots(2m-1)]}{\dfrac{m!\times 2^m\,(1\cdot 3\cdot 5\cdots(2m-1)}{2^m\,[1\cdot 3\cdot 5\cdots(2m-1)]}}$$

$$=\frac{2^{2m}\,[1\cdot 3\cdot 5\cdots(2m-1)]}{\dfrac{2m!}{1\cdot 3\cdot 5\cdots(2m-1)}}$$

$$=\frac{2^{2m}\,[1\cdot 3\cdot 5\cdots(2m-1)]}{(2m)\,(2m-2)\,(2m-4)\,(2m-6)\cdots 4,2)}$$

$$=\frac{2^{2m}\,[1\cdot 3\cdot 5\cdots(2m-1)]}{2\cdot 4\cdot 6\cdots(2m)}$$

$$=\frac{2^n\,[1\cdot 3\cdot 5\cdots(n-1)]}{2\cdot 4\cdot 6\cdots n}\qquad[\because n=2m]$$

21. (b) Given, pth, $(p+1)$th and $(p+2)$th term in the expansion of $(1+x)^n$ are in AP.

$\therefore$ ${}^nC_{p-1}$, ${}^nC_p$ and ${}^nC_{p+1}$ are in AP.

$\Rightarrow$ $2\,{}^nC_p={}^nC_{p-1}+{}^nC_{p+1}$

[$\because a, b$ and c are in AP, $2b=a+c$]

$$\Rightarrow 2\frac{n!}{p!\,(n-p)!}=\frac{n!}{(p-1)!\,(n-p+1)!}+\frac{n!}{(p+1)!\,(n-p-1)!}$$

$$2\frac{n!}{p!(n-p)!}=\frac{n!}{(p-1)!(n-p+1)(n-p)(n-p-1)!}+\frac{n!}{(p+1)p\,(p-1)!\,(n-p-1)!}$$

$$\frac{2n!}{p\,(p-1)!\,(n-p)\,(n-p-1)!}=\frac{n!}{(p-1)!\,(n-p-1)!}\left[\frac{1}{(n-p)\,(n-p+1)}+\frac{1}{p\,(p+1)}\right]$$

$$\frac{2}{p\,(n-p)}=\frac{1}{(n-p)\,(n-p+1)}+\frac{1}{p\,(p+1)}$$

$$\frac{2}{p\,(n-p)}=\frac{p\,(p+1)+(n-p)\,(n-p+1)}{p\,(p+1)\,(n-p)\,(n-p+1)}$$

$\Rightarrow$ $2\,(p+1)\,(n-p+1)=p\,(p+1)+(n-p)\,(n-p+1)$

$\Rightarrow$ $2\,(np-p^2+p+n-p+1)$
$=p^2+p+n^2-np+n-np+p^2-p$

$\Rightarrow 2np-2p^2+2p+2n-2p+2$
$=p^2+p+n^2-np+n-np+p^2-p$

$\Rightarrow$ $2np-2p^2+2n+2=2p^2-2np+n+n^2$

$\Rightarrow$ $4p^2-4np-n+n^2-2=0$

$\Rightarrow n^2-n\,(4p+1)+4p^2-2=0$

22. (c) It is given that 6th term in the expansion of $\left(\frac{1}{x^{8/3}}+x^2\log_{10}x\right)^8$ is 5600.

i.e. $T_6=5600$

$\Rightarrow$ ${}^8C_5\,(x^2\log_{10}x)^5\left(\frac{1}{x^{8/3}}\right)^3=5600$

$\Rightarrow$ $56\,x^{10}\,(\log_{10}x)^5\cdot\frac{1}{x^8}=5600$

$\Rightarrow$ $x^2\,(\log_{10}x)^5=100$

$\Rightarrow$ $x^2\,(\log_{10}x)^5=10^2\,(\log_{10}10)^5$ $\quad[\because \log_a a=1]$

$\therefore$ $x=10$

23. (b) Let a, b and c, be coefficient of rth, $(r+1)$th and $(r+2)$th term in the expansion of $(1+x)^n$.

$\Rightarrow$ ${}^nC_{r-1}=a$ ...(i)

$\Rightarrow$ ${}^nC_r=b$ ...(ii)

$\Rightarrow$ ${}^nC_{r+1}=c$ ...(iii)

On dividing Eq. (i) by Eq. (ii), we get

$$\frac{{}^nC_{r-1}}{{}^nC_r}=\frac{a}{b}$$

$$\Rightarrow \frac{\dfrac{n!}{(r-1)!\,(n-r+1)!}}{\dfrac{n!}{r!\,(n-r)!}}=\frac{a}{b}$$

$$\Rightarrow \frac{1}{n-r+1}\times\frac{r}{1}=\frac{a}{b}$$

$$\Rightarrow \frac{r}{n-r+1}=\frac{a}{b}$$

$\Rightarrow$ $rb=an-ar+a$

$\Rightarrow$ $r\,(b+a)=an+a$

$$r=\frac{a\,(n+1)}{b+a}\qquad\text{...(iv)}$$

On dividing Eq. (iii) by Eq. (ii), we get

$$\frac{^nC_{r+1}}{^nC_r} = \frac{c}{b}$$

$$\Rightarrow \quad \frac{\dfrac{n!}{(r+1)!\,(n-r-1)!}}{\dfrac{n!}{r!\,(n-r)!}} = \frac{c}{b}$$

$$\Rightarrow \quad \frac{1}{(r+1)!\,(n-r-1)!} \times r! \times (n-r)! = \frac{c}{b}$$

$$\Rightarrow \frac{1}{(r+1)\,r!\,(n-r-1)!} \times r!\,(n-r)\,(n-r-1)! = \frac{c}{b}$$

$$\Rightarrow \quad \frac{(n-r)}{r+1} = \frac{c}{b}$$

$$\Rightarrow \quad b\,(n-r) = c\,(r+1)$$

$$\Rightarrow \quad bn - b\,r = cr + c$$

$$\Rightarrow \quad r\,(c+b) = bn - c$$

$$\Rightarrow \quad r = \frac{bn-c}{c+b} \quad \ldots(\text{v})$$

$$\frac{a\,(n+1)}{b+a} = \frac{bn-c}{c+b}$$

$$a\,(c+b)\,(n+1) = (bn-c)\,(b+a)$$

$$(ac+ab)\,(n+1) = b^2n - bc + abn - ac$$

$$\Rightarrow \quad acn + abn + ac + ab = b^2n - bc + abn - ac$$

$$\Rightarrow \quad acn - b^2n + 2ac + ab + bc = 0$$

$$\Rightarrow n\,(ac - b^2) + 2ac + ab + bc = 0$$

$$\therefore \quad n = \frac{ab + bc + 2ac}{b^2 - ac}$$

24. (b) Given, $(5^{1/2} + 7^{1/8})^{1024}$

To find Number of integral terms in the expansion of $(5^{1/2} + 7^{1/8})^{1024}$.

The general term T_{r+1} in the expansion of $(5^{1/2} + 7^{1/8})^{1024}$ is given by

$$T_{r+1} = {}^{1024}C_r\,(5^{1/2})^{1024-r}\,(7^{1/8})^r$$

$$T_{r+1} = {}^{1024}C_r\,(5)^{512-r/2} \times 7^{r/8}$$

$$T_{r+1} = {}^{1024}C_r\,(5)^{512-r+\frac{r}{2}}\,(7)^{r/8}$$

$$T_{r+1} = {}^{1024}C_r\,(5)^{512-r}\,(5^{r/2}7^{r/8})$$

$$T_{r+1} = {}^{1024}C_r\,(5)^{512-r}\,(5^4 \times 7)^{r/8}$$

Clearly, T_{r+1} will be an integer, iff $r/8$ is an integer such that $0 \le r \le 1024$.

$\Rightarrow$ r is multiple of satisfying $0 \le r \le 1024$.

$\Rightarrow \quad r = 0, 8, 16, 24, 32, \ldots, 1024.$

Therefore, we have to find number of values of r.

$\therefore$ Let r have n values excluding '0'

i.e. 8, 16, 24, 32, 40, ..., 1024.

This is an AP.

Here, $a = 8, d = 8$ and $l = 1024$ $\quad [\because l = a + (n-1)d]$

$$\Rightarrow \quad 1024 = 8 + (n-1)\,8$$

$$\Rightarrow \quad 1024 = 8 + 8n - 8$$

$$\Rightarrow \quad 1024 = 8n$$

$$\Rightarrow \quad n = \frac{1024}{8}$$

$$n = 128$$

Including '0', r can assume 129 values.

25. (c) We have,

$$99^{50} + 100^{50} - 101^{50}$$

$$= (100-1)^{50} + 100^{50} - (100+1)^{50}$$

$$= (100-1)^{50} - (100+1)^{50} + 100^{50}$$

$$= -\{(100+1)^{50} - (100-1)^{50}\} + 100^{50}$$

$$= -\{(1+100)^{50} - (1-100)^{50}\} + 100^{50}$$

$$= -2\,\{^{50}C_1 \times 100 + {}^{50}C_3 \times (100)^3 + \cdots + {}^{50}C_{49} \times 100^{49}\} + 100^{50}$$

$$[\because (1+x)^n - (1-x)^n = 2\,(^nC_1x + {}^nC_3x^3 + \cdots)]$$

$$= -2\,[^{50}C_1\,100 + {}^{50}C_3\,(100)^3 + \cdots + {}^{50}C_{47} \times (100)^{47}] < 0$$

$$\Rightarrow \quad 99^{50} + 100^{50} - 101^{50} < 0$$

$$\Rightarrow \quad 99^{50} + 100^{50} < 101^{50}$$

26. (a) Let $S = 1 - {}^nC_1\left(\dfrac{1+x}{1+nx}\right) + {}^nC_2\left(\dfrac{1+2x}{(1+nx)^2}\right) + \cdots$

$$= \sum_{r=0}^{n} (-1)^r\,{}^nC_r\,\frac{1+rx}{(1+nx)^r}$$

$$= \sum_{r=0}^{n} (-1)^r\left[\frac{^nC_r}{(1+nx)^r} + \frac{^nC_r rx}{(1+nx)^r}\right]$$

$$= \sum_{r=0}^{n} C_r\left(-\frac{1}{1+nx}\right)^r + x\sum_{r=0}^{n} \frac{n.^{n-1}C_{r-1}}{(1+nx)^r}(-1)^r$$

$$= \left(1 - \frac{1}{1-nx}\right)^n - \left(\frac{nx}{1+nx}\right)\sum_{r=0}^{n} {}^{n-1}C_{r-1}\left(-\frac{1}{1+nx}\right)^{r-1}$$

$$= \left(1 - \frac{1}{1+nx}\right)^n - \left(\frac{nx}{1+nx}\right)\left(1 - \frac{1}{1+nx}\right)^{n-1}$$

$$= \left(1 - \frac{1}{1+nx}\right)^{n-1}\left(1 - \frac{1}{1+nx} - \frac{nx}{1+nx}\right)$$

$$= \left(1 - \frac{1}{1+nx}\right)^{n-1}\left(\frac{nx}{1+nx} - \frac{nx}{1+nx}\right) = 0$$

27. (d) Given, $(1 + x - 2x^2)^6 = a_0 + a_1x + a_2x^2 + \cdots + a_{12}x^{12}$ $\quad \ldots$(i)

Put $x = -1$ in Eq. (i), we get

$$\Rightarrow \quad (-2)^6 = a_0 - a_1 + a_2 - a_3 + \cdots + a_{12} \quad \ldots(\text{ii})$$

Again, put $x = 1$ in Eq. (i), we get

$$0 = a_0 + a_1 + a_2 + a_3 + \cdots + a_{12} \quad \ldots(\text{iii})$$

Put $x = 0$ in Eq. (i), we get

$$1 = a_0 \quad \ldots(\text{iv})$$

On adding Eqs. (ii) and (iii), we get

$$2\,(a_0 + a_2 + a_4 + a_6 + \cdots + a_{12}) = 2^6$$

$$\Rightarrow \quad a_0 + a_2 + a_4 + a_6 + \cdots + a_{12} = 2^5$$

$$\Rightarrow \quad a_2 + a_4 + a_6 + \cdots + a_{12} = 2^5 - 1$$

$$\Rightarrow \quad a_2 + a_4 + a_6 + \cdots + a_{12} = 31$$

28. (b) Given, $(1 + x + x^2)^n = (a_0 + a_1x + a_2x^2 + \cdots + a_{2n}x^{2n})$

Now, we have

$$(1 + x + x^2)^n = (a_0 + a_3x^3 + a_6x^6 + \cdots) + x\,(a_1 + a_4x^3 + \cdots) + x^2\,(a_2 + a_5x^3 + \cdots) \ldots(\text{i})$$

Put $x = 1$ in Eq. (i), we get

$$3^n = (a_0 + a_3 + a_6 + \cdots) + (a_1 + a_4 + \cdots) + (a_2 + a_5 + \cdots) \ldots(\text{ii})$$

Put $x = \omega$ in Eq. (i), we get

$$0 = (a_0 + a_3 + a_6 + \cdots) + \omega(a_1 + a_4 + \cdots) + \omega^2\,(a_2 + a_5 + \cdots) \quad \ldots(\text{iii})$$

Put $x = \omega^2$ in Eq. (i), we get

$$0 = (a_0 + a_3 + a_6 + \cdots) + \omega^2(a_1 + a_4 + \cdots) + \omega(a_2 + a_5 + \cdots) \quad \ldots(iv)$$

On adding Eqs. (ii), (iii) and (iv), we get

$$3^n = 3(a_0 + a_3 + a_6 + \cdots)$$

$$\Rightarrow \quad a_0 + a_3 + a_6 + \cdots = 3^{n-1}$$

29. (b) Given, $T_0, T_1, T_2, T_3, \ldots, T_n$ represents terms in the expansion of $(x + a)^n$, then

$$(x + a)^n = T_0 + T_1 + T_2 + T_3 + \cdots + T_n \quad \ldots(i)$$

On replacing a by ia in Eq. (i), we get

$$(x + ai)^n = (T_0 - T_2 + T_4 - T_6 + \cdots) + i(T_1 - T_3 + T_5 \cdots) \quad \ldots(ii)$$

On replacing a by $-ai$ in Eq. (i), we get

$$(x - ai)^n = (T_0 - T_2 + T_4 - T_6 + \cdots) - i(T_1 - T_3 + T_5 + \cdots) \quad \ldots(iii)$$

On multiplying Eqs. (ii) and (iii), we get

$$(x + ai)^n(x - ai)^n = (T_0 - T_2 + T_4 - T_6 \cdots)^2 + (T_1 - T_3 + T_5 - \cdots)^2$$

$$\Rightarrow \quad (x^2 + a^2)^n = (T_0 - T_2 + T_4 - T_6 + \cdots)^2 + (T_1 - T_3 + T_5 - \cdots)^2$$

30. (c) Given,

$$(3 + x^{2008} + x^{2009})^{2010} = a_0 + a_1x + a_2x^2 + \cdots + a_nx^n \quad \ldots(i)$$

Put $x = \omega$ in Eq. (i), we get

$$[3 + (\omega)^{2008} + (\omega)^{2009}]^{2010} = a_0 + a_1\omega + a_2\omega^2 + a_3\omega^3 + \cdots a_n\omega^n$$

$$\Rightarrow [3 + (\omega^3)^{669} \cdot \omega + (\omega^3)^{669}\omega^2]^{2010}$$

$$= a_0 + a_1\omega + a_2\omega^2 + a_3 + \cdots + a_n\omega^n \quad [\because \omega^3 = 1]$$

$$[3 + \omega + \omega^2]^{2010}$$

$$= a_0 + a_1\frac{(-1 + \sqrt{3}i)}{2} + a_2\left(\frac{-1 - \sqrt{3}i}{2}\right) + \cdots + a_n\omega^n$$

$$\left[\because \omega = \frac{-1 + \sqrt{3}i}{2}, \omega^2 = \frac{-1 - \sqrt{3}i}{2}\right]$$

$$= \left(a_0 - \frac{a_1}{2} - \frac{a_2}{2} + a_3 - \frac{a_4}{2} - \frac{a_5}{2} + a_6 \cdots\right) + i\left(\frac{\sqrt{3}}{2}a_1 - \frac{\sqrt{3}}{2}a_2 + \frac{\sqrt{3}}{2}a_4 + \cdots\right)$$

$$\Rightarrow (2 + 1 + \omega + \omega^2)^{2010}$$

$$= \left(a_0 - \frac{a_1}{2} - \frac{a_2}{2} + a_3 - \frac{a_4}{2} - \frac{a_5}{2} + a_6 \cdots\right) + \frac{\sqrt{3}}{2}i(a_1 - a_2 + a_4 \cdots)$$

$$\Rightarrow (2)^{2010}$$

$$= \left(a_0 - \frac{a_1}{2} - \frac{a_2}{2} + a_3 - \frac{a_4}{2} - \frac{a_5}{2} + a_6 \cdots\right) + \frac{\sqrt{3}}{2}i(a_1 - a_2 + a_4 - a_5 + \cdots)$$

$$[\because 1 + \omega + \omega^2 = 0]$$

On equating real part on both side, we get

$$a_0 - \frac{a_1}{2} - \frac{a_2}{2} + a_3 - \frac{a_4}{2} - \frac{a_5}{2} + a_6 - \cdots = (2)^{2010}$$

DPP-2 Divisibility, Remainder, Last Digits, Comparision of Numbers and $(A + \sqrt{B})^n$

1. (i) To show $9^n + 7$ is divisible by 8.

We have,

$$9^n = (1 + 8)^n = 1 + n8 + \frac{n(n-1)}{2!}(8)^2 + \frac{n(n-1)(n-2)}{3!}(8)^3 + \cdots$$

$$\left[\because (1 + x)^n = 1 + nx + \frac{n(n-1)}{2!}x^2 + \frac{n(n-1)(n-2)}{3!}(x)^3 + \cdots\right]$$

$$\Rightarrow 9^n = 1 + 8n + \frac{n(n-1)}{2!}(8)^2 + \frac{n(n-1)(n-2)}{3!}(8)^3 + \cdots$$

On adding 7 on both side, we get

$$\Rightarrow 9^n + 7 = 8 + 8n + \frac{n(n-1)}{2!}(8)^2 + \frac{n(n-1)(n-2)}{3!}(8)^3 + \cdots$$

$$\Rightarrow 9^n + 7 = 8\left[1 + n + \frac{n(n-1)}{2!} \times 8 + \frac{n(n-1)(n-2)}{3!}(8)^2 + \cdots\right]$$

$$\Rightarrow \quad 9^n + 7 = 8\lambda$$

$$\text{Where, } \lambda = \left(1 + n + \frac{n(n-1)}{2!} \times (8) + \frac{n(n-1)(n-2)}{3!}(8)^2 + \cdots\right)$$

$\Rightarrow$ $9^n + 7$ is divisible by 8.

(ii) $3^{2n+1} + 2^{n+2}$ is divisible by 7.

We have, $3^{2n+1} + 2^{n+2} \Rightarrow 3^{2n}3 + 2^n2^2 = 9^n3 + 2^n4$

$$\Rightarrow (2 + 7)^n3 + 2^n4$$

$$\Rightarrow 3 \cdot [{}^nC_02^n + {}^nC_12^{n-1}7 + {}^nC_22^{n-2}7^2 + \cdots] + 2^24$$

$$\Rightarrow 3\,{}^nC_02^n + 3\,{}^nC_12^{n-1}7 + 3\,{}^nC_22^{n-2}7^2 + \cdots + 2^n4$$

$$\Rightarrow (3 + 4)2^n + 21\,{}^nC_12^{n-1} + 147\,{}^nC_22^{n-2} + \cdots$$

$$\Rightarrow 72^n + 21\,{}^nC_12^{n-1} \cdots$$

$$\Rightarrow 7(2^n + 3\,{}^nC_12^{n-1} + 21\,{}^nC_22^{n-2} + \cdots$$

$$\Rightarrow 7\lambda$$

Where, $\lambda = 2^n + 3\,{}^nC_1 \cdot 2^{n-1} + 21\,{}^nC_22^{n-2} + \cdots$

$\therefore$ $3^{2n+1} + 2^{n+2}$ is divisible by 7.

(iii) To show $6^{2n} - 35n - 1$ is divisible by 1225.

So, $6^{2n} - 35n - 1 = (6^2)^n - 35n - 1 = (36)^n - 35n - 1$

$$\text{Now}(36)^n = (1 + 35)^n = 1 + 35n + \frac{n(n-1)}{2!}(35)^2 + \frac{n(n-1)(n-2)}{3!}(35)^3 + \cdots$$

$$\left[\because (1 + x)^n = 1 + nx + \frac{n(n-1)}{2!}(x)^2 + \frac{n(n-1)(n-2)}{3!} \cdot (x)^3 + \cdots\right]$$

$$\Rightarrow (36)^n = 1 + 35n + \frac{n(n-1)}{2!}(35)^2 + \frac{n(n-1)(n-2)}{3!}(35)^3 + \cdots$$

$$\Rightarrow (36)^n - 35n - 1 = (35)^2\frac{n(n-1)}{2!} + (35)^3\frac{n(n-1)(n-2)}{3!} + \cdots$$

$$\Rightarrow (36)^n - 35n - 1 = (35)^2\left[\frac{n(n-1)}{2!} + (35)\frac{n(n-1)(n-2)}{3!} + \cdots\right]$$

$\Rightarrow (36)^n - 35n - 1$

$$= 1225\left[\frac{n(n-1)}{2!} + (35)\frac{n(n-1)(n-2)}{3!} + \cdots\right]$$

$\Rightarrow (36)^n - 35n - 1 = 1225\lambda$

Where, $\lambda = \frac{n(n-1)}{2!} + (35)\frac{n(n-1)(n-2)}{3!} + \cdots$

$\Rightarrow \quad (6)^{2n} - 35n - 1 = 1225\lambda$

$\Rightarrow \quad (6)^{2n} - 35n - 1$ is divisible by 1225.

2. (b) Let $E = 2^{100} = 2\cdot 2^{99} = 2\cdot(2^3)^{33} = 2\cdot(8)^{33}$

$= 2\cdot(1+7)^{33}$

So, $E = 2^{100} = 2\cdot(1+7)^{33} = 2\cdot\left(1 + 33\times 7 + \frac{33\times 32}{2}(7)^2 + \cdots\right)$

$E = 2 + 7\cdot 2\left(33 + \frac{33\times 32}{2}\times 7 + \cdots\right)$

$E = 2 + 7k$

Where, $k = 2\cdot\left(33 + \frac{33\times 32}{2}\times 7 + \cdots\right)$

Now, $\frac{E}{7} = \frac{7k+2}{7} = k + \frac{2}{7}$

Hence, remainder is 2.

3. (d) Here, $5^2 = 25$ which is close to $26 = 13\times 2$

Here, $E = 5^{99} = 5\times 5^{98} = 5\times(5^2)^{49} = 5(26-1)^{49}$

$\Rightarrow E = 5\,[{}^{49}C_0(26)^{49} - {}^{49}C_1(26)^{48} + {}^{49}C_2(26)^{47} + \cdots {}^{49}C_{48}(26) - {}^{49}C_{49}]$

$[\because (x-a)^n = {}^nC_0x^n - {}^nC_1x^{n-1}a + {}^nC_2x^{n-2}a^2 + \cdots(-1)^n\,{}^nC_na^n]$

$\Rightarrow E = 5\times 26\,[{}^{49}C_0(26)^{48} - {}^{49}C_1(26)^{47} + {}^{49}C_2(26)^{46} + \cdots {}^{49}C_{48}] - 5$

$E = 5\times 26k - 5$

Where, $k = [{}^{49}C_0(26)^{48} - {}^{49}C_1(26)^{47} + \cdots {}^{49}C_{48}]$

Now, $\frac{E}{13} = 5\times\frac{26k}{13} - \frac{5}{13} = 10k - \frac{5}{13} = 10k - 1 + \frac{8}{13}$

Hence, the remainder is 8.

4. (d) Let $E = 7^{103}$

$= 7\cdot 7^{102} = 7\cdot(7^2)^{51} = 7\cdot(49)^{51} = 7\,(50-1)^{51}$

$\therefore \quad E = 7\cdot[{}^{51}C_0(50)^{51} - {}^{51}C_1(50)^{50} + {}^{51}C_2(50)^{49} + \cdots + {}^{51}C_{50}(50) - {}^{51}C_{51}]$

$[\because (x-1)^n = {}^nC_0x^n - {}^nC_1x^{n-1} + {}^nC_2x^{n-2} + \cdots(-1)^n\cdot{}^nC_n]$

$\Rightarrow \quad E = 7\times[{}^{51}C_0(50)^{51} - {}^{51}C_1(50)^{50} + {}^{51}C_2(50)^{49} + \cdots + {}^{51}C_{50}(50)] - 7$

$E = 7\times 50\,[{}^{51}C_0(50)^{50} - {}^{51}C_1(50)^{49} + {}^{51}C_2(50)^{48} + \cdots + {}^{51}C_{50}] - 7$

$E = 7\times 50k - 7$

Where, $k = [{}^{51}C_0(50)^{50} - {}^{51}C_1(50)^{49} + {}^{51}C_2(50)^{48} + \cdots + {}^{51}C_{50}]$

Now, $\frac{E}{25} = \frac{7\times 50k}{25} - \frac{7}{25} = 14k - \frac{7}{25} = 14k - 1 + \frac{18}{25}$

Hence, remainder is 18.

5. (a) We have,

$(17)^{10} = (17^2)^5 = (289)^5 = (290-1)^5$

$(17)^{10} = (290-1)^5$

$= {}^5C_0(290)^5 - {}^5C_1(290)^4 + {}^5C_2(290)^3 - {}^5C_3(290)^2 + {}^5C_4(290) - {}^5C_5$

$[\because (x-1)^n = {}^nC_0x^n - {}^nC_1x^{n-1} + {}^nC_2x^{n-2}\ldots(-1)^n\cdot{}^nC_n]$

$(17)^{10} = [(290)^5 - 5\times(290)^4 + 10\times(290)^3 - 10\times(290)^2 + 5\times 290 - 1]$

$= [(290)^5 - 5\times(290)^4 + 10\times(290)^3 - 10\times(290)^2] + 1450 - 1$

$= (290)^2[(290)^3 - 5\times(290)^2 + 10\times 290 - 10] + 1449$

$= 84100\,[24389000 - 420500 + 2900 - 10] + 1449$

$= 1000\,[841\times 2438900 - 841\times 42050 + 290 - 1] + 1449$

$= 1000m + 1449$

Where, $m = [841\times 2438900 - 841\times 42050 + 290 - 1]$

Hence, last two digit in $(17)^{10}$ is 49.

6. (a) We have,

$(1.01)^{1000000} = (1 + 0.01)^{1000000}$

$(1.01)^{1000000} = {}^{1000000}C_0 + {}^{1000000}C_1(0.01) + \cdots + {}^{1000000}C_{1000000}(0.01)^{1000000}$

$[\because (1+x)^n + {}^nC_0 + {}^nC_1x + {}^nC_2x^2 + {}^nC_3x^3 + \cdots + {}^nC_nx^n]$

$(1.01)^{1000000} = 1 + 1000000\times(0.01) + {}^{1000000}C_2\,(0.01)^2 + \cdots + (0.01)^{1000000}$

$(1.01)^{1000000} = 1 + 10000 + {}^{1000000}C_2\,(0.01)^2 + \cdots + (0.01)^{1000000}$

$\Rightarrow (1.01)^{1000000} = 10000 + (1 + {}^{1000000}C_2\,(0.01)^2 + \cdots + (0.01)^{1000000})$

$\Rightarrow \quad (1.01)^{1000000} = 10000 + k, k \in I$

$\therefore \quad (1.01)^{1000000} > 10000$

7. (a) Given, $\quad (7+4\sqrt{3})^n = \alpha + \beta$

Let $\quad (7-4\sqrt{3})^n = G$

Now, $\quad \alpha + \beta + G = (7+4\sqrt{3})^n + (7-4\sqrt{3})^n$

$= 2\,[{}^nC_0(7)^n + {}^nC_2(7)^{n-2}(4\sqrt{3})^2 + {}^nC_4(7)^{n-4}(4\sqrt{3})^4 + \cdots]$

$\alpha + \beta + G$ = an even integer

$\Rightarrow \beta + G$ = an even integer $- \alpha$

$\Rightarrow \quad \beta + G$ = an integer $\quad [\because \alpha$ is an integer$]$

$\Rightarrow \quad \beta + G = 1 \quad \left[\begin{array}{l}\because 0 \le \beta < 1, 0 < G < 1\\ \therefore 0 < \beta + G < 2\\ \Rightarrow \beta + G = 1\end{array}\right]$

$\Rightarrow \quad G = (1-\beta)$

$\therefore \quad (1-\beta)(\alpha+\beta) = G\cdot(\alpha+\beta)$

$= (7-4\sqrt{3})^n\,(7+4\sqrt{3})^n$

$= [(7)^2 - (4\sqrt{3})^2]^n = (49-48)^n = 1$

8. (a) Let $(7+4\sqrt{3})^n = I + f$, where $I \in N$ and $0 < f < 1$ and $g = (7-4\sqrt{3})^n$.

Then, $\quad I + f + g = (7+4\sqrt{3})^n + (7-4\sqrt{3})^n$

$\Rightarrow I + f + g = 2\,[{}^nC_0(7)^n + {}^nC_2(7)^{n-2}(4\sqrt{3})^2 + \cdots]$

$\Rightarrow I + f + g$ = an even integer.

$\Rightarrow \quad f + g$ = (an even integer $-I$)

$\Rightarrow \quad f + g$ = an integer $\quad [\because I$ is an integer$]$

$\Rightarrow \quad f + g = 1 \quad \left[\begin{array}{l}\because 0 \le f < 1, 0 < g < 1\\ \Rightarrow 0 < f + g < 2 \Rightarrow f + g = 1\end{array}\right]$

Now, since $I + f + g$ = an even integer

and $\quad f + g = 1$

$\Rightarrow \quad I$ = an even integer -1

$\Rightarrow \quad I$ = an odd integer.

9. (b) $I + f = (5\sqrt{2} + 7)^{2n+1}$

$g = (5\sqrt{2} - 7)^{2n+1}$

Then, $I + f - g = 2\cdot\{{}^{2n+1}C_1\cdot(5\sqrt{2})^{2n}\cdot 7^1 + {}^{2n+1}C_3\cdot(5\sqrt{2})^{2n-2}\cdot 7^3 \cdots\}$

$I + f - g$ = even integer. $\quad \ldots$(i)

Here, $\quad 0 < f < 1, 0 < g < 1 \Rightarrow -1 < f - g < 1$

As, $(f-g)$ is an integer and between -1 to 1

$\therefore \quad f - g = 0 \quad \ldots$(ii)

From Eqs. (i) and (ii), I = even integer.

10. (d) **To find** Last digit of $(3^P + 2)$.

Where, $P = 3^{4n}$

As, 3^{4n} where $n \in N$.

$\Rightarrow \quad (81)^n \Rightarrow (80 + 1)^n \Rightarrow 80\lambda + 1.$

$\therefore \quad 3^P + 2 = 3^{80\lambda + 1} + 2 = (3^4)^{20\lambda} \cdot 3 + 2$

$= 3 \cdot (81)^{20\lambda} + 2 = 3\,[80 + 1]^{20\lambda} + 2$

$= 3\,[80\,K + 1] + 2$

$= 240\,K + 3 + 2 = 240\,K + 5$

$\therefore$ Last digit of $3^P + 2$, is 5.

11. (b) $(99)^n + 1 = (100 - 1)^n + 1$

$= [{}^nC_0(100)^n - {}^nC_1(100)^{n-1} + {}^nC_2(100)^{n-2} - \dots - {}^nC_n] + 1$ [$\because n$ is odd]

$[\because (x-1)^n = {}^nC_0x^n - {}^nC_1x^{n-1} + {}^nC_2x^{n-2} - \dots {}^nC_n]$

$= {}^nC_0(100)^n - {}^nC_1(100)^{n-1} + {}^nC_2(100)^{n-2} - \dots + {}^nC_{n-1}100 - 1 + 1$

$= {}^nC_0(100)^n - {}^nC_1(100)^{n-1} + {}^nC_2(100)^{n-2} - \dots + {}^nC_{n-1}100$

$= 100\,[{}^nC_0(100)^{n-1} - {}^nC_1(100)^{n-2} + {}^nC_2(100)^{n-3} - \dots + {}^nC_{n-1}]$

$= 100 \times$ An integer whose units digit is different from zero [$\because n$ is odd]

Hence, the number of digits at the end of the sum $99^n + 1$ is 2.

12. (c) Let $E = (6)^{83} + (8)^{32}$

$E = (7-1)^{83} + (7+1)^{32}$

$E = ({}^{83}C_0(7)^{83} - {}^{83}C_1(7)^{82} + {}^{83}C_2(7)^{81} - \dots + {}^{83}C_{82}(7) - {}^{83}C_{83})$
$+ ({}^{32}C_0(7)^{32} + {}^{32}C_1(7)^{31} + {}^{32}C_2(7)^{30} + \dots + {}^{32}C_{31}(7) + {}^{32}C_{32})$

$E = {}^{83}C_0(7)^{83} - {}^{83}C_1(7)^{82} + {}^{83}C_2(7)^{81} -$
$\dots - {}^{83}C_{81}(7)^2 + {}^{83}C_{82}(7) - 1 + {}^{32}C_0(7)^{32}$
$+ {}^{32}C_1(7)^{31} + \dots + {}^{32}C_{30}(7)^2 + {}^{32}C_{31}(7) + 1$

$E = (7)^2\,({}^{83}C_0(7)^{81} - {}^{83}C_1(7)^{80} + {}^{83}C_2(7)^{79} - \dots - {}^{83}C_{81})$
$+ 7 \times 83 + (7)^2\,({}^{32}C_0(7)^{30} + {}^{32}C_1(7)^{29} + \dots + {}^{32}C_{30}) + 32 \times 7$

$E = 49\lambda + 7 \times 83 + 49 \times \mu + 32 \times 7$

Where,

$\lambda = {}^{83}C_0(7)^{81} - {}^{83}C_1(7)^{80} + \dots - {}^{83}C_{81}$

$\mu = {}^{32}C_0(7)^{30} + {}^{32}C_1(7)^{29} + \dots + {}^{32}C_{30}$

$E = 49(\lambda + \mu) + 7(83 + 32)$

$E = 49k + 7(7 \times 11 + 6 + 7 \times 4 + 4)$ where, $k \in I$.

$E = 49k + 7\,(7 \times 15 + 10)$

$E = 49k + 7(7 \times 15 + 7 \times 1 + 3)$

$E = 49k + 7(7 \times 16 + 3)$

$E = 49k + 49 \times 16 + 21$

Now, $\dfrac{E}{49} = \dfrac{49k}{49} + \dfrac{49 \times 16}{49} + \dfrac{21}{49} = k + 16 + \dfrac{21}{49}$

Hence, remainder is 21.

13. (a) Let $E = (11)^{27} + (21)^{27}$

$E = (16-5)^{27} + (16+5)^{27}$

$E = 2\,[{}^{27}C_0(16)^{27} + {}^{27}C_2(16)^{25}(5)^2 + {}^{27}C_4(16)^{23}(5)^4 + \dots + {}^{27}C_{26}(16)\,(5)^{26}]$

$[\because (a-b)^n + (a+b)^n = 2\,[{}^nC_0a^n + {}^nC_2a^{n-2}b^2 + \dots]$

$E = 2 \times 16\,[\,{}^{27}C_0(16)^{26} + {}^{27}C_2(16)^{24}(5)^2 + {}^{27}C_4(16)^{22}(5)^4 + \dots + {}^{27}C_{26}(5)^{26}]$

$E = 2 \times 16 \times k,\ k \in I$

Now, $\dfrac{E}{16} = \dfrac{2 \times 16 \times k}{16} = 2k + 0$

Hence, remainder is zero, when $(11)^{27} + (21)^{27}$ is divided by 16.

14. (d) Given, $N = 7^{100} - 3^{100}$

$N = (5+2)^{100} - (5-2)^{100}$

$\Rightarrow N = 2\,[{}^{100}C_1(5)^{99}(2) + {}^{100}C_3(5)^{97}(2)^3 + \dots + {}^{100}C_{99}(5)(2)^{99}]$

$[\because (a+b)^n - (a-b)^n = 2\,({}^nC_1(a)^{n-1}b + {}^nC_3(a)^{n-3}(b)^3 + \dots)]$

$\Rightarrow N = [{}^{100}C_1(5)^{99}(2)^2 + {}^{100}C_3(5)^{97}(2)^4 + \dots] + 100 \times (5 \times 2) \times 2^{99}$

$N = k + 2^{99} \times 1000, \quad k \in I$

Hence, last three digit of the number is 000.

15. (c) Let $E = [(15)^{23} + (23)^{23}]$

$E = [(19-4)^{23} + (19+4)^{23}]$

$\Rightarrow \quad E = 2\,[{}^{23}C_0(19)^{23} + {}^{23}C_2(19)^{21}(4)^2 + {}^{23}C_4(19)^{19}(4)^4 + \dots + {}^{23}C_{22}(19)(4)^{22}]$

$[\because (a-b)^n + (a+b)^n = 2\,[{}^nC_0a^n + {}^nC_2a^{n-2}b^2 + \dots]$

$\Rightarrow \quad E = 2 \times 19\,[{}^{23}C_0(19)^{22} + {}^{23}C_2(19)^{20}(4)^2 + {}^{23}C_3(19)^{19}(4)^3 + \dots + {}^{23}C_{22}(4)^{22}]$

$\Rightarrow \quad E = 2 \times 19\,k,\ k \in I$

Hence, when $(15)^{23} + (23)^{23}$ is divided by 19 remainder is zero.

16. (a) Given, $3^{400} = (3^4)^{100}$

$= (81)^{100} = (1 + 80)^{100}$

$= 1 + \dfrac{100}{1!}(80) + \dfrac{100\,(99)}{2!}(80)^2 + \dots + (80)^{100}$

$= 1 + 8000 + 100\,[\lambda]$

$= 8001 + 100\lambda.$

$\therefore$ Last two digit is 01.

17. (b) Given, $(5 + 2\sqrt{6})^n = p + f$

Where, $n \in N$, $p \in N$ and $0 < f < 1$.

Let $\quad g = (5 - 2\sqrt{6})^n$

Then, $\quad p + f + g = (5 + 2\sqrt{6})^n + (5 - 2\sqrt{6})^n$

$= 2\,[{}^nC_0(5)^n + {}^nC_2(5)^{n-2}(2\sqrt{6})^2 + \dots]$

$p + f + g =$ an even integer

$\Rightarrow \quad f + g =$ an integer $[\because p \in I]$

$\Rightarrow \quad f + g = 1$ $[\because 0 < f < 1, 0 < g < 1]$

$0 < f + g < 2$

$\Rightarrow \quad f + g = 1)$

Now, $\quad p + f = (5 + 2\sqrt{6})^n$

$p = (5 + 2\sqrt{6})^n - f$

$\Rightarrow \quad p = \dfrac{1}{(5 - 2\sqrt{6})^n} - f$

$\Rightarrow \quad p = \dfrac{1}{1 - f} - f$

$\Rightarrow \quad (1 - f)p = 1 - f\,(1 - f)$

$\therefore \quad f^2 - f + pf - p = -1$

18. (b) Let $E = 21^{100} = (20 + 1)^{100}$

$\Rightarrow \quad E = {}^{100}C_0(20)^{100} + {}^{100}C_1(20)^{99} + {}^{100}C_2(20)^{98} + \dots + {}^{100}C_{99}(20) + {}^{100}C_{100}$

$\Rightarrow \quad E = (20)^{100} + {}^{100}C_1(20)^{99} + {}^{100}C_2(20)^{98} + \dots + 100 \times 20 + 1$

$\Rightarrow \quad E = (20)^{100} + {}^{100}C_1(20)^{99} + {}^{100}C_2(20)^{98} + \dots + 2000 + 1$

$\Rightarrow \quad E = (20)^{100} + {}^{100}C_1(20)^{99} + {}^{100}C_2(20)^{98} + \dots + {}^{100}C_{98}(20)^2 + 2001$

Hence, the last two digit in $(21)^{100}$ is 01.

19. (a) To find remainder when
$1 + 2 + 2^2 + \cdots + 2^{1999}$ is divided by 5.

Let $E = 1 + 2 + 2^2 + \cdots + 2^{1999}$.

Which is a GP with first term 1 and common ratio 2.

$$E = \frac{1(2^{2000} - 1)}{2 - 1} \qquad \left[\because S_n = \frac{a(r^n - 1)}{r - 1}, r > 1\right]$$

$\Rightarrow \quad E = 2^{2000} - 1 \quad \Rightarrow \quad E = (2^4)^{500} - 1$

$E = (16)^{500} - 1 \quad \Rightarrow \quad E = (15 + 1)^{500} - 1$

$E = {}^{500}C_0(15)^{500} + {}^{500}C_1(15)^{499} + \cdots + {}^{500}C_{499}(15) + 1 - 1$

$E = {}^{500}C_0(15)^{500} + {}^{500}C_1(15)^{499} + \cdots + {}^{500}C_{499}(15)$

$E = 15\,[{}^{500}C_0(15)^{499} + {}^{500}C_1(15)^{498} + \cdots + {}^{500}C_{499}]$

$E = 15 \times \lambda$, where $\alpha \in I$

Now, $\frac{E}{5} = \frac{15 \times \lambda}{5} + 0 \Rightarrow \frac{E}{5} = 3\lambda + 0$

Hence, remainder is zero when $1 + 2 + 2^2 + \cdots + 2^{1999}$ is divided by 5.

20. (d) To find remainder when
$(8)^{2n} - (62)^{2n+1}$ is divided by 9.

Now, $8^{2n} - (62)^{2n+1} = (1 + 63)^n - (63 - 1)^{2n+1}$

$= (1 + 63)^n + (1 - 63)^{2n+1}$

Let $E = 8^{2n} - (62)^{2n+1} = (1 + 63)^n + (1 - 63)^{2n+1}$

$E = (1 + {}^nC_1 63 + {}^nC_2(63)^2 + \cdots + (63)^n] + (1 - {}^{2n+1}C_1(63) + {}^{2n+1}C_2(63)^2 + \cdots + (-1)(63)^{2n+1})$

$E = 2 + 63\,({}^nC_1 + {}^nC_2(63) + \cdots + (63)^n - {}^{2n+1}C_1 + {}^{2n+1}C_2(63) + \cdots - (63)^{2n}]$

$E = 2 + 63\lambda$, where, $\lambda \in I$.

Now, $\frac{E}{9} = \frac{63\lambda}{9} + \frac{2}{9}$

$= 7\lambda + \frac{2}{9}$

Hence, remainder is 2.

DPP-3 **Properties of Binomial Coefficients Based on Summation**

1. We have,

$$(1 + x)^n = {}^nC_0 + {}^nC_1x + {}^nC_2x^2 + \cdots + {}^nC_nx^n \quad \ldots(i)$$

Put $x = 1$ in Eq. (i), we get

$$(1 + 1)^n = {}^nC_0 + {}^nC_1 + {}^nC_2 + {}^nC_3 + \cdots + {}^nC_n \quad \ldots(i)$$

(i) ${}^nC_0 + {}^nC_1 + {}^nC_2 + \cdots + {}^nC_n = 2^n \quad \ldots(ii)$

Put $x = -1$ in Eq. (i), we get

$$0 = {}^nC_0 - {}^nC_1 + {}^nC_2 - {}^nC_3 + \cdots \quad \ldots(iii)$$

On adding Eqs. (ii) and (iii), we get

$2({}^nC_0 + {}^nC_2 + {}^nC_4 + \cdots) = 2^n$

(ii) ${}^nC_0 + {}^nC_2 + {}^nC_4 + \cdots = 2^{n-1}$

On subtracting Eq. (iii) from Eq. (ii), we get

$2({}^nC_1 + {}^nC_3 + {}^nC_5 + \cdots) = 2^n$

(iii) ${}^nC_1 + {}^nC_3 + {}^nC_5 + \cdots = 2^{n-1}$

(iv) Put $x = 3$ in Eq. (i), we get

${}^nC_0 + 3\cdot{}^nC_1 + 3^2\cdot{}^nC_2 + \cdots + 3^n\,{}^nC_n = 4^n$

(v) Put $x = -1$, we get

${}^nC_0 - {}^nC_1 + {}^nC_2 - {}^nC_3 + \cdots + (-1)^n\cdot{}^nC_n = 0$

(vi) $\sum_{r=0}^{n} (-1)^r \cdot r \cdot {}^nC_r = 0$

$\Rightarrow \quad \sum_{r=0}^{n} (-1)^r \cdot r\,\frac{n}{r}\,{}^{n-1}C_{r-1} \qquad [\because {}^nC_r = {}^{n-1}C_{r-1}]$

$\Rightarrow \quad n\sum_{r=0}^{n} (-1)^r\,{}^{n-1}C_{r-1}$

$n[{}^{n-1}C_0 - {}^{n-1}C_1 + {}^{n-1}C_2 - \cdots (-1)^{n-1}\,{}^{n-1}C_{n-1}]$

$n \times 0 = 0$

2. We have, $(1 + x)^n = {}^nC_0 + {}^nC_1x + {}^nC_2x^2 + \cdots + {}^nC_nx^n \quad \ldots(i)$

(i) Differentiate Eq. (i) w.r.t x both side, we get

$n(1 + x)^{n-1} = {}^nC_1 + 2{}^nC_2x + 3\,{}^nC_3x^2 + \cdots + n\,{}^nC_nx^{n-1} \quad \ldots(ii)$

Put $x = 1$ in Eq. (ii), we get

${}^nC_1 + 2\,{}^nC_2 + 3\,{}^nC_3 + \cdots + n\,{}^nC_n = n2^{n-1}$

(ii) In Eq. (i) replace x by x^2, we get

$(1 + x^2)^n = {}^nC_0 + {}^nC_1x^2 + {}^nC_2x^4 + \cdots + {}^nC_nx^{2x} \quad \ldots(iii)$

On multiplying both side by x, we get

$x(1 + x^2)^n = {}^nC_0x + {}^nC_1x^3 + {}^nC_2x^5 + \cdots + {}^nC_nx^{2n+1}$

On differentiating both side w.r.t x, we get

$(1 + x^2)^n + 2nx^2(1 + x^2)^{n-1} = {}^nC_0 + 3{}^nC_1x^2 + 5{}^nC_2x^4 + \cdots + (2n + 1)\,{}^nC_nx^{2x}$

Put $x = 1$ on both side, we get

$2^n + 2n\cdot 2^{n-1} = {}^nC_0 + 3{}^nC_1 + 5{}^nC_2 + \cdots + (2n + 1){}^nC_n$

$\Rightarrow {}^nC_0 + 3\,{}^nC_1 + 5{}^nC_2 + \cdots + (2n + 1)\,{}^nC_n = (n + 1)2^n$

(iii) On differentiating Eq. (i) both side w.r.t. x, we get

$n(1 + x)^n = {}^nC_1 + 2\,{}^nC_2x + \cdots + n\,{}^nC_nx^{n-1}$

On multiplying both side by x, we get

$nx(1 + x)^{n-1} = {}^nC_1x + 2\,{}^nC_2x^2 + 3\,{}^nC_3x^3 + \cdots + n\,{}^nC_nx^n$

Again, differentiating both side w.r.t, we get

$\Rightarrow n[(1 + x)^{n-1} + (n - 1)x(1 + x)^{n-2}]$

$= {}^nC_1 + 2^2\,{}^nC_2x + 3^2\,{}^nC_3x^2 - n^2\,{}^nC_nx^{n-1}]$

Put $x = 1$ on both side, we get

$\Rightarrow n\,[2^{n-1} + (n - 1)2^{n-2}]$

$= {}^nC_1 + 2^2\,{}^nC_2 + 3^2\,{}^nC_3 + \cdots + n^2\,{}^nC_n$

$\Rightarrow 1^2\,{}^nC_1 + 2^2\,{}^nC_2 + 3^2\,{}^nC_3 + \cdots + n^2\,{}^nC_n = n(n + 1)2^{n-2}$

(iv) ${}^nC_0 - 2\cdot{}^nC_1 + 3\cdot{}^nC_2 + \cdots + (-1)^n\cdot(n + 1)\cdot{}^nC_n$

Here, $(1 + x)^n = {}^nC_0 + {}^nC_1\,x + {}^nC_2x^2 + \cdots + {}^nC_n\cdot x^n$

Multiplying by x,

$\Rightarrow x(1 + x)^n = {}^nC_0\cdot x + {}^nC_1\cdot x^2 + {}^nC_2\cdot x^3 + \cdots + {}^nC_n\cdot x^{n+1}$

On differentiation both sides

$x\cdot n(1 + x)^{n-1} + (1 + x)^n = {}^nC_0 + 2\cdot{}^nC_1\cdot x + 3\cdot{}^nC_2\cdot x^2 + \cdots + (n + 1)\cdot{}^nC_n\cdot x^n.$

put, $x = -1$

$\Rightarrow 0 = {}^nC_0 - 2\cdot{}^nC_1 + 3\cdot{}^nC_2 + \cdots + (-1)^n\cdot(n + 1)\cdot{}^nC_n$

(v) Replace x by x^4 in Eq. (i), we get

$(1 + x^4)^n = {}^nC_0 + {}^nC_1x^4 + {}^nC_2x^8 + {}^nC_3x^{12} + \cdots + {}^nC_nx^{4n}$

On multiplying x both side, we get

$x(1 + x^4)^n = {}^nC_0x + {}^nC_1x^5 + {}^nC_2x^9 + {}^nC_3x^{13} + \cdots + {}^nC_nx^{4n+1}$

On differentiating both side w.r.t., we get

$$(1+x^4)^n + 4nx^4(1+x)^{n-1} = {}^nC_0 + 5\,{}^nC_1x^4 + 9\,{}^nC_2x^8 + \cdots + (4n+1)\,{}^nC_nx^{4n}$$

Put $x = 1$ both side, we get

$$2^n + 4n2^{n-1} = {}^nC_0 + 5\,{}^nC_1 + 9\,{}^nC_2 + \cdots + (4n+1)\,{}^nC_n$$

$$\Rightarrow {}^nC_0 + 5\,{}^nC_1 + 9\,{}^nC_2 + \cdots + (4n+1)\,{}^nC_n = (2n+1)2^n$$

(vi) We have,

$$(1+x)^n = {}^nC_0 + {}^nC_1x + {}^nC_2x^2 + \cdots + {}^nC_nx^n$$

$$\Rightarrow (1+x)^{5n} = {}^{5n}C_0 + {}^{5n}C_1x + {}^{5n}C_2x^2 + \cdots + {}^{5n}C_{5n}x^{5n}$$

$$\Rightarrow [(1+x)^n]^5 = {}^{5n}C_0 + {}^{5n}C_1x + {}^{5n}C_2x^2 + \cdots + {}^{5n}C_{5n}x^{5n}$$

$$\Rightarrow [1 + {}^nC_1x + {}^nC_2x^2 + \cdots + {}^nC_nx^n]^5 = {}^{5n}C_0 + {}^{5n}C_1x + \cdots + {}^{5n}C_{5n}x^{5n}$$

Put $x = 1$ both side, we get

$$\Rightarrow (1 + {}^nC_1 + {}^nC_2 + \cdots + {}^nC_n)^5 = {}^{5n}C_0 + {}^{5n}C_1 + {}^{5n}C_2 + \cdots + {}^{5n}C_{5n}$$

(vii) We have,

$$a - (a-1)\,{}^nC_1 + (a-2)\,{}^nC_2 - (a-3)\,{}^nC_3 \cdots (-1)^n(a-n)\,{}^nC_n$$

$$\Rightarrow a(1 - {}^nC_1 + {}^nC_2 - {}^nC_3 + \cdots (-1)^n\,{}^nC_n) + ({}^nC_1 - 2\,{}^nC_2 + 3\,{}^nC_3 \cdots (-1^n)(-n)\,{}^nC_n]$$

$$= a(0) + 0 = 0$$

$[\because {}^nC_0 - {}^nC_1 + {}^nC_2 - \cdots (-1)^n\,{}^nC_n = 0$ and ${}^nC_1 - 2\,{}^nC_2 + 3\,{}^nC_3 - \cdots (-1)^n - n\,{}^nC_n = 0]$

(viii) Integrating between the limits 0 and 1 in Eq. (i), we get

$$\int_0^1 (1+x)^n dx = \int_0^1 ({}^nC_0 + {}^nC_1x + {}^nC_2x^2 + \cdots + {}^nC_nx^n)dx$$

$$\left[\frac{(1+x)^{n+1}}{n+1}\right]_0^1 = \left[{}^nC_0x + \frac{{}^nC_1x^2}{2} + \frac{{}^nC_2x^3}{3} + \cdots + \frac{{}^nC_nx^{n+1}}{n+1}\right]_0^1$$

$$\frac{2^{n+1}}{n+1} - \frac{1}{n+1} = {}^nC_0 + \frac{{}^nC_1}{2} + \frac{{}^nC_2}{3} + \cdots + \frac{{}^nC_n}{n+1}$$

$$\frac{2^{n+1}-1}{n+1} = {}^nC_0 + \frac{{}^nC_1}{2} + \frac{{}^nC_2}{3} + \cdots + \frac{{}^nC_n}{n+1}$$

(ix) Integrating between the limits 0 and 2, we get

$$\int_0^2 (1+x)^n dx = \int_0^2 ({}^nC_0 + {}^nC_1x + {}^nC_2x^2 + \cdots + {}^nC_nx^n)dx$$

$$\left[\frac{(1+x)^{n+1}}{n+1}\right]_0^2 = \left[{}^nC_0x + \frac{{}^nC_1x^2}{2} + \frac{{}^nC_2x^3}{3} + \cdots + \frac{{}^nC_nx^{n+1}}{n+1}\right]_0^2$$

$$\frac{3^{n+1}-1}{n+1} = 2\,{}^nC_0 + 2^2\cdot\frac{{}^nC_1}{2} + \frac{2^{3}\,{}^nC_2}{3} + \cdots + \frac{2^{n+1}\,{}^nC_n}{n+1}$$

(x) We have,

$$a\,{}^nC_0 + (a-b)\,{}^nC_1 + (a-2b)\,{}^nC_2 + \cdots + (a-nb)\,{}^nC_n$$

$$\Rightarrow a({}^nC_0 + {}^nC_1 + {}^nC_2 + \cdots + {}^nC_n) - b({}^nC_1 + 2\,{}^nC_2 + 3\,{}^nC_2 + \cdots + n\,{}^nC_n)$$

$$= a(2^n) - b(n\cdot 2^{n-1})$$

$[\because {}^nC_0 + {}^nC_1 + {}^nC_2 + \cdots + {}^nC_n = 2^n]$

and ${}^nC_1 + 2\,{}^nC_2 + \cdots + n\,{}^nC_n = n\cdot 2^{n-1}] = 2^{n-1}[2a - nb]$

3. (a) Sum of the given binomial coefficient is 256.

$$\therefore \quad 2^n = 256$$

$$\Rightarrow \quad n = 8$$

$$\Rightarrow \left(2x + \frac{1}{x}\right)^8$$

$$T_{r+1} = {}^8C_r(2x)^{8-r}\left(\frac{1}{x}\right)^r = {}^8C_r2^{8-r}x^{8-2r}$$

For constant term,

$$8 - 2r = 0$$

$$\Rightarrow \quad r = 4$$

Hence, constant term is ${}^8C_42^4 = \dfrac{8!}{4!4!} \times 16 = 1120$

Hence, option (a) is correct.

4. (b) We have,

$$\frac{1}{1\cdot(n-1)!} + \frac{1}{3!(n-3)!} + \frac{1}{5!(n-5)!} + \cdots + \frac{1}{(n-1)!1!}$$

On multiplying and dividing by $n!$, we get

$$\Rightarrow \frac{1}{n!}\left[\frac{n!}{1!(n-1)!} + \frac{n!}{3!(n-3)!} + \frac{n!}{5!(n-5)!} + \cdots + \frac{n!}{(n-1)!\,1!}\right]$$

$$\Rightarrow \frac{1}{n!}[{}^nC_1 + {}^nC_3 + {}^nC_5 + \cdots + {}^nC_{n-1}]$$

$$\Rightarrow \frac{2^{n-1}}{n!} \qquad [\because {}^nC_1 + {}^nC_3 + {}^nC_5 + \cdots = 2^{n-1}]$$

Hence, option (b) is correct.

5. (a) We have, $(2x^2 - 3x + 1)$

Sum of all coefficient for the given expansion, we put $x = 1$

$$\therefore \quad (2 - 3 + 1)^{11} = 0$$

Hence, option(a) is correct.

6. (b) We have,

$$(2x^2 - 3x + 1)^{11}$$

Let $(2x^2 - 3x + 1)^{11} = a_0 + a_1x + a_2x^2 + a_3x^3 + \cdots$ ...(i)

Put $x = 1$ in Eq. (i), we get

$$0 = a_0 + a_1 + a_2 + a_3 + \cdots \quad \text{...(ii)}$$

Put $x = -1$ in Eq. (i), we get

$$6^{11} = a_0 - a_1 + a_2 - a_3 + \cdots \quad \text{...(iii)}$$

For the sum of even power of x, on adding Eqs. (ii) and (iii), we get

$$6^{11} = 2(a_0 + a_2 + a_4 + \cdots)$$

$$\Rightarrow a_0 + a_2 + a_4 + \cdots = \frac{6\cdot 6^{10}}{2} = 3\cdot 6^{10}$$

Hence, option (b) is correct.

7. (b) We have,

$$(1 + x - 3x^2)^{2145} = a_0 + a_1x + a_2x^2 + \cdots$$

Put $x = -1$, we get

$$a_0 - a_1 + a_2 - a_3 \cdots = (1 - 1 - 3)^{2145} = (-3)^{2145}$$

$$(-3)^{4\times 526} \times (-3) = (81)^{526} \times (-3)$$

$$= (1)^{526}(-3)$$

$$= -(\cdots 3)$$

$\therefore$ End digit is 3.

Hence, option (b) is correct.

8. (c) We have,
$$(1+x)^n = {}^nC_0 + {}^nC_1x + {}^nC_2x^2 + \cdots + {}^nC_nx^n$$
$$\frac{{}^nC_1}{{}^nC_0} + 2\frac{{}^nC_2}{{}^nC_1} + 3\frac{{}^nC_3}{{}^nC_2} + \cdots + n\frac{{}^nC_n}{{}^nC_{n-1}}$$
$$\Rightarrow \quad n + 2\frac{(n-1)}{2} + 3\frac{(n-2)}{3} - \frac{n(1)}{n} \quad \left[\because \frac{{}^nC_r}{{}^nC_{r-1}} = \frac{n+1-r}{r}\right]$$
$$\Rightarrow \quad n + (n-1) + (n-2) + (n-3) + \cdots + 1$$
$$\Rightarrow \quad 1 + 2 + 3 + 4 + \cdots + n$$
$$\Rightarrow \quad \frac{n(n+1)}{2}$$
Hence, option (c) is correct.

9. (c) We have, $\sum_{r=0}^{10} {}^{20}C_r$
$$\Rightarrow \quad {}^{20}C_0 + {}^{20}C_1 + {}^{20}C_2 + \cdots + {}^{20}C_{10}$$
We know that,
$$2^{20} = {}^{20}C_0 + {}^{20}C_1 + {}^{20}C_2 + \cdots + {}^{20}C_{19} + {}^{20}C_{20}$$
$$\Rightarrow \quad 2^{20} = ({}^{20}C_0 + {}^{20}C_{20}) + ({}^{20}C_1 + {}^{20}C_{19}) + \cdots + ({}^{20}C_{10})$$
$$\Rightarrow \quad 2^{20} = 2({}^{20}C_0 + {}^{20}C_1 + {}^{20}C_2 + \cdots + {}^{20}C_9) + {}^{20}C_{10}$$
$$[\because {}^nC_r = {}^nC_{n-r}]$$
$$\Rightarrow \quad {}^{20}C_0 + {}^{20}C_1 + {}^{20}C_2 + \cdots + {}^{20}C_{10} = 2^{19} + \frac{1}{2}\,{}^{20}C_{10}$$
Hence, option (c) is correct.

10. (c) We have,
$$\left(1+\frac{{}^nC_1}{{}^nC_0}\right)\left(1+\frac{{}^nC_2}{{}^nC_1}\right)\left(1+\frac{{}^nC_3}{{}^nC_2}\right)\cdots\left(1+\frac{{}^nC_n}{{}^nC_{n-1}}\right)$$
$$\Rightarrow \quad \left(1+\frac{n}{1}\right)\left(1+\frac{n-1}{2}\right)\left(1+\frac{n-2}{3}\right)\cdots\left(1+\frac{1}{n}\right)$$
$$\Rightarrow \quad \frac{(1+n)}{1}\frac{(1+n)}{2}\frac{(1+n)}{3}\cdots\frac{(1+n)}{n} \quad \left[\because \frac{{}^nC_r}{{}^nC_{r-1}} = \frac{n+1-r}{r}\right]$$
$$\Rightarrow \quad \frac{(n+1)^n}{n!}$$
Hence, option (c) is correct.

11. (c) We have, $(\alpha^2x^2 - 2\alpha x + 1)^{51}$

The sum of the coefficient of the expansion is vanishes iff $\alpha^2 - 2\alpha + 1 = 0$
$$\Rightarrow \quad (\alpha - 1)^2 = 0$$
$$\Rightarrow \quad \alpha = 1$$
Hence, option (c) is correct.

12. (c) We have,
$$S_1 = \sum_{j=1}^{10} j(j-1)\,{}^{10}C_j$$
$$\Rightarrow \quad S_1 = \sum_{j=1}^{10} j(j-1)\frac{10!}{j(j-1)(j-2)!(10-j)!}$$
$$\Rightarrow \quad S_1 = 90\sum_{j=2}^{10} \frac{8!}{(j-2)![8-(j-2)]!}$$
$$= 90\cdot 2^8$$
and
$$S_2 = \sum_{j=1}^{10} j\frac{10!}{j(j-1)![9-(j-1)]!}$$
$$= 10\sum_{j=1}^{10} \frac{9!}{(j-1)![9-(j-1)]!} = 10\cdot 2^9$$
Also,
$$S_3 = \sum_{j=1}^{10} [j(j-1)+j]\frac{10!}{j!(10-j)!}$$
$$= \sum_{j=1}^{10} j(j-1)\,{}^{10}C_j + \sum_{j=1}^{10} j\,{}^{10}C_j$$
$$= 90\cdot 2^8 + 10\cdot 2^9$$
$$= 110\cdot 2^8$$
$$= 55\cdot 2^9$$
Hence, Statement I is correct, Statement II is incorrect.

13. (a) Since, $\sum_{r=0}^{n} {}^nC_r\cdot x^r = (1+x)^n$

On multiplying by x, we get
$$\sum_{r=0}^{n} {}^nC_r\cdot x^{r+1} = x(1+x)^n$$
On differentiating w.r.t x, we get
$$\sum_{r=0}^{n} (r+1)\,{}^nC_rx^r = (1+x)^n + nx(1+x)^{n-1}$$
Statement II is correct.

If we put $x = 1$, then
$$\sum_{r=1}^{n} (r+1)\,{}^nC_r = 2^n + n2^{n-1}$$
$$= (n+2)2^{n-1}$$
∴ Hence, Statement I is also correct and Statement II is a correct explanation of Statement I.

14. (b) We have,
$$(1+x)^n = \sum_{r=0}^{n} a_rx^r$$
i.e.
$$a_r = {}^nC_r$$
and
$$b_r = 1 + \frac{a_r}{a_{r-1}} = 1 + \frac{{}^nC_r}{{}^nC_{r-1}}$$
$$\Rightarrow \quad b_r = 1 + \frac{n+1-r}{r} \quad \left[\because \frac{{}^nC_r}{{}^nC_{r-1}} = \frac{n+1-r}{r}\right]$$
$$\Rightarrow \quad b_r = \frac{n+1}{r}$$
$$\prod_{r=1}^{n} b_r = \left(\frac{n+1}{1}\right)\left(\frac{n+1}{2}\right)\left(\frac{n+1}{3}\right)\cdots\left(\frac{n+1}{n}\right)$$
$$= \frac{(n+1)^n}{n!}$$
If $n = 100$, then
$$\sum_{r=1}^{100} b_r = \frac{(101)^{100}}{100!}$$
and
$$\sum_{r=0}^{n} \frac{1}{b_r} = \sum_{r=0}^{n} \frac{r}{n+1}$$
$$= \frac{1+2+3+\cdots+n}{n+1}$$
$$\sum_{r=0}^{n} \frac{1}{b_r} = \frac{n(n+1)}{2(n+1)} = \frac{n}{2}$$
if $n = 100$,
$$\sum_{r=0}^{100} \frac{1}{b_r} = \frac{100}{2} = 50$$
Hence, Statement I is correct and also Statement II is correct but not the correct explanation of Statement I.

15. (d) We have, $\sum_{r=0}^{15} r \cdot {}^{15}C_r p^r q^{15-r}$

$$\Rightarrow \sum_{r=0}^{15} r \cdot \frac{15}{r} {}^{14}C_{r-1} p^r q^{15-r} \quad \left[\because {}^nC_r = \frac{n}{r} {}^{n-1}C_{r-1}\right]$$

$$\Rightarrow 15 \sum_{r=0}^{15} {}^{14}C_{r-1} p^r q^{15-r}$$

$$\Rightarrow 15p \sum_{r=1}^{15} {}^{14}C_{r-1} p^{r-1} q^{15-r}$$

$$\Rightarrow 15p(p+q)^{14}$$

$$\Rightarrow 15 \times \frac{2}{3}(1)^{14} = 10 \quad \left[\begin{array}{l}\because p+q=1\\ q=\frac{1}{3} \text{ and } p=\frac{2}{3}\end{array}\right]$$

Statement I is incorrect.

Now, $\sum_{r=0}^{n} r \cdot {}^nC_r p^r q^{n-r}$

$$\Rightarrow \sum_{r=0}^{n} r \cdot \frac{n}{r} {}^{n-1}C_{r-1} p^r q^{n-r}$$

$$\Rightarrow np \sum_{r=1}^{n} {}^{n-1}C_{r-1} p^{r-1} q^{(n-1)-(r-1)}$$

$$\Rightarrow np(p+q)^{n-1} = np \quad [\because p+q=1]$$

Statement II is correct.

Hence, option (d) is correct.

16. (c) $(1-x)^n = {}^nC_0 - {}^nC_1 x + {}^nC_2 x^2 - {}^nC_3 x^3 + \cdots (-1)^n {}^nC_n x^n$

On multiplying both side by x^{k-1}, we get

$$x^{k-1}(1-x)^n = {}^nC_0 x^{k-1} - {}^nC_1 x^k + {}^nC_2 x^{k+1} - \cdots (-1)^n {}^nC_n x^{n+k-1}$$

On integrating both sides between limit 0 and 1

$$\int_0^1 x^{k-1}(1-x)^n dx = \left[\frac{{}^nC_0 x^k}{k} - \frac{{}^nC_1 x^{k+1}}{k+1} + \cdots (-1)^n \frac{{}^nC_n x^{n+k}}{n+k}\right]_0^1$$

To evaluate $\int_0^1 x^{k-1}(1-x)^n dx$

Let
$$I_{k-1,n} = \int_0^1 x^{k-1}(1-x)^n dx$$

$$= \left[\frac{-x^{k-1}(1-x)^{n+1}}{n+1}\right]_0^1 + \frac{k-1}{n+1}\int_0^1 x^{k-2}(1-x)^{n+1} dx$$

$$I_{k-1,n} = \frac{k-1}{n+1}\int_0^1 x^{k-2}(1-x)^{n+1} dx$$

$$I_{k-1,n} = \frac{k-1}{n+1} I_{k-2,n+1}$$

$$I_{k-1,n} = \frac{(k-1)(k-2)(k-3)\cdots 1}{(n+1)(n+2)\cdots(n+k)}$$

$$I_{k-1,n} = \frac{n!}{k(k+1)(k+2)\cdots(n+k)}$$

$$\therefore \left[\frac{{}^nC_0 x^k}{k} - \frac{{}^nC_1 x^{k+1}}{k+1} + \cdots (-1)^n \frac{{}^nC_n x^{n+k}}{n+k}\right]_0^1$$

$$= \frac{n!}{k(k+1)(k+2)\cdots(n+k)} \quad \frac{{}^nC_0}{k} - \frac{{}^nC_1}{k+1} + \frac{{}^nC_2}{k+2} - \cdots \frac{(-1)^n {}^nC_n}{n+k}$$

$$= \frac{n!}{k(k+1)(k+2)\cdots(n+k)}$$

Put $k = \frac{1}{2}$ on both side, we get

$$\frac{2\,{}^nC_0}{1} - \frac{2\,{}^nC_1}{3} + \frac{2\,{}^nC_2}{5} - \cdots \frac{(-1)^n 2\,{}^nC_n}{2n+1} = \frac{n!\,2^n}{1\cdot 3\cdot 5\cdots(2n+1)}$$

$$\Rightarrow {}^nC_0 - \frac{{}^nC_1}{3} - \frac{{}^nC_2}{5}\cdots \frac{(-1)^n {}^nC_n}{2n+1} = \frac{n!\,2^{n-1}}{1\cdot 3\cdot 5\cdots(2n+1)}$$

$$= \frac{n!\,2^{2n}\cdot n!}{1\cdot 2\cdot 3\cdot 4\cdots 2n\cdot(2n+1)} = \frac{2^{2n}(n!)^2}{(2n+1)!}$$

Hence, option (c) is correct.

17. (c) We have, $\frac{{}^nC_0}{n(n+1)} - \frac{{}^nC_1}{(n+1)(n+2)} + \cdots + \frac{(-1)^n {}^nC_n}{2n(2n+1)}$

$$\left(\frac{{}^nC_0}{n} - \frac{{}^nC_0}{n+1}\right) - \left(\frac{{}^nC_1}{n+1} - \frac{{}^nC_1}{n+2}\right) + \cdots + (-1)^n\left(\frac{{}^nC_n}{2n} - \frac{{}^nC_n}{2n+1}\right)$$

$$= \left(\frac{{}^nC_0}{n} - \frac{{}^nC_0}{n+1} + \frac{{}^nC_2}{n+2}\cdots + (-1)^n\frac{{}^nC_n}{2n}\right)$$

$$- \left(\frac{{}^nC_0}{n+1} - \frac{{}^nC_1}{n+2}\cdots + (-1)^n\frac{{}^nC_n}{2n+1}\right)$$

$$= \int_0^1 x^{n-1}(1-x)^n dx - \int_0^1 x^n(1-x)^n dx$$

$$= \int_0^1 x^{n-1}[(1-x)^n(1-x)]dx$$

$$= \int_0^1 x^{n-1}(1-x)^{n+1} dx$$

Hence, option (c) is correct.

18. (d) General term of given series is

$$T_r = (-1)^{r-1}\left(1 + \frac{1}{2} + \frac{1}{3} + \cdots + \frac{1}{r}\right){}^nC_r$$

Now, $(-1)^{r-1}(1 + x + x^2 + \cdots + x^{r-1}){}^nC_r$

$$\Rightarrow \left(\frac{1-x^r}{1-x}\right)(-1)^{r-1}\,{}^nC_r$$

$$= \frac{1}{1-x}\sum_{r=1}^{n}[(-1)^{r-1}\,{}^nC_r - (-1)^{r-1}x^r\,{}^nC_r]$$

$$= \frac{1}{1-x}[(-1)\sum_{r=1}^{n}(-1)^r\cdot{}^nC_r + \sum_{r=1}^{n}(-1)^r\cdot x^r\,{}^nC_r]$$

$$= \frac{1}{1-x}[(-1)\{(1-1)^n - 1\} + (1-x)^n - 1]$$

$$= \frac{(1-x)^n}{1-x} = (1-x)^{n-1}$$

$$\therefore \sum_{r=1}^{n}(-1)^{r-1}(1 + x + x^2 + \cdots + x^{r-1}){}^nC_r = (1-x)^{n-1}$$

On integrating between limit 0 and 1,

$$\int_0^1 \sum_{r=1}^{n}(-1)^{r-1}(1 + x + x^2 + \cdots + x^{r-1}){}^nC_r dx = \int_0^1 (1-x)^{n-1}dx$$

$$\Rightarrow \sum_{r=1}^{n}(-1)^{r-1}\,{}^nC_r\int_0^1(1 + x + x^2 + \cdots + x^{r-1})dx = \left[-\frac{(1-x)^n}{n}\right]_0^1$$

$$\Rightarrow \sum_{r=1}^{n}(-1)^{r-1}\,{}^nC_r\left[x + \frac{x^2}{2} + \frac{x^3}{3} + \cdots + \frac{x^r}{r}\right]_0^1 = \frac{1}{n}$$

$$\Rightarrow \sum_{r=1}^{n}(-1)^{r-1}\,{}^nC_r\left[1 + \frac{1}{2} + \frac{1}{3} + \cdots + \frac{1}{r}\right] = \frac{1}{n}$$

Hence, option (d) is correct.

19. (b) We have, $(1+x)^n = a_0 + a_1 x + a_2 x^2 + \cdots + a_n x^n$

put $x = i$,

$$(1+i)^n = a_0 + a_1 i + a_2 i^2 + \cdots + a_n(i)^n \quad \left[2^{\frac{1}{2}}\left(\cos\frac{\pi}{4} + i\sin\frac{\pi}{4}\right)\right]^n$$

$$= a_0 + a_1 i - a_2 - a_3 i + a_4 \cdots$$

$$2^{n/2}\left(\cos\frac{n\pi}{4} + i\sin\frac{n\pi}{4}\right) = a_0 + a_1 i - a_2 - a_3 i + a_4 \cdots$$

On equating real and imaginary part both sides

$$a_0 - a_2 + a_4 - a_6 + \cdots = 2^{n/2}\cos\frac{n\pi}{4}$$

Hence, option (b) is correct.

20. (a) $a_1 - a_3 + a_5 - a_7 + \cdots = 2^{n/2}\cdot\sin\frac{n\pi}{4}$

Hence, option (a) is correct.

DPP-4 Product of Binomial Coefficients and Double Summation

1. (i) ${}^nC_n + {}^{n+1}C_n + {}^{n+2}C_n + \cdots + {}^{n+K}C_n$

$\Rightarrow \underbrace{{}^{n+1}C_{n+1} + {}^{n+1}C_n} + {}^{n+2}C_n + \cdots + {}^{n+K}C_n$

$\Rightarrow {}^{n+2}C_{n+1} + {}^{n+2}C_n + {}^{n+3}C_n + \cdots + {}^{n+K}C_n,$

[using ${}^{n+1}C_{n+1} + {}^{n+1}C_n = {}^{n+2}C_{n+1}$]

$\Rightarrow {}^{n+3}C_{n+1} + {}^{n+3}C_n + \cdots + {}^{n+K}C_n$

$\Rightarrow {}^{n+4}C_{n+1} + \cdots + {}^{n+K}C_n$

$\vdots \quad \vdots \quad \vdots$

$\Rightarrow {}^{n+K+1}C_{n+1}.$

(ii) LHS

${}^mC_1 + {}^{m+1}C_2 + {}^{m+2}C_3 + \cdots + {}^{m+n-1}C_m$

$\Rightarrow [(1+m) + {}^{m+1}C_2 + {}^{m+2}C_3 + \cdots + {}^{m+n-1}C_m] - 1$

$\Rightarrow ({}^{m+1}C_1 + {}^{m+1}C_2 + {}^{m+2}C_3 + \cdots + {}^{m+n-1}C_m) - 1$

$\Rightarrow [{}^{m+2}C_2 + {}^{m+2}C_3 + \cdots + {}^{m+n-1}C_m] - 1$

$[{}^{m+3}C_3 + \cdots + {}^{m+n-1}C_m - 1]$

$\Rightarrow {}^{m+n}C_{m+1} - 1$

RHS

${}^nC_1 + {}^{n+1}C_2 + {}^{n+2}C_3 + \cdots + {}^{n+m-1}C_m$

$\Rightarrow [{}^{n+1}C_1 + {}^{n+1}C_2 + {}^{n+2}C_3 + \cdots + {}^{n+m-1}C_m] - 1$

$\Rightarrow ({}^{n+2}C_2 + {}^{n+2}C_3 + \cdots + {}^{n+m-1}C_m) - 1$

$\Rightarrow ({}^{n+3}C_3 + \cdots + {}^{n+m-1}C_m) - 1$

$\vdots \quad \vdots \quad \vdots$

$\Rightarrow {}^{n+m}C_{m+1} - 1$

$\therefore$ LHS = RHS

2. (i) Given, $(1+x)^n = C_0 + C_1x + C_2x^2 + \cdots + C_rx^r + \cdots + {}^nC_nx^n$ ...(i)

Replacing x by $\frac{1}{x}$ in Eq. (i), we get

$$\left(1+\frac{1}{x}\right)^n = {}^nC_0 + {}^nC_1\frac{1}{x} + {}^nC_2\frac{1}{x^2} + \cdots + {}^nC_n\frac{1}{x^n} \quad \text{...(ii)}$$

On multiplying Eqs. (i) and (ii), we get

$$\frac{(1+x)^{2n}}{x^n} = ({}^nC_0 + {}^nC_1x + {}^nC_2x^2 + \cdots)$$
$$\left({}^nC_0 + {}^nC_1\frac{1}{x} + {}^nC_2\frac{1}{x^2} + \cdots\right)_n^0$$

On comparing coefficient of x^r both side, we get

$${}^nC_0{}^nC_r + {}^nC_1{}^nC_{r+1} + \cdots + {}^nC_{n-r}{}^nC_n = {}^{2n}C_{n-r} \quad \text{...(iii)}$$

Put $r = 0$ in Eq. (iii), we get

$${}^nC_0{}^nC_0 + {}^nC_1{}^nC_1 + {}^nC_2{}^nC_2 + \cdots + {}^nC_n{}^nC_n = {}^{2n}C_n$$

Hence proved.

(ii) Put $r = 1$ in Eq. (iii) of part (i), we get

$${}^nC_0{}^nC_1 + {}^nC_1{}^nC_2 + {}^nC_2{}^nC_3 + \cdots + {}^nC_{n-1}{}^nC_n = {}^{2n}C_{n-1}$$

(iii) By part (i) of this question, we have

$${}^nC_0{}^nC_0 + {}^nC_1{}^nC_1 + \cdots + {}^nC_n{}^nC_n = {}^{2n}C_n$$

$\Rightarrow {}^nC_0{}^nC_n + {}^nC_1{}^nC_{n-1} + \cdots + {}^nC_n{}^nC_{n-n} = {}^{2n}C_n$

[$\because {}^nC_r = {}^nC_{n-r}$]

$\Rightarrow {}^nC_0{}^nC_n + {}^nC_1{}^nC_2 + \cdots + {}^nC_n{}^nC_0 = {}^{2n}C_n$

(iv) Put $r = 3$.

$$\sum_{r=0}^{n} {}^nC_r \cdot {}^nC_{r+3} = {}^{2n}C_{n-3}$$

(v) To show,

$${}^mC_r{}^nC_0 + {}^mC_{r-1}{}^nC_1 + {}^mC_{r-2}{}^nC_2 + \cdots + {}^mC_0{}^nC_r = {}^{m+n}C_r$$

$$\text{LHS} = {}^mC_r{}^nC_0 + {}^mC_{r-1}{}^nC_1 + {}^mC_{r-2}{}^nC_2 + \cdots + {}^mC_0{}^nC_r$$

Hence, sum of lower suffices in each term is constant.

Now, $(1+x)^n = {}^nC_0 + {}^nC_1x + {}^nC_2x^2 + \cdots + {}^nC_{r-1}x^{r-1} + {}^nC_rx^r + \cdots + {}^nC_nx^n$ ...(i)

and $(1+x)^m = {}^mC_0 + {}^mC_1x + {}^mC_2x^2 + \cdots + {}^mC_{r-1}x^{r-1} + {}^mC_rx^r + {}^mC_mx^m$ ...(ii)

On multiplying Eqs. (i) and (ii), we get

$({}^nC_0 + {}^nC_1x + {}^nC_2x^2 + \cdots + {}^nC_{r-1}x^{r-1} + {}^nC_rx^r + \cdots + {}^nC_nx^n)$

$\times ({}^mC_0 + {}^mC_1x + {}^mC_2x^2 + \cdots + {}^mC_{r-1}x^{r-1} + {}^mC_rx^r + \cdots + {}^mC_mx^m)$

$(1+x)^{m+n} = {}^{m+n}C_0 + {}^{m+n}C_1x + {}^{m+n}C_2x^2 + \cdots + {}^{m+n}C_rx^r + \cdots + {}^{m+n}C_{m+n}x^{m+n}$

On equating coefficient of x^r, we get

$${}^mC_r{}^nC_0 + {}^mC_{r-1}{}^nC_1 + {}^mC_{r-2}{}^nC_2 + \cdots + {}^mC_0{}^nC_r = {}^{m+n}C_r$$

3. (i) To show

$${}^nC_0{}^nC_0 - {}^nC_1{}^nC_1 + {}^nC_2{}^nC_2 - \cdots (-1)^n{}^nC_n{}^nC_n = \begin{cases} 0, \text{ if } n \text{ is odd} \\ (-1)^{n/2}\, {}^nC_{n/2}, \text{ if } n \text{ is even} \end{cases}$$

$\text{LHS} = {}^nC_0{}^nC_0 - {}^nC_1{}^nC_1 + {}^nC_2{}^nC_2 - \cdots (-1)^n\, {}^nC_n{}^nC_n$

${}^nC_0{}^nC_{n-0} - {}^nC_1{}^nC_{n-1} + {}^nC_2{}^nC_{n-2} \cdots (-1)^n{}^nC_n{}^nC_{n-n}$

$= {}^nC_0{}^nC_n - {}^nC_1{}^nC_{n-1} + {}^nC_2{}^nC_{n-2} - \cdots (-1)^n{}^nC_n{}^nC_0$

= coefficient of x^n in $(1+x)^n(1-x)^n$

= coefficient of x^n in $(1-x^2)^n$.

Now, general term in $(1-x^2)^n$ is

$T_{r+1} = {}^nC_r\,(-x^2)^r$

$= (-1)^r\, {}^nC_rx^{2r}$

To get x^n put $2r = n$

$\Rightarrow r = \frac{n}{2}$

$\Rightarrow$ n must be even

If n is odd, r is not an integer or we can say when n is odd x^n term does not occur in $(1-x^2)^n$ which is obvious.

When r is even, $r = n/2$

Hence, $T_{n/2+1} = {}^nC_{n/2}(-x^2)^{n/2}$

$= (-1)^{n/2}\, {}^nC_{n/2}x^n$

$$\Rightarrow {}^nC_0{}^nC_0 - {}^nC_1{}^nC_1 + {}^nC_2{}^nC_2 - \cdots (-1)^n\, {}^nC_n{}^nC_n = \begin{cases} 0, n \text{ is odd} \\ (-1)^{n/2}\, {}^nC_{n/2}, \text{ if } n \text{ is even.} \end{cases}$$

(ii) Using part (a) of this question, we have

$^nC_0{}^nC_0 - {}^nC_1{}^nC_1 + {}^nC_2{}^nC_2 - \cdots + (-1)^n\, {}^nC_n{}^nC_n = (-1)^{n/2}\, {}^nC_{n/2}$

when n is even.

On replacing n by $2n$, we get

$$^{2n}C_0{}^{2n}C_0 - {}^{2n}C_1{}^{2n}C_1 + {}^{2n}C_2{}^{2n}C_2 - \cdots (-1)^{2n}\, {}^{2n}C_{2n}{}^{2n}C_{2n} = (-1)^n \cdot {}^{2n}C_n$$

$$\Rightarrow {}^{2n}C_0^2 - {}^{2n}C_1^2 + {}^{2n}C_2^2 - \cdots (-1)^{2n}\, {}^{2n}C_{2n}^2 = (-1)^n\, {}^{2n}C_n$$

Hence proved.

4. We have,

$$(1 + 2x + 2x^2)^n = C_0 + C_1x + C_2x^2 + \cdots + C_{2n} \cdot x^{2n} \quad \ldots(i)$$

Replacing x by $-x$, we get

$$(1 - 2x + 2x^2)^n = C_0 - C_1x + C_2x^2 - \cdots + C_{2n} \cdot x^{2n} \quad \ldots(ii)$$

On multiplying Eqs. (i) and (ii), we get

$$\Rightarrow (C_0 + C_1x + C_2x^2 + \cdots + C_{2n}x^{2n})(C_0 - C_1x + C_2x^2 \cdots + C_{2n} \cdot x^{2n})$$

$$= (1 + 2x + 2x^2)^n \cdot (1 - 2x + 2x^2)^n$$

$$= [(1 + 2x^2) - 4x^2]^n$$

Equating coefficient of x^{2n} in both sides, we get

$$\Rightarrow C_0C_{2n} - C_1C_{2n-1} + C_2C_{2n-2} + \cdots + C_{2n} \cdot C_0$$

$= $ Coefficient of x^{2n} in $(1 + 4x^4)^n$

Here, $(r + 1)$th term of $(1 + 4x^4)^n$

i.e. $t_{r+1} = {}^nC_r\,(4x^4)^r = {}^nC_r \cdot 4^r \cdot x^{4r}$,

$$\{4r = 2n \Rightarrow r = n/2\}$$

$$\therefore C_0C_{2n} - C_1 \cdot C_{2n-1} + C_2 \cdot C_{2n-2} \cdots + C_{2n} \cdot C_0 = \begin{cases} 0, & n \text{ is odd} \\ 4^{n/2} \cdot {}^nC_{n/2} & n \text{ is even} \end{cases}$$

5. (a) We have,

$$\left(x - \frac{^{50}C_1}{^{50}C_0}\right)\left(x - 2^2 \cdot \frac{^{50}C_2}{^{50}C_1}\right)\left(x - 3^2 \cdot \frac{^{50}C_3}{^{50}C_2}\right) \cdots \left(x - 50^2 \cdot \frac{^{50}C_{50}}{^{50}C_{49}}\right)$$

$$\Rightarrow x^{50} - x^{49}\left\{\frac{^{50}C_1}{^{50}C_0} + 2^2\frac{^{50}C_2}{^{50}C_1} + 3^2\frac{^{50}C_3}{^{50}C_2} + \cdots + 50^2\frac{^{50}C_{50}}{^{50}C_{49}}\right\} + \cdots$$

$\therefore$ Coefficient of x^{49}

$$= -\left\{\frac{^{50}C_1}{^{50}C_0} + 2^2\frac{^{50}C_2}{^{50}C_1} + 3^2\frac{^{50}C_3}{^{50}C_2} + \cdots + 50^2\frac{^{50}C_{50}}{^{50}C_{49}}\right\}$$

$$= -\left(\sum_{r=0}^{49}(r+1)^2\frac{^{50}C_{r+1}}{^{50}C_r}\right) = -\left(\sum_{r=0}^{49}(r+1)^2\frac{^{50}C_{r+1}}{^{50}C_r}\right)$$

$$= -\left[\sum_{r=0}^{49}(r+1)^2\frac{50!}{(r+1)!\,(50-r-1)!} \times \frac{r!\,(50-r)!}{50!}\right]$$

$$= -\left[\sum_{r=0}^{49}(r+1)^2\frac{(50-r)}{(r+1)}\right] = -\left[\sum_{r=0}^{49}(r+1)(50-r)\right]$$

$$= -\left[\sum_{r=0}^{49}(49r + 50 - r^2)\right]$$

$$= -\left[49\sum_{r=0}^{49}r + \sum_{r=0}^{49}50 - \sum_{r=0}^{49}r^2\right]$$

$$= -\left[49 \times \frac{49 \times 50}{2} + 50 \times 50 - \frac{49 \times 50 \times 99}{6}\right]$$

$= -[49 \times 49 \times 25 + 2500 - 49 \times 25 \times 33]$

$= -[25 \times 49\,(49 - 33) + 2500]$

$= -[25 \times 49 \times 16 + 2500]$

$= -22100$

Hence, option (a) is correct

6. (c) Given sum $\sum_{i=0}^{m}\binom{10}{i}\binom{20}{m-i}$, where $\binom{p}{q} = 0$, if $p < q$

To find Value of m, when the given sum is maximum

Now, $\sum_{i=0}^{m} {}^{10}C_i\, {}^{20}C_{m-i}$

$= {}^{10}C_0{}^{20}C_m + {}^{10}C_1{}^{20}C_{m-1} + {}^{10}C_2{}^{20}C_{m-2} + \cdots + {}^{10}C_m{}^{20}C_0$

$=$ Coefficient of x^m in the expansion of product $(1 + x)^{10}(1 + x)^{20}$

$=$ Coefficient of x^m in the expansion of $(1 + x)^{30} = {}^{30}C_m$

Now, ${}^{30}C_m$ will be maximum, when $m = 15$

Hence, option (c) is correct.

7. (a) Given,

$$\binom{30}{0}\binom{30}{10} - \binom{30}{1}\binom{30}{11} + \binom{30}{2}\binom{30}{12} - \cdots + \binom{30}{20}\binom{30}{30}$$

$= {}^{30}C_0{}^{30}C_{10} - {}^{30}C_1{}^{30}C_{11} + {}^{30}C_2{}^{30}C_{12} - \cdots + {}^{30}C_{20}{}^{30}C_{30}$

$= {}^{30}C_0{}^{30}C_{30-10} - {}^{30}C_1{}^{30}C_{30-11} + {}^{30}C_2{}^{30}C_{30-12} \cdots + {}^{30}C_{20}{}^{30}C_{30-30}$ $[\because {}^nC_r = {}^nC_{n-r}]$

$= {}^{30}C_0{}^{30}C_{20} - {}^{30}C_1{}^{30}C_{19} + {}^{30}C_2{}^{30}C_{18} - \cdots + {}^{30}C_{20}{}^{30}C_0$

$=$ Coefficient of x^{20} in the expansion of $(1 + x)^{30}(1 - x)^{30}$

$=$ Coefficient of x^{20} in the expansion of $(1 - x^2)^{30}$

Now, let $(r + 1)$th term contains x^{20} in $(1 - x^2)^{30}$

$\therefore \quad T_{r+1} = {}^{30}C_r(-x^2)^r = (-1)^r\, {}^{30}C_r x^{2r}$

To get x^{20} put $\quad 2r = 20$

$\Rightarrow \quad r = 10$

$\therefore$ Coefficient of x^{20} is ${}^{30}C_{10}$

Hence, ${}^{30}C_0{}^{30}C_{20} - {}^{30}C_1{}^{30}C_{19} + {}^{30}C_2{}^{30}C_{18} - \cdots + {}^{30}C_{20}{}^{30}C_0$

$$= {}^{30}C_{10} = \binom{30}{10}$$

Hence, option (a) correct.

8. (d) Since, A_r, B_r and C_r denote the coefficient of rth term in the expansion of $(1 + x)^{10}$, $(1 + x)^{20}$ and $(1 + x)^{30}$, respectively.

$\therefore \quad A_r = {}^{10}C_r,\ B_r = {}^{20}C_r$ and $C_r = {}^{30}C_r$

$$\therefore \sum_{r=1}^{10}A_r\,(B_{10}B_r - C_{10}A_r) = B_{10}\left(\sum_{r=1}^{10}A_rB_r\right) - C_{10}\left(\sum_{r=1}^{10}(A_r)^2\right)$$

$$= B_{10}\left(\sum_{r=1}^{10}{}^{10}C_r{}^{20}C_r\right) - C_{10}\left(\sum_{r=1}^{10}({}^{10}C_r)^2\right)$$

$[\because A_r = {}^{10}C_r,\ B_r = {}^{20}C_r$ and $C_r = {}^{30}C_r]$

$$= B_{10}\left(\sum_{r=1}^{10}{}^{10}C_r{}^{20}C_{20-r}\right) - C_{10}\left(\sum_{r=0}^{10}({}^{10}C_r)^2 - 1\right)$$

$[\because {}^nC_r = {}^nC_{n-r}$ and ${}^nC_0 = 1]$

$$= B_{10}\left(\sum_{r=0}^{10}{}^{10}C_r{}^{20}C_{20-r} - 1\right) - C_{10}\left(\sum_{r=0}^{10}({}^{10}C_r)^2 - 1\right)$$

$= B_{10} \times$ (coefficient of x^{20} in $(1 + x)^{10}(x + 1)^{20} - 1$)

$$- C_{10}\{{}^{20}C_{10} - 1\}\left[\because \sum_{r=0}^{10}{}^{10}C_r^2 = {}^{20}C_{10}\right]$$

$= B_{10} \times \{{}^{30}C_{20} - 1\} - C_{10}\{{}^{20}C_{10} - 1\}$

$= B_{10}\{{}^{30}C_{10} - 1\} - C_{10}\{{}^{20}C_{10} - 1\}$ $\quad [\because {}^nC_r = {}^nC_{n-r}]$

$= B_{10}(C_{10} - 1) - C_{10}(B_{10} - 1)$

$= B_{10}C_{10} - B_{10} - C_{10}B_{10} + C_{10} = C_{10} - B_{10}$

Hence option (d) is correct

9. (c) Given, $(1+x)^n = {}^nC_0 + {}^nC_1x + {}^nC_2x^2 + \cdots + {}^nC_nx^n$ ...(i)

Put $x = 1$ in Eq. (i), we get

$${}^nC_0 + {}^nC_1 + {}^nC_2 + \cdots + {}^nC_n = 2^n$$

$$\Rightarrow \sum_{r=0}^{n} C_r = 2^n \quad \text{...(ii)}$$

We have, $\sum_{r=0}^{n}\sum_{s=0}^{n} {}^nC_r \cdot {}^nC_s = \left(\sum_{r=0}^{n} {}^nC_r\right)\left(\sum_{s=0}^{n} C_s\right)$

$$= \sum_{r=0}^{n} {}^nC_r \cdot 2^n \quad \text{[using Eq. (ii)]}$$

$$= 2^n \cdot 2^n \quad \text{[using Eq. (iii)]}$$

$$= (2)^{2n}$$

Hence, option (c) is correct.

10. (a) Given,

$$(1+x)^n = {}^nC_0 + {}^nC_1x + {}^nC_2x^2 + \cdots + {}^nC_nx^n \quad \text{...(i)}$$

Put $x = 1$ in Eq. (i), we get

$$2^n = {}^nC_0 + {}^nC_1 + {}^nC_2 + \cdots + {}^nC_n$$

$$2^n = \sum_{r=0}^{n} C_r \quad \text{...(ii)}$$

Now, $\sum_{r=0}^{n}\sum_{s=0}^{n} C_rC_s = \left(\sum_{r=0}^{n} C_r\right)\left(\sum_{s=0}^{n} C_s\right) = 2^n \cdot 2^n$

$$\sum_{r=0}^{n}\sum_{s=0}^{n} C_rC_s = (2)^{2n} \quad \text{...(iii)}$$

We have, $\sum_{r=0}^{n}\sum_{s=0}^{n} C_rC_s = \left(\sum_{r=0}^{n} C_r^2\right) + 2 \sum\sum_{0 \le r < s \le n} C_rC_s$

$$\Rightarrow (2)^{2n} = {}^{2n}C_n + 2 \sum\sum_{0 \le r < s \le n} C_rC_s$$

[using Eq. (iii) and $\sum_{r=0}^{n} C_r^2 = {}^{2n}C_n$]

$$\Rightarrow \sum\sum_{0 \le r < s \le n} C_rC_s = \frac{(2)^{2n} - {}^{2n}C_n}{2}$$

Hence, option (a) is correct

11. (a) Given, $(1+x)^n = {}^nC_0 + {}^nC_1x + {}^nC_2x^2 + \cdots + {}^nC_nx^n$ ...(i)

Put $n = 1$ in Eq. (i), we get

$$\sum_{r=0}^{n} {}^nC_r = 2^n \quad \text{...(ii)}$$

We have,

$$\sum_{r=0}^{n}\sum_{s=0}^{n} ({}^nC_r + {}^nC_s) = \sum_{r=0}^{n}\sum_{s=0}^{n} {}^nC_r + \sum_{r=0}^{n}\sum_{s=0}^{n} {}^nC_s$$

$$= \sum_{s=0}^{n}\left(\sum_{r=0}^{n} {}^nC_r\right) + \sum_{r=0}^{n}\left(\sum_{s=0}^{n} {}^nC_s\right)$$

$$= \sum_{s=0}^{n} 2^n + \sum_{r=0}^{n} 2^n \quad \text{[using Eq. (ii)]}$$

$$= (n+1)\,2^n + (n+1)2^n$$

$$= (2n+2)2^n = (n+1)2^{n+1}$$

Hence, option (a) is correct.

12. (d) Given, $(1+x)^n = {}^nC_0 + {}^nC_1x + {}^nC_2x^2 + \cdots + {}^nC_nx^n$ ...(i)

We have,

$$\sum_{r=0}^{n}\sum_{s=0}^{n} (C_r + C_s) = \sum_{r=0}^{n} (C_r + C_r) + 2 \sum\sum_{0 \le r < s \le n} (C_r + C_s)$$

$$(n+1)2^{n+1} = 2\sum_{r=0}^{n} (C_r) + 2 \sum\sum_{0 \le r < s \le n} (C_r + C_s)$$

[using the result obtained in Q. 11]

$$\Rightarrow (n+1)2^{n+1} = 2 \cdot 2^n + 2 \sum\sum_{0 \le r < s \le n} (C_r + C_s) \quad \left[\because \sum_{r=0}^{n} C_r = 2^n\right]$$

$$\Rightarrow (n+1)2^{n+1} - 2 \cdot 2^n = 2 \sum\sum_{0 \le r < s \le n} (C_r + C_s)$$

$$\Rightarrow n \cdot 2^{n+1} = 2 \sum\sum_{0 \le r < s \le n} (C_r + C_s)$$

$$\Rightarrow \sum\sum_{0 \le r < s \le n} (C_r + C_s) = n \cdot 2^n$$

Hence, option (d) is correct.

13. (a) Here, $\sum_{r=0}^{s}\sum_{s=1}^{n} {}^nC_s \cdot {}^sC_r$

$$\Rightarrow \sum_{s=1}^{n} {}^nC_s\,[{}^sC_0 + {}^sC_1 + {}^sC_2 + \cdots + {}^sC_s] \Rightarrow \sum_{s=1}^{n} {}^nC_s \cdot 2^s$$

$$\Rightarrow (1+2)^n - 1 \Rightarrow 3^n - 1.$$

14. (c) Given,

$$(1 + x + x^2)^n = a_0 + a_1x + a_2x^2 + \cdots + a_{2n}x^{2n}$$

$$= \sum_{r=0}^{2n} a_rx^r \quad \text{...(i)}$$

Put of $x = 1$ in Eq. (i), we get

$$\sum_{r=0}^{2n} a_r = 3^n \quad \text{...(ii)}$$

On replacing x by $\frac{1}{x}$, we have

$$\left(1 + \frac{1}{x} + \frac{1}{x^2}\right)^n = \sum_{r=0}^{2n} \frac{a_r}{x^r}$$

$$\Rightarrow \frac{(1+x+x^2)^n}{x^{2n}} = \sum_{r=0}^{2n} \frac{a_r}{x^r}$$

$$\Rightarrow (1+x+x^2)^n = \sum_{r=0}^{2n} \frac{a_r}{x^r} \cdot x^{2n}$$

$$\Rightarrow (1+x+x^2)^n = \sum_{r=0}^{2n} a_rx^{2n-r} \quad \text{...(iii)}$$

From Eqs. (i) and (iii), we get

$$\sum_{r=0}^{2n} a_rx^r = \sum_{r=0}^{2n} a_rx^{2n-r}$$

Equating coefficient of x^{2n-r} on both side, we have

$$\Rightarrow a_{2n-r} = a_r \text{ for } 0 \le r \le 2n$$

$$\Rightarrow \sum_{r=0}^{n-1} a_r = \sum_{r=0}^{n-1} a_{2n-r}$$

$$\Rightarrow a_0 + a_1 + a_{n-1} = a_{2n} + a_{2n-1} + \cdots + a_{n+1}$$

On adding both side $a_0 + a_1 + a_2 + a_3 + \cdots + a_n$, we have

$$\Rightarrow 2(a_0 + a_1 + a_2 + \cdots + a_{n-1}) + a_n = a_0 + a_1 + a_2 + a_3 + \cdots + a_{2n}$$

$$\Rightarrow 2(a_0 + a_1 + a_2 + \cdots + a_{n-1}) + a_n = \sum_{r=0}^{n} a_r$$

$$\Rightarrow 2(a_0 + a_1 + a_2 + \cdots + a_{n-1}) + a_n = 3^n \quad \text{[using Eq. (iii)]}$$

$$\Rightarrow a_0 + a_1 + a_2 + \cdots + a_{n-1} = \frac{3^n - a_n}{2}$$

Hence, option (c) is correct.

15. (a,b,c,d) Given,

$$1 + (1+x) + (1+x)^2 + \cdots + (1+x)^n = \sum_{k=0}^{n} a_k x^k$$

$$\Rightarrow \quad \frac{[(1+x)^{n+1} - 1]}{1+x-1} = \sum_{k=0}^{n} a_k x^k$$

$$\left[\because 1 + r + r^2 + r^3 + \cdots + r^n = \frac{r^{n+1}-1}{r-1}\right]$$

$$\Rightarrow \quad \frac{[(1+x)^{n+1} - 1]}{x} = \sum_{k=0}^{n} a_k x^k \quad \ldots(i)$$

Put $x = 1$ in Eq. (i), we get

$$\Rightarrow \quad 2^{n+1} - 1 = \sum_{k=0}^{n} a_k$$

Hence, option (a) is correct.

Also, coefficient of $x^k \Rightarrow a_k = {}^{n+1}C_{k+1}$

$$\therefore \quad a_{n-2} = {}^{n+1}C_{n-1} = \frac{n(n+1)}{2}$$

$\Rightarrow$ (b) is correct.

Also, $a_p > a_{p-1} \Rightarrow p < \frac{n}{2}$

$\therefore$ (c) is correct.

Also, $(a_9)^2 - (a_8)^2 = {}^{n+2}C_{10}({}^{n+1}C_{10} - {}^{n+1}C_9)$

$\Rightarrow$ (d) is correct.

16. (a, b, d) Given,

$$f(m) = \sum_{i=0}^{m} {}^{30}C_{30-i}\, {}^{20}C_{m-i}$$

$$= \sum_{i=0}^{m} \binom{30}{30-i}\binom{20}{m-i}$$

$$= {}^{30}C_{30}\, {}^{20}C_m + {}^{30}C_{29}\, {}^{20}C_{m-1} + \cdots + {}^{30}C_{30-m} \times {}^{20}C_0$$

$$= {}^{30}C_0\, {}^{20}C_m + {}^{30}C_1\, {}^{20}C_{m-1} + \cdots + {}^{30}C_m\, {}^{20}C_0 \quad [\because {}^nC_r = {}^nC_{n-r}]$$

= Coefficient of x^m in the expansion of product $(1+x)^{30}(x+1)^{20}$

= Coefficient of x^m in the expansion of $(1+x)^{50} = {}^{50}C_m$

Now, ${}^{50}C_m$ will be maximum when $m = 25$

Also, $f(0) + f(1) + f(2) + \cdots + f(50)$

$$= {}^{50}C_0 + {}^{50}C_1 + {}^{50}C_2 + \cdots + {}^{50}C_{50} = 2^{50} \quad \left[\therefore \sum_{r=0}^{n} C_r = 2^n\right]$$

Also, ${}^{50}C_m$ is not divisible by 50 for any m as 50 is not a prime number.

$$\sum_{i=0}^{m} [f(m)]^2 = ({}^{50}C_0)^2 + ({}^{50}C_1)^2 + ({}^{50}C_2)^2 + \cdots + ({}^{50}C_{50})^2 = {}^{100}C_{50}$$

$$[\because C_0^2 + C_1^2 + C_2^2 + \cdots + C_n^2 = {}^{2n}C_n]$$

Hence, option (a), (b) and (d) is correct.

17. (a, b, c, d)

$${}^{1000}C_{50} + {}^{999}C_{49} + {}^{998}C_{48} + \cdots + {}^{950}C_0$$

$\Rightarrow$ Coefficient of x^{950} in

$$\{(1+x)^{950} + (1+x)^{951} + \cdots + (1+x)^{1000}\}$$

$\Rightarrow$ Coefficient of x^{950} in

$$(1+x)^{950} \cdot \frac{\{(1+x)^{51} - 1\}}{(1+x) - 1}$$

$\Rightarrow$ Coefficient of x^{951} in

$$\{(1+x)^{1001} - (1+x)^{950}\}$$

$$\Rightarrow \quad {}^{1001}C_{951} = {}^{1001}C_{50} = {}^{1002}C_{951} - {}^{1001}C_{51}$$

$$= {}^{1002}C_{51} - {}^{1001}C_{950}$$

18. (b) $C_0 + C_1x + C_2x^2 + \cdots + C_nx^n = (1+x)^n$

On differentiate w.r.t. x, we get

$\Rightarrow C_1 + 2\cdot C_2x + 3\cdot C_3x^2 + \cdots + n\cdot C_nx^{n-1} = n(1+x)^{n-1}$

Put, $x = i$

$(C_1 - 3\cdot C_3 + 5\cdot C_5 - \cdots) + i(2\cdot C_2 - 4\cdot C_4 + \cdots) = n(1+i)^{n-1}$

Taking modulus on both sides and squaring, we get

$$(C_1 - 3C_3 + 5C_5 \ldots)^2 + (2\cdot C_2 - 4\cdot C_4 \ldots)^2 = n^2\cdot 2^{n-1}$$

19. (c) We have,

$$\frac{{}^nC_0}{n(n+1)} - \frac{{}^nC_1}{(n+1)(n+2)} + \frac{{}^nC_2}{(n+2)(n+3)} - \cdots \text{ upto } (n+1) \text{ term}$$

$$= \frac{{}^nC_0}{n(n+1)} - \frac{{}^nC_1}{(n+1)(n+2)} + \frac{{}^nC_2}{(n+2)(n+3)} - \cdots + (-1)^n \frac{{}^nC_n}{2n(2n+1)}$$

$$= \left(\frac{{}^nC_0}{n} - \frac{{}^nC_0}{n+1}\right) - \left(\frac{{}^nC_1}{n+1} - \frac{{}^nC_1}{n+2}\right) + \cdots (-1)^n \left(\frac{{}^nC_n}{2n} - \frac{{}^nC_n}{2n+1}\right)$$

$$= \frac{{}^nC_0}{n} - \frac{{}^nC_1}{n+1} + \frac{{}^nC_2}{n+2} - \cdots (-1)^n \frac{{}^nC_n}{2n}$$

$$- \left[\frac{{}^nC_0}{n+1} - \frac{{}^nC_1}{n+2} + \cdots (-1)^n \frac{{}^nC_n}{2n+1}\right] \quad \ldots(i)$$

Now,

$$\therefore \quad (1-x)^n = {}^nC_0 - {}^nC_1x + {}^nC_2x^2 - \cdots + (-1)^n\, {}^nC_nx^n$$

On multiplying both side by x^{n-1}, we get

$${}^nC_0x^{n-1} - {}^nC_1x^n + {}^nC_2x^{n+1} - \cdots + (-1)^n\, {}^nC_nx^{2n-1} = x^{n-1}(1-x)^n$$

On integrating both side between the limit 0 to 1, we get

$$\left[{}^nC_0\frac{x^n}{n} - {}^nC_1\frac{x^{n+1}}{n+1} + C_2\frac{x^{n+2}}{n+2} - \cdots (-1)^n\, {}^nC_n\frac{x^{2n}}{2n}\right]_0^1$$

$$= \int_0^1 x^{n-1}(1-x)^n dx \quad \ldots(ii)$$

Using Eq. (ii) in Eq. (i), we have

$$\frac{{}^nC_0}{n(n+1)} - \frac{{}^nC_1}{(n+1)(n+2)} + \cdots (-1)^n \frac{{}^nC_n}{2n(2n+1)}$$

$$= \int_0^1 x^{n-1}(1-x)^n dx - \int_0^1 x^n(1-x)^n dx$$

$$= \int_0^1 x^{n-1}(1-x)^n(1-x)dx$$

$$= \int_0^1 x^{n-1}(1-x)^{n+1}dx$$

20. (a) $\displaystyle\sum_{0 \le i < f \le n}\sum \left(\frac{1}{{}^nC_i} + \frac{1}{{}^nC_j}\right) = n \cdot \sum_{r=0}^{n} \frac{1}{{}^nC_r} = na$

DPP-5 Greatest Term, Binomial Theorem for any Index and Multinomial Theorem

1. (i) Let, r th term be the greatest term in the expansion of $(2 + 3x)^9$

Now, $T_r = {}^9C_{r-1}(2)^{9-r+1}(3x)^{r-1}$

$= {}^9C_{r-1}(2)^{10-r}(3x)^{r-1}$

and $T_{r+1} = {}^9C_r(2)^{9-r}(3x)^r$

$T_{r-1} = {}^9C_{r-2}(2)^{11-r}(3x)^{r-2}$

Since, r th term is greatest term

$$\therefore \quad \left|\frac{T_r}{T_{r+1}}\right| \geq 1$$

$$\therefore \quad \left|\frac{{}^9C_{r-1}2^{10-r}(3x)^{r-1}}{{}^9C_r 2^{9-r}(3x)^r}\right| \geq 1$$

$$\Rightarrow \quad \left|\frac{r}{10-r}\right| \geq \left|\frac{3x}{2}\right|$$

$$\Rightarrow \quad \frac{r}{10-r} \geq \frac{3}{2} \times \frac{3}{2} \qquad \left[\because x = \frac{3}{2}\right]$$

$$\Rightarrow \quad 4r \geq 90 - 9r$$

$$\Rightarrow \quad 13r \geq 90$$

$$r \geq \frac{90}{13} = 6\frac{12}{13} \quad \text{and} \quad \left|\frac{T_r}{T_{r-1}}\right| \geq 1$$

$$\therefore \quad \left|\frac{{}^9C_{r-1}2^{10-r}(3x)^{r-1}}{{}^9C_{r-2}(2)^{11-r}(3x)^{r-2}}\right| \geq 1$$

$$\Rightarrow \quad \left|\frac{11-r}{r-1}\right| \geq \left|\frac{2}{3x}\right|$$

$$\Rightarrow \quad \frac{11-r}{r-1} \geq \frac{4}{9} \qquad [\because x = 3/2]$$

$$\Rightarrow \quad 99 - 9r \geq 4r - 4$$

$$\Rightarrow \quad 103 \geq 13r$$

$$\Rightarrow \quad r \leq \frac{103}{13} = 7\frac{12}{13}$$

$$\therefore \quad 6\frac{12}{13} \leq r \leq 7\frac{12}{13}$$

$$r = 7$$

Hence, 7th term is the greatest term

$$\therefore \quad T_7 = {}^9C_6 2^3 \left(\frac{9}{2}\right)^6 = \frac{9 \times 8 \times 7}{1 \times 2 \times 3} \times \frac{3^{12}}{2^3} = \frac{7 \cdot 3^{13}}{2}$$

(ii) Let, r th term be the greatest term of the expansion $(3 - 5x)^{15}$.

Now, $T_r = {}^{15}C_{r-1}(3)^{16-r}(-5x)^{r-1}$

and $T_{r+1} = {}^{15}C_r(3)^{15-r}(-5x)^r$

Since, r th term is greatest

$$\therefore \quad \left|\frac{T_r}{T_{r+1}}\right| \geq 1$$

$$\Rightarrow \quad \left|\frac{{}^{15}C_{r-1}(3)^{16-r}(-5x)^{r-1}}{{}^{15}C_r(3)^{15-r}(-5x)^r}\right| \geq 1$$

$$\Rightarrow \quad \left|\frac{r}{16-r}\right| \geq \left|\frac{-5x}{3}\right|$$

$$\Rightarrow \quad \left|\frac{r}{16-r}\right| \geq \left|\frac{-1}{3}\right| \qquad \left[\because x = \frac{1}{5}\right]$$

$$\Rightarrow \quad 3r \geq 16 - r$$

$$4r \geq 16$$

$$r \geq 4$$

and $\left|\frac{T_r}{T_{r-1}}\right| \geq 1 \Rightarrow \left|\frac{{}^{15}C_{r-1}(3)^{16-r}(-5x)^{r-1}}{{}^{15}C_{r-2}(3)^{17-r}(-5x)^{r-2}}\right| \geq 1$

$$\Rightarrow \quad \left|\frac{17-r}{r-1}\right| \geq \left|\frac{3}{5x}\right|$$

$$\Rightarrow \quad \left|\frac{17-r}{r-1}\right| \geq 3 \qquad \left[\because x = \frac{1}{5}\right]$$

$$\Rightarrow \quad 17 - r \geq 3r - 3 \Rightarrow 4r \leq 20$$

$$r \leq 5$$

$\therefore$ The greatest term be T_5 or T_4

$$\therefore \quad T_5 = {}^{15}C_4(3)^{11}(-5x)^4 = \frac{15!}{4!\,11!}(3)^{11}(-1)^4 \qquad \left[\because x = \frac{1}{5}\right]$$

$$= \frac{15 \times 14 \times 13 \times 12}{1 \times 2 \times 3 \times 4}(3)^{11} = 455 \times 3^{12}$$

2. (i) The coefficient of $x^2y^3z^4$ in the expansion of $(ax - by + cz)^9$ is

$$\frac{9!}{2!\,3!\,4!}a^2(-b)^3(c)^4 = -1260a^2b^3c^4$$

(ii) The coefficient of $a^2b^3c^4d$ in the expansion of $(a - b - c + d)^{10}$ is

$$\frac{10!}{2!\,3!\,4!\,1!}(1)^2(-1)^3(-1)^4(1)^1 = -12600$$

3. (i) We have, $(1 + x + x^2 + x^3)^{11}$

$\Rightarrow [(1 + x)(1 + x^2)]^{11} = (1 + x)^{11}(1 + x^2)^{11}$

$\Rightarrow [1 + {}^{11}C_1x + {}^{11}C_2x^2 + {}^{11}C_3x^3 + {}^{11}C_4x^4 + \cdots]$

$[1 + {}^{11}C_1x^2 + {}^{11}C_2x^4 + \cdots]$

Coefficient of x^4 in the above expansion is

$${}^{11}C_2 + {}^{11}C_2 \cdot {}^{11}C_1 + {}^{11}C_4 = 55 + 55 \times 11 + 330 = 990$$

(ii) We have, $(2 - x + 3x^2)^6$

General term of the expansion is

$$\frac{6!}{a!\,b!\,c!}(2)^a(-x)^b(3x^2)^c$$

$$\Rightarrow \quad \frac{6!}{a!\,b!\,c!}2^a(-1)^b(3)^c x^{b+2c}$$

Coefficient of x^4, if $b + 2c = 4$ and $a + b + c = 6$

(a) $a = 4, b = 0, c = 2$

(b) $a = 3, b = 2, c = 1$

(c) $a = 2, b = 4, c = 0$

$\therefore$ Coefficient of x^4 is

$$\frac{6!}{4!\,0!\,2!}(2)^4 \cdot 3^2 + \frac{6!}{3!\,2!\,1!}(2)^3 \cdot (3)^1 + \frac{6!}{2!\,4!\,0!}(2)^2(3)^0$$

$$= 2160 + 1440 + 60 = 3660$$

4. (a) We have, $(ab + bc + ca)^8$

The general term is $\frac{8!}{x!\,y!\,z!}(ab)^x(bc)^y(ca)^z$

$$\Rightarrow \quad \frac{8!}{x!\,y!\,z!}a^{x+z}b^{x+y}c^{y+z}$$

Coefficient of $a^3b^4c^7$ in the above expansion is possible if

$$x + z = 3,\ x + y = 4,\ y + z = 7$$

$$\therefore \quad x = 0,\ y = 4,\ z = 3$$

$\therefore$ Coefficient of $(ab + bc + ca)^8$ is

$$\frac{8!}{0!\,4!\,3!} = 280$$

Hence, option (a) is correct.

5. (a) We know that, the total number of terms in the expansion $(x_1 + x_2 + x_3 + \cdots + x_m)^n$ is ${}^{m+n-1}C_n$

$\therefore$ The number of terms in the expansion of $(a + b + c + d + e + f)^n$ is ${}^{6+n-1}C_n = {}^{n+5}C_n$

Hence, option (a) is correct.

6. (d) The expansion of $\dfrac{1}{(4-3x)^{1/2}}$ is valid iff $\left|\dfrac{3x}{4}\right| < 1$

$$\Rightarrow \quad |3x| < 4$$
$$|x| < \frac{4}{3}$$

Hence, option (d) is correct.

7. (b) We have, $\dfrac{1}{(6-3x)^{1/3}}$

$$\Rightarrow \quad (6-3x)^{-1/3}$$

$$\Rightarrow \quad 6^{-1/3}\left[1 - \frac{3x}{6}\right]^{-1/3}$$

$$\Rightarrow 6^{-1/3}\left[1 + \frac{1}{3} \times \frac{3}{6}x + \frac{\frac{-1}{3}\left(\frac{-1}{3} - 1\right)}{2!}\left(\frac{3x}{6}\right)^2 - \frac{\left(\frac{-1}{3}\right)\left(\frac{-1}{3} - 1\right)\left(\frac{-1}{3} - 2\right)}{3!}\left(\frac{3x}{6}\right)^3 + \cdots\right]$$

$$\Rightarrow \quad 6^{-1/3}\left[1 + \frac{x}{6} + \frac{2x^2}{6^2} + \cdots\right]$$

Hence, option (b) is correct.

8. (a) The general term of the expansion $(1-3x)^{\frac{-1}{3}}$ is

$$\frac{\frac{-1}{3}\left(\frac{-1}{3} - 1\right)\left(\frac{-1}{3} - 2\right)\cdots\left(\frac{-1}{3} - (r-1)\right)}{r!}(-3x)^r$$

$$= \frac{(-1)^r\left[\frac{1}{3} \times \frac{4}{3} \times \frac{7}{3} \times \cdots \times \left(\frac{3r-2}{3}\right)\right]}{r!}(-1)^r\, 3^r\, x^r$$

$$\Rightarrow \quad \frac{1 \cdot 4 \cdot 7 \cdots (3r-2)}{3^r \cdot r!}(3)^r\, x^r$$

$$\Rightarrow \quad \frac{1 \cdot 4 \cdot 7 \cdots (3r-2)}{r!}x^r$$

Hence, option (a) is correct.

9. (b) We have,

$$(1 + 3x)^{1/2}\,(1 - 2x)^{-1/3}$$

$$\Rightarrow \left[1 + \frac{1}{2} \times 3x + \frac{\left(\frac{1}{2}\right)\left(\frac{1}{2} - 1\right)(3x)^2}{2!} + \cdots\right]\left[1 + \frac{1}{3}(2x) + \frac{\frac{1}{3}\cdot\frac{4}{3}}{2!}(-2x)^2 - \cdots\right]$$

$\Rightarrow$ Coefficient of x^2 is

$$\frac{4 \times 4}{9 \times 2!} + \frac{1}{2} \times 3 \times \frac{1}{3} \times 2 + \frac{\left(\frac{1}{2}\right)\left(\frac{-1}{2}\right)}{2!} \times 9 = \frac{16}{18} + 1 - \frac{9}{8}$$

$$= \frac{64 + 72 - 81}{72}$$

$$= \frac{55}{72}$$

Hence, option (b) is correct.

10. (a) The general term of the expansion $\left(\frac{1}{2}y^{1/3} + y^{-1/5}\right)^8$

$$T_{r+1} = {}^8C_r\left(\frac{1}{2}y^{1/3}\right)^{8-r}(y)^{-r/5} = {}^8C_r\left(\frac{1}{2}\right)^{8-r}(y)^{\frac{8-r}{3} - \frac{r}{5}}$$

The term is independent of y

if $$\frac{8-r}{3} - \frac{r}{5} = 0$$

$$\Rightarrow \quad 40 - 5r - 3r = 0$$

$$\Rightarrow \quad r = 5$$

$\therefore$ The term independent of y is 6th term

Hence, option (a) is correct.

11. (d) We have, $\quad 1 + 2x + 3x^2 + 4x^3 + \cdots$

$$\Rightarrow \quad (1-x)^{-2}$$

$\therefore$ Square root of $(1-x)^{-2}$, i.e.

$$[(1-x)^{-2}]^{1/2} = (1-x)^{-1}$$

Hence, option (d) is correct.

12. (c) We have, $\quad (1 + 2x + 3x^2 + 4x^3 + \cdots)^{1/2}$

$$\Rightarrow \quad [(1-x)^{-2}]^{1/2} = (1-x)^{-1}$$

$$(1-x)^{-1} = 1 + x + x^2 + x^3 + \cdots$$

$\therefore$ Coefficient of x^n is 1.

Hence, option (c) is correct.

13. (d) We have, $\quad (1 + x + x^2 + x^3 + \cdots)^2$

$$\Rightarrow \quad = [(1-x)^{-1}]^2$$

$$\Rightarrow \quad = (1-x)^{-2}$$

$$\Rightarrow \quad 1 + 2x + 3x^2 + 4x^3 + \cdots + (n+1)\,x^n + \cdots$$

Hence, coefficient of x^n is $(n+1)$.

Hence, option (d) is correct.

14. (b) We have, $(1-x)^{-4}$

The $(r+1)$ th term is

$${}^{r+4-1}C_{4-1}\,x^r \Rightarrow {}^{r+3}C_3\,x^r$$

$$\Rightarrow \quad \frac{(r+3)!}{3!\,(r+3-3)!}x^r = \frac{(r+3)\,(r+2)\,(r+1)}{6}x^r$$

Hence, option (b) is correct.

15. (a) We have, $\left(\dfrac{1+x}{1-x}\right)^2$

$$\Rightarrow \quad (1+x)^2\,(1-x)^{-2}$$

$$\Rightarrow [1 + 2x + x^2]\,[1 + 2x + 3x^2 + \cdots + (n-1)\,x^{n-2} + nx^{n-1} + (n+1)\,x^n + \cdots]$$

The coefficient of x^n is

$$(n+1) + 2(n) + (n-1) = 4n$$

Hence, option (a) is correct.

16. (a) We have, $\quad 1 + \dfrac{1}{4} + \dfrac{1 \cdot 3}{4 \cdot 8} + \dfrac{1 \cdot 3 \cdot 5}{4 \cdot 8 \cdot 12} + \cdots$

Given, series is $$1 + \frac{\left(\frac{-1}{2}\right)}{1!}\left(\frac{-1}{2}\right) + \frac{\left(\frac{-1}{2}\right)\left(\frac{-1}{2} - 1\right)}{2!}\left(-\frac{1}{2}\right)^2 + \frac{\left(\frac{-1}{2}\right)\left(\frac{-1}{2} - 1\right)\left(\frac{-1}{2} - 2\right)}{3!}\left(-\frac{1}{2}\right)^3 + \cdots$$

$$\Rightarrow \quad \left(1 - \frac{1}{2}\right)^{-1/2} = \left(\frac{1}{2}\right)^{-1/2} = (2)^{1/2} = \sqrt{2}$$

Hence, option (a) is correct.

17. (c) We have, $y = \frac{2}{5} + \frac{1\cdot 3}{2!}\left(\frac{2}{5}\right)^2 + \frac{1\cdot 3\cdot 5}{3!}\left(\frac{2}{5}\right)^3 + \cdots$

$$\Rightarrow \quad 1 + y = 1 + \left(\frac{2}{5}\right) + \frac{1\cdot 3}{2!}\left(\frac{2}{5}\right)^2 + \frac{1\cdot 3\cdot 5}{3!}\left(\frac{2}{5}\right)^3 + \cdots$$

Let $1 + y = (1 + x)^n = 1 + nx + \frac{n(n-1)}{2!}x^2 + \cdots$

On equating the given series, we get

$$nx = \frac{2}{5} \quad \ldots(i)$$

and $$\frac{n(n-1)}{2}x^2 = \frac{1\cdot 3}{2!}\left(\frac{2}{5}\right)^2 \quad \ldots(ii)$$

On solving Eqs. (i) and (ii), we get

$$x = -\frac{4}{5} \text{ and } n = -\frac{1}{2}$$

$$\therefore \quad 1 + y = \left(1 - \frac{4}{5}\right)^{-1/2}$$

$$\Rightarrow \quad (y + 1) = (5)^{1/2}$$

On squaring both sides, we get

$$y^2 + 2y + 1 = 5$$

$$\Rightarrow \quad y^2 + 2y - 4 = 0$$

$\therefore$ The value of $y^2 + 2y - 4 = 0$

Hence, option (c) is correct.

18. (b) We have, $\dfrac{\sqrt{1+x} + \sqrt[3]{(1-x)^2}}{1 + x + \sqrt{1+x}}$

$$\Rightarrow \quad \frac{(1+x)^{1/2} + (1-x)^{2/3}}{1 + x + (1+x)^{1/2}}$$

$$\Rightarrow \quad \frac{\left(1 + \frac{1}{2}x\right) + \left(1 - \frac{2}{3}x\right)}{1 + x + 1 + \frac{1}{2}x}$$

[$\because x^2$ and higher powers of x neglected]

$$\Rightarrow \quad \left[\frac{2 - \frac{1}{6}x}{2 + \frac{3}{2}x}\right] \Rightarrow \frac{1 - \frac{1}{12}x}{1 + \frac{3}{4}x}$$

$$\Rightarrow \quad \left(1 - \frac{1}{12}x\right)\left(1 + \frac{3}{4}x\right)^{-1} \Rightarrow \left(1 - \frac{1}{12}x\right)\left(1 - \frac{3}{4}x\right)$$

$$\Rightarrow \quad 1 - \frac{1}{12}x - \frac{3}{4}x \Rightarrow 1 - \frac{5x}{6}$$

Hence, option (b) is correct.

19. (c) We have, $\dfrac{1}{(1-ax)(1-bx)}$

$$\Rightarrow \quad (1-ax)^{-1}(1-bx)^{-1}$$

$$\Rightarrow (1 + ax + a^2x^2 \cdots)(1 + bx + b^2x^2 \cdots)$$

Hence, a_n is the coefficient of x^n of $(1-ax)^{-1}(1-bx)^{-1}$

$$= a^0b^n + a^1b^{n-1} + \cdots + a^nb^0$$

$$= a^0b^n\left[1 + \frac{a}{b} + \left(\frac{a}{b}\right)^2 + \cdots + \left(\frac{a}{b}\right)^n\right]$$

$$\Rightarrow \quad a^0b^n \frac{\left[\left(\frac{a}{b}\right)^{n+1} - 1\right]}{\frac{a}{b} - 1}$$

$$\Rightarrow \quad \frac{b^n(a^{n+1} - b^{n+1})}{a - b} \times \frac{b}{b^{n+1}}$$

$$\Rightarrow \quad \frac{a^{n+1} - b^{n+1}}{a - b} \Rightarrow \frac{b^{n+1} - a^{n+1}}{b - a}$$

Hence, option (c) is correct.

20. (a, b) 4th term of the given expansion $\left(2 + \frac{3x}{8}\right)^{10}$

i.e. $$T_4 = {}^{10}C_3(2)^7\left(\frac{3x}{8}\right)^3$$

T_4 is the maximum numerical value

$$\therefore \quad T_4 \geq T_5$$

$$\left|\frac{T_4}{T_5}\right| \geq 1 \Rightarrow \left|\frac{{}^{10}C_3 2^7 \left(\frac{3x}{8}\right)^3}{{}^{10}C_4 2^6 \left(\frac{3x}{8}\right)^4}\right| \geq 1$$

$$\frac{4}{7} \geq \left|\frac{3x}{16}\right| \Rightarrow |x| \leq \frac{64}{21} \quad \ldots(i)$$

and $$\left|\frac{T_4}{T_3}\right| \geq 1$$

$$\Rightarrow \left|\frac{{}^{10}C_3 2^7\left(\frac{3x}{8}\right)^3}{{}^{10}C_2 2^8\left(\frac{3x}{8}\right)^2}\right| \geq 1 \Rightarrow \frac{8}{3} \geq \left|\frac{2 \times 8}{3x}\right|$$

$$|x| \geq 2 \quad \ldots(ii)$$

From Eqs. (i) and (ii), $2 \leq |x| \leq \frac{64}{21}$

Hence, value of x be

$$x \in \left[\frac{-64}{21}, -2\right] \cup \left[2, \frac{64}{21}\right]$$

Hence, options (a) and (b) are correct.

21. (a, b, c, d) We have, $(x + y + z)^{10}$

(a) The total number of terms is ${}^{3+10-1}C_{10} = {}^{12}C_{10} = \frac{12!}{10!\,2!} = 66$

(b) Coefficient of $x^3y^2z^5$ of the given expansion is $\frac{10!}{3!\,2!\,5!}$.

(c) The sum of the radical of $x^2y^3z^4$ is $2 + 3 + 4 = 9$ is not equal to 10.

Since, $x^2y^3z^4$ cannot be term of the $(x + y + z)^{10}$.

(d) $\because (x + y + z)^{10} = \sum_{\alpha+\beta+\gamma=10} \frac{10!}{\alpha!\,\beta!\,\gamma!} x^\alpha y^\beta z^\gamma$

Put $x = y = z = 1$ both sides, we have

$$\Rightarrow \quad \sum_{\alpha+\beta+\gamma=10} \frac{10!}{\alpha!\,\beta!\,\gamma!} = 3^{10}$$

Hence, (d) is correct.

22. (a,b,d)

(a) We have, $\dfrac{(1+2x)^n + (1+3x)^n}{1-x}$

$$\Rightarrow (1+2x)^n(1-x)^{-1} + (1+3x)^n(1-x)^{-1}$$

$$\Rightarrow [1 + {}^nC_1(2x) + {}^nC_2(2x)^2 + \cdots][1 + x + x^2 + \cdots]$$
$$+ [1 + {}^nC_1(3x) + {}^nC_2(3x)^2 + \cdots][1 + x + x^2 + \cdots]$$

Coefficient of x^n is

$$(1 + {}^nC_1 2 + {}^nC_2 2^2 + {}^nC_3 2^3 + \cdots + {}^nC_n 2^n)$$
$$+ (1 + {}^nC_1 3 + {}^nC_2 3^2 + \cdots + {}^nC_n 3^n)$$

$$\Rightarrow \quad (1+2)^n + (1+3)^n = 3^n + 4^n$$

(a) is true.

(b) We have, $\sum_{r=0}^{n} (-1)^r\, {}^nC_r \left[\frac{1}{2^r} + \frac{3^r}{2^{2r}} + \frac{7^r}{2^{3r}} + \cdots \infty\right]$

$$\Rightarrow \sum_{r=0}^{n} (-1)^r\, {}^nC_r\left(\frac{1}{2}\right)^r + \sum_{r=0}^{n}(-1)^r\, {}^nC_r\left(\frac{3}{4}\right)^r$$
$$+ \sum_{r=0}^{n}(-1)^r\, {}^nC_r\left(\frac{7}{8}\right)^r + \cdots$$

$$\Rightarrow \left(1-\frac{1}{2}\right)^n + \left(1-\frac{3}{4}\right)^n + \left(1-\frac{7}{8}\right)^n + \cdots \infty$$

$$\Rightarrow \left(\frac{1}{2}\right)^n + \left(\frac{1}{4}\right)^n + \left(\frac{1}{8}\right)^n + \cdots \quad \left[\because \sum_{r=0}^{n} (-1)^r\, {}^nC_r x^r = (1-x)^n\right]$$

$$\Rightarrow \frac{1}{2^n} + \frac{1}{2^{2n}} + \frac{1}{2^{3n}} + \cdots \Rightarrow \frac{\frac{1}{2^n}}{1-\frac{1}{2^n}} = \frac{1}{2^n - 1}$$

Hence, (b) is true.

(c) We have, $(1 + x + x^2)^{10} \Rightarrow \left(\frac{1-x^3}{1-x}\right)^{10}$

$$\Rightarrow (1-x^3)^{10}(1-x)^{-10}$$

$$\Rightarrow (1 - {}^{10}C_1 x^3 + {}^{10}C_2 x^6 + \cdots)(1-x)^{-10}$$

Coefficient of x^2 is depend on $(1-x)^{-10}$.

$\therefore$ Coefficient of x^2 of $(1-x)^{-10}$ is

$${}^{2+10-1}C_{10-1} = {}^{11}C_9$$

$${}^{11}C_9 \neq {}^{12}C_{10}$$

Hence, (c) is false.

(d) Coefficient of x^2 in $(1 + x + x^2)^{10}$ is ${}^{11}C_9 = {}^{11}C_2$

$\therefore$ (d) is true.

23. (b) We have, $(1 + x + x^2)^n = a_0 + a_1 x + a_2 x^2 + \cdots + a_{2n} x^{2n}$

Put $x = 1$

$$3^n = a_0 + a_1 + a_2 + a_3 + \cdots + a_{2n}$$

$$3^n = (a_0 + a_{2n} + a_1 + a_{2n-1} + \cdots + a_{n-1} + a_{n+1}) + a_n$$

$$3^n = 2(a_0 + a_1 + a_2 + \cdots + a_{n-1}) + a_n$$

$$[\because a_0 = a_{2n} \text{ and } a_1 = a_{2n-1}]$$

$\therefore$ Hence, option (b) is correct.

24. (a) Put $x = \frac{-1}{x}$ in place of x

$$\left(1 - \frac{1}{x} + \frac{1}{x^2}\right)^n = a_0 - \frac{a_1}{x} + \frac{a_2}{x^2} - \cdots + \frac{a_{2n}}{x^{2n}} \quad \text{...(i)}$$

$$\Rightarrow (x^2 - x + 1)^n = a_0 x^{2n} - a_1 x^{2n-1} + a_2 x^{2n-2} + \cdots + a_{2n} \text{...(ii)}$$

On multiplying Eqs. (i) and (ii), we get

$$(1 + x^2 + x^4)^n = (a_0 + a_1 x + a_2 x^2 + \cdots + a_{2n} x^{2n})$$
$$(a_0 x^{2n} - a_1 x^{2n-1} + \cdots + a_{2n}) \quad \text{...(iii)}$$

On putting x^2 in place of x, we get

$$(1 + x^2 + x^4)^n = a_0 + a_1 x^2 + a_2 x^4 + \cdots + a_n x^{2n} + \cdots + a_{2n} x^{4n}$$

Coefficient of x^{2n} in $(1 + x^2 + x^4)^n$ is a_n.

On equating the coefficient of x^{2n}, we get

$$a_0^2 - a_1^2 + a_2^2 - a_1^2 - \cdots + a_{2n}^2 = a_n$$

Hence, $k = 1$

25. (b) We have, $(1 + x + x^2)^n = \sum_{r=0}^{2n} a_r x^r$

On dividing both sides by x^n, we get

$$\left(1 + x + \frac{1}{x}\right)^n = \sum_{r=0}^{2n} a_r x^{r-n}$$

Thus, a_n = constant term in the expansion of $\left(1 + x + \frac{1}{x}\right)^n$.

But $\left(1 + x + \frac{1}{x}\right)^n = {}^nC_0 + {}^nC_1\left(x + \frac{1}{x}\right)$

$$+ {}^nC_2\left(x + \frac{1}{x}\right)^2 + \cdots + {}^nC_n\left(x + \frac{1}{x}\right)^n$$

Thus, $a_n = {}^nC_0 + {}^nC_2({}^2C_1) + {}^nC_4({}^4C_2) + \cdots$

$$a_n = 1 + a \Rightarrow a_n - a = 1$$

Hence, option (b) is correct.

Revisal Problems (JEE Main)

1. (b) We have, $(2n+1)(2n+3)(2n+5)\cdots(4n-1)$

$$\Rightarrow \frac{1\cdot2\cdot3\cdot4\cdots2n(2n+1)(2n+2)(2n+3)\cdots(4n-1)4n}{1\cdot2\cdot3\cdot4\cdots2n(2n+2)(2n+4)(2n+6)\cdots4n}$$

$$\Rightarrow \frac{(4n)!}{(2n)!2^n(n+1)(n+2)(n+3)\cdots(2n)}$$

$$\Rightarrow \frac{4n!\cdot1\cdot2\cdot3\cdot4\cdots n}{(2n)!2^n(1\cdot2\cdot3\cdots n(n+1)(n+2)\cdots(2n)}$$

$$\Rightarrow \frac{(4n)!\,n!}{(2n)!2^n(2n)!} = \frac{(4n)!\,n!}{2^n(2n)!(2n)!}$$

Hence, option (b) is correct.

2. (a) We have, $\left(\frac{3}{\sqrt[3]{84}} + \sqrt{3}\log_e x\right)^9$

7th term of the expansion is 729.

$$\therefore T_7 = {}^9C_6\left(\frac{3}{84^{1/3}}\right)^3 \times (\sqrt{3}\log_e x)^6 = 729$$

$$\Rightarrow \frac{9!}{3!6!}\frac{3^3}{84} \times (3)^3(\log_e x)^6 = 729$$

$$\Rightarrow (\log_e x)^6 = \frac{729 \times 3!\,6! \times 84}{3^3\cdot9! \times 3^3}$$

$$\Rightarrow (\log_e x)^6 = 1$$

$$\Rightarrow \log_e x = 1$$

$$x = e$$

Hence, option (a) is correct.

3. (d) General term of the expansion $(2^{1/2} + 3^{1/5})^{10}$.

$$\therefore \quad T_{r+1} = {}^{10}C_r(2)^{\frac{10-r}{2}}(3)^{r/5}$$

This term is rational iff $r = 0$ and $r = 10$, if $r = 0$, then

$$T_1 = 2^5 = 32$$

If $r = 10$, then $\quad T_{11} = 3^2 = 9$

Sum of the rational terms i.e. $T_1 + T_{11} = 32 + 9 = 41$

Hence, option (d) is correct.

4. (c) We have, ${}^{18}C_{r-2} + 2\,{}^{18}C_{r-1} + {}^{18}C_r$

$$\Rightarrow {}^{18}C_{r-2} + {}^{18}C_{r-1} + {}^{18}C_{r-1} + {}^{18}C_r$$

$$\Rightarrow {}^{19}C_{r-1} + {}^{19}C_r \quad [\because {}^nC_{r-1} + {}^nC_r = {}^{n+1}C_r]$$

$$\Rightarrow {}^{20}C_r \Rightarrow {}^{20}C_r \geq {}^{20}C_{13}$$

$$13 + r = 20 \quad [\because {}^nC_r = {}^nC_{n-r}]$$

$$r = 7$$

Hence, option (c) is correct.

5. (b) The general term of the expansion

$$(\sqrt{2} + \sqrt[4]{3})^{100} \Rightarrow T_{r+1} = {}^{100}C_r(2)^{\frac{100-r}{2}}(3)^{\frac{r}{4}}$$

The rational terms of the expansion is, iff r is multiple of 4.

$\therefore \quad r = 0, 4, 8, 12, \cdots, 100.$

Total number of terms 26.

Hence, option (b) is correct.

6. (c) We have, $15^{23} + 23^{23}$

$\Rightarrow (19-4)^{23} + (19+4)^{23}$

$= 2(19^{23} + {}^{23}C_2\, 19^{21}\cdot 4^2 + \cdots + {}^{23}C_{22} 19\cdot 4^{22})$

$= 2\times 19\,(19^{22} + {}^{23}C_2 19^{20} 4^2 + \cdots + {}^{23}C_{22} 4^{22})$

$= 2\times 19\,(k) \quad [\because 19^{22} + {}^{23}C_2\, 19^{20}4^2 + \cdots + {}^{23}C_{22}4^{22} = k]$

This is divided by 19, then remainder is 0. Hence, option (c) is correct.

7. (b) We have, $(\alpha+p)^{m-1} + (\alpha+p)^{m-2}(\alpha+q) + (\alpha+p)^{m-3}(\alpha+q)^2 + \cdots + (\alpha+q)^{m-1}$

$$\Rightarrow (\alpha+p)^{m-1}\left[\frac{\left(1-\left(\frac{\alpha+q}{\alpha+p}\right)^m\right)}{\left(1-\frac{\alpha+q}{\alpha+p}\right)}\right]$$

$$\Rightarrow (\alpha+p)^{m-1}\left[\frac{((\alpha+p)^m-(\alpha+q)^m)(\alpha+p)}{(\alpha+p)^m[(\alpha+p)-(\alpha+q)]}\right]$$

$$\Rightarrow \frac{(\alpha+p)^m-(\alpha+q)^m}{p-q}$$

The coefficient of α^t in $\frac{(\alpha+p)^m-(\alpha+q)^m}{p-q}$ is

$${}^mC_t\left(\frac{p^{m-t}-q^{m-t}}{p-q}\right)$$

Hence, option (b) is correct.

8. (d) Given, $\left(9x - \frac{1}{3\sqrt{x}}\right)^{18}$

$$T_{r+1} = {}^{18}C_r(9)^{18-r}(x)^{18-r}\left(-\frac{1}{3}\right)^r\left(\frac{1}{x}\right)^{\frac{r}{2}}$$

$$= {}^{18}C_r\cdot 9^{18-r}\left(-\frac{1}{3}\right)^r\cdot x^{18-r-\frac{r}{2}}$$

The term is independent of x,

if $18 - r - \frac{r}{2} = 0 \Rightarrow r = 12$

$\therefore \quad T_{12+1} = {}^{18}C_{12}9^6\left(-\frac{1}{3}\right)^{12} = {}^{18}C_{12}$

$\therefore$ According to the problem,

$${}^{18}C_{12} = \alpha\, {}^{18}C_{12}$$

$\therefore \quad \alpha = 1$

Hence, option (d) is correct.

9. (d) We have, ${}^{99}C_{97} + {}^{98}C_{96} + {}^{97}C_{95} + \cdots + {}^3C_1 + {}^2C_0$

$\Rightarrow {}^{99}C_2 + {}^{98}C_2 + {}^{97}C_2 + \cdots + {}^3C_2 + {}^2C_2 \quad [\because {}^nC_r = {}^nC_{n-r}]$

$= {}^2C_2 + {}^3C_2 + {}^4C_2 + \cdots + {}^{99}C_2$

$= {}^{99+1}C_{2+1} \quad [\because {}^nC_n + {}^{n+1}C_n + \cdots + {}^{n+k}C_n = {}^{n+k+1}C_{n+1}]$

$= {}^{100}C_3 = {}^{100}C_{97}$

Hence, option (d) is correct.

10. (b) We have, $\sum_{r=1}^{10} r(r-1)\,{}^{10}C_r$

$$\Rightarrow \sum_{r=1}^{10} r(r-1)\frac{10!}{r(r-1)(r-2)!(10-r)!} \Rightarrow \sum_{r=1}^{10}\frac{10\times 9\times 8!}{(r-2)!\,[8-(r-2)]!}$$

$$\Rightarrow 90\sum_{r=2}^{10} {}^8C_{r-2} = 90\times 2^8 = 45\times 2^9$$

$\therefore \quad k = 45$

Hence, option (b) is correct.

11. (a) We have, $\sum_{k=0}^{4}\frac{5^{4-k}}{(4-k)!}\cdot\frac{x^k}{k!} = \frac{8}{3}$

$$\Rightarrow \frac{5^4}{4!} + \frac{5^3x}{3!1!} + \frac{5^2x^2}{2!2!} + \frac{5x^3}{1!3!} + \frac{x^4}{4!} = \frac{8}{3}$$

$$\Rightarrow \frac{1}{4!}[{}^4C_05^4 + {}^4C_15^3x + {}^4C_25^2x^2 + {}^4C_35x^3 + {}^4C_4x^4) = \frac{8}{3}$$

$\Rightarrow \quad (x+5)^4 = \frac{8\times 4!}{3}$

$\Rightarrow \quad (x+5)^4 = 64$

$\Rightarrow \quad x + 5 = \pm 2\sqrt{2}$

$\Rightarrow \quad x = -5 \pm 2\sqrt{2}$

Then, the largest real value of x is $2\sqrt{2} - 5$.

Hence, option (a) is correct.

12. (a) $(1+x)^{20} = {}^{20}C_0 + {}^{20}C_1x + {}^{20}C_2x^2 + \cdots + {}^{20}C_{10}x^{10} + {}^{20}C_{11}x^{11} + \cdots + {}^{20}C_{20}x^{20}$

Put $x = -1$

$(1-1)^{20} = {}^{20}C_0 - {}^{20}C_1 + {}^{20}C_2 - {}^{20}C_3 + \cdots + {}^{20}C_{10} - {}^{20}C_{11} + \cdots + {}^{20}C_{20}$

$0 = ({}^{20}C_0 + {}^{20}C_{20}) - ({}^{20}C_1 + {}^{20}C_{19}) + ({}^{20}C_2 + {}^{20}C_{18}) \cdots ({}^{20}C_9 + {}^{20}C_{11}) + {}^{20}C_{10}$

$0 = 2[{}^{20}C_0 - {}^{20}C_1 + {}^{20}C_2 - \cdots - {}^{20}C_9] + {}^{20}C_{10}$

$\Rightarrow {}^{20}C_{10} = 2[{}^{20}C_0 - {}^{20}C_1 + {}^{20}C_2 - \cdots - {}^{20}C_9 + {}^{20}C_{10}]$

$[\therefore {}^nC_r = {}^nC_{n-r}]$

$\Rightarrow \quad {}^{20}C_0 - {}^{20}C_1 + {}^{20}C_2 - \cdots + {}^{20}C_{20} = \frac{1}{2}\cdot {}^{20}C_{10}$

Hence, option (a) is correct.

13. (c) We have,

The sum of 5th and 6th terms of the binomial expansion. $(a-b)^x, x \geq 5$ is zero.

$\therefore {}^nC_4\, a^{n-4}b^4 + {}^nC_5\, a^{n-5}(-b)^5 = 0$

$\Rightarrow \quad {}^nC_4a^{n-4}b^4 = -{}^nC_5\, a^{n-5}(-b)^5$

$\Rightarrow \quad \frac{a}{b} = \frac{{}^nC_5}{{}^nC_4} \quad \left[\because \frac{{}^nC_r}{{}^nC_{r-1}} = \frac{n+1-r}{r}\right]$

$\Rightarrow \quad \frac{a}{b} = \frac{n-4}{5}$

Hence, option (c) is correct.

14. (b) The coefficient of rth, $(r+1)$th and $(r+2)$th terms in the expansion of $(1+y)^m$ are in AP.

$\therefore \quad 2T_{r+1} = T_r + T_{r+2}$

$\Rightarrow \quad 2\,{}^mC_r = {}^mC_{r-1} + {}^mC_{r+1}$

$$\Rightarrow \frac{2m!}{r!(m-r)!} = \frac{m!}{(r-1)!\,(m+1-r)!} + \frac{m!}{(r+1)!(m-r-1)!}$$

$$\Rightarrow \frac{2}{r(m-r)} = \frac{1}{(m+1-r)(m-r)} + \frac{1}{r(r+1)}$$

$$\Rightarrow \frac{2}{r(m-r)} = \frac{r(r+1)+(m+1-r)(m-r)}{r(r+1)(m+1-r)(m-r)}$$

$\Rightarrow 2(mr - r^2 + r + m - r + 1) = r^2 + r + m^2 + r^2 - 2mr + m - r$

$\Rightarrow \quad 4r^2 - 4mr - m - 2 + m^2 = 0$

$\Rightarrow \quad 4r^2 - m(4r+1) + m^2 - 2 = 0$

$\Rightarrow \quad m^2 - m(4r+1) + 4r^2 - 2 = 0$

Hence, option (b) is correct.

15. (a) We have, $(1 - x - x^2 + x^3)^6 \Rightarrow [(1-x) - x^2(1-x)]^6$

$\Rightarrow [(1-x)(1-x^2)]^6 \Rightarrow (1-x)^6(1-x^2)^6$

$\Rightarrow (1 - 6x + 15x^2 - 20x^3 + 15x^4 - 6x^5 + x^6)$

$(1 - 6x^2 + 15x^4 - 20x^6 + 15x^8 - 6x^{10} + x^{12})$

Now, coefficient of x^7 in the given expansion is

$(-6)(-20) + (-20)(15) + (-6)(-6)$

$\Rightarrow 120 - 300 + 36 = -144$

Hence, coefficient of x^7 in $(1 - x - x^2 + x^3)^6$ is -144.

Hence, option (a) is correct.

16. (b) We have, $(\sqrt{3}+1)^{2n} - (\sqrt{3}-1)^{2n}$

$\Rightarrow 2({}^{2n}C_1(\sqrt{3})^{2n-1} + {}^{2n}C_3(\sqrt{3})^{2n-3} + \dots + {}^{2n}C_{2n-1}(\sqrt{3})]$

$\Rightarrow 2\sqrt{3}\,[{}^{2n}C_1\sqrt{3}^{2n-2} + {}^{2n}C_3\sqrt{3}^{2n-4} + \dots + {}^{2n}C_{2n-1}]$

$[\because (a+b)^n - (a-b)^n = 2({}^nC_1 a^{n-1}b + {}^nC_3 a^{n-3}b^3 + \dots + {}^nC_{n-1}ab^{n-1}]$

$= 2\sqrt{3}\,[{}^{2n}C_1 3^{n-1} + {}^{2n}C_3 3^{n-2} + \dots + {}^{2n}C_{2n-1}]$

$= 2\sqrt{3}$ (rational number)

Hence, $(\sqrt{3}+1)^{2n} - (\sqrt{3}-1)^{2n}$ is an irrational number.

17. (d) We have, ${}^{n-1}C_r = (k^2 - 3)\,{}^nC_{r+1}$

$\Rightarrow k^2 - 3 = \dfrac{{}^{n-1}C_r}{{}^nC_{r+1}} = \dfrac{r+1}{n}$

$0 < \dfrac{r+1}{n} \le 1 \quad \left[\because n \ge r+1 \Rightarrow \dfrac{r+1}{n} \le 1\right]$

$\therefore 0 < k^2 - 3 \le 1 \Rightarrow 3 < k^2 \le 4$

$\sqrt{3} < |K| \le 2$

Hence, $|k| \in (\sqrt{3}, 2]$

Hence, option (d) is correct.

18. (d) Given, $y = 3x + 6x^2 + 10x^3 + \cdots$ ...(i)

or $xy = 3x^2 + 6x^3 + 10x^4 + \cdots$ ...(ii)

On subtracting Eq. (ii) from Eq. (i), we get

$\Rightarrow y(1-x) = 3x + 3x^2 + 4x^3 + 5x^4 + \cdots$

$\Rightarrow y(1-x) = -1 + 1 + x + 2x + 3x^2 + 4x^3 + \cdots$

$\Rightarrow y(1-x) = -1 + x + (1 + 2x + 3x^2 + 5x^4 + \cdots)$

$\Rightarrow y(1-x) = -1 + x + (1-x)^{-2}$

$\Rightarrow y(1-x) = -1 + x + \dfrac{1}{(1-x)^2}$

$\Rightarrow y(1-x) = -\left[(1-x) - \dfrac{1}{(1-x)^2}\right]$

$\Rightarrow y(1-x) = -\left[\dfrac{(1-x)^3 - 1}{(1-x)^2}\right]$

$\Rightarrow y(1-x)^3 = -(1-x)^3 + 1$

$\Rightarrow (1-x)^3 = \dfrac{1}{1+y} \Rightarrow x = 1 - (1+y)^{-1/3}$

Hence, option (d) is correct.

19. (b) We have, $\dfrac{\left(1+\dfrac{3x}{4}\right)^{-4}(\sqrt{16-3x})}{(8+x)^{2/3}} = P + Qx$

$\Rightarrow \left(1+\dfrac{3x}{4}\right)^{-4} \cdot 4\left(1-\dfrac{3x}{16}\right)^{1/2} \cdot 8^{-2/3}\,(1 + x/8)^{-2/3}$

$\Rightarrow \left(1+\dfrac{3x}{4}\right)^{-4}\left(1-\dfrac{3x}{16}\right)^{1/2}\left(1+\dfrac{x}{8}\right)^{-2/3}$

$\Rightarrow (1 - 3x + \cdots)\left(1 - \dfrac{3x}{32} + \cdots\right)\left(1 - \dfrac{x}{12} + \cdots\right)$

Constant term is 1 and coefficient of x is

$\left(-3 - \dfrac{3}{32} - \dfrac{1}{12}\right) = -\dfrac{305}{96}$

$\therefore P = 1$ and $Q = -\dfrac{305}{96}$

Hence, (b) is correct option.

20. (c) We have, $(1 + 3x + 6x^2 + 10x^3 + \cdots)^2$

Let $S = 1 + 3x + 6x^2 + 10x^3 + \cdots$

or $xS = x + 3x^2 + 6x^3 + \cdots$

$\Rightarrow (1-x)S = 1 + 2x + 3x^2 + \cdots$

$\Rightarrow (1-x)S = (1-x)^{-2} \Rightarrow S = (1-x)^{-3}$

Now, $(1 + 3x + 6x^2 + 10x^3 + \cdots)^2 \Rightarrow [(1-x)^{-3}]^2 = (1-x)^{-6}$

Coefficient of x^r in the expansion of $(1-x)^{-6}$ is ${}^{r+6-1}C_{6-1}$.

$[\because$ coefficient of x^r in $(1-x)^{-n}$ is ${}^{r+n-1}C_{n-1}]$

$= {}^{r+5}C_5$

$\Rightarrow \dfrac{(r+5)!}{5!(r+5-5)!} \Rightarrow \dfrac{(r+5)(r+4)(r+3)(r+2)(r+1)}{5!}$

Hence, option (c) is correct.

21. (a) $S_\infty = 1 + n\left(1-\dfrac{1}{x}\right) + \dfrac{n(n+1)}{2!}\left(1-\dfrac{1}{x}\right)^2 + \cdots \infty$

Put, $1 - \dfrac{1}{x} = t$

$\therefore S_\infty = 1 + nt + \dfrac{n(n+1)}{2!}t^2 + \cdots \infty$

$= (1-t)^{-n} = \left[1 - \left(1 - \dfrac{1}{x}\right)\right]^{-n} = \left[\dfrac{1}{x}\right]^{-n} = x^n.$

22. (b) We have, $(1 - 9x + 20x^2)^{-1} \Rightarrow (1 - 5x - 4x + 20x^2)^{-1}$

$\Rightarrow (1-5x)^{-1}(1-4x)^{-1}$

$\Rightarrow [1 + 5x + 5^2x^2 + 5^3x^3 + \cdots]\quad[1 + 4x + 4^2x^2 + 4^3x^3 + \cdots]$

Coefficient of x^n in the above expansion is,

$4^n + 5\cdot 4^{n-1} + 5^2\cdot 4^{n-2} + \cdots + 5^n$

$4^n\left[1 + \dfrac{5}{4} + \left(\dfrac{5}{4}\right)^2 + \cdots + \left(\dfrac{5}{4}\right)^n\right],\ 4^n\left[\dfrac{(5/4)^{n+1} - 1}{5/4 - 1}\right]$

$4^n\left[\dfrac{5^{n+1} - 4^{n+1}}{4^n}\right] \Rightarrow 5^{n+1} - 4^{n+1}$

Hence, coefficient of x^n is $5^{n+1} - 4^{n+1}$.

Hence, option (b) is correct.

23. (c) As, x is nearly equal to 1.

$\therefore x = 1 + h$, where h is very small, so that

$h^2 = h^3 = h^4 = \ldots = 0$

Thus, $\dfrac{px^q - qx^p}{x^q - x^p} = \dfrac{p(1+h)^q - q(1+h)^p}{(1+h)^q - (1+h)^p}$

$= \dfrac{p(1+qh) - q(1+ph)}{(1+qh) - (1+ph)} = \dfrac{p-q}{h(q-p)} = \dfrac{1}{-h} = \dfrac{1}{1-x}$

Hence, option (c) is the correct.

24. (c) Since, $(r+1)$th term in the expansion of $(1+x)^{\frac{27}{5}}$

$= \dfrac{\dfrac{27}{5}\left(\dfrac{27}{5}-1\right)\left(\dfrac{27}{5}-2\right)\cdots\left(\dfrac{27}{5}-r+1\right)}{r!}x^r$

Now, this term will be negative, if the last factor in numerator is the only one negative factor.

$\Rightarrow \dfrac{27}{5} - r + 1 < 0 \Rightarrow \dfrac{32}{5} < r \Rightarrow r > 6.4$

$\Rightarrow$ Least value of r is 7. Thus, the first negative term will be 8th.

Hence, option (c) is correct.

25. (b) We have, $\dfrac{(1+x)^{3/2}-\left(1+\dfrac{x}{2}\right)^3}{(1-x)^{1/2}}$

$$=\frac{\left(1+\frac{3}{2}x+\frac{\frac{3}{2}\cdot\frac{1}{2}}{2!}x^2+\cdots\right)-\left(1+\frac{3x}{2}+\frac{3\cdot 2}{2!}\frac{x^2}{4}+\cdots\right)}{(1-x)^{1/2}}$$

[neglecting higher power of x]

$$=-\frac{3x^2}{8}(1-x)^{-1/2}=-\frac{3x^2}{8}\left(1+\frac{1}{2}x+\frac{\frac{1}{2}\times\frac{3}{2}}{2}x^2\right)=-\frac{3x^2}{8}$$

Hence, option (b) is correct.

26. (a) Let, $f=(\sqrt{5}+2)^n$, $g=(\sqrt{5}-2)^n$.

Then, $f-g=(\sqrt{5}+2)^n-(\sqrt{5}-2)^n$

$=2\,[{}^nC_1(\sqrt{5})^{n-1}\cdot 2+{}^nC_3(\sqrt{5})^{n-3}\cdot 2^3+\cdots+2^n]$

$f-g=2\,[{}^nC_1(\sqrt{5})^{n-1}\cdot 2+{}^nC_3(\sqrt{5})^{n-3}\cdot 2^3+\cdots+{}^nC_{n-2}\cdot(\sqrt{5})^2\cdot 2^{n-2}\}+2^{n+1}$

$\therefore\quad f-2^{n+1}=20\,[\lambda]+g$

$\Rightarrow [f-2^{n+1}]=20\cdot n$, as $0<g<1$

$\therefore [(\sqrt{5}+2)^n-2^{n+1}]$ is divisible by 20 n.

27. (c) We have,

Statement I $11^{25}+12^{25}\ \Rightarrow\ (11+12-12)^{25}+12^{25}$

$\Rightarrow (23-12)^{25}+12^{25}$

$\Rightarrow 23^{25}-{}^{25}C_1\,(23)^{24}\cdot 12+{}^{25}C_2(23)^{23}\cdot 12^2+\cdots-12^{25}+12^{25}.$

$\Rightarrow 23^{25}-{}^{25}C_1 23^{24}\cdot 12+\cdots+{}^{25}C_{24}23\cdot 12^{24}$

$\Rightarrow 23[23^{24}-{}^{25}C_1 23^{23}\cdot 12+\cdots+{}^{25}C_{24}\cdot 12^{24}]=23\,k$

Hence, $11^{25}+12^{25}$ is divisible by 23.

Statement II a^n+b^n

$(a+b-b)^n+b^n$

$\Rightarrow\quad (a+b)^n-{}^nC_1(a+b)^n b+\cdots+(-1)^n b^n+b^n$

a^n+b^n is divisible by $a+b$, if n is an odd.

$\therefore$ Statement I is correct and Statement II is incorrect.

28. (a) $\quad(1+x)^n={}^nC_0+{}^nC_1x+{}^nC_2x^2+\cdots+{}^nC_rx^r+\cdots+{}^nC_nx^n$

and $(1+x)^m={}^mC_0+{}^mC_1x+{}^mC_2x^2+\cdots+{}^mC_rx^r+\cdots+{}^mC_mx^m$

$(1+x)^n\cdot(1+x)^m={}^nC_0{}^mC_0+{}^nC_0{}^mC_1x+\cdots+({}^nC_0{}^mC_r+{}^nC_1{}^mC_{m-1}\cdots)x^r$

$\Rightarrow\quad (1+x)^{n+m}={}^nC_0{}^mC_0x+{}^nC_0{}^mC_1x+\cdots$

Coefficient of x^r in $(1+x)^{n+m}={}^{n+m}C_r$

$\therefore\qquad {}^{n+m}C_r={}^nC_0{}^mC_r+{}^nC_1{}^mC_{r-1}+\cdots+{}^nC_r\cdot{}^mC_0$

Statement I is correct.

Statement II is ${}^nC_0{}^mC_r+{}^nC_1{}^mC_{r-1}+\cdots+{}^nC_r{}^mC_0.$

Hence, Statement I is correct and Statement II is correct explanation of Statement I.

29. (b) We have, $T_n=\sum_{r=1}^{n}\dfrac{1}{r}\ \Rightarrow\ T_n=1+\dfrac{1}{2}+\dfrac{1}{3}+\dfrac{1}{4}+\cdots+\dfrac{1}{n}$

and $\qquad S_n=\sum_{r=1}^{r}{}^nC_r(-1)^{r-1}\cdot\dfrac{1}{r}$

$\Rightarrow\qquad S_n=C_1-\dfrac{1}{2}C_2+\dfrac{1}{2}C_3+\cdots+\dfrac{(-1)^{n-1}\,{}^nC_n}{n}$

$(1-x)^n=1-{}^nC_1x+{}^nC_2x^2-{}^nC_3x^3+\cdots+(-1)^n\,{}^nC_nx^n$

$\Rightarrow\quad \dfrac{1-(1-x)^n}{x}=\dfrac{1}{x}[{}^nC_1x-{}^nC_2x^2+\cdots+(-1)^{n-1}\cdot{}^nC_nx^n]$

$\Rightarrow\quad \dfrac{1-(1-x)^n}{x}={}^nC_1-{}^nC_2x+{}^nC_3x^2+\cdots+(-1)^{n-1}\,{}^nC_nx^{n-1}$

On integrating between the limits 0 and 1, we get

$$\int_0^1\frac{1-(1-x)^n}{x}dx=\left[{}^nC_1x-\frac{{}^nC_2x^2}{2}+\frac{{}^nC_3x^3}{3}+\cdots+(-1)^{n-1}\frac{{}^nC_nx^n}{n}\right]_0^1$$

$$\int_0^1(1+x+x^2+x^3+\cdots+x^{n-1})dx={}^nC_1-\frac{{}^nC_2}{2}+\frac{{}^nC_3}{3}-\cdots+\frac{(-1)^{n-1}\,{}^nC_n}{n}$$

$$\Rightarrow\left[x+\frac{x^2}{2}+\frac{x^3}{3}+\cdots+\frac{x^n}{n}\right]_0^1={}^nC_1-\frac{{}^nC_2}{2}+\frac{{}^nC_3}{3}+\cdots+\frac{(-1)^{n-1}\,{}^nC_n}{n}$$

$$1+\frac{1}{2}+\frac{1}{3}+\cdots+\frac{1}{n}={}^nC_1-\frac{{}^nC_2}{2}+\frac{{}^nC_3}{3}+\cdots+(-1)^{n-1}\cdot\frac{{}^nC_n}{n}.$$

$\Rightarrow\qquad T_n=S_n$

Statement I is correct.

$$S_{n+1}-S_n=\left(1-\frac{{}^nC_2}{2}+\frac{1}{3}\,{}^nC_3-\cdots+(-1)^n\frac{{}^{n+1}C_{n+1}}{n+1}\right)-\left(1-\frac{{}^nC_2}{2}+\frac{1}{3}\,{}^nC_3+\cdots+(-1)^{n-1}\frac{{}^nC_n}{n}\right)=\frac{1}{n+1}$$

$$T_{n+1}-T_n=\left(1+\frac{1}{2}+\frac{1}{3}+\cdots+\frac{1}{n+1}\right)-\left(1+\frac{1}{2}+\frac{1}{3}+\cdots+\frac{1}{n}\right)=\frac{1}{n+1}$$

$\therefore\qquad S_{n+1}-S_n=T_{n+1}-T_n$

Statement II is correct.

30. (a) $\sum_{r=0}^{1}(-1)^r\,{}^nC_r={}^nC_0-{}^nC_1+{}^nC_2-\cdots+(-1)^n\,{}^nC_n=0$

$\sum_{r=0}^{n}(-1)^r r\,{}^nC_r=\sum_{r=0}^{n}(-1)^n\,n^{n-1}C_{r-1}=-n\sum_{r=0}^{n}(-1)^{r-1}\,{}^{n-1}C_{r-1}$

$=-n[{}^{n-1}C_0-{}^{n-1}C_1+{}^{n-1}C_2-\cdots+(-1)^{n-1}\,{}^{n-1}C_{n-1}]$

$=-n(1-1)^{n-1}=0$

$\sum_{r=0}^{n}(-1)^r r^2\,{}^nC_r=\sum_{r=0}^{n}(-1)^r\,nr^{n-1}C_{r-1}$

$=n\sum_{r=0}^{n}(-1)^r r\cdot^{n-1}C_{r-1}=n\sum_{r=0}^{n}(-1)^r(r+1-1)^{n-1}C_{r-1}$

$=n\left[\sum_{r=0}^{n}(-1)^r(r-1)^{n-1}C_{r-1}+\sum_{r=1}^{n}(-1)^r\,{}^{n-1}C_{r-1}\right]$

$=n\left[\sum_{r=2}^{n}(-1)^r(n-1)\,{}^{n-2}C_{r-2}+({}^{n-1}C_0-{}^{n-1}C_1+{}^{n-1}C_2-\cdots+{}^{n-1}C_{n-1})\right]$

$=n(n-1)[{}^{n-2}C_0+{}^{n-2}C_1+\cdots]+0=n(n-1)(1-1)^{n-2}=0$

Similarly, $\sum_{r=0}^{n}(-1)^r r^P\,{}^nC_r=0$

Statement II is correct.

Now, $\sum_{r=0}^{n}(-1)^r\,{}^nC_r(a-r)(b-r)(c-r)$

$\Rightarrow\sum_{r=0}^{n}(-1)^r\,{}^nC_r\,[abc+(a+b+c)r^2-(ab+bc+ca)r-r^3]$

$\Rightarrow abc\sum_{r=0}^{n}(-1)^r\,{}^nC_r+(a+b+c)\sum_{r=0}^{n}(-1)^r r^2\,{}^nC_r-(ab+bc+ca)\sum_{r=0}^{n}(-1)^r\,r\,{}^nC_r-\sum_{r=0}^{n}(-1)^r\,r^3\,{}^nC_r$

$=abc(0)+(a+b+c)\,0-(ab+bc+ca)\Rightarrow(0)-(0).$

Statement I is correct and Statement II is correct explanation of Statement I.

Revisal Problems (JEE Advanced)

1. (c) We have, $1 + (1+x) + (1+x)^2 + \cdots + (1+x)^{16}$

$$= \frac{1-(1+x)^{17}}{1-(1+x)} = \frac{1}{x}[(1+x)^{17} - 1]$$

Thus, the coefficient of x^k in $\frac{1}{x}[(1+x)^{17} - 1] = {}^{17}C_{k+1}$

and the coefficient of x^{2k} in $\frac{1}{x}[(1+x)^{17} - 1] = {}^{17}C_{2k+1}$

Coefficient of x^k and x^{2k} are equal

$\therefore \quad {}^{17}C_{k+1} = {}^{17}C_{2k+1}$

or $\quad k + 1 + 2k + 1 = 17$

$\Rightarrow \quad k = 5$

Hence, option (c) is correct.

2. (c) Here, $(\sqrt{5} - 2)$ is a positive proper fraction.

$\therefore (\sqrt{5} - 2)^p$ is also a positive proper fraction say f.

Now, consider

$(\sqrt{5} + 2)^p - (\sqrt{5} - 2)^p$

$$= 2\,[{}^pC_1(\sqrt{5})^{p-1}\cdot 2 + {}^pC_3(\sqrt{5})^{p-3}\cdot 2^3 + \cdots]$$

$\{(p-1)$ is even$\}$

$I + f - f' = 2\,[{}^pC_1\cdot(\sqrt{5})^{p-1}\cdot 2 + {}^pC_3(\sqrt{5})^{p-3}\cdot 2^3 + \cdots + {}^pC_p\cdot 2^p]$

Here, $\quad -1 < f - f' < 1$

$\Rightarrow \quad f - f' = 0.$

$$\therefore I - 2^{p+1} = 2p[(\sqrt{5})^{p-1}\cdot 2 + \frac{(p-1)(p-2)}{3!}(\sqrt{5})^{p-3}\cdot 2^3 + \cdots + {}^pC_{p-2}\cdot(\sqrt{5})^2\cdot 2^{p-2}]$$

But, $\quad I = [(2 + \sqrt{5})^p]$

Hence, $\quad [(2 + \sqrt{5})^p] - 2^{p+1} = 2p \quad$ [an integer]

So, must be divisible by $(2p)$.

3. (a) We have,

$\Rightarrow \quad (7)^{1995} \Rightarrow (7^{2\times 997})\cdot 7$

$\Rightarrow \quad (49)^{997}\cdot 7 \Rightarrow 7(50 - 1)^{997}$

$\Rightarrow \quad 7\,[50^{997} - {}^{997}C_1 50^{996} + \cdots {}^{997}C_{996}\,50 - (1)^{997}]$

$\Rightarrow 7 \times 50^2[50^{995} - {}^{997}C_1\,50^{994} + \cdots - {}^{997}C_{995}] + 7\,[997 \times 50 - 1]$

$\Rightarrow \quad 100k + [348950 - 7]$

$\Rightarrow \quad 100k + 348943$

$\Rightarrow \quad 100k + 348900 + 43$

$\Rightarrow \quad 100[k + 3489] + 43$

$\therefore$ Where, 7^{1995} is divided by 100 leaves the remainder is 43.

Hence, option (a) is correct.

4. (d) We have,

$(x+1)(x+2)(x+3)\ldots(x+10)$
$+ (x+2)(x+3)\ldots(x+11) + \cdots$
$+ (x+11)(x+12)\ldots(x+20)$

Coefficient of x^9 in the given polynomial is

$(1 + 2 + 3 + \cdots + 10) + (2 + 3 + 4 + \cdots + 11)$
$\cdots (11 + 12 + 13 + \cdots + 20)$

$55 + 65 + 75 + \cdots + 155$

$$= \frac{11}{2}(55 + 155) = \frac{11}{2} \times 210 = 1155$$

Hence, option (d) is correct.

5. (c) We have, $(1+x)^n(1+y)^n(1+z)^n$

$$= \left(\sum_{p=0}^{n} {}^nC_p x^p\right)\left(\sum_{q=0}^{n} {}^nC_q y^q\right)\left(\sum_{s=0}^{n} {}^nC_s z^s\right)$$

$$= \sum_{0 \le p,q,s \le n} ({}^nC_p)({}^nC_q)({}^nC_s)\,x^p y^q z^s$$

For the coefficient of degree r, we must have $p + q + s = r$ where, p, q, s are integers with $p, q, s \ge 0$.

Sum of such coefficient $= \sum_{p,q,s\ge 0} ({}^nC_p)({}^nC_q)({}^nC_s) = {}^{3n}C_r$

$[\because p + q + s = r]$

Hence, option (c) is correct.

6. (c) We have, $\sum_{r=0}^{n} (-1)^r \left(\frac{{}^nC_r}{{}^{r+3}C_r}\right) = \sum_{r=0}^{n} (-1)^r \frac{n!\,(r!)\,(r+3-r)!}{r!\,(n-r)!\,(r+3)!}$

$$= 3!\sum_{r=0}^{n} (-1)^r \frac{n!}{(n-r)!\,(r+3)!}$$

$$= \frac{3!}{(n+1)(n+2)(n+3)}\sum_{r=0}^{n} (-1)^r \frac{(n+3)!}{(n-r)!\,(r+3)!}$$

$$= \frac{3!}{(n+1)(n+2)(n+3)}\sum_{r=0}^{n} (-1)^r\,{}^{n+3}C_{r+3}$$

$$= \frac{3!\,(-1)^3}{(n+1)(n+2)(n+3)}\sum_{s=3}^{n+3} (-1)^s\,{}^{n+3}C_s$$

$$= \frac{-3!}{(n+1)(n+2)(n+3)}\left[\left(\sum_{s=0}^{n+3} (-1)^s\,{}^{n+3}C_s\right) - ({}^{n+3}C_0 - {}^{n+3}C_1 + {}^{n+3}C_2)\right]$$

$$= \frac{3!}{(n+1)(n+2)(n+3)} \times \left[-(n+2) + \frac{(n+3)(n+2)}{2!}\right]$$

$$= \frac{3!}{(n+1)(n+2)(n+3)}\left[\frac{(n+1)(n+2)}{2}\right] = \frac{3!}{2(n+3)} = \frac{3}{(n+3)}$$

Hence, option (c) is correct.

7. (c) We have,

$$\frac{3^{2n}}{8} = \frac{9^n}{8} = \frac{1}{8}(8+1)^n = \frac{1}{8}[8^n + {}^nC_1 8^{n-1} + \cdots + {}^nC_{n-1}8 + 1]$$

$$= \frac{1}{8} \times 8\,[8^{n-1} + {}^nC_1 8^{n-2} + \cdots + {}^nC_{n-1}] + \frac{1}{8}$$

$$= [8^{n-1} + {}^nC_1 8^{n-2} + \cdots + {}^nC_{n-1}] + \frac{1}{8} = k + \frac{1}{8} \quad [k \text{ is an integer}]$$

$\therefore$ Fractional part of $\frac{3^{2n}}{8}$ is $\frac{1}{8}$.

Hence, option (c) is correct.

8. (a) We have, $13^n + 7^n - 3^n = 13^n - 3^n + 7^n = 10k + 7^n$

$[\because 13^n - 3^n$ is divisible by $13 - 3 = 10]$

$$= 10k + (7^4)^{\left(\frac{n}{4} - \frac{1}{4}\right)}\cdot 7 = 10k + (2401)^{\frac{n-1}{4}}\cdot 7$$

$$= 10k + (2400 + 1)^{\frac{n-1}{4}}\cdot 7 = 10k + 100k + 7$$

The unit digit is 7, if $\frac{n-1}{4} = k$

$n = 4k + 1, \quad k \in I$

Hence, option (a) is correct.

9. (d) We have, $\quad S = \sum_{r=0}^{n-1} \frac{1}{(n-r)^2}\left(\frac{{}^nC_{r+1}}{{}^nC_r}\right)^2$

$$S = \sum_{r=0}^{n-1} \frac{1}{(n-r)^2}\left[\frac{n-r}{r+1}\right]^2 \quad \left[\because \frac{{}^nC_{r+1}}{{}^nC_r} = \frac{n-r}{r+1}\right]$$

$$S = \sum_{r=0}^{n-1} \frac{1}{(r+1)^2} = \frac{1}{1^2} + \frac{1}{2^2} + \frac{1}{3^2} + \cdots + \frac{1}{n^2}$$

Hence, option (d) is correct.

10. (c) Let coefficient of y^m in $e^{xy}(e^y - 1)^m$

= Coefficient of y^m in $e^{xy}\{e^{my} - {}^mC_1 e^{(m-1)y} + \cdots + (-1)^m \cdot {}^mC_m\}$

$\Rightarrow$ Coefficient of y^m in $e^{xy}\left\{\left(1 + \frac{y}{1!} + \frac{y^2}{2!} + \cdots\right) - 1\right\}^m$

= Coefficient of y^m in

$\{e^{xy}[e^{my} - {}^mC_1 e^{(m-1)y} + \cdots + (-1)^m \cdot {}^mC_m]\}$

$\Rightarrow$ Coefficient of y^m in $e^{xy} y^m \left(1 + \frac{y}{2!} + \frac{y^2}{3!} + \cdots\right)^m$

= Coefficient of y^m in

$\{e^{(x+m)y} - {}^mC_1 \cdot e^{y(m+x-1)} + \cdots + (-1)^m \cdot {}^mC_m \cdot e^{xy}\}$

$\Rightarrow \quad 1 = \frac{(x+m)^m}{m!} - {}^mC_1 \cdot \frac{(x+m-1)^m}{m!} + \cdots + (-1)^m \frac{{}^mC_m \cdot x^m}{m!}$

$\Rightarrow \quad m! = (x+m)^m - {}^mC_1 \cdot (x+m-1)^m$

$+ {}^mC_2 (x+m-2)^m + \cdots + (-1)^m \cdot {}^mC_m$

11. (b) $(2010)^n - 2(1010)^n - 4(510)^n + 5(10)^n$

$\Rightarrow (2000+10)^n - 2(1000+10)^n - 4(500+10)^n + 5(10)^n$.

Here, $(2000+10)^n = {}^nC_0(2000)^n + {}^nC_1(2000)^{n-1} \cdot (10) + \cdots$

$+ {}^nC_{n-1}(2000)(10)^{n-1} + {}^nC_n 10^n$

= Except last term the remaining terms are divisible by 10^3.

$= N_1 10^3 + 10^n$ ...(i)

Similarly, $2(1010)^n = 2N_2 \cdot 10^3 + 2(10)^n$ ...(ii)

$4(510)^n = N_3 \cdot 10^3 + 4(10)^n$ ...(iii)

From Eqs. (i), (ii) and (iii), we get

$= (2010)^n - 2(1010)^n - 4(510)^n + 5(10)^n$

$\Rightarrow N_1 \cdot 10^3 + 10^n - 2N_2 \cdot 10^3 - 2(10)^n$

$- N_3 \cdot (10)^3 - 4(10)^n + 5(10)^n$

$\Rightarrow 10^3(N_1 - 2N_2 - N_3)$ where, $N_1, N_2, N_3 \in N$.

$\therefore$ The given number is always divisible by 10^3.

12. (a, c, d) We have, $(9+\sqrt{80})^n = I + f$

Clearly, $0 < 9 - \sqrt{80} < 1$

$\therefore \quad 0 < (9-\sqrt{80})^n < 1$

Let $(9-\sqrt{80})^n = F, \; 0 < F < 1$

Now, $I + f = (9+\sqrt{80})^n$

$= 9^n + {}^nC_1 9^{n-1}\sqrt{80} + {}^nC_2 9^{n-2}(\sqrt{80})^2 + \cdots + (\sqrt{80})^n$ ...(i)

and $F = (9-\sqrt{80})^n$

$= 9^n - {}^nC_1 9^{n-1}\sqrt{80} + {}^nC_2 9^{n-2}(\sqrt{80})^2 \cdots (-1)^n \sqrt{80}^n$ ...(ii)

On adding Eqs. (i) and (ii), we get

$+ f + F = 2[9^n + {}^nC_2 9^{n-2}(\sqrt{80})^2 + \cdots]$

$= 2[9^n + {}^nC_2 9^{n-2} \cdot 80 + \cdots]$

$\therefore \quad I + f + F$ = an even number

$\Rightarrow \quad f + F$ = an even number $- I$ = an integer

$\because \quad 0 < f < 1$

$\therefore \quad 0 < f + F < 2$

$\therefore \quad f + F = 1 \Rightarrow F = 1 - f = (9-\sqrt{80})^n$

$I + 1$ = an even number

$\therefore \quad I$ = an even number $- 1$ = an odd number

$(I+f)(I-f) = (I+f)(F)$

$= (9+\sqrt{80})^n (9-\sqrt{80})^n = (81-80)^n = 1$

Hence, option (a), (c) and (d) are correct.

13. (a, c) We have,

${}^{100}C_6 + 4 \cdot {}^{100}C_7 + 6 \cdot {}^{100}C_8 + 4 \cdot {}^{100}C_9 + {}^{100}C_{10}$

$\Rightarrow {}^4C_0\,{}^{100}C_{94} + {}^4C_1\,{}^{100}C_{93} + {}^4C_2\,{}^{100}C_{92}$

$+ {}^4C_3\,{}^{100}C_{91} + {}^4C_4\,{}^{100}C_{90}$

$(1+x)^4 = {}^4C_0 + {}^4C_1 x + {}^4C_2 x^2 + {}^4C_3 x^3 + {}^4C_4 x^4$

and $(1+x)^{100} = {}^{100}C_0 + {}^{100}C_1 x^2 + {}^{100}C_3 x^3$

$\cdots + {}^{100}C_{94}x^{94} + {}^{100}C_{95}x^{95} + \cdots + x^{100}]$

$(1+x)^4(1+x)^{100} = {}^4C_0\,{}^{100}C_0 + {}^4C_1\,{}^{100}C_1 x^3 + \cdots$

$(1+x)^{104} = [{}^4C_0\,{}^{100}C_0 + {}^4C_1\,{}^{100}C_1 x^3 + \cdots]$

Coefficient of x^{94}, we get

${}^{104}C_{94} = {}^4C_0\,{}^{100}C_{94} + {}^4C_1\,{}^{100}C_{93}$

$+ {}^4C_2\,{}^{100}C_{92} + {}^4C_3\,{}^{100}C_{91} + {}^4C_4\,{}^{100}C_{90}$

$\therefore \quad x + y = 104 + 94 = 198$

Also, ${}^{100}C_6 + 4 \cdot {}^{100}C_7 + 6 \cdot {}^{100}C_8 + 4 \cdot {}^{100}C_9 + {}^{100}C_{10}$

$\Rightarrow {}^4C_4\,{}^{100}C_6 + {}^4C_3\,{}^{100}C_7 + {}^4C_2\,{}^{100}C_8 + {}^4C_1\,{}^{100}C_9 + {}^4C_{10}\,{}^{100}C_{10}$

Coefficient of x^{10} in $(1+x)^{104}$ is

${}^{104}C_{10} = {}^4C_4\,{}^{100}C_6 + {}^4C_3\,{}^{100}C_7$

$+ {}^4C_2\,{}^{100}C_8 + {}^4C_1\,{}^{100}C_9 + {}^4C_{10}\,{}^{100}C_{10}$

$\therefore \quad x + y = 104 + 10 = 114$

Hence, option (a) and (c) is correct.

14. (b, c) we have, $\left(x^2 + \frac{1}{x^2} - 2\right)^{10} \Rightarrow \left(x - \frac{1}{x}\right)^{20}$

$T_{r+1} = {}^{20}C_r (x)^{20-r}(-x)^{-r}$

$T_{r+1} = {}^{20}C_r (x)^{20-2r}(-1)^r$

The value of constant term in the binomial expansion $\left(x - \frac{1}{x}\right)^{10}$, iff

$20 - 2r = 0 \Rightarrow r = 10$

$T_{10+1} = {}^{20}C_{10}(-1)^{10} = {}^{20}C_{10}$

(a) Number of different dissimilar term in the expansion

$(x_1 + x_2 + x_3 + \cdots + x_{10})^{10} = {}^{10+10-1}C_{10-1} = {}^{19}C_9$

(b) $({}^{10}C_0)^2 + ({}^{10}C_1)^2 + \cdots + ({}^{10}C_{10})^2 = \frac{(2 \times 10)!}{10!\,10!} = \frac{20!}{10!\,10!} = {}^{20}C_{10}$

$\left[\because ({}^nC_1)^2 + ({}^nC_2)^2 + \cdots + ({}^nC_n)^2 = \frac{2n!}{n!\,n!}\right]$

(c) Coefficient of x^{10}, in $(1-x)^{20} = {}^{20}C_{10}(-1)^{10} = {}^{20}C_{10}$

(d) Number of linear arrangement of 20 things of which 10 alike of one kind and rest all are different, taken all at a time $= \frac{20!}{10!}$.

Hence, option (b) and (c) is correct.

15. (a, d) We have, $\sum_{k=0}^{100} {}^{100}C_k (x-2)^{100-k} 3^k$

$(x - 2 + 3)^{100} = \sum_{k=0}^{100} {}^{100}C_k (x-2)^{100-k} 3^k$

Coefficient of x^{50} in $(x+1)^{100} = {}^{100}C_{50} = \frac{100!}{50!\,50!}$

(a) Number of ways in which 50 identical books can be distributed in 100 students, if each student can get atmost one book is, coefficient of x^{50} in $(x^0 + x^1)^{100}$

i.e. coefficient of x^{50} in $(1+x)^{100} = {}^{100}C_{50} = \frac{100!}{50!\,50!}$

(b) Number of ways in which 100 different white balls and 50 identical red balls can be arranged in a circle, if no two red balls are together is $99! \, {}^{99}C_{50}$.

(c) Number of dissimilar terms in $(x_1 + x_2 + x_3 + \cdots + x_{50})^{50}$, is

$${}^{50+50-1}C_{50-1} = {}^{99}C_{49}$$

(d) $$\frac{2\cdot 6\cdot 10\cdot 14 \cdots 198}{50!} = \frac{2^{50}(1\cdot 3\cdot 5\cdot 7 \cdots 99)}{50!}$$

$$= \frac{2^{50}[1\cdot 2\cdot 3\cdot 4\cdot 5 \cdots 99\cdot 100]}{(2\cdot 4\cdot 6\cdot 8 \cdots 100)(50!)}$$

$$= \frac{2^{50}(100!)}{2^{50}[1\cdot 2\cdot 3 \cdots 50](50!)}$$

$$= \frac{100!}{50!\,50!} = {}^{100}C_{50}$$

∴ Hence, option (a) and (d) is correct.

16. (a) $(1-x)^{200} = {}^{200}C_0 - {}^{200}C_1 x + {}^{200}C_2 x^2 - {}^{200}C_3 x^3 + \cdots$ ...(i)

and $\left(\frac{1-x^{149}}{1-x}\right) = (x^{148} + x^{147} + x^{146} + \cdots + x + 1)$ ...(ii)

On multiplying and equating coefficient of x^{148}. is

${}^{200}C_0 - {}^{200}C_1 + {}^{200}C_2 \cdots + {}^{200}C_{148}$

$= \text{Coefficient of } x^{148} \text{ in} \left[(1-x)^{200}\cdot\frac{(1-x^{149})}{(1-x)}\right]$

$= \text{Coefficient of } x^{148} \text{ in } [(1-x)^{199}(1-x^{149})]$

$= \text{Coefficient of } x^{148} \text{ in } (1-x)^{199} = {}^{199}C_{148}$ or ${}^{199}C_{51}$.

17. (b) We have, $(1+px+x^2)^n = 1 + a_1x + a_2x^2 + \cdots + a_{2n}x^{2n}$

Differentiate with respect to x, we get

$n(1+px+x^2)^{n-1}(p+2x) = a_1 + 2a_2x + 3a_2x^2 + \cdots + 2na_{2n}x^{2n-1}$

Multiply by $(1+px+x^2)$ both side, we get

$$n(1+px+x^2)^n(p+2x) = (1+pn+x^2)(a_1 + 2a_2x + 3a_3x^2 + \cdots + 2na_{2n}x^{2n-1})$$

$$n(p+2x)(1+a_1x+a_2x^2+\cdots+a_{2n}x^{2n}) = (1+px+x^2)$$
$$(a_1 + 2a_2x + 3a_3x^2 + \cdots + (r-1)a_{r-1}x^{r-1} + ra_r x^{r-1} + (r+1)a_{r+1}x^r)$$

Equating the coefficient of x^r, we get

$$nPa_r + 2na_{r-1} = (r+1)a_{r+1} + pra_r + (r-1)a_{r-1}$$
$$npa_r - pra_r = (r+1)a_{r+1} + (r-1)a_{r-1} - 2na_{r-1}$$
$$(np - pr)a_r = (r+1)a_{r+1} + (r-1-2n)a_{r-1}$$

Hence, option (b) is correct.

18. (c) We have, $(1+px+x^2)^n = 1 + a_1x + a_2x^2 + a_3x^3 + \cdots + a_{2n}x^{2n}$

On replacing x by x^4, we get

$$(1+px^4+x^8)^n = 1 + a_1x^4 + a_2x^8 + a_3x^{12} + \cdots + a_{2n}x^{8n}$$

On dividing by x^3 both side, we get

$$\frac{(1+px^4+x^8)^n}{x^3} = \frac{1}{x^3} + a_1x + a_2x^5 + a_3x^9 + \cdots + a_{2n}x^{8n-3}$$

[On differentiate w.r.t x, we get]

$$\frac{nx^3(4px^3+8x^7)(1+px^4+x^8)^{n-1} - 3x^2(1+px^4+x^8)^n}{x^6}$$
$$= \frac{-3}{x^4} + a_1 + 5a_2x^4 + 9a_3x^8 + \cdots + (8n-3)a_{2n}x^{8n-4}$$

put $x = 1$ both side, we get

$$n(4p+8)(2+p)^{n-1} - 3(2+p)^n = -3 + a_1 + 5a_2 + 9a_3 + \cdots + (8n-3)a_{2n}$$
$$(p+2)^n\cdot(4n-3) + 3 = a_1 + 5a_2 + 9a_3 + \cdots + (8n-3)a_{2n}$$

when $a_1 + 5a_2 + 9a_3 + \cdots + (8n-3)a_{2n}$ is divide by $(p+2)$ the remainder is 3.

Hence, option (c) is correct.

19. (d) We have, $(1+px+x^2)^n = 1 + a_1x + a_2x^2 + a_3x^3 + \cdots + a_{2n}x^{2n}$

On replacing x by x^2, we get

$$(1+px^2+x^4)^n = 1 + a_1x^2 + a_2x^4 + a_3x^6 + \cdots + a_{2n}x^{4n}$$

On dividing by x, we get

$$\frac{(1+px^2+x^4)^n}{x} = \frac{1}{x} + a_1x + a_2x^3 + a_3x^5 + \cdots + a_{2n}x^{4n-1}$$

On differentiating w.r.t lo x, we get

$$\frac{nx(2px+4x^3)(1+px^2+x^4)^{n-1} - (1+px^2+x^4)^n}{x^2}$$
$$= \frac{-1}{x^2} + a_1 + 3a_2x^2 + 5a_3x^4 + \cdots + (4n-1)\cdot x^{4n-2}\cdot a_{2n}$$

put $p = -3$ and $x = 1$

$$\Rightarrow n(-6+4)(1-3+1)^{n-1} - (1-3+1)^n = -1 + a_1 + 3a_2 + 5a_3 + \cdots + (4n-1)a_{2n}$$
$$\Rightarrow -2n(-1)^{n-1} - (-1)^n + 1 = a_1 + 3a_2 + 5a_3 + \cdots + (4n-1)a_{2n}$$
$$\Rightarrow 2n(-1)^n - (-1)^n + 1 = a_1 + 3a_2 + 5a_3 + \cdots + (4n-1)a_{2n}$$
$$\Rightarrow 2n - 1 + 1 = a_1 + 3a_2 + 5a_3 + \cdots + (4n-1)a_{2n}$$
$$\Rightarrow a_1 + 3a_2 + 5a_3 + \cdots + (4n-1)a_{2n} = 2n \quad [\because (-1)^n = 1, n \text{ is even}]$$

Hence, option (d) is correct.

20. (b) We have, $(1+x+x^2)^{100} = \sum_{r=0}^{200} a_r x^r$

$$\Rightarrow (1+x+x^2)^{100} = a_0 + a_1x + a_2x^2 + \cdots + a_{200}x^{200} \quad ...(i)$$

put $\frac{1}{x}$ in place of x, we get

$$(x^2+x+1)^{100} = a_0x^{200} + a_1x^{199} + a_2x^{198} + \cdots + a_{200} \quad ...(ii)$$

From Eqs. (i) and (ii), we get

$$a_0 = a_{200},\ a_1 = a_{199} \cdots a_r = a_{200-r}$$

∴ $a_{56} = a_{200-56} = a_{144}$

Since, $a_{56} = a_{144}$

Hence, option (b) is correct.

21. (c) Put x = 1 in Eq. (i), we get

$$3^{100} = a_0 + a_1 + a_2 + a_3 + \cdots + a_{100} + a_{101} + \cdots + a_{200}$$
$$3^{100} = 2(a_0 + a_1 + a_2 + \cdots + a_{99}) + a_{100}$$

$[\because a_0 = a_{200}, a_r = a_{200-r}]$

$$\Rightarrow a_0 + a_1 + a_2 + \cdots + a_{99} = \frac{3^{100} - a_{100}}{2}$$

Hence, option (c) is correct.

22. (a) We have,

$(1+x+x^2)^{100} = a_0 + a_1x + a_2x^2 + a_3x^3 + \cdots + a_{200}x^{200}$

On differentiate w.r.t x, we get

$100(1+x+x^2)^{99}(1+2x) = a_1 + 2a_2x + 3a_3x^2 + \cdots + 200a_{200}x^{199}$

On multiply by $(1+x+x^2)$ we, get

$$100(1+x+x^2)^{100}(1+2x) = (1+x+x^2)(a_1 + 2a_2x + 3a_3x^2 + \cdots + 200a_{200}x^{199})$$

$$100(1+2x)(a_0 + a_1x + a_2x^2 + \cdots + a_{200}x^{200}) = (1+x+x^2)$$
$$(a_1 + 2a_2x + 3a_3x^2 + \cdots + 200a_{200}x^{199})$$

Equating the coefficient of x^r, we get

$$100(a_r + 2a_{r-1}) = (r+1)a_{r+1} + ra_r + (r-1)a_{r-1}$$
$$100a_r + 200a_{r-1} = (r+1)a_{r+1} + ra_r + (r-1)a_{r-1}$$
$$(r+1)a_{r+1} = (100-r)a_r + (200-r+1)a_{r-1}$$

put $r = 36$, we get

$$37a_{37} = 64a_{36} + 165a_{35}$$

Hence, option (a) is correct.

23. (d) We have,

$$(1+x)^n = {}^nC_0 + {}^nC_1x + {}^nC_2x^2 + \cdots + {}^nC_nx^n$$

Let $$P = \sum\sum_{0 \le r < s \le n} (r+s)({}^nC_r + {}^nC_s) \quad \ldots(i)$$

On replacing r by $(n-r)$ and s by $(n-s)$, then

$$P = \sum\sum_{0 \le r < s \le n} (n-r+n-s)({}^nC_{n-r} + {}^nC_{n-s})$$

$$P = \sum\sum_{0 \le r < s \le n} (2n-r+s)({}^nC_r + {}^nC_s) \quad \ldots(ii)$$

On adding Eqs. (i) and (ii), we get

$$2P = \sum\sum_{0 \le r < s \le n} 2n({}^nC_r + {}^nC_s)$$

$$P = n\sum\sum_{0 \le r < s \le n} ({}^nC_r + {}^nC_s)$$

$$P = n(n2^n) = n^2 \cdot 2^n$$

Hence, option (d) is correct.

24. (b) $({}^nC_0 + {}^nC_1 + {}^nC_2 + \cdots + {}^nC_n)^2$

$$= \sum_{r=0}^{n} ({}^nC_r)^2 + 2\sum\sum_{0 \le r < s \le n} {}^nC_r\, {}^nC_s$$

$$\Rightarrow \quad (2^n)^2 = {}^{2n}C_n + 2\sum\sum_{0 \le r < s \le n} {}^nC_r\, {}^nC_s$$

$$\sum\sum_{0 \le r < s \le n} {}^nC_r\, {}^nC_s = 2^{2n-1} - \frac{1}{2}\,{}^{2n}C_n$$

We have to find

$$\sum\sum_{0 \le r < s \le n} (r+s)({}^nC_r + {}^nC_s + {}^nC_r \cdot {}^nC_s)$$

$$= \sum\sum_{0 \le r < s \le n} (r+s)({}^nC_r + {}^nC_s) + \sum\sum_{0 \le r < s \le n} (r+s)\, {}^nC_r\, {}^nC_s$$

$$= n^2 \cdot 2^n + \left[n \cdot 2^{2n-1} - \frac{n}{2} \cdot {}^{2n}C_n\right]$$

$$= n^2 \cdot 2^n + \frac{n}{2}[2^{2n} - {}^{2n}C_n]$$

Hence, option (b) is correct.

25. (d) We have, $$\sum_{r=0}^{n}\sum_{s=0}^{n}\sum_{t=0}^{n}\sum_{u=0}^{n}(1)$$

$$\sum_{r=0}^{n}\sum_{s=0}^{n}\sum_{t=0}^{n}(n+1)$$

Similarly, $$\sum_{r=0}^{n}\sum_{s=0}^{n}(n+1)^2$$

$$\Rightarrow \quad \sum_{r=0}^{n}(n+1)^3$$

$$\Rightarrow \quad (n+1)^4$$

Hence, option (d) is correct.

26. (b) $\sum\sum\sum\sum_{0 \le r < s < t < u \le n} (2) \Rightarrow 2 \cdot {}^{n+1}C_4.$

27. (3) We have, $$\sum_{k=0}^{100}\left(\frac{k}{k+1}\right){}^{100}C_k = \frac{a(2^{100})+b}{c}$$

$$\Rightarrow \quad \sum_{k=0}^{100}\left(\frac{k+1-1}{k+1}\right){}^{100}C_k = \frac{a(2)^{100}+b}{c}$$

$$\Rightarrow \quad \sum_{k=0}^{100}\left(1-\frac{1}{k+1}\right){}^{100}C_k = \frac{a(2^{100})+b}{c}$$

$$\sum_{k=0}^{100} {}^{100}C_k - \sum_{k=0}^{100}\frac{{}^{100}C_k}{k+1} = a\frac{(2^{100})+b}{c}$$

$$\Rightarrow \quad 2^{100} - \frac{2^{101}-1}{101} = \frac{a(2^{100})+b}{c}$$

$$\left[\because \sum_{r=0}^{n}\frac{{}^nC_r}{r+1} = \frac{2^{n+1}-1}{n+1}\right]$$

$$\Rightarrow \quad \frac{101(2^{100}) - 2^{101} + 1}{101} = \frac{a(2^{100})+b}{c}$$

$$\Rightarrow \quad \frac{2^{100}(101-2)+1}{101} = \frac{a(2^{100})+b}{c}$$

$$\Rightarrow \quad \frac{99(2^{100})+1}{101} = \frac{a(2^{100})+b}{c}$$

On comparing, we get

$$a = 99,\ b = 1 \text{ and } c = 101$$

$\therefore \quad a+b+c = 99+1+101 = 201$

$\therefore$ Number of digits is 3.

28. (2) $$\lim_{n\to\infty}\sum_{r=0}^{n}\frac{{}^nC_r}{n^r(r+3)} = \lim_{n\to\infty}\sum_{r=0}^{n}\frac{{}^nC_r}{n^r}\cdot\int_0^1 x^{r+2}\,dx.$$

$$= \int_0^1\left(x^2\lim_{x\to\infty}\sum_{r=0}^{n}{}^nC_r\left(\frac{x}{n}\right)^r\right)dx$$

$$= \int_0^1\left(x^2\cdot\lim_{n\to\infty}\left(1+\frac{x}{n}\right)^n\right)dx = \int_0^1 x^2\cdot e^x\cdot dx = e-2$$

But $$e-2 = e-x$$

$\therefore$ $$x = 2$$

29. (3) We heve, $$\lim_{x\to 0}\left[\sum_{r=0}^{n}(-1)^r\,{}^nc_r\left(\sum_{k=0}^{n-r}{}^{n-r}C_k\cdot x^k\cdot 2^k\right)(x^2-x)^r\right]^{1/x} = e^{n\lambda}$$

$$\Rightarrow \quad \lim_{x\to 0}\left[\sum_{r=0}^{n}(-1)^r\,{}^nC_r(1+2x)^{n-r}(x^2-x)^r\right]^{1/x}.$$

$$\Rightarrow \quad \lim_{x\to 0}[(1+2x-x^2+x)^n]^{1/x} \Rightarrow \lim_{x\to 0}(1+3x-x^2)^{n/x}$$

$$\Rightarrow \quad e^{\lim_{x\to 0}\frac{(3x-x^2)\cdot n}{x}} \Rightarrow e^{3n} \qquad \left[\because \lim_{x\to 0}[1+f(x)]^{g(x)} = e^{\lim_{x\to 0}f(x)\cdot g(x)}\right]$$

$$\Rightarrow \quad e^{3n}$$

$\because$ $$\lambda = 3$$

30. (9) ${}^nC_0 - {}^nC_1 + {}^nC_2 - {}^nC_3 + \cdots + (-1)^r \cdot {}^nC_r$

$$\Rightarrow {}^{n-1}C_0 - ({}^{n-1}C_0 + {}^{n-1}C_1) + ({}^{n-1}C_1 + {}^{n-1}C_2) - ({}^{n-1}C_2 + {}^{n-1}C_3) + \cdots + (-1)^r \cdot ({}^{n-1}C_{r-1} + {}^{n-1}C_r)$$

$$\Rightarrow \quad (-1)^r \cdot {}^{n-1}C_r$$

$\therefore \quad (-1)^r \cdot {}^{n-1}C_r = 28$, r must be even

$$\Rightarrow \quad {}^{n-1}C_r = 28 = 7\times 4 = \frac{7\times 8}{2} = {}^8C_2$$

$\therefore \quad n-1 = 8$ and $r = 2$

$$\Rightarrow \quad n = 9.$$

31. (2) Here, $$S = \sum_{K=1}^{\infty}\frac{1}{K^2} = 1^2 + \left(\frac{1}{2}\right)^2 + \left(\frac{1}{3}\right)^2 + \cdots +$$

$$t = \sum_{K=1}^{\infty}\frac{(-1)^{K+1}}{K^2} = (1)^2 - \left(\frac{1}{2}\right)^2 + \left(\frac{1}{3}\right)^2 - \cdots$$

$$S - t = 2\cdot\left\{\left(\frac{1}{2}\right)^2 + \left(\frac{1}{4}\right)^2 + \left(\frac{1}{6}\right)^2 + \cdots +\right\} = 2\cdot\frac{S}{4} = \frac{S}{2}$$

$$\Rightarrow \quad \frac{S}{2} = t$$

$$\therefore \quad \frac{S}{t} = 2$$

JEE Main & AIEEE Archive

1. (d) In the expansion of $(1 + ax + bx^2)(1 - 2x)^{18}$,

Coefficient of x^3 in $(1 + ax + bx^2)(1 - 2x)^{18}$

= Coefficient of x^3 in $(1 - 2x)^{18}$ + Coefficient of x^2 in $a(1 - 2x)^{18}$ + Coefficient of x in $b(1 - 2x)^{18}$

$$= -{}^{18}C_3 \cdot 2^3 + a\,{}^{18}C_2 \cdot 2^2 - b\,{}^{18}C_1 \cdot 2$$

$$\because -{}^{18}C_3 \cdot 2^3 + a\,{}^{18}C_2 \cdot 2^2 - b\,{}^{18}C_1 \cdot 2 = 0$$

$$\Rightarrow \frac{18 \times 17 \times 16}{3 \times 2} \cdot 8 + a \cdot \frac{18 \times 17}{2} 2^2 - b \cdot 18 \cdot 2 = 0$$

$$\Rightarrow 17a - b = \frac{34 \times 16}{3} \quad \ldots(i)$$

Similarly, coefficient of x^4

$${}^{18}C_4 \cdot 2^4 - a \cdot {}^{18}C_3 2^3 + b \cdot {}^{18}C_2 \cdot 2^2 = 0$$

$$\therefore \quad 32a - 32b = 240 \quad \ldots(ii)$$

On solving Eqs. (i) and (ii), we get

$$a = 16 \text{ and } b = \frac{272}{3}$$

2. (c) Consider $\left[\dfrac{x+1}{x^{2/3} - x^{1/3} + 1} - \dfrac{(x-1)}{x - x^{1/2}}\right]^{10}$

$$= \left[\frac{(x^{1/3})^3 + 1^3}{x^{2/3} - x^{1/3} + 1} - \frac{\{(\sqrt{x})^2 - 1\}}{\sqrt{x}(\sqrt{x} - 1)}\right]^{10}$$

$$= \left[\frac{(x^{1/3} + 1)(x^{2/3} + 1 - x^{1/3})}{x^{2/3} - x^{1/3} + 1} - \frac{(\sqrt{x} - 1)(\sqrt{x} + 1)}{\sqrt{x}(\sqrt{x} - 1)}\right]^{10}$$

$$= \left[(x^{1/3} + 1) - \frac{(\sqrt{x} + 1)}{\sqrt{x}}\right]^{10} = (x^{1/3} - x^{-1/2})^{10}$$

$\therefore$ The general term is

$$T_{r+1} = {}^{10}C_r (x^{1/3})^{10-r} (-x^{-1/2})^r = {}^{10}C_r (-1)^r x^{\frac{10-r}{3} - \frac{r}{2}}$$

For independent of x, put $\dfrac{10 - r}{3} - \dfrac{r}{2} = 0$

$$\Rightarrow \quad 20 - 2r - 3r = 0$$

$$\Rightarrow \quad 20 = 5r \Rightarrow r = 4$$

$$\therefore \quad T_5 = {}^{10}C_4 = \frac{10 \times 9 \times 8 \times 7}{4 \times 3 \times 2 \times 1} = 210$$

3. (a) $(\sqrt{3} + 1)^{2n} = {}^{2n}C_0(\sqrt{3})^{2n} + {}^{2n}C_1(\sqrt{3})^{2n-1} + {}^{2n}C_2(\sqrt{3})^{2n-2} + \cdots + {}^{2n}C_{2n}(\sqrt{3})^{2n-2n}$

$(\sqrt{3} - 1)^{2n} = {}^{2n}C_0(\sqrt{3})^{2n}(-1)^0 + {}^{2n}C_1(\sqrt{3})^{2n-1}(-1)^1 + {}^{2n}C_2(\sqrt{3})^{2n-2}(-1)^2 + \cdots + {}^{2n}C_{2n}(\sqrt{3})^{2n-2n}(-1)^{2n}$

Adding both the binomial expansions above, we get

$(\sqrt{3} + 1)^{2n} - (\sqrt{3} - 1)^{2n} = 2[{}^{2n}C_1(\sqrt{3})^{2n-1} + {}^{2n}C_3(\sqrt{3})^{2n-3} + {}^{2n}C_5(\sqrt{3})^{2n-5} + \cdots + {}^{2n}C_{2n-1}(\sqrt{3})^{2n-(2n-1)}]$

Which is most certainly an irrational number because of odd powers of $\sqrt{3}$ in each of the terms.

4. (b) Here, $(1 - x - x^2 + x^3)^6$

$$= \{(1 - x) - x^2(1 - x)\}^6 = \{(1 - x)(1 - x^2)\}^6$$

$$= (1 - x)^6 \cdot (1 - x^2)^6 = \left\{\sum_{r=0}^{6} (-1)^r\, {}^6C_r \cdot x^r\right\}\left\{\sum_{s=0}^{6} (-1)^s\, {}^6C_s \cdot x^{2s}\right\}$$

$$= \sum_{r=0}^{6}\sum_{s=0}^{6} (-1)^{r+s} \cdot {}^6C_r \cdot {}^6C_s \cdot x^{r+2s}$$

For coefficient of x^7, we have

$r + 2s = 7$ i.e. $s = 1, r = 5$

or $s = 2, r = 3$

or $s = 3, r = 1$

$\therefore$ Coefficient of x^7 is

$$\{(-1)^{5+1} \cdot {}^6C_5 \cdot {}^6C_1\} + \{(-1)^{3+2} \cdot {}^6C_3 \cdot {}^6C_2\} + \{(-1)^{1+3} \cdot {}^6C_1 \cdot {}^6C_3\}$$

$$\Rightarrow \quad (36) - (20)(15) + 6(20)$$

$$\Rightarrow \quad 36 - 300 + 120 = -144$$

5. (d) Here, we use $\sum_{r=0}^{n} {}^nC_r = 2^n$

$$\because \quad S_1 = \sum_{j=1}^{10} j(j-1)\frac{10!}{j(j-1)(j-2)!(10-j)!}$$

$$= 90\sum_{j=2}^{10}\frac{8!}{(j-2)![8-(j-2)]!} = 90 \cdot 2^8$$

$$\text{and } S_2 = \sum_{j=1}^{10} j\frac{10!}{j(j-1)![9-(j-1)]!}$$

$$= 10\sum_{j=1}^{10}\frac{9!}{(j-1)![9-(j-1)]!} = 10 \cdot 2^9$$

$$\text{Also, } S_3 = \sum_{j=1}^{10}[j(j-1) + j]\frac{10!}{j!(10-j)!}$$

$$= \sum_{j=1}^{10} j(j-1)\,{}^{10}C_j = \sum_{j=1}^{10} j\,{}^{10}C_j$$

$$= 90 \cdot 2^8 + 10 \cdot 2^9 = 90.2^8 + 20.2^8$$

$$= 110 \cdot 2^8 = 55 \cdot 2^9$$

Hence, Statement I is correct and Statement II is incorrect.

6. (b) $8^{2n} - (62)^{2n+1} = (1 + 63)^n - (63 - 1)^{2n+1}$

$$= (1 + 63)^n + (1 - 63)^{2n+1}$$

$$= (1 + {}^nC_1\, 63 + {}^nC_2 (63)^2 + \cdots + (63)^n + [1 - {}^{(2n+1)}C_1\, 63 + {}^{(2n+1)}C_2 (63)^2 + \cdots + (-1)(63)^{(2n+1)}]$$

$$= 2 + 63\,[{}^nC_1 + {}^nC_2(63) + \cdots + (63)^{n-1} + {}^{(2n+1)}C_2(63) - \cdots - (63)^{(2n)}]$$

Hence, the remainder is 2.

7. (b) Since, $\sum_{r=0}^{n} {}^nC_r \cdot x^r = (1 + x)^n$

On multiplying by x, we get

$$\sum_{r=0}^{n} {}^nC_r \cdot x^{r+1} = x(1 + x)^n$$

On differentiating w.r.t. x, we get

$$\sum_{r=0}^{n} (r + 1) \cdot {}^nC_r \cdot x^r = (1 + x)^n + nx(1 + x)^{n-1}$$

Hence, Statement II is correct.

If $x = 1$, then $\sum_{r=0}^{n} (r + 1) \cdot {}^nC_r = 2^n + n(2)^{n-1} = (n + 2)2^{n-1}$

Hence, Statements I and II are correct and Statement II is a correct explanation of Statement I.

8. (d) Since, in a binomial expansion of $(a - b)^n, n \geq 5$, the sum of 5th and 6th terms is equal to zero.

$$\therefore \quad {}^nC_4\, a^{n-4}(-b)^4 + {}^nC_5\, a^{n-5}(-b)^5 = 0$$

$$\Rightarrow \quad \frac{n!}{(n-4)!4!} a^{n-4} \cdot b^4 - \frac{n!}{(n-5)!5!} a^{n-5} b^5 = 0$$

$$\Rightarrow \quad \frac{n!}{(n-5)!4!} a^{n-5} \cdot b^4 \left(\frac{a}{n-4} - \frac{b}{5}\right) = 0$$

$$\Rightarrow \quad \frac{a}{b} = \frac{n-4}{5}$$

9. (b) We know that,

$$(1+x)^{20} = {}^{20}C_0 + {}^{20}C_1x + \cdots + {}^{20}C_{10}x^{10} + \cdots + {}^{20}C_{20}x^{20}$$

On putting $x = -1$ in the above expansion, we get

$$0 = {}^{20}C_0 - {}^{20}C_1 + \cdots - {}^{20}C_9 + {}^{20}C_{10} - {}^{20}C_{11} + \cdots + {}^{20}C_{20}$$

$$\Rightarrow \quad 0 = {}^{20}C_0 - {}^{20}C_1 + \cdots - {}^{20}C_9 + {}^{20}C_{10} - {}^{20}C_9 + \cdots + {}^{20}C_0$$

$$\Rightarrow \quad 0 = 2({}^{20}C_0 - {}^{20}C_1 + \cdots - {}^{20}C_9) + {}^{20}C_{10}$$

$$\Rightarrow \quad {}^{20}C_{10} = 2({}^{20}C_0 - {}^{20}C_1 + \cdots + {}^{20}C_{10})$$

$$\Rightarrow \quad {}^{20}C_0 - {}^{20}C_1 + \cdots + {}^{20}C_{10} = \frac{1}{2}\,{}^{20}C_{10}$$

10. (b) Now, $(1-ax)^{-1}(1-bx)^{-1}$

$$= (1 + ax + a^2x^2 + \cdots)(1 + bx + b^2x^2 + \cdots)$$

Hence, a_n = Coefficient of x^n in $(1-ax)^{-1}(1-bx)^{-1}$

$$= a^0b^n + ab^{n-1} + \cdots + a^nb^0$$

$$= a^0b^n\left(1 + \frac{a}{b} + \left(\frac{a}{b}\right)^2 + \cdots\right)$$

$$= a^0b^n\left[\frac{\left(\frac{a}{b}\right)^{n+1} - 1}{\frac{a}{b} - 1}\right]$$

$\left[\text{sum of } n\text{th terms of a GP with common ratio } \frac{a}{b}\right]$

$$= \frac{b^n(a^{n+1} - b^{n+1})}{a-b}\cdot\frac{b}{b^{n+1}}$$

$$= \frac{a^{n+1} - b^{n+1}}{a-b}$$

11. (c) $(1-y)^m(1+y)^n = 1 + a_1y + a_2y^2 + a_3y^3 + \cdots$

On differentiating w.r.t. y, we get

$$-m(1-y)^{m-1}(1+y)^n + (1-y)^m n(1+y)^{n-1}$$

$$= a_1 + 2a_2y + 3a_3y^2 + \cdots \quad \ldots(\text{i})$$

On putting $y = 0$ in Eq. (i), we get

$$-m + n = a_1 = 10 \quad \ldots(\text{ii})$$

$[\because a_1 = 10, \text{ given}]$

On differentiating Eq. (i), we get

$$-m[-(m-1)(1-y)^{m-2}(1+y)^n + (1-y)^{m-1}n(1+y)^{n-1}]$$

$$+ n[-m(1-y)^{m-1}(1+y)^{n-1} + (1-y)^m(n-1)(1+y)^{n-2}]$$

$$= 2a_2 + 6a_3y + \cdots \quad \ldots(\text{iii})$$

On putting $y = 0$ in Eq. (iii), we get

$$-m[-(m-1) + n] + n[-m + (n-1)] = 2a_2 = 20$$

$$\Rightarrow \quad m(m-1) - mn - mn + n(n-1) = 20$$

$$\Rightarrow \quad m^2 + n^2 - m - n - 2mn = 20$$

$$\Rightarrow \quad (m-n)^2 - (m+n) = 20$$

$$\Rightarrow \quad 100 - (m+n) = 20$$

$$\therefore \quad m + n = 80 \quad \ldots(\text{iv})$$

On solving Eqs. (ii) and (iv), we get

$$m = 35 \quad \text{and} \quad n = 45$$

12. (a) Since, we know that ${}^nC_r + {}^nC_{r-1} = {}^{n+1}C_r$.

$$\therefore \quad {}^{50}C_4 + {}^{55}C_3 + {}^{54}C_3 + {}^{53}C_3 + {}^{52}C_3 + {}^{51}C_3 + {}^{50}C_3$$

$$= {}^{50}C_3 + {}^{50}C_4 + {}^{51}C_3 + {}^{52}C_3 + {}^{53}C_3 + {}^{54}C_3 + {}^{55}C_3$$

$$= {}^{51}C_4 + {}^{51}C_3 + {}^{52}C_3 + {}^{53}C_3 + {}^{54}C_3 + {}^{55}C_3$$

$[\because {}^nC_r + {}^nC_{r-1} = {}^{n+1}C_r]$

$$= {}^{52}C_4 + {}^{52}C_3 + {}^{53}C_3 + {}^{54}C_3 + {}^{55}C_3$$

$$= {}^{53}C_4 + {}^{53}C_3 + {}^{54}C_3 + {}^{55}C_3 = {}^{54}C_4 + {}^{54}C_3 + {}^{55}C_3$$

$$= {}^{55}C_4 + {}^{55}C_3 = {}^{56}C_4$$

13. (b) Since, the coefficient of given terms are ${}^mC_{r-1}, {}^mC_r, {}^mC_{r+1}$, respectively and they also in AP.

$$\therefore \quad {}^mC_{r-1} + {}^mC_{r+1} = 2\cdot{}^mC_r$$

$$\Rightarrow \quad \frac{m!}{(r-1)!(m-r+1)!} + \frac{m!}{(r+1)!(m-r-1)!} = 2\,\frac{m!}{r!(m-r)!}$$

$$\Rightarrow \quad \frac{1}{(m-r+1)(m-r)} + \frac{1}{(r+1)r} = \frac{2}{r(m-r)}$$

$$\Rightarrow \quad \frac{r(r+1) + (m-r+1)(m-r)}{r(r+1)(m-r+1)(m-r)} = \frac{2}{r(m-r)}$$

$$\Rightarrow \quad r^2 + r + m^2 + r^2 - 2mr + m - r$$

$$= 2(mr - r^2 + r + m - r + 1)$$

$$\Rightarrow \quad 4r^2 - 4mr - m - 2 + m^2 = 0$$

$$\Rightarrow \quad m^2 - m(4r+1) + 4r^2 - 2 = 0$$

14. (a) Let x^7 be contained in $(r+1)$th term in the expansion of $\left(ax^2 + \frac{1}{bx}\right)^{11}$.

$$\therefore \quad T_{r+1} = {}^{11}C_r(ax^2)^{11-r}\left(\frac{1}{bx}\right)^r = {}^{11}C_r\,\frac{a^{11-r}}{b^r}\cdot x^{22-3r}$$

For coefficient of x^7, put

$$22 - 3r = 7 \quad \Rightarrow r = 5$$

$$\therefore \quad T_6 = {}^{11}C_5\,\frac{a^6}{b^5}\cdot x^7$$

So, the coefficient of x^7 in the expansion of $\left(ax^2 + \frac{1}{bx}\right)^{11}$ is ${}^{11}C_5\,\frac{a^6}{b^5}$.

Similarly, coefficient of x^{-7} in the expansion of $\left(ax - \frac{1}{bx^2}\right)^{11}$ is ${}^{11}C_6\,\frac{a^5}{b^6}$.

According to the given condition,

$${}^{11}C_5\,\frac{a^6}{b^5} = {}^{11}C_6\,\frac{a^5}{b^6} \quad \Rightarrow \quad \frac{a^6}{b^5} = \frac{a^5}{b^6}$$

$$\therefore \quad ab = 1$$

15. (b)

$$\frac{(1+x)^{3/2} - \left(1 + \frac{1}{2}x\right)^3}{(1-x)^{1/2}}$$

$$= \frac{\left(1 + \frac{3}{2}x + \frac{\frac{3}{2}\cdot\frac{1}{2}}{2}x^2\right) - \left(1 + \frac{3x}{2} + \frac{3\cdot 2}{2}\cdot\frac{x^2}{4}\right)}{(1-x)^{1/2}}$$

[neglecting higher powers of x]

$$= -\frac{3x^2}{8}(1-x)^{-1/2} = -\frac{3x^2}{8}\left(1 + \frac{1}{2}x + \frac{\frac{1}{2}\cdot\frac{3}{2}}{2}\cdot x^2\right) = -\frac{3x^2}{8}$$

[since, higher powers of x can be neglected]

16. (c) The coefficient of the middle term in powers of x of $(1+\alpha x)^4 = {}^4C_2\alpha^2$.

The coefficient of the middle term in powers of x of $(1-\alpha x)^6 = {}^6C_3(-\alpha)^3$.

According to given condition,

$${}^4C_2\alpha^2 = {}^6C_3(-\alpha)^3$$

$$\Rightarrow \quad \frac{4!}{2!2!}\alpha^2 = -\frac{6!}{3!\,3!}\alpha^3$$

$$\Rightarrow \quad 6\alpha^2 = -20\alpha^3 \quad \Rightarrow \quad \alpha = -\frac{6}{20}$$

$$\therefore \quad \alpha = -\frac{3}{10}$$

17. (b) The coefficient of x^n in the expansion of $(1+x)(1-x)^n$

$= $ Coefficient of x^n in $(1-x)^n +$ Coefficient of x^{n-1} in $(1-x)^n$

$[\because (1+x)(1-x)^n = (1+x)^n + x(1-x)^n]$

$= (-1)^n\,{}^nC_n + (-1)^{n-1}\,{}^nC_{n-1}$

$= (-1)^n\left(1-\dfrac{n!}{1!(n-1)!}\right) = (-1)^n(1-n)$

18. (a) Given that, $S_n = \sum_{r=0}^{n}\dfrac{1}{{}^nC_r}$

$S_n = \sum_{r=0}^{n}\dfrac{1}{{}^nC_{n-r}}$ $\quad [\because {}^nC_r = {}^nC_{n-r}]$

$\Rightarrow \quad nS_n = \sum_{r=0}^{n}\dfrac{n}{{}^nC_{n-r}}$

$\Rightarrow \quad nS_n = \sum_{r=0}^{n}\left(\dfrac{n-r}{{}^nC_{n-r}} + \dfrac{r}{{}^nC_{n-r}}\right)$

$\Rightarrow \quad nS_n = \sum_{r=0}^{n}\dfrac{n-r}{{}^nC_{n-r}} + \sum_{r=0}^{n}\dfrac{r}{{}^nC_r}$

$\Rightarrow \quad nS_n = \left(\dfrac{n}{{}^nC_n} + \dfrac{n-1}{{}^nC_{n-1}} + \dots + \dfrac{1}{{}^nC_1}\right) + \sum_{r=0}^{n}\dfrac{r}{{}^nC_r}$

$\Rightarrow \quad nS_n = t_n + t_n \quad \left[\because t_n = \sum_{r=0}^{n}\dfrac{r}{{}^nC_r}, \text{given}\right]$

$\Rightarrow \quad nS_n = 2t_n \Rightarrow \dfrac{t_n}{S_n} = \dfrac{n}{2}$

19. (b) The general term of $(\sqrt{3}+\sqrt[8]{5})^{256}$ is

$T_{r+1} = {}^{256}C_r(3)^{(256-r)/2}(5)^{r/8}$

For integral terms, $\dfrac{256-r}{2}$ and $\dfrac{r}{8}$ are both positive integers.

i.e. $r = 0, 8, 16, 24, 32, \dots, 256$

Hence, total number of terms are 33.

20. (c) Since, $(r+1)$th term in the expansion of $(1+x)^{27/5}$.

$= \dfrac{\frac{27}{5}\left(\frac{27}{5}-1\right)\dots\left(\frac{27}{5}-r+1\right)}{r!}x^r$

Now, this term will be negative, if the last factor in numerator is the only one negative factor.

$\Rightarrow \quad \dfrac{27}{5} - r + 1 < 0 \Rightarrow \dfrac{32}{5} < r$

$\Rightarrow \quad 6.4 < r \Rightarrow$ Least value of r is 7.

Thus, first negative term will be 8th.

21. (d) $(1+2x+3x^2+\dots)^{-3/2} = [(1-x)^{-2}]^{-3/2} = (1-x)^3$

So, coefficient of x^5 in $(1+2x+3x^2+\dots)^{-3/2}$

$=$ Coefficient of x^5 in $(1-x)^3 = 0$

22. (d)

$(1+x+x^2+x^3+\dots)^2 = [(1-x)^{-1}]^2 = (1-x)^{-2}$

Coefficient of x^n in $(1+x+x^2+\dots)^2 =$ Coefficient of x^n in $(1-x)^{-2}$

$= {}^{n+2-1}C_{2-1} = {}^{n+1}C_1 = n+1$

JEE Advanced & IIT JEE Archive

1. (c) Coefficient of x^r in $(1+x)^n$ is ${}^nC_r$.

In this type of questions, we find different composition of terms where product will give us x^{11}.

Coefficient of x^{11} in $(1+x^2)^4(1+x^3)^7(1+x^4)^{12}$

Now, consider the following cases for x^{11} in $(1+x^2)^4(1+x^3)^7(1+x^4)^{12}$

Coefficient of $x^0\,x^3\,x^8$; $\quad$ Coefficient of $x^2\,x^9\,x^0$

Coefficient of $x^4\,x^3\,x^4$; $\quad$ Coefficient of $x^8\,x^3\,x^0$

$= {}^4C_0\times{}^7C_1\times{}^{12}C_2 + {}^4C_1\times{}^7C_3\times{}^{12}C_0 + {}^4C_2\times{}^7C_1\times{}^{12}C_1 + {}^4C_4\times{}^7C_1\times{}^{12}C_0$

$= 462 + 140 + 504 + 7 = 1113$

2. (6) Let the three consecutive terms in $(1+x)^{n+5}$ be t_r, t_{r+1}, t_{r+2}.

Having coefficients ${}^{n+5}C_{r-1}, {}^{n+5}C_r, {}^{n+5}C_{r+1}$

given, ${}^{n+5}C_{r-1} : {}^{n+5}C_r : {}^{n+5}C_{r+1} = 5:10:14$

$\therefore \quad \dfrac{{}^{n+5}C_r}{{}^{n+5}C_{r-1}} = \dfrac{10}{5}$ and $\dfrac{{}^{n+5}C_{r+1}}{{}^{n+5}C_r} = \dfrac{14}{10}$

$\Rightarrow \quad \dfrac{n+5-(r-1)}{r} = 2$ and $\dfrac{n-r+5}{r+1} = \dfrac{7}{5}$

$\Rightarrow \quad n-r+6 = 2r$ and $5n-5r+25 = 7r+7$

$\Rightarrow \quad n+6 = 3r$ and $5n+18 = 12r$

$\therefore \quad \dfrac{n+6}{3} = \dfrac{5n+18}{12}$

$\Rightarrow \quad 4n+24 = 5n+18 \Rightarrow n = 6$

3. (d) $A_r =$ Coefficient of x^r in $(1+x)^{10} = {}^{10}C_r$

$B_r =$ Coefficient of x^r in $(1+x)^{20} = {}^{20}C_r$

$C_r =$ Coefficient of x^r in $(1+x)^{30} = {}^{30}C_r$

$\therefore \quad \sum_{r=1}^{10}A_r(B_{10}B_r - C_{10}A_r) = \sum_{r=1}^{10}A_rB_{10}B_r - \sum_{r=1}^{10}A_rC_{10}A_r$

$= \sum_{r=1}^{10}{}^{10}C_r\,{}^{20}C_{10}\,{}^{20}C_r - \sum_{r=1}^{10}{}^{10}C_r\,{}^{30}C_{10}\,{}^{10}C_r$

$= \sum_{r=1}^{10}{}^{10}C_{10-r}\,{}^{20}C_{10}\,{}^{20}C_r - \sum_{r=1}^{10}{}^{10}C_{10-r}\,{}^{30}C_{10}\,{}^{10}C_r$

$= {}^{20}C_{10}\sum_{r=1}^{10}{}^{10}C_{10-r}\cdot{}^{20}C_r - {}^{30}C_{10}\sum_{r=1}^{10}{}^{10}C_{10-r}\,{}^{10}C_r$

$= {}^{20}C_{10}({}^{30}C_{10}-1) - {}^{30}C_{10}({}^{20}C_{10}-1)$

$= {}^{30}C_{10} - {}^{20}C_{10} = C_{10} - B_{10}$

4. (c) Let $A = \binom{30}{0}\binom{30}{10} - \binom{30}{1}\binom{30}{11} + \binom{30}{2}\binom{30}{12} - \dots + \binom{30}{20}\binom{30}{30}$

$\therefore \quad A = {}^{30}C_0\cdot{}^{30}C_{10} - {}^{30}C_1\cdot{}^{30}C_{11} + {}^{30}C_2\cdot{}^{30}C_{12} - \dots + {}^{30}C_{20}\cdot{}^{30}C_{30}$

$=$ coefficient of x^{20} in $(1+x)^{30}(1-x)^{30}$

$=$ coefficient of x^{20} in $(1-x^2)^{30}$

$=$ coefficient of x^{20} in $\sum_{r=0}^{30}(-1)^r\,{}^{30}C_r(x^2)^r$

$= (-1)^{10}\,{}^{30}C_{10}$ (for coefficient of x^{20}, put $r = 10$) $= {}^{30}C_{10}$

2. Permutation and Combination

DPP-1 Factorial and Exponent of Prime Value p in $(n!)$

1. (i) Consider $\text{LHS} = \frac{1}{4!} + \frac{1}{5!} + \frac{1}{6!}$

$$= \frac{1}{4!}\left(1 + \frac{1}{5} + \frac{1}{6 \times 5}\right)$$

$$= \frac{1}{4!}\left(1 + \frac{1}{5} + \frac{1}{6 \times 5}\right)$$

$$= \frac{1}{4!}\left(\frac{30 + 6 + 1}{6 \times 5}\right)$$

$$= \frac{37}{6!} = \text{RHS}$$

(ii) Consider,

$$\text{LHS} = \frac{1}{9!} + \frac{1}{10!} + \frac{1}{11!} + \frac{1}{12!}$$

$$= \frac{1}{9!}\left(1 + \frac{1}{10} + \frac{1}{11 \times 10} + \frac{1}{12 \times 11 \times 10}\right)$$

$$= \frac{1}{9!}\left(\frac{1320 + 132 + 12 + 1}{12 \times 11 \times 10}\right)$$

$$= \frac{1465}{12!}$$

(iii) Consider,

$$\text{LHS} = n(n-1)(n-2)\cdots(n-r+1)$$

$$= n(n-1)(n-2)\cdots[n-(r-1)]$$

$$= n(n-1)(n-2)\cdots(n-(r-1))(n-r)!$$

[On multiplying numerator and denominator by $(n-r)!$]

$$= \frac{n!}{(n-r)!} = \text{RHS}$$

(iv) Consider,

$$\text{LHS} = 1 \cdot 3 \cdot 5 \cdots (2n-1)(2n+1) \cdot 2^n$$

$$= \frac{1 \cdot 2 \cdot 3 \cdot 4 \cdots (2n-1)(2n)(2n+1)}{2 \cdot 4 \cdot 6 \cdots (2n)} \cdot 2^n$$

[On multiplying Nr and Dr by $2 \cdot 4 \cdot 6 \cdots (2n)$]

$$= \frac{(2n+1)!}{2^n (1 \cdot 2 \cdot 3 \cdots n)} \cdot 2^n$$

$$= \frac{(2n+1)!}{n!}$$

$$= \text{RHS}$$

(v) Consider $2 \cdot 6 \cdot 10 \cdot 14 \cdots$ to n factors

$$= 2^n (1 \cdot 3 \cdot 5 \cdot 7 \cdots \text{ to } n \text{ factors})$$

$$= \frac{2^n [1 \cdot 2 \cdot 3 \cdot 4 \cdot 5 \cdot 6 \cdots (2n)]}{2 \cdot 4 \cdot 6 \cdots \text{ to } n \text{ factors}}$$

[On multiplying Nr and Dr by first n even numbers]

$$= \frac{2^n (2n)!}{2^n (1 \cdot 2 \cdot 3 \cdots \text{ to } n \text{ factors})}$$

$$= \frac{2^n (2n)!}{2^n \cdot n!}$$

$$= \frac{(2n)!}{n!} \quad \ldots\text{(i)}$$

Now, consider

$$= (n+1)(n+2)\cdots(2n)$$

$$= \frac{(1 \cdot 2 \cdot 3 \cdot 4 \cdots n)(n+1)(n+2)\cdots(2n)}{(1 \cdot 2 \cdot 3 \cdot 4 \cdots n)}$$

$$= \frac{(2n)!}{n!} \quad \ldots\text{(ii)}$$

From Eqs. (i) and (ii), we get

$$2 \cdot 6 \cdot 10 \cdot 14 \ldots \text{ to } n \text{ factors} = \frac{(2n)!}{(n)!}$$

$$= (n+1)(n+2)\cdots(2n).$$

(vi) Consider, $\text{LHS} = (100)!$

$$= 100 \cdot 99 \cdot 98 \cdot 97 \cdots 5 \cdot 4 \cdot 3 \cdot 2 \cdot 1$$

$$= (100 \cdot 98 \cdot 96 \cdots 4 \cdot 2)(99 \cdot 97 \cdots 5 \cdot 3 \cdot 1)$$

$$= 2^{50}(50 \cdot 49 \cdot 48 \cdots 2 \cdot 1)(99 \cdot 97 \cdots 5 \cdot 3 \cdot 1)$$

$$= 2^{50} \cdot (50)!\,(1 \cdot 3 \cdot 5 \cdot 7 \cdots 99)$$

$$= \text{RHS}$$

2. (i) We have, $(x)! = 132(x-2)!$

$\Rightarrow \quad x(x-1)(x-2)! = 132(x-2)!$

$\Rightarrow \quad x(x-1) = 132$

$\Rightarrow \quad x(x-1) = 12 \times (12-1)$

$\therefore \quad x = 12$

(ii) We have, $(x+2)! = 870(x)!$

$\Rightarrow \quad (x+2)(x+1)(x)! = 870(x)!$

$\Rightarrow \quad (x+2)(x+1) = 870$

$\Rightarrow \quad (x+2)(x+1) = 30 \times 29 = (28+2)(28+1)$

$\therefore \quad x = 28$

(iii) We have, $\dfrac{(x+3)!\,(2x-1)!}{(2x+1)!\,(x+2)!} = \dfrac{7}{72}$

$\Rightarrow \quad \dfrac{(x+3)(x+2)!\,(2x-1)!}{(2x+1)(2x)(2x-1)!\,(x+2)!} = \dfrac{7}{72}$

$\Rightarrow \quad \dfrac{x+3}{(2x+1)(2x)} = \dfrac{7}{72}$

$\Rightarrow \quad 72x + 216 = 7(4x^2 + 2x)$

$\Rightarrow \quad 72x + 216 = 28x^2 + 14x$

$\Rightarrow \quad 28x^2 - 58x - 216 = 0$

$\Rightarrow \quad 14x^2 - 29x - 108 = 0$

$\Rightarrow \quad 14x^2 - 56x + 27x - 108 = 0$

$\Rightarrow \quad 14x(x-4) + 27(x-4) = 0$

$\therefore \quad (x-4)(14x+27) = 0$

$\therefore \quad x = 4$ [since, x cannot be negative]

(iv) We have, $\dfrac{x!}{2!\,(x-2)!} : \dfrac{x!}{4!\,(x-4)!} = \dfrac{2}{1}$

$\Rightarrow \quad \dfrac{x!}{2!\,(x-2)!} \times \dfrac{4!\,(x-4)!}{x!} = \dfrac{2}{1}$

$\Rightarrow \quad \dfrac{4 \cdot 3 \cdot 2!\,(x-4)!}{2!\,(x-2)(x-3)(x-4)!} = \dfrac{2}{1}$

$\Rightarrow \quad \dfrac{12}{(x-2)(x-3)} = \dfrac{2}{1}$

$\Rightarrow \quad (x-2)(x-3) = 6$

$\Rightarrow \quad x^2 - 5x + 6 = 6$

$\Rightarrow \quad x^2 - 5x = 0$

$\therefore \quad x = 5$ [since, x cannot be less than 4]

3. $E_2(65!) = \left[\dfrac{65}{2}\right] + \left[\dfrac{65}{2^2}\right] + \left[\dfrac{65}{2^3}\right] + \left[\dfrac{65}{2^4}\right] + \left[\dfrac{65}{2^5}\right] + \left[\dfrac{65}{2^6}\right]$

$$= 32 + 16 + 8 + 4 + 2 + 1 = 63$$

4. $E_3(100)! = \left[\dfrac{100}{3}\right] + \left[\dfrac{100}{3^2}\right] + \left[\dfrac{100}{3^3}\right] + \left[\dfrac{100}{3^4}\right]$

$$= 33 + 11 + 3 + 1 = 48$$

5. $E_2(10!) = \left[\frac{10}{2}\right] + \left[\frac{10}{2^2}\right] + \left[\frac{10}{2^3}\right] = 8 = p$

$E_3(10!) = \left[\frac{10}{3}\right] + \left[\frac{10}{3^2}\right] = 4 = q$

$E_5(10!) = \left[\frac{10}{5}\right] = 2 = r$

$E_7(10!) = \left[\frac{10}{7}\right] = 1 = s$

6. $x = 20 \times 19 \times 18 \times 17 \times \cdots \times 11$

$= \frac{\{20 \times 19 \times 18 \times \cdots \times 11\} \cdot (10!)}{(10!)} = \frac{(20)!}{(10)!}$

$E_2(20!) = \left[\frac{20}{2}\right] + \left[\frac{20}{2^2}\right] + \left[\frac{20}{2^3}\right] + \left[\frac{20}{2^4}\right]$

$= 10 + 5 + 2 + 1 = 18$

$E_2(10!) = \left[\frac{10}{2}\right] + \left[\frac{10}{2^2}\right] + \left[\frac{10}{2^3}\right]$

$= 5 + 2 + 1 = 8$

$\therefore \quad E_2\left\{\frac{(20)!}{(10)!}\right\} = 18 - 8 = 10$

7. $E_2(50!) = \left[\frac{50}{2}\right] + \left[\frac{50}{2^2}\right] + \left[\frac{50}{2^3}\right] + \left[\frac{50}{2^4}\right] + \left[\frac{50}{2^5}\right]$

$= 25 + 12 + 6 + 3 + 1$
$= 47$

$E_3(50!) = \left[\frac{50}{3}\right] + \left[\frac{50}{3^2}\right] + \left[\frac{50}{3^3}\right]$

$= 16 + 5 + 1 = 22$

$E_{2^2}(50!) = \left[\frac{47}{2}\right] = 23$

$\therefore E_{12}(50!) = \text{min. } \{22, 23\} = 22$

8. $E_2(60!) = \left[\frac{60}{2}\right] + \left[\frac{60}{2^2}\right] + \left[\frac{60}{2^3}\right] + \left[\frac{60}{2^4}\right] + \left[\frac{60}{2^5}\right]$

$= 30 + 15 + 7 + 3 + 1 = 56$

$E_5(60!) = \left[\frac{60}{5}\right] + \left[\frac{60}{5^2}\right]$

$= 12 + 2 = 14$

$\therefore E_{10}(60!) = 14$

9. (d) Clearly, $105! + 2$ is divisible by 2, as

$105! + 2 = (1 \cdot 2 \cdot 3 \cdots 105) + 2 = 2((1 \cdot 3 \cdot 4 \cdots 105) + 1)$

Similarly, the numbers $105! + 3$, $105! + 4, \cdots 105! + 105$ are divisible by 3, 4, 5 $\cdots$ 105, respectively.

So, none of the given numbers are prime, as prime number has factors only 1 and itself.

10. (c) Since, perfect square of every natural number ends with {0,1, 4, 5, 6, 9}, but square of any natural number cannot have unit's digit {2, 3, 7, 8} ...(i)

Also, $1! + 2! + 3! + 4! + 5! + \dots + x!$, when $x \geq 4$ units place ends with 3.

Thus, $x \in \{1, 2, 3\}$

Now, if $x = 1 \Rightarrow 1! = 1$ [perfect square]

if $x = 2 \Rightarrow 1! + 2! = 3$ [cannot be perfect square]

if $x = 3 \Rightarrow 1! + 2! + 3! = 9$ [perfect square]

Thus, x can take value 1 and 3.

$\therefore$ Integral solution for x is two.

DPP-2 Fundamental Principle of Counting

1. (i) In order to drive from A to C, one can drive A to B in 6 ways and B to C in 4 ways. Thus, the number ways to drive from A to C is $6 \times 4 = 24$ ways.

(ii) Clearly, there are 24 ways to drive from A to C.

Now, in order to drive from C to A, one can drive C to B in 4 ways and B to A in 6 ways, i.e. one can drive from C to A in 24 ways.

Thus, the number of ways to drive from A to C and back to A

$= 24 \times 24$

$= 576$ ways

(iii) Clearly, one can drive from A to B in 6 ways and B to C in 4 ways, i.e. one can drive from A to B in 24 ways. But to come back one can drive from C to B in 3 ways, as we cannot choose the way from which we came and one can drive from B to A in 5 ways, i.e. one can drive from C to A in 15 ways.

Thus, the number of ways to drive the circular trip without using the same road $= 24 \times 15 = 360$

2. (i)

4 Ways	5 Ways	6 Ways

Clearly, units's place can be filled in 6 ways. Now, tenth place can be filled with any of remaining 5 digits, i.e. in 5 ways. Now, the hundred's place can be filled with in any of the remaining 4 digits, i.e. in 4 ways. Thus, the total number of required numbers

$= 6 \times 5 \times 4$

$= 120$

(ii) For number to be less than 400, hundred's place cannot be greater than 3. So, hundred's place can be filled with any of 2 digits 2, 3 i.e. 2 ways. Now, tenth place can be filled with any of the remaining 5 digits, i.e.in 5 ways and unit place can be filled with any of the remaining 4 digit, i.e. in 4 ways.

Thus, the total number of required numbers

$= 2 \times 5 \times 4 = 40$

(iii) For number to be even, unit's place is 2 or 6 i.e. unit's place can be filled in 2 ways. Now, tenth place can be filled with any of the remaining 5 digits, i.e. in 5 ways and hundred's place can be filled with any of the remaining 4 digits, i.e. in 4 ways.

4 Ways	5 Ways	2 Ways

Thus, the number of required numbers $= 4 \times 5 \times 2 = 40$

(iv) Since, total numbers are 120. and even numbers are 40.

$\therefore$ Number of required numbers $= 120 - 40 = 80$

(v) For number to be multiple of 5, unit's place is 5, i.e. unit's place can be filled in one way. Now, tenth place can be filled with any of the remaining 5 digits, i.e. in 5 ways and hundred's place can be filled with any of the remaining 4 digits, i.e. in ways.

4 Ways	5 Ways	1 Ways

Thus, the number of required numbers $= 4 \times 5 \times 1 = 20$

3. (i) Clearly, there are 7 letters in the word HISTORY. Now, first place can be filled by any of these letters, i.e. in 7 ways. Second place can be filled up by any of the remaining 6 letters, i.e. in 6 ways. Third place can be filled up any of the remaining 5 letters, i.e. in 5 ways and fourth place can be filled by any of the remaining 4 letters, i.e. in 4 ways.

I	II	III	IV
7 Ways	6 Ways	5 Ways	4 Ways

Thus, required number of words $= 4 \times 5 \times 6 \times 7 = 840$

(ii) Clearly, there are 5 consonants in the word HISTORY, *viz.* H, S, T, R, Y.

Now, first place can be filled by 5 ways,

second place can be filled by 4 ways,

third place can be filled by 3 ways,

and fourth place can be filled by 2 ways.

Thus, required number of words $= 5 \times 4 \times 3 \times 2 = 120$

(iii) Clearly, first place can be filled by 5 ways,

fourth place can be filled by 4 ways,

second place can be filled by any of remaining 5 letters (3 consonants and 2 vowels), i.e. in 5 ways.

and third place can be filled by 4 ways.

Thus, the required number of words $= 5 \times 5 \times 4 \times 4 = 400$

I	II	III	IV
5 Ways	5 Ways	4 Ways	4 Ways

(iv) Clearly, there are 2 vowels in the word HISTORY, *viz.* I, O.

Now, first place can be filled by 2 ways,

second place can be filled by any of the remaining 6 letters, i.e. in 6 ways.

Third place can be filled by 5 ways.

and fourth place can be filled by 4 ways.

Thus, required number of words

$$= 2 \times 6 \times 5 \times 4 = 240$$

(v) Let us consider the letter *Y* at first place

Y			
	6 Ways	5 Ways	4 Ways

Now, second place can be filled by 6 ways,

Third place can be filled by 5 ways

and fourth place can be filled by 4 ways.

Thus, the total number of words having Y at first place

$$= 6 \times 5 \times 4 = 120$$

Similarly, if we consider *Y* at second, third and fourth place respectively, then we get the same number of words.

Hence, the number of words contain the letter Y

$$= 4 \times 120 = 480$$

(vi) Clearly, first place can be filled by one way (i.e. T).

T			I or 0
	5 Ways	4 Ways	2 Ways

Fourth place can be filled by 2 ways (either by I or 0).

Second place can be filled by any of remaining 5 letters, i.e. in 5 ways and third place can be filled by 4 ways.

Hence, the required number of words

$$= 1 \times 5 \times 4 \times 2 = 40$$

(vii) Clearly, first place can be filled by one way (i.e. T).

Let us consider *S* at second place.

T	S		

Now, third place can be filled by any of the remaining 5 letters, i.e. in 5 ways and fourth place can be filled by 4 ways.

Thus, the number of words having T is beginning and S at second place

$$= 5 \times 4 = 20$$

Similarly, if we consider S at third and fourth place, we get same number of words.

Hence, the number of words begin with T and also contain

$$S = 20 \times 3 = 60$$

(viii) Clearly, there are two vowels in the word HISTORY, *viz.*, I, O.

Let us consider I and O at first and second place.

Now, third place can be filled with any of the remaining 5 letters, i.e. in 5 ways and fourth place can be filled by 4 ways.

$\therefore$ Number of words $= 5 \times 4 = 20$

Now, if we interchange the position of I and O, then again we get the same number of words = 20

Thus, the number of words having vowels at first two places $= 20 + 20 = 40$

Similarly, if we consider the vowels at (II, III), (III, IV), (IV, I), (I, III) and (II, IV) places, then we get the same number of words.

Hence, required number of words $= 6 \times 40 = 240$

4. (a) Let R_1, R_2, and R_3 be three railway passes that are allotted to 5 students.

Clearly, R_1 can be allotted to any of the 5 students, then R_2 can be allotted to any of the rest 4 students and R_3 can be allotted to any of the rest 3 students.

Hence, by fundamental principle of counting the number of ways of allotting 3 different railway passes to 5 students are $5 \times 4 \times 3 = 60$ ways.

5. (b) Let *A* denote jacket, *B* denote shirt and *C* denote a pair of slack.

Clearly, *A* can occur in 3 ways,

B can occur in 10 ways and *C* can occur in 5 ways.

Now, *A*, *B* and *C* together can occur in $3 \times 10 \times 5 = 150$ ways. Thus, the number of outfits, that the man can make is 150.

6. (b) Clearly, each letter can be mailed in 3 ways.

So, the number of ways in which 5 letters can be mailed

$$= 3 \times 3 \times 3 \times 3 \times 3 = 3^5 = 243$$

7. (c) We know that, a number is divisible by 2, if digit on unit place is 0, 2, 4, 6, 8, therefore unit place can be filled in 5 ways.

Clearly, tenth and hundred's place can be filled in 10 ways respectively, as repetition is allowed.

And thousand's place can be filled in 9 ways, as 0 cannot be inserted in the start of number and digit can be repeated.

9 Ways	10 Ways	10 Ways	5 Ways

Thus, the number of four-digit numbers are $= 9 \times 10 \times 10 \times 5$

$$= 4500$$

8. (a) Clearly, there are 1-digit, 2-digits, 3-digits, and 4-digits number from 1 to 1000.

In 1-digit number no digit is repeated, therefore the number of natural number from 1 to 9, which have none of their digits repeated is 9.

In 2-digit number, tenth place can be filled with any of the 9 digits 1, 2, 3, ...,9 . So, there are 9 ways of filling the tenth place. Now, 9 digits are left including 0. So, units place can be filled with any of the remaining 9 digits in 9 ways. Thus, there are $9 \times 9 = 81$ numbers which have none of their digits repeated.

↓	↓
9 Ways	9 Ways

In 3-digit number, hundred's place can be filled with any of the 9 digits 1, 2, 3, ..., 9. So, there are 9 ways to fill the hundred's place. Now, 9 digits are left including 0, so tenth's place can be filled in 9 ways. Now, the units place can be filled with any of the remaining 8 digits. So, there are 8 ways of filling the unit's place. Thus, there are $9 \times 9 \times 8 = 648$ number which have none of their digits repeated.

Hence, the total number of required numbers

$$= 9 + 81 + 648 = 738$$

9. (c) Clearly, a number between 100 and 1000 has three digits and there are 900 numbers between them.

Now, let us first find number of numbers of three digit having none of digit as 7.

Clearly, unit place can be filled by any of digits (0, 1, 2, 3, 4, 5, 6, 8, 9, 10) i.e. in 9 ways, tenth place can be filled by any of digits (0, 1, 2, 3, 4, 5, 6, 8, 9, 10) i.e. in 9 ways.

and hundred place can be filled by any of digits (1,2,3,4,5,6,7,8,9,10) i.e. in 8 ways.

↓	↓	↓
8 Ways	9 Ways	9 Ways

∴ Number of 3-digits numbers having none of the digit as 7

$$= 8 \times 9 \times 9 = 648$$

Hence, the number having at least one of their digit is 7

$$= 900 - 648 = 252$$

10. (d) For number to be odd, unit's place is either 1 or 3 i.e. unit's place can be filled in 2 ways. Now, ten thousand's place can be filled by any of remaining 3 digits (excluding 0) i.e. in 3 ways, thousand's place can be filled by any of remaining 3 digits (including 0) i.e. in 3 ways, hundred's place can be filled by any of the remaining 2 digits i.e. in 2 ways and tenth place can be filled by one way.

∴ Number of required numbers $= 3 \times 3 \times 2 \times 1 \times 2 = 36$

↓	↓	↓	↓	↓
3 Ways	3 Ways	2 Ways	1 Way	2 Ways

11. (c) Consider the following box, where green colour is in the middle.

			Green colour			
VII	VI	V	IV	III	II	I

Clearly, I box can be filled by any of 6 remaining colours i.e. in 6 ways.

Similarly, II box can be filled in 5 ways,
III box can be filled in 4 ways,
V box can be filled in 3 ways,
VI box can be filled in 2 ways,
VII box can be filled in 1 ways.

Hence, the number of ways in which 7 different colours in a rainbow can be arranged.

Considering green colour in middle $= 1 \times 2 \times 3 \times 4 \times 5 \times 6 = 6!$

12. (b) Since, each ring have 10 different letters, therefore total number of attempts are $10 \times 10 \times 10 = 1000$.

Clearly, out of 1000 attempts one will be the code of lock.

∴ Number of unsuccessful attempts $= 1000 - 1 = 999$

13. (a)

I	II	III	IV	V	VI	VII	VIII	X	XI
↓	↓	↓	↓	↓	↓	↓	↓	↓	↓
5 Ways	4 Ways	4 Ways	4 Ways	4 Ways	4 Ways	4 Ways	4 Ways	4 Ways	4 Ways

Clearly, first box can be filled by any of the digits (1, 3, 5, 7, 9) i.e. in 5 ways, second box can be filled by any of the remaining 4 digit, i.e. in 4 ways, third box can be filled by any of the remaining 4 digit (including digit of first box) i.e. in 4 ways, fourth box can be filled by any of the remaining 4 digit (including digit of second box) i.e. in 4 ways and so on.

Thus, the required number of ways $= 5 \times 4^9$.

14. (a)

	W		W		W		W	
1	2	3	4	5	6	7	8	9

Clearly, for 4 women four places are available and for 5 men five places are available. Now, 2nd place can be filled by any of the 4 women, i.e. in 4 ways, 4th place can be filled by any of the remaining 3 women, i.e. in 3 ways, 6th place can be filled by any of the remaining 2 women, i.e. in 2 ways and the 8th place can be filled by one way.

∴ Number of ways in which 4 women sit at the even places

$$= 4 \times 3 \times 2 \times 1 = 4!$$

Similarly, number of ways in which 5 men sit at odd places

$$= 5 \times 4 \times 3 \times 2 \times 1 = 5!$$

Thus, required number of ways $= 5! \times 4! = 2880$

15. (c) Clearly, a plaindrome is a word, which is same while read from backward or forward.

∴ 5 lettered plaindrome is a word in which first letter is same as last letter and second letter in same as second last letter.

I	II	III	II	I

Clearly, first place can be filled in 26 ways.
Second place can be filled in 26 ways.
and third place can be filled in 26 ways.

∴ Required number of 5 letter palin dromes $= (26)^3$

16. (b) Clearly, there are three vowels in the word MIRACLE, *viz.*, I, A, E. Now, as vowels can occupy only old places.

Vowel		Vowel		Vowel		Vowel
1	2	3	4	5	6	7

Therefore, there are 4 places available for 3 vowels and consonants can be arrange on the remaining 4 places.

∴ Required number of ways $= {}^4P_3 \times 4!$

$$= 4! \times 4! = 24 \times 24$$

$$= 576$$

17. (c) Clearly, there are 2 vowels in word AKSHI, viz. A and I.

***Case* I** When each word begins with vowel.

Vowel				
I	II	III	IV	V

Clearly, first place can be filled by any of vowels (A to I) i.e. in 2 ways, second place can be filled by any of remaining 4 letters, i.e. in 4 ways, third place can be filled by 3 ways, fourth place can be filled by 2 ways and fifth place can be filled by 1 way.

$\therefore$ Number of word begins with vowels $= 2 \times 4 \times 3 \times 2 \times 1 = 48$

***Case* II** When each word terminates with vowel.

1 Way	2 Ways	3 Ways	4 Ways	Vowel
I	II	III	IV	V → 2 Ways

By similar argument as in *Case* I, we get 48 words that terminates with vowel.

Note that in above two cases, we have some words which begin and terminate with vowel.

***Case* III** When each word begin and terminate with vowel.

A				I

Let us consider A in begin and I is last, then second place can be filled by 3 ways, third can be filled by 2 ways and fourth place can be filled by 1 way.

$\therefore$ Number of words $= 3 \times 2 = 6$

Similarly, if we consider A in last and I in begin, we get same number of words $= 6$

Hence, required number of words $= 48 + 48 - 6 - 6 = 84$

18. (c) Clearly, first passenger can be accommodated in any of the three railway carriages i.e. in 3 ways, second passenger can be accommodated in any of the three railway carriages i.e. in 3 ways.

Similarly, third and fourth passenger can be accommodated in any of the three railway carriages.

Hence, the number of ways in which four passengers be accommodate in three railway carriage

$$= 3 \times 3 \times 3 \times 3$$
$$= 3^4 = 81$$

19. (a) Clearly, first chair can be accommodated by any of the four persons, i.e. in 4 ways. Second chair can be accommodated by any of remaining 3 persons, i.e. in 3 ways and their chain can be accomodated by any of the remaining 2 persons, i.e. in 2 ways.

Thus, number of ways in which four persons can be accommodated in 3 different chairs

$$= 4 \times 3 \times 2 = 24 \text{ ways}$$

20. (d) Let $x, y \in N$, such that $x^3 + y^3 =$ natural number. ...(i)

Here, we need to find number of possible order pair (x, y) such that it satisfy (i).

Clearly, we can find infinite ordered pair (x, y) which satisfy Eq. (i), as cubes of natural number is natural number and sum of two natural number is natural number.

Hence, there are infinite natural number that can be written as sum of cubes of two natural numbers.

21. (d) Clearly, the word SACHIN starting with S. Now, let us first find the number of words started with A, C, H, I and N.

The number of words started with A $= 5! = 120$

A (SCHIN) → 5 ! Way

As remaining five letters can be arrange in 5! ways.

Similarly, if we start the word with C or H or I or N, we get same number of words. Now, we have $120 \times 5 = 600$ words before the word start with S.

Clearly, the first word that start with S is SACHIN.

Thus, the word SACHIN appears at 601 serial number.

22. (d) Clearly, the first word start with A and we get 24 words starting with A, as rest letters can be arrange in 4! ways.

Similarly, number of words starting with E are 24, number of words starting with H are 24.

Now, if we start the word with M, then again we get 24 words, i.e. we have 96 words before the words start with S.

Thus, the 86th, word will be start with M.

Now, if we start the word with M, then next letter will be A and remaining 3 letters can be arrange in 3! ways, i.e. we get 6 words.

Now, if we take letter E after M, then again we get 6 words.

Thus, over all we have $= 72 + 6 + 6 = 84$ words

Now, next two words are M H A E S and M H A S E.

Hence, the last letter of the 86th word in the list is E.

23. (b) Clearly, first prize can be given to any of the 5 students, i.e. first prize can be distributed in 5 ways. Second prize can be given to any of the remaining 4 students, i.e. second prize can be distributed in 4 ways. Similarly, third prize can be distributed in 3 ways and fourth prize can be distributed in 2 ways.

Thus, the required number of ways $= 5 \times 4 \times 3 \times 2 = 120$

24. (a) Clearly, each prize can be given to any of the 5 students, i.e. each prize can be distributed in 5 ways.

Thus, the required number of ways $= 5 \times 5 \times 5 \times 5$
$= 625$

25. (b) As there are only five students, therefore there are only five possible cases in which student getting all the prizes because may be first student get all prizes or second student get all prizes etc.

$\therefore$ Required number of ways $= 625 - 5$
$= 620$

26. (d) If the equal side is unity and x be the third side, then $x < 2$ must satisfied. Now, $x < 2$ has only one solution namely $x = 1$.

Similarly, for equal side to be 3, $x < 6$ and 4 solutions and for equal side to be 4, $x < 8$ and 4 solutions will be there.

$\Rightarrow$ Total triangles $= 1 + 3 + 4 + 4 = 12$.

27. (c) The number of equal side can be vary from 1 to $2c$. The number of Δ's can be formed by finding integer solutions $x < 2$, $x < 4, x < 6, \cdots, x < 2c, x < 2c + 2, x < 2c + 4, \cdots, x < 4c$.

$\therefore$ Total number of solutions $= \{1 + 3 + 5 + \cdots + (2c - 1)\}$
$+ \{(2c + 2c + \cdots + \text{upto } c \text{ times})\}$
$= 3c^2$

DPP-3 Application of ^nP_r

1. (i) Consider, $^{50}P_3 = \dfrac{50!}{(50-3)!}$ $\left[\because {}^nP_r = \dfrac{n!}{(n-r)!}\right]$

$$= \frac{50!}{47!} = \frac{50 \times 49 \times 48 \times 47!}{47!}$$

$$= 117600$$

(ii) $^6P_4 + {}^8P_2 = \dfrac{6!}{2!} + \dfrac{8!}{6!}$

$$= 6 \times 5 \times 4 \times 3 + 8 \times 7$$

$$= 360 + 56$$

$$= 416$$

2. (c) We have,

$$^nP_8 = 12 \cdot {}^nP_6$$

$\Rightarrow \dfrac{n!}{(n-8)!} = 12 \cdot \dfrac{n!}{(n-6)!}$

$\Rightarrow (n-6)(n-7) = 12$

$\Rightarrow n^2 - 13n + 42 - 12 = 0$

$\Rightarrow n^2 - 13n + 30 = 0$

$\Rightarrow n^2 - 10n - 3n + 30 = 0$

$\Rightarrow (n-10)(n-3) = 0$

$\therefore n = 10$

[since, n cannot be less than 8]

3. (c) We have, $^{2n}P_4 = 396 \times {}^nP_2$

$\Rightarrow \dfrac{(2n)!}{(2n-4)!} = 396 \times \dfrac{n!}{(n-2)!}$

$\Rightarrow \dfrac{(2n)(2n-1)(2n-2)(2n-3)(2n-4)!}{(2n-4)!} = 396 \times \dfrac{n(n-1)(n-2)!}{(n-2)!}$

$\Rightarrow 2^2 \cdot n(n-1)(2n-1)(2n-3) = 396 \times n(n-1)$

$\Rightarrow (2n-1)(2n-3) = 99$

$\Rightarrow 4n^2 - 8n - 96 = 0$

$\Rightarrow n^2 - 2n - 24 = 0$

$\Rightarrow n^2 - 6n + 4n - 24 = 0$

$\Rightarrow (n-6)(n+4) = 0$

$\therefore n = 6$

[since, n cannot be negative]

4. (a) Consider, $^{n-1}P_r + r \cdot {}^{n-1}P_{r-1}$

$$= \frac{(n-1)!}{(n-1-r)!} + r \cdot \frac{(n-1)!}{(n-r)!} = \frac{(n-1)!}{(n-r-1)!}\left[1 + \frac{r}{n-r}\right]$$

$$= \frac{(n-1)!}{(n-r-1)!}\left[\frac{n}{(n-r)}\right] = \frac{n!}{(n-r)!} = {}^nP_r$$

5. (b) We have, $^{l+m}P_2 = 56$ and $^{l-m}P_2 = 12$

Consider, $^{l+m}P_2 = 56$

$\Rightarrow \dfrac{(l+m)!}{(l+m-2)!} = 56$

$\Rightarrow \dfrac{y!}{(y-2)!} = 56$, where $y = l + m$

$\Rightarrow y(y-1) = 56$

$\Rightarrow y^2 - y - 56 = 0$

$\Rightarrow (y-8)(y+7) = 0$

$\Rightarrow y = 8$

[$\because y = l + m$, cannot be negative]

$\Rightarrow l + m = 8$ ...(i)

Now, consider $^{l-m}P_2 = 12$

$\Rightarrow \dfrac{(l-m)!}{(l-m-2)!} = 12$

$\Rightarrow \dfrac{x!}{(x-2)!} = 12$, where $x = l - m$

$\Rightarrow x(x-1) = 12$

$\Rightarrow x^2 - x - 12 = 0$

$\Rightarrow (x-4)(x+3) = 0$

$\Rightarrow x = 4$

[$\because x = l - m$, cannot be negative]

$\Rightarrow l - m = 4$ ...(ii)

On solving Eqs. (i) and (ii), we get

$$l = 6 \text{ and } m = 2$$

$\therefore l + m = 8$

Alternate Method

Clearly, by hit and trial method we can get the result, as

$^6P_2 = 30$ [taking $l + m = 6$]

$^8P_2 = 56$ [taking $l + m = 8$]

$^{10}P_2 = 90$

and $^{12}P_2 = 132$

6. (c) Consider, the given series

$$1 + 1 \cdot {}^1P_1 + 2 \cdot {}^2P_2 + 3 \cdot {}^3P_3 + \cdots + n \cdot {}^nP_n$$

$$= 1 + 1 \cdot 1! + 2 \cdot 2! + 3 \cdot 3! + \cdots + n \cdot n!$$

[$\because {}^nP_n = n!, \forall \in N$]

$= 1! + 1 \cdot 1! + 2 \cdot 2! + 3 \cdot 3! + \cdots + n \cdot n!$ [$\because 1! = 1$]

$= 1!(1 + 1) + 2 \cdot 2! + 3 \cdot 3! + \cdots + n \cdot n!$

$= 2! + 2 \cdot 2! + 3 \cdot 3! + \cdots + n \cdot n!$

[$\because n! \cdot (n+1) = (n+1)!$]

$= 2!(1 + 2) + 3 \cdot 3! + \cdots + n \cdot n!$

$= 3! + 3 \cdot 3! + \cdots + n \cdot n! = n! + n \cdot n!$

$= n!(1 + n) = (n+1)!$

7. (b) Clearly, a bundle of 2 specific things (say P and Q) can be put in r places in $(r-1)$ ways.

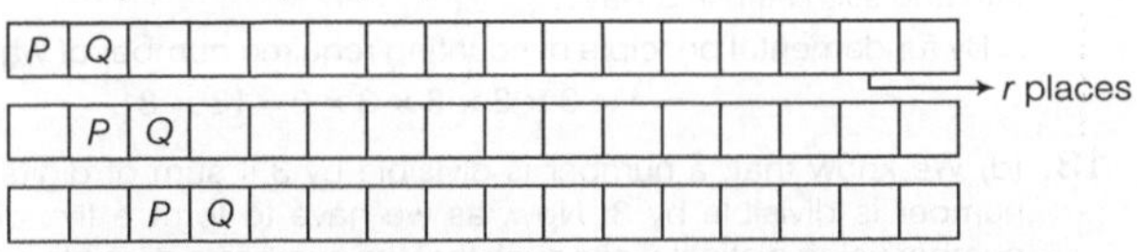

Also, 2 things in the bundle can be arranged themselves into 2! ways. Now, number of permutation of $(n-2)$ things taking $(r-2)$ things at a time is $^{n-2}P_{r-2}$.

Thus, using the fundamental principle of counting, the required number of permutations will be $2!(r-1)\,{}^{n-2}P_{r-2}$

$$= \frac{(n-2)!}{(n-r)!} \cdot 2 \cdot (r-1)$$

8. (a) We first arrange the m men. This can be done in $m!$ ways. After m men have taken their seats, we can arrange the n women on $(m+1)$ seats marked with cross.

$$\times \underset{\text{1st}}{M} \times \underset{\text{2nd}}{M} \times \underset{\text{3rd}}{M} \times \underset{\text{4th}}{M} \times \cdots \times \underset{\text{mth}}{m} \times$$

This can be done in $^{m+1}P_n$ ways.

Thus, the required number of ways $= m! \times {}^{m+1}P_n$

$$= \frac{m!(m+1)!}{(m-n+1)!}$$

9. (a) Consider, $P(n, r) = \frac{n!}{(n-r)!}$

$$= n(n-1)(n-2)(n-3)\cdots[n-(r-1)]$$

Clearly, $P(n, r)$ is prime, if $(n-1)(n-2)\cdots[n-(r-1)]$ is 1 as product of continuous numbers cannot be prime.

Also, $(n-1)(n-2)\cdots(n-(r-1))$ is 1 only if $r = 1$,

$\therefore$ We have, $P(n, r) = n$, where $r = 1$

Now, for $P(n, r)$ to be prime, n should be prime, i.e. n can be 2, 3, 5, 7.

$\therefore$ Possible ordered pairs are (2, 1), (3, 1), (5, 1) and (7, 1).

10. (a) Let us first find the number of three-digit numbers having none of the digits even.

Clearly, there are 5 odd digits, *viz.* 1, 3, 5, 7, 9.

$\therefore$ Number of three-digit numbers having all digits odd $= 5^3 = 125$

$\therefore$ Required number $= 900 - 125 = 775$

11. (b) Clearly, there are 9000 numbers from 1000 to 9999 and each number is of four digits.

Now, let us first find the number of four-digit numbers that have all digit different.

Number of four digit numbers having none of the digit same

= Number of four digits number using digits 0 to 9 – Number of four digit number having 0 at thousand's place $= {}^{10}P_4 - {}^9P_3$

$$= 5040 - 504 = 4536$$

$\therefore$ Required number $= 9000 - 4536 = 4464$

12. (a)

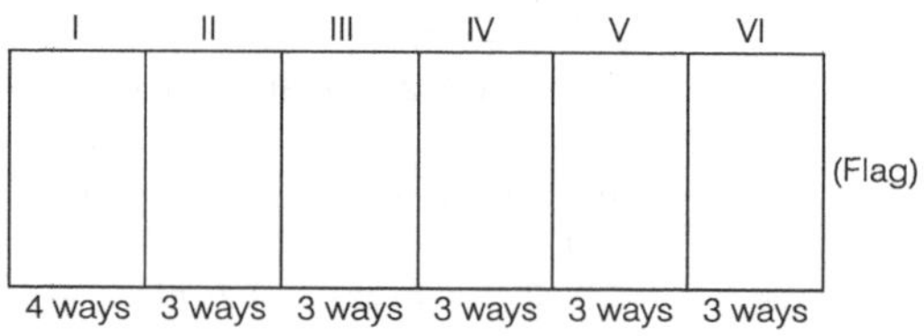

Clearly, first strip can be filled by any of the four colours, i.e. in 4 ways, second strip can be filled by any of the remaining 3 colours, i.e. in 3 ways, third strip can be filled by any of the remaining 3 colours (including colour of first strip) i.e. in 3 ways, fourth strip can be filled by any of the remaining 3 colours (including colour of 2nd strip) i.e. in 3 ways. Similarly, we can fill fifth and sixth strip in 3 ways.

$\therefore$ By fundamental principle of counting required number of ways

$$= 4 \times 3 \times 3 \times 3 \times 3 \times 3 = 12 \times 81$$

13. (d) We know that, a number is divisible by 3 if sum of digits of number is divisible by 3. Now, as we have to form a five-digit number using distinct digits such that the number is divisible by 3, therefore sum of digits can be 12 or 15.

***Case* I** Sum of digits is 15.

In this case, digits are 1, 2, 3, 4, 5 and required number of five-digit number $= {}^5P_5 = 5! = 120$

***Case* II** Sum of digits is 12.

In this case, digits are 0, 1, 2, 4, 5 and required number of five-digit number

$$= {}^5P_5 - {}^4P_4 = 120 - 4! = 96$$

Thus, total number of five-digit number which are divisible by 3 and having all digits distinct $= 120 + 96 = 216$

14. (b) We know that, a number is divisible by 9, if sum of digits of number is divisible by 9. Now, as we have to form a 9-digit number using distinct digits such that the number is divisible by 3, therefore sum of digits can be 36 or 45.

***Case* I** Sum of digits is 36.

In this case digits are 0, 1, 2, 3, 4, 5, 6, 7, 8 and required number of 9-digit number

$$= {}^9P_9 - {}^8P_8 = 9! - 8!$$

***Case* II** Sum of digits is 45.

In this case digits are 1, 2, 3, 4, 5, 6, 7, 8, 9 and required number of 9-digit number $= {}^9P_9 = 9!$

Thus, total number of 9-digit number which are divisible by 9 and having all digits distinct $= 2(9!) - 8!$

$\Rightarrow \quad k(8!) = 2(9!) - 8!$

$\Rightarrow \quad (k+1)8! = 2(9!)$

$\Rightarrow \quad k + 1 = 18$

$\therefore \quad k = 17$

15. (d) Clearly, the numbers which are less 1000 are of 1 digit, 2 digits and 3 digits. Also, we know that a number is divisible by 5 if unit is 0 or 5.

Now, number of required 1-digit number = 1,

Number of required 2-digit number $= {}^9P_1 + {}^8P_1 = 17$

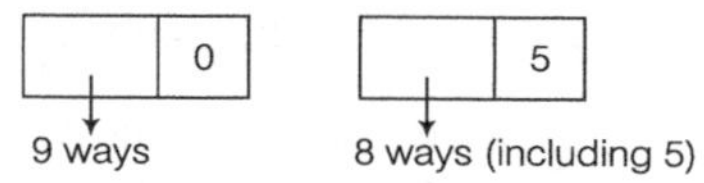

and number of required 3-digit number

$$= 8 \times 8 + 9 \times 8$$
$$= 64 + 72 = 136$$

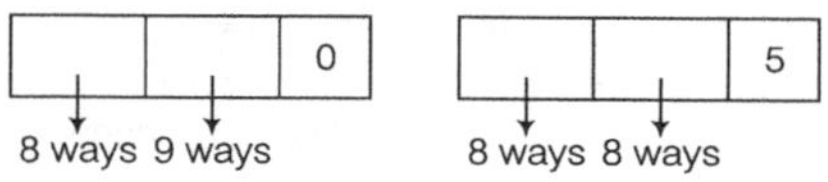

$\therefore$ Total number of natural numbers less than 1000 and divisible by $5 = 136 + 17 + 1 = 154$

16. (d) Number of ways in which 5 different books can be distributed among 10 people such that each people get at most one book is same as number of permutations of 10 people taking 5 at a time (without repetition) *i.e.* ${}^{10}P_5$.

17. (d) Product of all odd positive integers less than 10000

$$= 1 \cdot 3 \cdot 5 \cdot 7 \cdots 9999$$
$$= \frac{1 \cdot 2 \cdot 3 \cdot 4 \cdot 5 \cdot 6 \cdot 7 \cdot 8 \cdots 9999 \cdot 10000}{2 \cdot 4 \cdot 6 \cdot 8 \cdots 10000}$$
$$= \frac{(10000)!}{2^{5000}(1 \cdot 2 \cdot 3 \cdot 4 \cdots 5000)} = \frac{(10000)!}{2^{5000}(5000)!}$$

18. (c) Clearly, each of the x arms can have any one of the $(x-1)$ position in $(x-1)$ ways.

But all the x arms will be in rest position in $\underbrace{1 \times 1 \times 1 \times 1 \times \cdots \times 1}_{x \text{ times}}$

= 1 way and in this case no signal will be made.

$\therefore$ Required number of signals $= \underbrace{(x-1)(x-1)\cdots(x-1)}_{x \text{ times}} - 1$

$$= (x-1)^x - 1$$

19. (c) Since, each digit can be filled by any of the two numbers 1 or 2, i.e. 2 ways.

$\therefore$ Number of n-digit numbers $= \underbrace{2 \times 2 \times 2 \times 2 \times \cdots \times 2}_{n \text{ times}} = 2^n$

$\Rightarrow \quad 2^n = 512 = 2^9$

$\therefore \quad n = 9$

20. (a) Let us first arrange the 5 lions. This can be done in 5! ways. After giving the position to 5 lions, tamer can bring 4 tigers at 6 positions marked with P.

$$P\ \underset{\text{1st}}{L}\ P\ \underset{\text{2nd}}{L}\ P\ \underset{\text{3rd}}{L}\ P\ \underset{\text{4th}}{L}\ P\ \underset{\text{5th}}{L}\ P$$

This can be done in ${}^6P_4$ ways.

Thus, the required number of ways $= 5! \times {}^6P_4$

DPP-4 Application of ^nC_r

1. (a) We have, $^{10}P_r = 604800$ and $^{10}C_r = 120$

$$^nP_r = r!\,^nC_r$$
$$\Rightarrow \quad ^{10}P_r = r!\,^{10}C_r$$
$$\Rightarrow \quad 604800 = r!\,120$$
$$\Rightarrow \quad r! = \frac{604800}{120} = 5040$$
$$\Rightarrow \quad r! = 7!$$
$$\therefore \quad r = 7$$

Hence, option (a) is correct.

2. (b) We have, $^nC_{12} = {}^nC_8$

$$\Rightarrow \quad n = 12 + 8 \quad [\because {}^nC_x = {}^nC_y, \text{ then } x + y = n]$$
$$\Rightarrow \quad n = 20$$
$$^{20}C_{17} = \frac{20!}{17!3!} = \frac{20 \times 19 \times 18}{1 \times 2 \times 3} = 1140$$

Hence, option (b) is correct.

3. (d) We have, $^{15}C_{3r} = {}^{15}C_{r+2}$

If $^nC_x = {}^nC_y$

Then, either $x + y = n$ or $x = y$

$$\therefore \quad 3r + r + 2 = 15 \quad \text{or} \quad 3r = r + 2$$
$$\Rightarrow \quad 4r = 13 \quad \text{or} \quad 2r = 2$$
$$\Rightarrow \quad r = \frac{13}{4} \quad \text{or} \quad r = 1$$
$$r \neq \frac{13}{4}$$
$$\therefore \quad r = 1$$

Hence, option (d) is correct.

4. (c) We have,

$$^{18}C_r = {}^{18}C_{r+2}$$
$$\Rightarrow \quad r + r + 2 = 18$$
$$\Rightarrow \quad r = 8$$

Now, $^rC_5 = {}^8C_5$

$$\Rightarrow \quad ^8C_5 = \frac{8!}{5!3!} = \frac{8 \times 7 \times 6}{1 \times 2 \times 3} = 56$$

Hence, option (c) is correct.

5. (a) We have,

$$^{47}C_4 + \sum_{j=1}^{5} + {}^{52-j}C_3$$
$$\Rightarrow \quad ^{47}C_4 + {}^{51}C_3 + {}^{50}C_3 + {}^{49}C_3 + {}^{48}C_3 + {}^{47}C_3$$
$$\Rightarrow \quad ^{47}C_3 + {}^{47}C_4 + {}^{48}C_3 + {}^{49}C_3 + {}^{50}C_3 + {}^{51}C_3$$
$$\Rightarrow \quad ^{48}C_4 + {}^{48}C_3 + {}^{49}C_3 + {}^{50}C_3 + {}^{51}C_3$$
$$[\because {}^nC_{r-1} + {}^nC_r = {}^{n+1}C_r]$$
$$\Rightarrow \quad ^{49}C_4 + {}^{49}C_3 + {}^{50}C_3 + {}^{51}C_3$$
$$\Rightarrow \quad ^{50}C_4 + {}^{50}C_3 + {}^{51}C_3$$
$$\Rightarrow \quad ^{51}C_4 + {}^{51}C_3 = {}^{52}C_4$$

Hence, option (a) is correct.

6. (b) We have, $^{15}C_8 + {}^{15}C_9 - {}^{15}C_6 - {}^{15}C_7$

$$\Rightarrow \quad ^{15}C_8 + {}^{15}C_9 - {}^{15}C_9 - {}^{15}C_8 \quad [\because {}^nC_r = {}^nC_{n-r}]$$
$$\Rightarrow \quad 0$$

Hence, option (b) is correct.

7. (b) We have, $^{10}C_{x-1} > 2 \cdot {}^{10}C_x$

$$\Rightarrow \quad \frac{10!}{(x-1)!\,(10-(x-1))!} > \frac{2 \cdot 10!}{x!\,(10-x)!}$$
$$\Rightarrow \quad \frac{1}{(x-1)!\,(11-x)!} > \frac{2}{x!\,(10-x)!}$$
$$\Rightarrow \quad x > 2\,(11 - x)$$
$$\Rightarrow \quad x > 22 - 2x$$
$$\Rightarrow \quad 3x > 22$$
$$\Rightarrow \quad x > \frac{22}{3} = 7.33$$

So, least positive integer is 8.

Hence, option (b) is correct.

8. (a) We have,

$$(^7C_0 + {}^7C_1) + (^7C_1 + {}^7C_2) + (^7C_2 + {}^7C_3) + \cdots + (^7C_6 + {}^7C_7)$$
$$\Rightarrow \quad ^8C_1 + {}^8C_2 + {}^8C_3 + \cdots + {}^8C_7 \quad [\because {}^nC_{r-1} + {}^nC_r = {}^{n+1}C_r]$$
$$\Rightarrow \quad (^8C_0 + {}^8C_1 + {}^8C_2 + \cdots + {}^8C_7 + {}^8C_8) - (^8C_8 + {}^8C_8)$$
$$= 2^8 - (1 + 1) = 2^8 - 2$$
$$[\because {}^nC_0 + {}^nC_1 + {}^nC_2 + \cdots + {}^nC_n = 2^n]$$

Hence, option (a) is correct.

9. (a) We have, $\frac{^{20}C_n}{^{25}C_m}$,

Numerator has greatest value, if $n = 10$ and

Denominator has greatest value, if $m = 12$ or 13

$$\therefore \quad \frac{^{20}C_{10}}{^{25}C_{12}} = \frac{20! \times 13!\,12!}{10!\,10! \times 25!} = \frac{143}{4025}$$

Hence, option (a) is correct.

10. (b) We have, $^nC_{r+1} + 2 \cdot {}^nC_r + {}^nC_{r-1}$

$$\Rightarrow \quad ^nC_{r+1} + {}^nC_r + {}^nC_r + {}^nC_{r-1}$$
$$\Rightarrow \quad ^{n+1}C_{r+1} + {}^{n+1}C_r$$
$$\Rightarrow \quad ^{n+2}C_{r+1}$$

Hence, option (b) is correct.

11. (a) We have, $^{50}C_4 + \sum_{r=1}^{6} {}^{56-r}C_3$

$$\Rightarrow \quad ^{50}C_4 + {}^{55}C_3 + {}^{54}C_3 + {}^{53}C_3 + {}^{52}C_3 + {}^{51}C_3 + {}^{50}C_3$$
$$\Rightarrow \quad ^{50}C_3 + {}^{50}C_4 + {}^{51}C_3 + {}^{52}C_3 + {}^{53}C_3 + {}^{54}C_3 + {}^{55}C_3$$
$$\Rightarrow \quad ^{51}C_4 + {}^{51}C_3 + {}^{52}C_3 + {}^{53}C_3 + {}^{54}C_3 + {}^{55}C_3$$
$$[\because {}^nC_{r-1} + {}^nC_r = {}^{n+1}C_r]$$
$$\Rightarrow \quad ^{52}C_4 + {}^{52}C_3 + {}^{53}C_3 + {}^{54}C_3 + {}^{55}C_3$$
$$\Rightarrow \quad ^{53}C_4 + {}^{53}C_3 + {}^{54}C_3 + {}^{55}C_3$$
$$\Rightarrow \quad ^{54}C_4 + {}^{54}C_3 + {}^{55}C_3$$
$$\therefore \quad ^{55}C_4 + {}^{55}C_3 = {}^{56}C_4$$

Hence, option (a) is correct.

12. (a) Total number of questions = 13

Student has to answer 10 out of 13 questions in which atleast 3 questions from first five questions.

$\therefore$ Total number of selection is

$$^5C_3\,^8C_7 + {}^5C_4\,^8C_6 + {}^5C_5\,^8C_5$$
$$\Rightarrow \quad 10 \times 8 + 5 \times 28 + 1 \times 56$$
$$\Rightarrow \quad 80 + 140 + 56 = 276$$

$\therefore$ Total number of ways is 276.

Hence, option (a) is correct.

13. (a) We have, given digits 1, 2 and 3 in which digit 2 occur exactly twice.

Digit 1	Digit 2	Digit 3	Number of arrangement
1	2	4	$\frac{7!}{2!4!}=105$
2	2	3	$\frac{7!}{2!2!3!}=210$
3	2	2	$\frac{7!}{3!2!2!}=210$
4	2	1	$\frac{7!}{4!2!1!}=105$
0	2	5	$\frac{7!}{2!5!}=21$
5	2	0	$\frac{7!}{5!2!}=21$

Total number of different seven-digit numbers that can be formed using digits 1, 2 and 3 and the digit 2 occurs exactly twice is $105 + 210 + 210 + 105 + 21 + 21 = 672$.

Hence, option (a) is correct.

14. (c) We have, 7 consonants and 4 vowels.

The number of ways of selecting 4 consonants out of 7 and 2 vowels out of 4

$$= {}^7C_4 \times {}^4C_2$$

$$= \left(\frac{7\times 6\times 5}{1\times 2\times 3}\times\frac{4\times 3}{1\times 2}\right) = 210$$

$\therefore$ The number of groups each containing 4 consonants and 2 vowels = 210

Since, each group contains 6 letters, which can be arranged amongst themselves in $6! = 720$ ways

Hence, the required number of words $= 210 \times 720 = 151200$

Hence, option (c) is correct.

15. (a) We have, give 5 mangoes and 6 oranges.

Total number of selection so as to take at least one mango and one orange is

$$({}^5C_1 + {}^5C_2 + {}^5C_3 + {}^5C_4 + {}^5C_5)$$

$$({}^6C_1 + {}^6C_2 + {}^6C_3 + {}^6C_4 + {}^6C_5 + {}^6C_6)$$

$$(2^5 - 1)(2^6 - 1) = (32 - 1)(64 - 1)$$

$$= 31 \times 63$$

Hence, option (a) is correct.

16. (c) We have, $\{a_1, a_2, a_3, \cdots, a_n\}$

Total number of elements = n

$\therefore$ Total number of selection 3 elements out of n elements in which a_3 is included is

$$^{n-1}C_{3-1} = {}^{n-1}C_2$$

Hence, option (c) is correct.

17. (b) We have,

$$^{n+1}C_{n-2} - {}^{n+1}C_{n-1} \le 100$$

$$\Rightarrow \frac{(n+1)!}{(n-2)!\,(n+1-n+2)!} - \frac{(n+1)!}{(n-1)!\,(n+1-n+1)!} \le 100$$

$$\Rightarrow \frac{(n+1)!}{(n-2)!3!} - \frac{(n+1)!}{(n-1)!2!} \le 100$$

$$\Rightarrow \frac{(n+1)!}{2!\,(n-2)!}\left[\frac{1}{3} - \frac{1}{n-1}\right] \le 100$$

$$\Rightarrow \frac{(n+1)!\,(n-4)}{3!\,(n-1)!} \le 100$$

$$\Rightarrow (n+1)(n)(n-4) \le 600$$

The number of positive integer satisfying the inequality, if $n \le 9$ and $n \ge 2$ i.e. $2 \le n \le 9$.

$\therefore$ Total number of positive integers = 8

Hence, option (b) is correct.

18. (b) We have, 8 different colours in which red, blue and green are included.

Total number of selection 5 colours out of 8 different colours in which blue and green are included.

$$^2C_2 \times {}^6C_3 = \frac{2!}{0!2!} \times \frac{6!}{3!\,3!}$$

$$= 20$$

Hence, option (b) is correct.

19. (c) Total number of selection 5 colours out of 8 colours in which red is always excluded

$$^{8-1}C_{3-1} = {}^7C_2 = \frac{7!}{2!\,5!} = 21$$

Hence, option (c) is correct.

20. (a) Total number of selection 5 colours out of 8 colours in which red and blue are included and green excluded is

$$^2C_2 \times {}^5C_3 = \frac{2!}{0!\,2!} \times \frac{5!}{2!\,3!} = 10$$

Hence, option (a) is correct.

DPP-5 **Restricted Arrangements**

1. (i) The 5 boys may occupy 5 places in ${}^5P_5 = 5!$. Since, no two girls are sit together, we may arrange the 5 girls at the 6 places, shown by cross marks, as given below.

X B X B X B X B X B X

Now, the 5 girls at 6 places my be seated in ${}^6P_5 = 6!$.

$\therefore$ The required number of ways $= 5! \times 6!$

(ii) All the girls sit together and all the boys sit together.

The number of ways of 5 girls sit together = 5!

and the number of ways of 5 boys sit together = 5!

The number of ways boys and girls among themselves = 2!

Hence, total number of arrangement $= 2! \times 5! \times 5! = 2!(5!)^2$

(iii) The total number of boys and girls $= 5 + 5 = 10$

The total number of arrangement of 10 persons = 10!

But when all the five girls sit together, regarding the five girls as one person, we have only $5 + 1 = 6$ persons.

These 6 person can be arranged in 6! ways.

But the five girls can be arranged among themselves in 5! ways.

$\therefore$ Number of ways when five girls sit together = 6! 5!

$\therefore$ Required number of ways in which all the five girls do not sit together $= 10! - 6!\,5!$.

2. The given word 'MISSISSIPPI' contains 11 letters out of which I is repeated 4 times, S is repeated 4 times , P is repeated 2 time and the rest are different.

$\therefore$ The required number of arrangement $= \dfrac{11!}{4!\,4!\,2!}$

3. There are 11 letters in the word 'MATHEMATICS'. Out of these M letters is repeated 2 times, A occurs twice, T occurs twice and rest are different.

$\therefore$ Required number of arrangement $= \dfrac{11!}{2!\,2!\,2!}$

$= 4989600$

The given word contains 4 vowels namely AEAI.

Treating these four vowels as one letters, we have to arrange 8 letters, out of which M occurs twice, T occurs twices and the rest are all different.

$\therefore$ Number of such arrangement $= \dfrac{8!}{2!\,2!}$

Corresponding to each such arrangement, the four vowels in which A occurs twice and the rest are all different can be arranged amongst themselves in $\dfrac{4!}{2!}$.

$\therefore$ Total number of arrangements in which vowels are together

$= \dfrac{8!}{2!\,2!} \times \dfrac{4!}{2!} = 120960$

4. (i) The given word 'HARYANA' contains 7letters out of which A is repeated 3 times and the rest are all different.

$\therefore$ The required number of different words formed

$= \dfrac{7!}{3!} = 7 \times 6 \times 5 \times 4 = 840$

(ii) Starting with H and end with N we have find the number of arrangement of the remaining 5 letters out of which A is repeated 3 times and the rest of letters are distinct.

$\therefore$ The required number of arrangement $= \dfrac{5!}{3!} = 5 \times 4 = 20$

(iii) When the H and N together, treated as one letter, we have to arrange 6 letters, out of which A occurs thrice and the rest are distinct.

$\therefore$ The number of arrangement $= \dfrac{6!}{3!} = 6 \times 5 \times 4 = 120$

Now, H and N can be arranged in 2! ways

Thus, total number of arrangements $= 120 \times 2 = 240$

5. (i) The given word 'DAUGHTER' contains 8 different letters when the vowels A, U, E are always together, treated as one letters.

We have to arrange 6 letters i.e. $6! = 720$ ways,

but corresponding to each of these arrangement, the vowels in the group (A, U, E) can be amongst themselves in $3! = 6$ ways.

$\therefore$ The total number of arrangement when vowels always together $= 720 \times 6 = 4320$

(ii) Clearly, the total number of words formed by using all the 8 letters of the given word is $8! = 40320$

$\therefore$ The number of words in which vowels are never together

= (total number of words)

– (number of words in which vowels are always together)

$= (40320 - 4320) = 36000$

6. (i) The total number of person $= 4 + 4 = 8$. These 8 person can be seated at the round table $= (8-1)! = 7! = 5040$ ways.

(ii) Since, there is no restriction on men, therefore first of all we arrange the 4 men.

Now, 4 men can be seated a round table $= 3! = 6$ ways

Now, there a 4 places for 4 women, therefore 4 women can be seated in $4!$ way $= 24$ ways.

$\therefore$ Required number of ways $= 6 \times 24 = 144$ ways

(iii) consider two particular women as one person, therefore total number of person = 7. These 7 person can be seated at the round table $= (7-1)! = 6! = 720$ ways.

Now, also two particular women be arrange $= 2!$ ways

$\therefore$ Required number of ways $= 720 \times 2 = 1440$ ways

(iv) Two particular women sit together =

(Total number of arrangement) – (Number of arrangement of particular women sit together)

$= 5040 - 1440 = 3600$

(v) Consider 4 women as one person, therefore total number of person = 5. These 5 person can be seated at round table $= (5-1)! = 4! = 24$ ways.

Now, also 4 women can be arrange $= 4! = 24$ ways

$\therefore$ Required number of ways $= 24 \times 24 = 576$ ways

(vi) Let P_1, P_2 be two particular women and H be the person sit between two particular women.

$\therefore$ Required number of ways $= 2! = 2$ ways

Now, one person out of six is between $P_1, P_2 = {}^6C_1 = 6$

Thus, we have 5 persons and 1 group (P_1, P_2, and H) that can be arranged in $= 5!$ ways

$\therefore$ Total number of ways $= 2 \times 6 \times 5! = 1440$

7. (a) The given word 'BANANA' contains 6 letters out of which A occur thrice, N occur twice and rest are distinct.

$\therefore$ Total number of arrangements $= \dfrac{6!}{3!\,2!} = 60$

When two N's are appear adjacently, treated as one letter.

We have to arrange 5 letters out of which A occur thrice and rest are distinct

$\therefore$ Required number of arrangement $= \dfrac{5!}{3!} = 20$

$\therefore$ The total number of word two N's do not appear adjacently = (Total number of words) – (Number of words in which two N's appear adjacently)

$= 60 - 20 = 40$

Hence, option (a) is correct.

8. (c) There are nine-digits numbers 223355888 out of which 5 are even number and 4 are odd numbers.

Let us mark out the position to filled up as follows

(1) (2) (3) (4) (5) (6) (7) (8) (9)

Now, the 4 odd number can be placed at any of the four place, marked (2, 4, 6, 8).

So, the number of ways of arranging the digit (3, 3, 5, 5) out of which 3 and 5 occur twice

$= \dfrac{4!}{2!\,2!} = 6$

Also, the 5 even numbers (2, 2, 8, 8, 8) out of which 2 occurs twice and 8 occurs thrice may be arranged $= \dfrac{5!}{2!\,3!} = 10$

$\therefore$ The required number of ways $= 6 \times 10 = 60$ ways

9. (a) There are 18 beads out of which 8 being blue, 5 red, 4 green and 1 yellow. In this arrangement anti-clockwise or clockwise are not different.

$\therefore$ Total number necklace formed $= \dfrac{(18-1)!}{2!\,8!\,5!\,4!} = \dfrac{17!}{2 \cdot 8! \cdot 5! \cdot 4!}$

Hence, option (a) is correct.

10. (c) There are 11 letters in word 'MISSISSIPPI' out of which I occurs 4 times, S occurs 4 times and P occurs twice and rest are different.

Consider four S as one letter. We have to arrange 8 letters out of which I occurs 4 times and P occur twice and rest are distinct.

$\therefore$ Total number of arrangement $= \frac{8!}{4!2!}$.

Hence, option (c) is correct.

11. (b) Since, at most one ring is to be worn in any four fingers.

$\therefore$ Total number of arrangement of worn in four fingers $^4P_3$

$$= \frac{4!}{(4-3)!} = 4! = 24 \text{ ways}$$

Hence, option (b) is correct.

12. (c) Since, atleast one ring is to be worn in each finger,

So, we first select four rings out of 5 given rings and then they are arranged in fingers. This can be done in $^5P_4 = 5! = 120$ ways.

Now, one ring is left which can be worn in anyone of the four finger in 4 ways.

$\therefore$ Required number of ways $= 120 \times 4 = 480$ ways.

Hence, option (c) is correct.

13. (b) There are five rings and four fingers,

$\therefore$ $x_1 + x_2 + x_3 + x_4 + x_5 = 4$

The total number of selection of 5 rings in 4 fingers

$$^{4+5-1}C_5 = {}^8C_5$$

$$= \frac{8!}{5!\,3!} = 56 \text{ ways}$$

Now, 5 rings arranged in one fingers is $5! = 120$ ways.

$\therefore$ Total number of arrangement $= 56 \times 120 = 6720$ ways.

Hence, option(b) is correct.

14. (a) Here, the job putting r different objects in n boxes.

Putting one object in one of the n boxes $= n$ ways

Putting second object in one of the n boxes $= n$ ways.

$\vdots \qquad \vdots$

Putting rth object in one of the n boxes $= n$ ways.

$\therefore$ Total number of putting r object in n different boxes is

$$\underbrace{n \times n \times n \times \cdots \times n}_{r \text{ times}} = n^r$$

Hence, option (a) is correct.

15. (c) Here, the job putting r different object in n boxes so that each contains only one object.

Putting one object in one of the n boxes.

Putting second object in $(n-1)$ boxes.

Putting rth object in $(n-(r-1))$ boxes.

$\therefore$ Total number of putting r object in n different boxes, so that each box contains only one object is $n(n-1)(n-2)\cdots(n-r+1) = {}^nP_r$

16. (d)The number of ways of selection of r object from n boxes, if a box may contain any number of object is $^{n+r-1}C_n$.

Now, total number of n boxes be arranged in $n!$ ways.

$\therefore$ Total number of arrangement $= {}^{n+r-1}C_n \times n! = {}^{n+r-1}P_n$

Hence, option (d) is correct.

17. (c) Set of three positive integers (each greater than or equal to) 1, whose sum is 5

$$\{3, 1, 1\}, \{2, 2, 1\}.$$

Elements of 1st can be arranged in $\frac{3!}{2!}$ ways $= 3$ ways.

Elements of 2nd can be arranged in $\frac{3!}{2!}$ ways $= 3$ ways

$\therefore$ Required number

$$= \frac{3!}{2!} \times {}^5C_3 \times {}^2C_1 \times {}^1C_1 + \frac{3!}{2!} \times {}^5C_2 \times {}^3C_2 \times {}^1C_1$$

$$= 60 + 90 = 150$$

Hence, option (c) is correct.

18. (b) The required number of ways is the number of ways of distributing 5 items among 3 person so that a person receives atleast one item $= {}^{5-1}C_{3-1} = {}^4C_2 = 6$ ways.

Hence, option (b) is correct.

19. (c) As, boxes are identical, therefore we can placed (1, 1 in any two boxes and 3 balls in rest box) or (1 ball in any box and 2, 2 balls in rest two boxes.)

$\therefore$ Required number of ways

$$= {}^5C_1 \times {}^4C_1 \times {}^3C_3 + {}^5C_1 \times {}^4C_2 \times {}^2C_2$$

$$= 20 + 30 = 50$$

20. (a) Since, boxes and balls are identical, therefore we can placed balls in 2 ways only.

We can placed any two balls in any two boxes ((1 and 1) and rest 3 balls in rest box) or (any 1 ball in any box and 2, 2 balls in rest two boxes.)

DPP-6 Geometrical Application of ^nC_r

1. (a) We know that, by joining the two points we can obtain a line, so number of lines joining vertices of octagon $= {}^8C_2 = 28$

$\therefore$ Required number of diagonals

$= 28 -$ (Number of sides of octagon)

$= 28 - 8 = 20$

2. (c) Required number of diagonals

= number of lines joining the vertices of polygon − number of sides of polygon.

$$= {}^nC_2 - n = \frac{n(n-1)}{2 \times 1} - n$$

$$= n\left(\frac{n-1}{2} - 1\right)$$

$$= n\left(\frac{n-3}{2}\right)$$

3. (b) Number of lines drawn through 6 points lying on a circle

$$= {}^6C_2 = \frac{6 \times 5}{2 \times 1} = 15$$

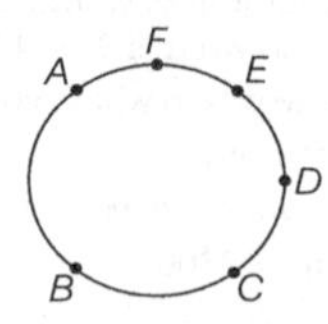

4. (b) Here, total number of points are $(m + n + k)$ which must give $^{m+n+k}C_3$ number of triangles, but m points on l, taking 3 points at a time gives $^mC_3$ combinations which produce no triangle. Similarly, $^nC_3$ and $^kC_3$ number of triangles can not be formed. Therefore, the required number of triangles are

$$^{m+n+k}C_3 - {}^mC_3 - {}^nC_3 - {}^kC_3$$

5. (b) Given, total number of points = 18 and number of collinear points = 5

We know that, one line has two ends points. Therefore, total number of lines = ${}^{18}C_2$

Since, 5 points are collinear, therefore total number of lines drawn from collinear points = ${}^5C_2 = 10$

Also, there is a line corresponding to the line of collinearity.

$\therefore$ Required number of lines = ${}^{18}C_2 - {}^5C_2 + 1$

$= 153 - 10 + 1 = 144$

6. (b) Since, two straight lines intersect at one point.

$\therefore$ Maximum number of points of intersections = ${}^8C_2$

$= \dfrac{8 \times 7}{2 \times 1} = 28$

7. (c) Let n be the number of sides of polygon

$\therefore$ Number of diagonals = ${}^nC_2 - n$

$\Rightarrow \quad \dfrac{n(n-1)}{2} - n = 44 \Rightarrow \dfrac{n(n-3)}{2} = 44$

$\Rightarrow \quad n^2 - 3n - 88 = 0$

$\Rightarrow n^2 - 11n + 8n - 88 = 0$

$\Rightarrow n(n-11) + 8(n-11) = 0$

$\Rightarrow \quad (n-11)(n+8) = 0$

$\therefore \quad n = 11$ [$\because n$ cannot be negative]

8. (c) Number of points required for the fixed circle is 3.

So, first select any three points from the 10 points in ${}^{10}C_3$ ways.

In these ways, circle with four concyclic points is selected in ${}^4C_3$ ways. But it should be taken once.

$\therefore$ Total number of circles = $({}^{10}C_3 - {}^4C_3) + 1$

$= \left(\dfrac{10 \times 9 \times 8}{3 \times 2 \times 1} - 4\right) + 1$

$= (120 - 4) + 1 = 117$

9. (b) We know that, two circles intersect at two distinct points, two straight lines intersect at one point, one circle and one line intersect at two distinct points, therefore total number of points of intersection are as follows.

Number of ways of selection	Point of intersection
Two straight lines : ${}^4C_2$	${}^4C_2 \times 1 = 6$
Two circles : ${}^4C_2$	${}^4C_2 \times 2 = 12$
One line and one circle : ${}^4C_1 \times {}^4C_1$	${}^4C_1 \times {}^4C_1 \times 2 = 32$

Thus, total number of point of intersection = 32 + 12 + 6 = 50

10. (c) Let number of sides of polygon = n

$\therefore \quad \dfrac{(n-2)\,180^\circ}{n} = 150$

[$\because$ interior angle of n sided polygon is given by $\dfrac{(n-2)}{n} 180^\circ$]

$\Rightarrow \quad (n-2)6 = 5n$

$\Rightarrow \quad 6n - 12 = 5n$

$\Rightarrow \quad n = 12$

$\therefore$ Number of diagonals = ${}^{12}C_2 - 12$

$= 6 \times 11 - 12$

$= 66 - 12 = 54$

11. (b) Clearly, there are three horizontal lines and four vertical lines.

As, we know that a parallelogram can be formed by two vertical lines and two horizontal lines.

$\therefore$ Required number of parallelogram = ${}^3C_2 \times {}^4C_2$

$= 3 \times \dfrac{4 \times 3}{2 \times 1} = 18$

12. (c) Let the given box is of size 4×3.

Now, on considering block of size 1×1, we get 4×3 = 12 squares (number of ways of selecting a unit length horizontally × number of ways of selecting a unit length vertically).

On considering block of size 2×2, we get 3×2 = 6 squares (number of ways of selecting a length of unit 2 horizontally × number of ways of selecting a length of unit 2 vertically)

On considering block of size 3×3, we get $2 \times 1 = 2$ squares

$\therefore$ Total number of squares = 12 + 6 + 2 = 20

13. (b) Given, n concurrent lines and another line parallel to one of them.

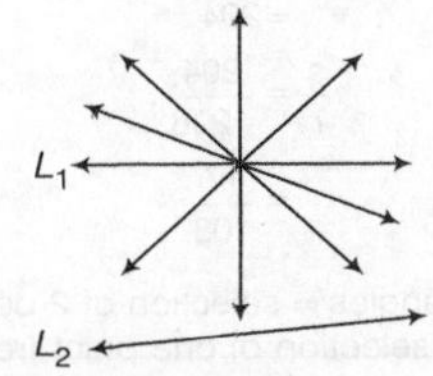

As, we have seen in figure if $L_1 \parallel L_2$, then any other two lines from congruent lines will intersect L_2 at two points and that form a triangle.

$\therefore$ Required number of triangle = number of ways of selecting two lines from $(n-1)$ concurrent lines

$= {}^{n-1}C_2 = \dfrac{(n-1)(n-2)}{2}$

14. (d) Total number of points are 18 which must give ${}^{18}C_3$ number of triangles, but on each side of triangles 7 points are lying.

Now, on taking 3 points at a time from any one side gives ${}^7C_3$ combinations which produce no triangle.

$\therefore$ Required number of triangle = ${}^{18}C_3 - {}^7C_3 - {}^7C_3 - {}^7C_3$

$= \dfrac{18 \times 17 \times 16}{3 \times 2 \times 1} - 3 \times \dfrac{7 \times 6 \times 5}{3 \times 2 \times 1}$

$= 816 - 105 = 711$

15. (a) The number of points of intersection of straight lines inside the polygon = number of intersection points of diagonals inside the polygon = ${}^9C_4 = 126$.

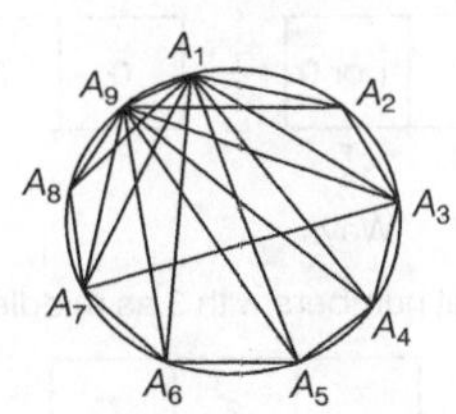

16. (b) Given, 9 horizontal lines and 9 vertical lines on an 8×8 chess board.

Clearly, a rectangle can be form by two vertical and two horizontal lines, therefore the number of rectangles

$(r) = {}^9C_2 \times {}^9C_2 = ({}^9C_2)^2 = (36)^2 = 1296.$

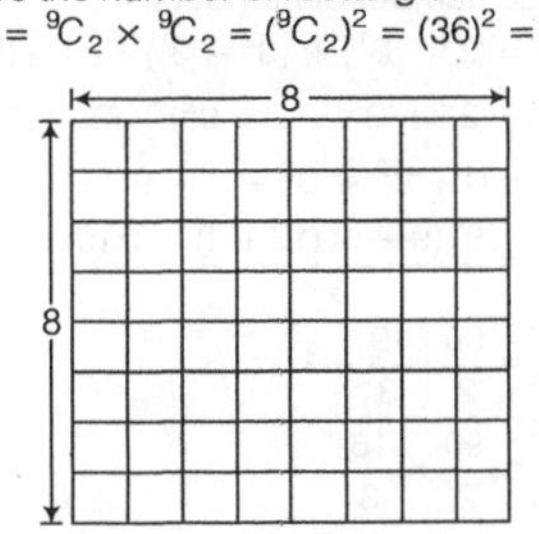

Chess board of size 8×8

Now, on considering block of sizes 1×1, we get $8 \times 8 = 8^2$

$= 64$ squares.

On considering block of size 2×2, we get

$$7 \times 7 = 7^2 = 49 \text{ squares}$$

$$\vdots \qquad \vdots$$

On considering block of size 8×8 we get

$$1 \times 1 = 1^2 = 1 \text{ square}$$

$\therefore$ Total number of square $= 8^2 + 7^2 + 6^2 + \cdots + 1^2$

$$= \frac{8(8+1)(2 \times 8 + 1)}{6}$$

$$= 204$$

$$\therefore \quad \frac{s}{r} = \frac{204}{1296} = \frac{17}{108}$$

17. (a) Number of triangles = selection of 2 points from AB and one point from AC + selection of one point from AB and two points from AC

$$= {}^mC_2 \cdot {}^nC_1 + {}^mC_1 \cdot {}^nC_2$$

$$= n \cdot {}^mC_2 + m \cdot {}^nC_2$$

18. (a) Let us find the number of triangle with one vertex as A.

Clearly, number of triangle having A as vertex = selection of one point from AB and one point from AC

$$= {}^mC_1 \cdot {}^nC_1$$

$$= mn$$

$\therefore$ Required number of triangles $= n \cdot {}^mC_2 + m \cdot {}^nC_2 + mn$

Explanations for Q. Nos. (19-20)

Total number of vertices of the polygon $= n$

Total number of selections of 3 vertices out of n vertices $= {}^nC_3$

$\therefore$ Each selection of 3 vertices gives one triangle.

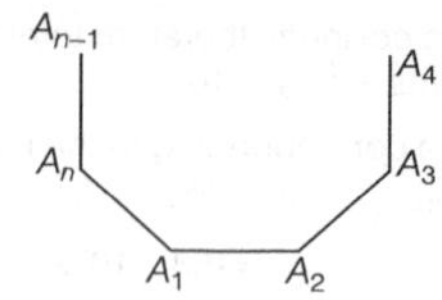

$\therefore$ Total number of triangles formed $= {}^nC_3 = \frac{n(n-1)(n-2)}{6}$

Now, number of triangles having two sides common with the sides of the polygon = number of selection of 3 consecutive vertices of the polygon $= n$

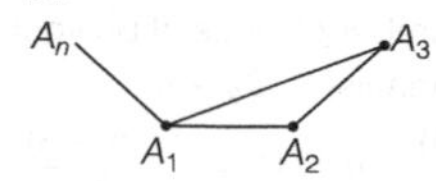

19. (a) Number of triangles having one sides common with the sides of the polygon (this is possible if $n \geq 5$) = number of selection of 3 vertices out of which only two are consecutive (i.e. if we select two consecutive vertices, A_1 and A_2, then the third vertex selected can not be A_3 or A_n).

$$= {}^nC_1 \times {}^{n-4}C_1 = n(n-4)$$

$\downarrow$

[number of ways of selecting consecutive vertices]

20. (b) Number of triangles that have none of sides common of polygon

$$= {}^nC_3 - n - n(n-4)$$

DPP-7 Problems on Arrangements and Selection

1. (b) The middle digit of 3-digit numbers can be 2 (when numbers are formed with digits 0, 1, 2), 3 (when 0, 1, 2, 3 are used),..., 9 (when 0, 1, 2, 3, 4, 5, 6, 7, 8, 9 are used).

$\therefore$ Number of 3-digit numbers with 2 as middle digit $= 2^2 - 2^1$

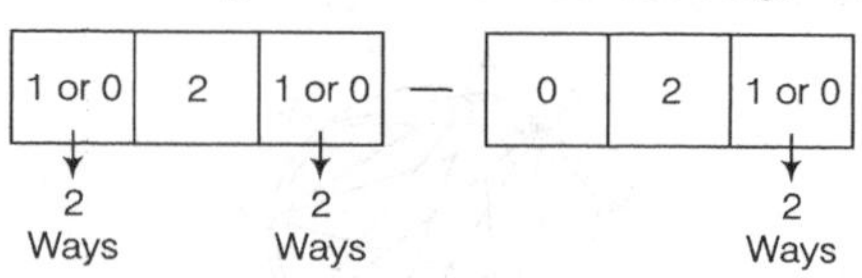

Number of 3-digit numbers with 3 as middle digit $= 3^2 - 3^1$

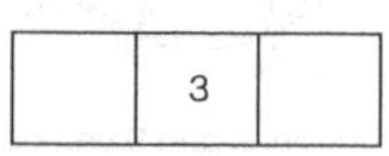

Number of 3-digit numbers with 4 as middle digit $= 4^2 - 4^1$ and so on.

Hence, required number of numbers

$$= (2^2 - 2^1) + (3^2 - 3^1) + \cdots + (9^2 - 9^1)$$

$$= (2^2 + 3^2 + \cdots + 9^2) - (2^1 + 3^1 + \cdots + 9^1)$$

$$= (1 + 2^2 + 3^2 + \cdots + 9^2) - (1 + 2 + 3 + \cdots + 9)$$

$$= \frac{9 \cdot (9+1)(18+1)}{6} - \frac{9(9+1)}{2}$$

$$= \frac{90}{2}\left[\frac{19}{3} - 1\right]$$

$$= \frac{90}{2} \times \frac{16}{3}$$

$$= 240$$

2. (b)

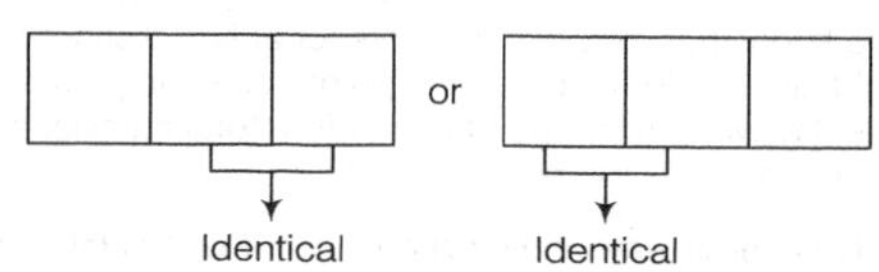

We have, the following digits 0, 1, 2, 3, 4, 5, 6, 7, 8, 9 to form 3-digit numbers.

***Case* I** When last two digits are identical.

In this case number of 3-digit numbers $= 9 \times 9 = 81$

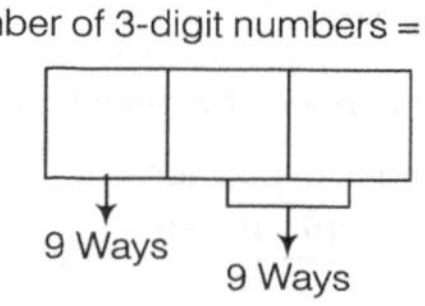

***Case* II** When first two digits are identical. In this case number of 3-digit numbers $= 9 \times 9$

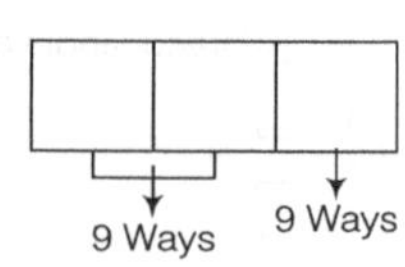

$\therefore$ Total number of 3-digits number

$$= 81 \times 2$$

$$= 162$$

3. (a) The digit at hundredth's place can be 1 (when numbers are formed with digits 0, 1), 2 (when numbers are formed digits 0, 1, 2), 3 (when 0, 1, 2, 3 are used),..., 9 (when 0, 1, 2, 3, 4, 5, 6, 7, 8, 9 are used).

$\therefore$ Number of 3-digit numbers having 1 at hundredth's place = 1

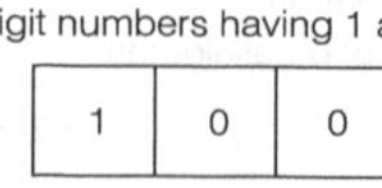

Number of 3-digit numbers having 2 at hundredth's place = 2^2

2	1 or 0	1 or 0
	2 Ways	2 Ways

Number of 3-digit numbers having 3 at hundredth's place = 3^2

3	0 or 1 or 2	0 or 1 or 2
	3 Ways	3 Ways

and so on.

$\therefore$ Total number of numbers $= 1 + 2^2 + 3^2 + \cdots + 9^2$

$$= \frac{9(9+1)(18+1)}{6} = \frac{9 \times 10 \times 19}{6} = 285$$

4. (b) ***Case I*** When successive digits are in their descending order.
The total number of arrangements of 10 digits 0, 1, 2, 3, 4,...,9 by taking 5 at a time

$= {}^{10}C_5 \times 5!$ or ${}^{10}P_5$

We observe that in every arrangement of 5 selected digits there is just one arrangement in which the digits are in descending order.

$\therefore$ The required number of numbers $(m) = \dfrac{{}^{10}C_5 \times 5!}{5!} = {}^{10}C_5$

Case II When successive digits are in their increasing order.
In this case we form 5-digit numbers from the digits 1, 2, 3,...,9, as if we select 5 digits from the digits 0, 1, 2, 3,...,9 and if it include 0 then, 0 is always at ten thousand's place which will not form a number.
Now, the number of arrangements of 1, 2, 3,...,9 by taking 5 at a time $= {}^9C_5 \times 5!$ or ${}^9P_5$
We observe that in every arrangement of 5 selected digits there is just one arrangement in which the digits are in increasing order.

$\therefore$ The required number of numbers $(n) = \dfrac{{}^9C_5 \times 5!}{5!} = {}^9C_5$

Thus, $m - n = {}^{10}C_5 - {}^9C_5$

$$= \frac{10 \times 9 \times 8 \times 7 \times 6}{5 \times 4 \times 3 \times 2 \times 1} - \frac{9 \times 8 \times 7 \times 6 \times 5}{5 \times 4 \times 3 \times 2 \times 1}$$

$$= \frac{9 \times 8 \times 7 \times 6}{5 \times 4 \times 3 \times 2 \times 1}(10-5) = \frac{9 \times 8 \times 7 \times 6 \times 5}{5 \times 4 \times 3 \times 2 \times 1} = {}^9C_5$$

5. (c) Number of ways in which each person may become the owner of one monkey = number of arrangement of m monkey taking n at a time

$= {}^mC_n \times n!$ or ${}^mP_n$

6. (c) We have, 7 identical green bottles (G) and 8 identical blue bottles (B).
Let us consider two green bottles GG (say) as single object.
Now, 8 blue bottles and (GG) can be arrange in $\dfrac{9!}{8!} = 9$ ways.

Now, we left with 5 green bottles and 8 places that can be filled in ${}^8C_5$ ways.

$\therefore$ Required number of ways $= 9 \times {}^8C_5 = 9 \times 56 = 504$

✓ ✓ ✓ ✓ ✓ ✓ ✓ ✓
GG B B B B B B B B
or
✓ ✓ ✓ ✓ ✓ ✓ ✓ ✓
B B B B B B B B GG
or
✓ ✓ ✓ ✓ ✓ ✓ ✓ ✓
B B B GG B B B B B

7. (d) We have, the following digits 0, 1, 2, 3, 4, 5, 6, 7, 8, 9 to form the 3-digit numbers.

Case I When exactly one 3 and one 2 present.

3	2		or		3	2	or	2		3
		8 Ways		7 Ways					8 Ways	

$\therefore$ Number of required numbers
$= 8 \times (2!) + 7 \times (2!) + 8 \times (2!) = 32 + 14 = 46$

Case II When two 3 and one 2 or two 2 and one 3 present.

3	3	2	or	2	2	3

$\therefore$ Number of required numbers $= \dfrac{3!}{2!} + \dfrac{3!}{2!} = 3 + 3 = 6$

$\therefore$ Total number of required numbers $= 46 + 6 = 52$

8. (b) Clearly, 10 red balls of different shades can be arrange in 10! ways.
Now, we left with 9 identical green balls and 11 places that can be filled in ${}^{11}C_9$ ways.

$\therefore$ Required number of ways $= (10!) \times ({}^{11}C_9)$

✓ ✓ ✓ ✓ ✓ ✓ ✓ ✓ ✓ ✓ ✓
R_1 R_2 R_3 R_4 R_5 R_6 R_7 R_8 R_9 R_{10}

9. (a) The number of ways in which the series can be won by India
$= {}^5C_0 + {}^5C_1 + {}^6C_2 + {}^7C_3 + {}^8C_4 = 126$

10. (a) The total number of four-digit numbers with the digits 0, 1, 2, 3
$= 3 \times 4 \times 4 \times 4$ or $4^4 - 4^3 = 192$

3 Ways	4 Ways	4 Ways	4 Ways

To find The sum of these 192 numbers, we will find the sum of digits at units, ten's, hundred's and thousand's place in all these number.
Clearly, occurrence of zero at any place will not affect the sum of digits at that place.
Now, number of numbers formed, having 1 at unit place $= 3 \times 4 \times 4 \times 1 = 48$

1 or 2 or 3	0 or 1 or 2 or 3	0 or 1 or 2 or 3	1
3 Ways	4 Ways	4 Ways	

Similarly, each of the digit 2, 3 will occur at unit place 48 times.
So, total for the digits in the unit's place in all the numbers $= (1 + 2 + 3) \times 48 = 288.$
Now, number of numbers formed having 1 at ten's place $= 3 \times 4 \times 1 \times 4 = 48$

1 or 2 or 3	0 or 1 or 2 or 3	1	0 or 1 or 2 or 3
3 Ways	4 Ways		4 Ways

Similarly, each of the digit 2, 3 will occur at ten's place 48 times.
So , sum of digit of ten's place in all numbers
$= (1+2+3) \times 48$
$= 288$

Similarly, we can get the same sum by adding digits of hundred's place.

Now, number of numbers formed, having 1 at thousand place

$= 1 \times 4 \times 4 \times 4 = 64$

1	0 or 1 or 2 or 3	0 or 1 or 2 or 3	0 or 1 or 2 or 3
	↓ 4 Ways	↓ 4 Ways	↓ 4 Ways

Similarly, each of digits 2, 3 will occur at thousand's place 64 times.

$\therefore$ Sum of digits at thousand place $= (1 + 2 + 3) \times 64 = 384$

$\therefore$ Sum of numbers formed

$= 288 \times 1 + 288 \times 10 + 288 \times 100 + 384 \times 1000 = 415968$

11. (a) There are 6 men and 6 women. 4 people may consists of all men and no women or all women and no men or 3 men and 1 women, where women is not among wives of 3 chosen men or 2 men and 2 women, where women are not among the wives of 2 chosen men or 1 men and 3 women none of whom is wife of chosen men.

$\therefore$ Required number of ways

$= {}^6C_4 + {}^6C_4 + {}^6C_3 \times {}^3C_1 + {}^6C_2 \times {}^4C_2 + {}^6C_3 \times {}^3C_1$

$= 15 \times 2 + 20 \times 3 + 15 \times 6 + 20 \times 3$

$= 30 + 60 + 90 + 60 = 240$

Alternate Method

Number of ways to select 4 groups out of six group of married coupes $= {}^6C_4$

Now, to select either a man or wowen $= 2^4$

$\therefore$ Number of ways $= {}^6C_4 \times 2^4 = 240$.

12. (a) 4 people may consists 1 men and 3 women (including the wife of selected man) or 2 men and 2 women (it include 2 cases, wife of 1st man and one of the wives of other 4 men (excluding 2nd man) or wife of 2nd man and one of the wives of other 4 men (excluding 1st man)) or 3 men and 1 women (one of wives of selected men)

$\therefore$ Required number of ways $= {}^6C_1 \times {}^1C_1 \times {}^5C_2 + {}^6C_2 \times {}^1C_1 \times {}^4C_1$
$+ {}^6C_2 \times {}^1C_1 \times {}^4C_1 + {}^6C_3 \times {}^3C_1$

$= 60 + 60 + 60 + 60 = 240$

13. (b) Required number of ways

= total number of selecting 4 people

– number of ways of selecting 4 people such that they do not form a couple

$= {}^{12}C_4 - 240$

$= 495 - 240 = 255$

14. (c) Required number of ways = number of ways of selecting 4 people so that they do not form a couple + number of ways of selecting 4 people so that they form exactly one couple

$= 240 + 240 = 480$

15. (9) Total number of digits = 9

Now, we can select 2 places for the digit 1 and 2 in ${}^9C_2$ ways, from the remaining 7 places select any two places for 3 and 4 in ${}^7C_2$ ways and from the remaining 5 places select any two for 5 and 6 in ${}^5C_2$ ways.

Now, we left with 3 digits 7, 8, 9 and 3 places that can be filled in 3! ways.

$\therefore$ Total number of ways $= {}^9C_2 \cdot {}^7C_2 \cdot {}^5C_2 \cdot 3! = 9 \cdot 7!$

$\therefore \quad k = 9$

16. (3) Here, we have to find the number of different natural numbers which are less than 2×10^8 and which can be written by means of digits 1 or 2 only.

Clearly, the required numbers are 1, 2, 11, 12, 21, 22,...,122222222.

Now, let us calculate how many numbers are these. There are 2 one digit such numbers. There are 2^2 two digit such numbers. And so on.

There are 2^8 eight digit such numbers. All the nine-digit numbers beginning with 1 and written by means of 1 or 2 are smaller than 2×10^8 and these are 2^8 in numbers.

Hence, required number of numbers

$= 2 + 2^2 + 2^3 + \cdots + 2^8 + 2^8$

$= \dfrac{2(2^8 - 1)}{2 - 1} + 2^8 = 2^9 - 2 + 2^8$

$= 2 \cdot 2^8 + 2^8 - 2 = 2^8(2 + 1) - 2 = 3 \cdot 2^8 - 2$

$\therefore \quad k = 3$

17. (9) Given, 5 Indian and 5 American couples meet at party and shake hand.

$\therefore$ Total number of possible hand shakes $= {}^{20}C_2$

Number of hand shakes, when wife shake hand with her own husband = 10 (5 Indian and 5 American)

Number of hand shakes when Indian wife shakes hand with a male $= {}^5C_1 \times {}^{10}C_1 = 50$ (it include the case where the Indian wife shake hand with her own husband)

$\therefore$ Total number of hand shakes that take place in the party

$= {}^{20}C_2 - 5 - 50 = 135 = 15 \times 9$

$\therefore \quad k = 9$

18. (4) Let us first locate vowels in alphabetical order at any 4 places out of 8 and this can be done in ${}^8C_4$ ways.

Now, we left with 4 letters R, B, N, D and 4 places that can be filled in 4! ways.

$\therefore$ Number of permutation of the word 'AUROBIND' in which vowels appear in an alphabetical order $= {}^8C_4 \times 4!$

$= \dfrac{8!}{4! \times 4!} \times 4! = \dfrac{8!}{4!}$

$\therefore \quad k = 4$

19. (5) All the possible numbers are ${}^9C_5$ (none containing the digit 0) = 126.

Total numbers starting with 1 $= {}^8C_4 = 70$ (using 2, 3, 4, 5, 6, 7, 8, 9)

1				

Total numbers starting with 23 $= {}^6C_3 = 20$ (using 4, 5, 6, 7, 8, 9)

2	3			

Total numbers starting with 245 $= {}^4C_2 = 6$ (using 6, 7, 8, 9)

2	4	5		

97th number =

2	4	6	7	8

20. (8) There are 3 choices for the first of n-letter and two choices for each subsequent letters.

Hence, using fundamental principle, number of good words

$= 3 \cdot 2^{n-1} = 384$

$= 2^{n-1} = 128 = 2^7$

$\therefore \quad n = 8$

DPP-8 Division of Objects into Groups of Equal and Unequal Groups, Number of Divisions and Sum of Divisions

1. (b) Here, 9 students are to be divided into 3 equal groups and the order of the groups are importants.

So, required number of ways $= \frac{9!}{(3!)^3 \cdot 3!} \times 3! = \frac{9!}{(3!)^3}$

Hence, option (b) is correct.

2. (c) We have, $9600 = 2^7 \times 3^1 \times 5^2$

The total number of divisors of 9600 including 1and 9600 is $(7+1)(1+1)(2+1) = 8 \times 2 \times 3 = 48$

Hence, option (c) is correct.

3. (b) First group can be given 8 different things out of 15 in ${}^{15}C_8$ ways second group can be given 4 different thing out remaining 7 in ${}^7C_4$ and 3 different things out of 3 remaining 3 things in ${}^3C_3$ ways.

Hence, the required number of ways (here, order of groups are not important) $= {}^{15}C_8 \times {}^7C_4 \times {}^3C_3 = \frac{15!}{8!\,7!} \times \frac{7!}{4!\,3!} \times \frac{3!}{3!} = \frac{15!}{8!\,4!\,3!}$

Hence, option (b) is correct.

4. (c) Here, there are 12 different balls and three different boxes the first box contain exactly 5 balls out of 12 ball is ${}^{12}C_5$ ways and the remaining 7 balls have two boxes for each of 7 balls

i.e. $2 \times 2 \times 2 \times 2 \times 2 \times 2 \times 2 = 2^7$ ways

$\therefore$ Required number of ways $= {}^{12}C_5 \times 2^7$ ways

Hence, option (c) is correct.

5. (b) There are 50 identical things distributed among 8 persons. In which 3 of them get 8 things each, two of them get 7 things each and remaining 3 get 4 thing each

$$8+8+8+7+7+4+4+4 = 50$$
$$3(8) + 2(7) + 3(4) = 50$$

Now, number of ways to distribute (3 + 2 + 3) items among 8 persons in the groups 3, 2, and 3 items

$$= \frac{(3+2+3)!}{3!\,2!\,3!} = \frac{8!}{(3!)^2(2!)}$$

Hence, option (b) is correct.

6. (c) Here, 10 balls can be divided between two boys one receiving two and other eight balls and the order of two boys is important.

So, required number of ways $= \frac{10!}{2!\,8!} \times 2! = 90$ ways

Hence, option (c) is correct

7. (c) We have, 'MISSISSIPPI' words contain I occurs 4 times, S occurs 4 times, P occurs twice and M are one time, M = 1, I = 4, S = 4, P = 2.

Zero or more 'I' can be selected out of 4I's = (4 + 1) = 5 ways

Zero or more 'S' can be selected out of 4S's = (4 + 1) = 5 ways

Zero or more 'P' can be selcted out of 2P's = (2 + 1) = 3 ways

Zero or more 'M' can be selected out of M's = (1 + 1) = 2 ways

So, total number of selection = 5 × 5 × 3 × 2 = 150 ways

But in one of these selection number of each type of letter is zero. Hence, this selection must be excluded.

$\therefore$ Total number of selection = 150 − 1 = 149 ways

Hence, option (c) is correct.

8. (d) We have, $a^p b^q c^r d^s$, where a, b, c and d are primes.

Total number of divisors $= (p+1)(q+1)(r+1)(s+1)$ this includes 1 and $a^p b^q c^r d^s$ as divisors. But the number of divisor excluding 1 and the number is $(p+1)(q+1)(r+1)(s+1) - 2$.

Hence, option (d) is correct.

9. (d) We have, given $2^5 \cdot 3^7 \cdot 5^3 \cdot 7^2$

The sum of the divisor is

$$\left(\frac{2^6-1}{2-1}\right)\left(\frac{3^8-1}{3-1}\right)\left(\frac{5^4-1}{5-1}\right)\left(\frac{7^3-1}{7-1}\right)$$
$$= \frac{(2^6-1)(3^8-1)(5^4-1)(7^3-1)}{2 \times 4 \times 6}$$

Hence, option (d) is correct.

10. (a) We have, $n = 21600 = 2^5 \times 3^3 \times 5^2$

A divisors of 21600 will be a multiple of 10 but not 15. We must have at least one 2 out of 5, 2's and at least one 5 out of 2 but 3 is not taken. Hence, the number of divisor which are multiple of 10 but not 15 is (5) (2) = 10 ways.

Hence, option (a) is correct.

11. (b) We have, $N = 27720 = 2^3 \times 3^2 \times 5 \times 7 \times 11$

Coprime mean consecutive prime.

Here, coprime are 2, 3, 5, 7, 11.

$n = 5$ (number of different factor in 27720)

Hence, number of ways in which a composite number 27720 can be resolved into two factors which are relative prime.

$$= 2^{5-1} = 2^4 = 16$$

Hence, option (b) is correct.

12. (b) We have, 2 white, 3 black and 4 red balls

Total number of ways of selection of three balls in which at least one black ball is

$${}^3C_1{}^6C_2 + {}^3C_2{}^6C_1 + {}^3C_3{}^6C_0 = 3 \times 15 + 3 \times 6 + 1 \times 1$$
$$= 45 + 18 + 1 = 64$$

$\therefore$ Total number of ways = 64

Hence, option (b) is correct.

13. (c) We have, total number of toys = 9

9 toys be distributed in 4 children in which 3 children get 2 toys each and younger children get 3 toys.

So, required number of ways $= \frac{9!}{(2!)^3 \times 3!}$

Hence, option (c) is correct.

14. (b) Here, 200 different things are to be divided in 100 equal groups.

$\therefore$ The total numbers of ways $= \frac{200!}{(2!)^{100} \times (100)!}$

$$= \frac{200!}{(100)!\,2^{100}}$$

Hence, option (b) is correct.

15. (b) Here, seven different coins are to be divided amongst three person in which no two persons receive the same number of coin but each receives at least one coin.

The possible number of coin distributed is (1, 2, 4).

$\therefore$ Total number of ways $= \frac{7!}{1!\,2!\,4!} \times 3! = 630$

Hence, option (b) is correct.

16. (a) Here, 8 different books are to be divided amongest 3 students, if each student receives at least 2 books. The possible number of distribution is (2, 2, 4) and (2, 3, 3).

$\therefore$ Total number of distribution is

$$\frac{8!}{2!\,2!\,4!} \times 3! + \frac{8!}{2!\,3!\,3!} \times 3! = 1260 + 1680 = 2940$$

Hence, option (a) is correct.

17. (a) Here, 10 men volunteers are to be divided into 2, 3 and 5 men.

$\therefore$ Required number of ways $= \dfrac{10!}{2!\,3!\,5!}$

Hence, option (a) is correct.

18. (c) Here, 20 different books have 4 are in single volume and other have 8, 5 and 3 volumes.

$\therefore$ Total number of book arranged in 7! ways.

Now, each set of volumes are together and in their due order be arranged in 2 ways.

$\therefore$ Total number of arrangement $= 7! \times 2 \times 2 \times 2 = 7! \times 8 = 8!$

Hence, option (c) is correct.

19. (b) We have, n identical red balls and m identicals green balls.

The number of different linear arrangement consisting n red balls but not necessarily all the green balls are

$$^nC_n + {}^{n+1}C_n + {}^{n+2}C_n + \cdots + {}^{n+m}C_n = {}^{m+n+1}C_{n+1}$$

$$\Rightarrow \quad {}^{m+n+1}C_m = {}^xC_y$$

$$\therefore \quad x = m + n + 1$$

$$[\because {}^nC_r + {}^{n-1}C_r + {}^{n-2}C_r + \cdots + {}^rC_r = {}^{n+1}C_{r+1}]$$

$$y = m$$

Hence, option (b) is correct.

20. (c) We have, 5'A' and 6 'B' can be arranged in arow B must be middle between 11 letters.

$\therefore$ Anyone left side or right $AABBB$ arranged.

Total number of arrangement $AABBB = \dfrac{5!}{2!\,3!} = 10$ ways

Hence, option (c) is correct.

21. (a) Total number of friends $= m + n$

m friends selected from $m + n$ friends $= {}^{m+n}C_m$

Now, arrangement of m friends sit in T_1 table $= {}^{m+n}C_m \times (m-1)!$

and n friends sit in T_2 table $= {}^nC_n(n-1)!$

$\therefore$ Required number of ways $= {}^{m+n}C_m \times (m-1)!\, {}^nC_n(n-1)!$

$$= \frac{(m+n)!}{m!\,n!}(m-1)! \times (n-1)!$$

But all the people have not same neighbour

$$= {}^{m+n}C_m \frac{(m-1)!}{2}\, {}^nC_n \frac{(n-1)!}{2} = \frac{1}{4}\frac{(m+n)!}{mn}$$

Hence, option (a) is correct.

22. (c) We have, $N = 1000$

$$\Rightarrow \quad N = 2^3 \times 5^3$$

The even divisor of 1000

2, 4, 8, 10, 20, 40, 50, 100, 200, 250, 500, 1000

the product of even divisor is

$2 \times 4 \times 8 \times 10 \times 20 \times 40 \times 50 \times 100 \times 200 \times 250 \times 500 \times 1000$

$2^{10} \times 10^{14} \times 5^4$

$$2^6 \times 2^4 \times 5^4 \times 10^{14} = 2^6 \times 10^{18} = 64 \times 10^{18}$$

Hence, option (c) is correct.

23. (b) We have, given digits 1, 3, 5, 7, 9.

Sum of number of one-digit $= 1 + 3 + 5 + 7 + 9 = 25$

Two-digit numbers are formed taken in pair

(1, 3) (1, 5) (1, 7) (1, 9) (3, 5) (3, 7) (3, 9) (5, 7) (5, 9) (7, 9)

Sum of two-digit numbers are $25 \times 4\,(11) = 1100$

Three-digit numbers are formed

(1, 3, 5) (1, 3, 7) (1, 3, 9) (1, 5, 7) (1, 5, 9) (1, 7, 9)

(3, 5, 7) (3, 5, 9) (3, 7, 9) (5, 7, 9).

Sum of three-digit numbers $= 25 \times 6 \times 2 \times (111) = 33300$

Four-digits numbers are

(1, 3, 5, 7) (1, 3, 5, 9) (1, 5, 7, 9) (3, 5, 7, 9) (1, 3, 7, 9)

Sum of four-digit numbers are $25 \times 4 \times 6 \times (1111) = 666600$

Sum of all the numbers from the digit 1, 3, 5, 7, 9 which are small than 10000, if repetition of digit is not allowed are

$$25 + 1100 + 33300 + 666600 = 701025$$
$$= 25\,(28041)$$

24. (b)

(i) Permutation are increasing in m symbols, therefore only one arrangement m object be selection from n object.

$\therefore$ Total number of ways $= {}^nC_m$

(ii) m men and n monkey. Number of ways in which every monkey has a master. First monkey has any m masters. Second monkey has also any m masters. nth monkey has also any m masters.

$\therefore$ Total number of ways $= \underbrace{m \times m \times m \times \cdots \times m}_{n \text{ times}} = m^n$

(iii) We have, n red balls and $(m-1)$ green balls

$$G \times G \times G \times G \times G \times \cdots \times G$$

There are m position from n red balls.

$\therefore$ Total number of arrangement are ${}^mC_n$. [$\because$ balls are identical]

(iv) First toy is given anyone children of n children.

Second toy is also given anyone children of n children.

nth toy is also given anyone children of n children

$\therefore$ Total number of arrangement $= \underbrace{n \times n \times n \times \cdots \times n}_{m \text{ times}} = n^m$

(i) $\to$ (r), (ii) $\to$ (s), (iii) $\to$ (q), (iv) $\to$ (p)

25. (a) We have, 4 oranges, 5 apples and 6 mangoes

Total number of selection when same types of fruits are identical

$$(4+1)(5+1)(6+1) - 1 = 210 - 1$$
$$= 209$$

Hence, option (a) is correct.

26. (b) Total number of selection when same types of fruits are different

$${}^{15}C_1 + {}^{15}C_2 + {}^{15}C_3 + \cdots + {}^{15}C_{15} = 2^{15} - 1$$

Hence, option (b) is correct.

27. (a) Here, 52 cards are to be divided into 4 groups and order is important.

$\therefore$ Required number of ways $= \dfrac{52!}{(13!)^4\,4!} \times 4!$

$$= \frac{52!}{(13!)^4}$$

Hence, option (a) is correct.

28. (b) Here, 52 cards are to be divided into 4 groups of 13 cards each but order is not important.

$\therefore$ Required number of ways $= \dfrac{52!}{(13!)^4\,4!}$

Hence, option (b) is correct.

29. (c) Here, 52 cards are to be divided into 4 sets contain 20, 15, 10, 7.

$\therefore$ Required number of ways $= \dfrac{52!}{20!\,15!\,10!\,7!}$

Hence, option (c) is correct.

30. (d) Here, 52 cards are to be divided into 4 groups in which three having 15 cards and fourth having 7 cards.

$\therefore$ Required number of ways $= \dfrac{52!}{(15!)^3 \times 3! \times 7!}$

Hence, option (d) is correct.

31. (a) We have, 16 players are divided into 4 groups and order is not important.

$\therefore$ Total number of ways

$$= \frac{16!}{(4!)^4 \times 4!} = \frac{16 \times 15 \times 14 \times \cdots \times 3 \times 2 \times 1}{(1 \times 2 \times 3 \times 4)^5}$$

$$= \frac{2^8 \times 1 \times 2 \times 3 \times 4 \times \cdots \times 8\,(1 \times 3 \times 5 \times 7 \times 9 \times \cdots \times 15)}{(1 \times 2 \times 3 \times 4)^4}$$

$$= \frac{35}{27}(1 \times 3 \times 5 \times 7 \times \cdots \times 15) = \frac{35}{27}\prod_{r=1}^{8}(2r-1)$$

Hence, option (a) is correct.

32. (c) Here, P_1, P_2, P_3 and P_4 are different groups.

Therefore, 12 players are divided into 4 groups equal.

$\therefore$ Total number of arrangement

$$= \frac{12!}{(3!)^4 \times 4!} \times 4! = \frac{12!}{(3!)^4} = \frac{12 \times 11!}{6 \times 6 \times 6 \times 6} = \frac{11!}{108}$$

Hence, option (c) is correct.

33. (d) We have, P_6 must be anyone of four equal groups.

In group P_6 there are 10 players are selected.

So, required number of ways of 3 players selected from 10 players is ${}^{10}C_3$.

Now, three equal groups of four player each.

$\therefore$ Total number of ways $= \dfrac{12!}{(4!)^3 \times 3!}$

Total number of ways P_6 is one of them is

$${}^{10}C_3 \frac{12!}{(4!)^3 \times 3!} = \frac{10 \times 9 \times 8}{1 \times 2 \times 3 \times 1 \times 2 \times 3} \times \frac{12!}{(4!)^3}$$

$$= 20 \times \frac{12!}{(4!)^3}$$

Since, value of $k = 20$

Hence, option (d) is correct.

DPP-9 Dearrangement, Multinominal, Selection and Arrangement from Identical Arrangement

1. (a) We know that, if there are n letters and n corresponding envelopes, then the number of ways in which letters can be placed in the evelopes, so that no letter is placed in the correct envelope is

$$n!\left[1 - \frac{1}{1!} + \frac{1}{2!} + \cdots + \frac{(-1)^n}{n!}\right]$$

Here, $n = 10$

$\therefore$ Required number of ways $= 10!\left[1 - \dfrac{1}{1!} + \dfrac{1}{2!} - \cdots + \dfrac{1}{10!}\right]$

2. (c) Let D_r denote the number of ways in which r things goes to r corresponding places wrongly.

$$\therefore \quad D_r = r!\left[\frac{1}{2!} - \frac{1}{3!} + \frac{1}{4!} - \frac{1}{5!} + \cdots + (-1)^r \frac{1}{r!}\right]$$

Case I Exactly two letters are placed in wrong envelopes.

$\therefore$ Required number of ways $= {}^5C_2 \cdot D_2$

Case II Exactly three letters are placed in wrong envelopes.

$\therefore$ Required number of ways $= {}^5C_3 \cdot D_3$

Case III Exactly four letters are placed in wrong envelopes.

$\therefore$ Required number of ways $= {}^5C_4 \cdot D_4$

Case IV All five letters are placed in wrong envelopes.

$\therefore$ Required number of ways $= D_5$

So, total number of required ways

$$= {}^5C_2 \cdot D_2 + {}^5C_3 \cdot D_3 + {}^5C_4 \cdot D_4 + D_5$$
$$= 10 \cdot 1 + 10 \cdot 2 + 5 \cdot 9 + 44$$
$$= 10 + 20 + 45 + 44 = 119$$

Alternate Method

Required number of ways

= Number of arrangement of 5 letters in 5 envelopes

– Number of ways of placing all letters in right envelope

$= {}^5P_5 - 1 = 5! - 1 = 120 - 1 = 119$

3. (a) Put $x = a - 1$, $y = b - 1$, $z = c - 1$, $t = d - 1$, then $a, b, c, d \ge 0$

and $\quad (a-1) + (b-1) + (c-1) + (d-1) = 20$

$\Rightarrow \quad a + b + c + d = 24$

which is equivalent to the number of ways of distributing 24 identical objects into 4 different boxes, if empty boxes are allowed which is

$${}^{24+4-1}C_{4-1} = {}^{27}C_3$$

4. (b) Let $x = p - 5$, $y = q - 5$ and $z = r - 5$, then $p, q, r \ge 0$ and the given equation reduces to $p + q + r = 15$ $\quad \ldots$(i)

Now, we have to find non- negative integral solution of Eq. (i). The total number of such solution is

$${}^{15+3-1}C_{3-1} = {}^{17}C_2 = \frac{17 \times 16}{2} = 136$$

5. (d) As $x, y, z \in N$, therefore $x, y, z > 0$.

Case I When $x + y + z = n$

Now, see k the positive integral solution of the equation $x + y + z + u = n$. The number of required $= {}^{n-1}C_{3-1} = {}^{n-1}C_2$

Case II When $x + y + z < n$

Let u be a variable such that $x + y + z + u = n$

Then, $\quad x + y + z < n \Rightarrow u > 0$

Now, see k the positive integral solution of the equation $x + y + z + u = n$. The number of required solution $= {}^{n-1}C_{4-1} = {}^{n-1}C_3$

$\therefore$ Total number of non-negative integral solution of $x + y + z \le n$ is ${}^{n-1}C_2 + {}^{n-1}C_3 = {}^nC_3$ $[\because {}^nC_r + {}^nC_{r-1} = {}^{n+1}C_r]$

6. (a) Let $x = 0$. Then, the equation reduces to $y + z = 33$.

Now, number non-negative integral solution of the equation $y + z = 33$ is ${}^{33+2-1}C_{2-1} = {}^{34}C_1$

Let $x = 1$. Then, the equation reduces to $y + z = 30$.

Now, number of non-negative integral solution of the equation $y + z = 30$ is ${}^{30+2-1}C_{2-1} = {}^{31}C_1$

Similarly, we have $x = 2, 3, 4, \ldots, 11$ and the corresponding number of non-negative integral solution are ${}^{28}C_1, {}^{25}C_1, \ldots, {}^1C_1$.

$\therefore$ Total number of integral solution $= {}^{34}C_1 + {}^{31}C_1 + \cdots + {}^1C_1$

$= 34 + 31 + 28 + \cdots + 1$

$= \dfrac{12}{2}[1 + 34]$ $\quad [\because$ sum of n terms of an AP $= \dfrac{n}{2}(a + l)]$

$= 6\,(35) = 210$

7. (c) Consider, $\quad 30 = 2 \times 3 \times 5 = 10 \times 3 \times 1 = 2 \times 15 \times 1$

$= 6 \times 5 \times 1 = 1 \times 1 \times 30$

Hence, the total number of positive integral solution of $abc = 30$ is

$$3! + 3! + 3! + 3! + \frac{3!}{2!} = 27$$

8. (c) Let $a = x_1 + x_2 + x_3$ and $b = y_1 + y_2$

$\therefore$ We have, $a, b > 0$ and $ab = 77$

Now, consider $77 = 7 \times 11 = 1 \times 77$

$\therefore$ Number of possible positive integral solution of $ab = 77$ is $2! + 2! = 4$.

But $a \neq 1$ and $b \neq 1$ as $x_1, x_2, x_3, y_1, y_2 \in N$.

$\therefore a = 7, b = 11$ or $a = 11; b = 7$

Case I When $a = 7; b = 11$

We have, $x_1 + x_2 + x_3 = 7$ and $y_1 + y_2 = 11$

Clearly, the number of positive integral solution of

$x_1 + x_2 + x_3 = 7$ is ${}^{7-1}C_{3-1} = {}^6C_2 = \frac{6 \times 5}{2 \times 1} = 15$

and the number of positive integral solution of $y_1 + y_2 = 11$ is

$${}^{11-1}C_{2-1} = {}^{10}C_1 = 10$$

$\therefore$ Required number of solutions $= 15 \times 10 = 150$

Case II When $a = 11; b = 7$

We have, $x_1 + x_2 + x_3 = 11$ and $y_1 + y_2 = 7$

Clearly, the number of positive integral solution of

$x_1 + x_2 + x_3 = 11$ is

$${}^{11-1}C_{3-1} = {}^{10}C_2 = \frac{10 \times 9}{2} = 45$$

and the number of positive integral solution of $y_1 + y_2 = 7$ is ${}^{7-1}C_{2-1} = {}^6C_1 = 6$.

Required number of solutions $= 45 \times 6 = 270$

$\therefore$ Total number of solutions $= 150 + 270 = 420$

9. (d) We have, $x_1 x_2 x_3 x_4 x_5 = 240$, where x_1, x_2, x_3, x_4, x_5 are natural numbers.

Now, by using prime factorisation of 240, we can write the given equation as

$$x_1x_2x_3x_4x_5 = 2^4 \times 3 \times 5$$

Now, we can assign 3 and 5 to any of 5 unknowns in 5×5 ways.

We can assign 2^4 to x_1, x_2, x_3, x_4, x_5 in

$$a + b + c + d + e = 4$$

$\Rightarrow$ ${}^{4+5-1}C_{5-1} = {}^8C_4$, ways

$\therefore$ Required number of ways $= {}^8C_4 \times 25 = 1750$

10. (c) Consider, $24 = 2^3 \times 3$

Now, consider three boxes x, y, z. 3 can be put in any of the three boxes.

Also, 2, 2, 2 can be distributed in three boxes in ${}^{3+3-1}C_{3-1} = {}^5C_2$ ways. Thus, total number of positive integral solutions is equal to the number of distibution which is given by $3 \times {}^5C_2 = 30$.

Now, to find the integral solution, where negative integers are also allowed, any two of the factors in each factorisation may be negative. Therefore, the number of ways to associate negative sign in each case is ${}^3C_2 = 3$.

Hence, total number of integral solutions is $30 + 3 \times 30 = 120$

Note If any box remains empty, say x, then $x = 1$

11. (a) Let x be the number of objects to the left of the first object chosen, y the number of objects between the first and the second, z the number of objects between the second and the third and u the number of objects to the right of the third object.

Then, $x, u \geq 0$, $y, z \geq 1$ and $x + y + z + u = n - 3$.

Let $y_1 = y - 1$ and $z_1 = z - 1$. Then, $y_1 \geq 0$, $z_1 \geq 0$ such that $x + y_1 + z_1 + u = n - 5$.

$\therefore$ Total number of non-negative integral solution of this equation is ${}^{n-5+4-1}C_{4-1} = {}^{n-2}C_3$.

$\therefore$ Required number of ways $= {}^{n-2}C_3$

12. (c) Consider the product

$$(x^0 + x^1 + x^2 + \dots + x^9)(x^0 + x^1 + x^2 + \dots + x^9)\dots 6 \text{ factors}$$

The number of ways in which the sum of the digits will be equal to 12 is equal to the coefficient of x^{12} in the above product.

So, required number of ways

$=$ Coefficient of x^{12} in $(x^0 + x^1 + x^2 + \dots + x^9)^6$

$=$ Coefficient of x^{12} in $\left(\frac{1 - x^{10}}{1 - x}\right)^6$

$=$ Coefficient of x^{12} in $(1 - x^{10})^6 (1 - x)^{-6}$

$=$ Coefficient of x^{12} in $(1 - x)^{-6} (1 - {}^6C_1 x^{10} + \dots)$

$=$ Coefficient of x^{12} in $(1 - x)^{-6} - {}^6C_1$ coefficient of x^2 in $(1 - x)^{-6}$

$= {}^{12+6-1}C_{6-1} - {}^6C_1\, {}^{2+6-1}C_{6-1}$

$= {}^{17}C_5 - 6 \times {}^7C_5 = 6062$

13. (c) Required number of ways is the number of positive integral solution of $x + y + z = 11$, where $1 \leq x, y, z \leq 6$.

$\therefore$ Required number of ways $=$ Coefficient of x^{11} in

$$(x + x^2 + x^3 + \dots + x^6)^3$$

$=$ Coefficient of x^8 in $(1 + x + x^2 + \dots + x^5)^3$

$=$ Coefficient of x^8 in $\left(\frac{1 - x^6}{1 - x}\right)^3$

$=$ Coefficient of x^8 in $(1 - x^6)^3 (1 - x)^{-3}$

$=$ Coefficient of x^8 in $(1 - {}^3C_1 x^6 + \dots)(1 - x)^{-3}$

$=$ Coefficient of x^8 in $(1 - x)^{-3} - {}^3C_1 \cdot$ Coefficient of x^2 in $(1 - x)^{-3}$

$= {}^{8+3-1}C_{3-1} - {}^3C_1 \cdot {}^{2+3-1}C_{3-1}$

$= {}^{10}C_2 - 3 \times {}^4C_2$

$= \frac{10 \times 9}{2} - 3 \times \frac{4 \times 3}{2}$

$= 45 - 18$

$= 27$

14. (b) We have, two classrooms A and B having capacity of 25 and $(n - 25)$ seats, respectively.

Now,

$A_n =$ Number of possible seating arrangement of room A, when n students are to be seated in these room.

$=$ Number of ways of arranging n students taking 25 at a time

$$= {}^nP_{25}$$

$\therefore$ $A_n - A_{n-1} = {}^nP_{25} - {}^{n-1}P_{25}$

$\Rightarrow$ ${}^nC_{25} \times (25!) - {}^{n-1}C_{25} \times (25!) = (25)!\,({}^{49}C_{25})$

$\Rightarrow$ ${}^nC_{25} - {}^{n-1}C_{25} = {}^{49}C_{25}$

$\Rightarrow$ ${}^{n-1}C_{24} + {}^{n-1}C_{25} - {}^{n-1}C_{25} = {}^{49}C_{25}$

$\Rightarrow$ ${}^{n-1}C_{24} = {}^{49}C_{25} = {}^{49}C_{24}$

$\Rightarrow$ $n - 1 = 49$

$\therefore$ $n = 50$

15. (c) Four letters can be selected in the following ways

(i) All different i.e. C, O, R, G

(ii) 2 like and 2 different

(iii) 3 like and 1 different i.e. three O and 1 from R, G,C.

The number of ways in (i) in ${}^4C_4 = 1$

The number of ways in (ii) is $1 \cdot {}^3C_2 = 3$

The number of ways in (iii) is $1 \times {}^3C_1 = 3$

Therefore, required number of ways $= 1 + 3 + 3$

$= 7$

16. (d) In 'MORADABAD' we have 6 different types of letters 3A's, 2D's and rest four different. We have, to form words of 4 letters.

Now, four letters word can be formed in following ways

(i) All different number of ways ${}^6P_4 = 6 \times 5 \times 4 \times 3 = 360$

(ii) Two different two alike number of ways ${}^2C_1 \times {}^5C_2 \times \frac{4!}{2!} = 240$

(iii) 3 alike 1 different number of ways ${}^1C_1 \times {}^5C_1 \times \frac{4!}{3!} = 20$

(iv) 2 alike of one type and 2 alike of other type, number of ways ${}^2C_2 \times \frac{4!}{2! \times 2!} = 6$

$\therefore$ Total numbers of words = 360 + 240 + 20 + 6 = 626

17. (a) In 'COMBINATION' we have 8 different types of letters 2 Is, 2 Os, 2 Ns and rest 5 different.

We have to form words by taking two same and two different letters and this can be done in ${}^3C_1 \times {}^7C_2 \times \frac{4!}{2!}$ ways

$= 3 \times 21 \times 12 = 756$

18. (a) Given letters are I, (N, N, N), (D, D), (E, E, E), P, T.

We have to select five letters from the given letters and this can be done in following ways.

(i) All the letters are distinct [different letters are I, N, D, E, P, T], number of ways = ${}^6C_5 = 6$

(ii) 3 distinct and 2 alike, number of ways = ${}^3C_1 \times {}^5C_3 = 30$

(iii) 2 distinct, 3 alike = ${}^2C_1 \times {}^5C_2 = 20$

(iv) 2 alike, 2 alike, 1 distinct = ${}^3C_2 \times {}^4C_1 = 12$

(v) 3 alike, 2 alike = ${}^2C_1 \times {}^2C_1 = 4$

$\therefore$ Total number of ways of selecting 5 letters

= 6 + 30 + 20 + 12 + 4 = 72

19. (a) We know that, the number of combination of n different things taken r at a time allowing repetition is ${}^{n+r-1}C_r$.

Here, $n = 8, r = 6$

$\therefore$ Required number of ways = ${}^{8+6-1}C_6 = {}^{13}C_6$

20. (b) We have to select 5 squares, taking at least one from each row and this can be done in following ways

$R_1(4)$	$R_2(2)$	$R_3(2)$	Number of selections
1	2	2	${}^4C_1 \times {}^2C_2 \times {}^2C_2 = 4$
2	1	2	${}^4C_2 \times {}^2C_1 \times {}^2C_2 = 6 \times 2 = 12$
2	2	1	${}^4C_2 \times {}^2C_2 \times {}^2C_1 = 6 \times 1 \times 2 = 12$
3	1	1	${}^4C_3 \times {}^2C_1 \times {}^2C_1 = 4 \times 2 \times 2 = 16$

$\therefore$ Total number of ways = 4 + 12 + 12 + 16 = 44

21. (c) Here, R_1 has 3 squares, R_2 has one square, R_3 has 3 squares, R_4 has one square and R_5 has 3 squares. We have to select 6 squares so that no row remains empty. The selection scheme is as follows

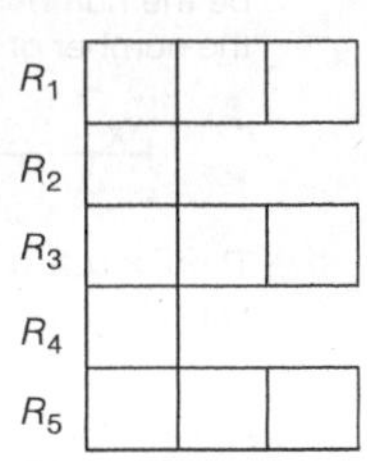

$R_1(3)$	$R_2(1)$	$R_3(3)$	$R_4(1)$	$R_5(3)$
2	1	1	1	1
1	1	1	1	2
1	1	2	1	1

Clearly, 2nd and 4th row has to be selected.

One row out of 1st, 3rd and 5th will contain two letters and remaining one one letter.

Selection of one row that contains two letters is done in ${}^3C_1$ ways i.e. 3 ways.

Now, two block can be selected from this row in ${}^3C_2$ ways i.e. 3 ways.

From each of remaining two rows, one blocks can be selected in ${}^3C_1$ ways = 3 ways

Hence, total selections = ${}^3C_1 \times {}^3C_2 \times {}^3C_1 \times {}^3C_1 = 81$

Number of ways of filling

A, A, A, B, B, B = $81 \times \frac{6!}{3!\,3!} = 81 \times 20 = 1620$

22. (b) As all the X's are identical, the question is of selection of 6 squares from 8 squares, so that no row remains empty. Here, R_1 has 2 squares, R_2 has 4 squares and R_3 has 2 squares. The selection scheme is as follows:

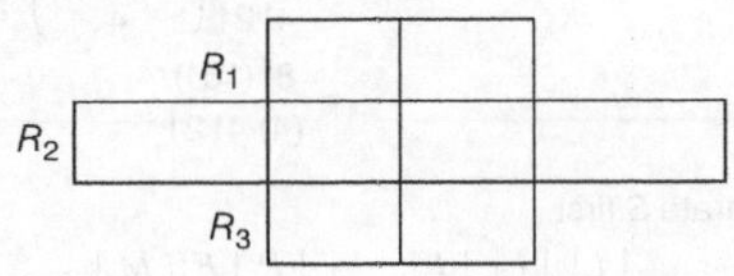

$R_1(2)$	$R_2(4)$	$R_3(2)$	Number of selections
1	4	1	${}^2C_1 \times {}^4C_4 \times {}^2C_1 = 2 \times 1 \times 2 = 4$
1	3	2	${}^2C_1 \times {}^4C_3 \times {}^2C_2 = 2 \times 4 \times 1 = 8$
2	3	1	${}^2C_2 \times {}^4C_3 \times {}^2C_1 = 1 \times 4 \times 2 = 8$
2	2	2	${}^2C_2 \times {}^4C_2 \times {}^2C_2 = 1 \times 6 \times 1 = 6$

$\therefore$ Total number of ways = 4 + 8 + 8 + 6 = 26

23. (b) Let S_1, S_2, S_3 and S_4 be the four intermediate stations, where the train stops.

Let a, b, c, d and e be the number of stations between A and S_1, S_1 and S_2, S_2 and S_3, S_3 and S_4 and S_4 and B, respectively.

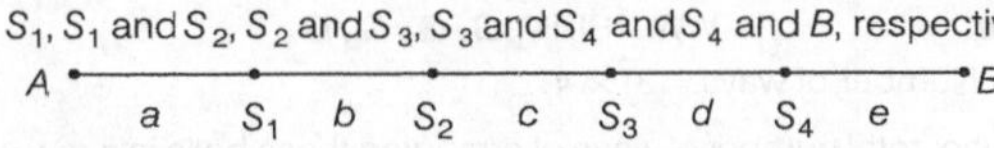

Clearly, $a + b + c + d + e = 8$, where

$a \ge 0, b \ge 1, c \ge 1, d \ge 1$ and $e \ge 0$.

Let $x = a,\ y = b - 1,\ z = c - 1, t = d - 1, w = e$

Now, $x + y + z + t + w = a + b + c + d + e - 3 = 8 - 3 = 5$

or $x + y + z + t + w = 5$.

Where, $x, y, z, t, w \ge 0$ ...(i)

$\therefore$ Required number

= number of non-negative integral solution of (i)

$= {}^{5+5-1}C_{5-1} = {}^9C_4$

24. (c) In 'MISSISSIPPI' we have 1 M, 4 I, 4 S and 2 P.

We have to select 5 letters from the given letters and this can be done in following ways:

(i) 3 distinct and 2 alike, number of ways = ${}^3C_1 \times {}^3C_3 = 3$

(ii) 2 distinct and 3 alike, number of ways = ${}^2C_1 \times {}^3C_2 = 6$

(iii) 1 distinct and 4 alike, number of ways = ${}^2C_1 \times {}^3C_1 = 6$

(iv) 2 alike, 2 alike different from first, 1 distinct = ${}^3C_2 \times {}^2C_1 = 6$

(v) 3 alike and 2 alike = ${}^2C_1 \times {}^2C_1 = 4$

$\therefore$ Total number of ways of selecting 5 letters

= 3 + 6 + 6 + 6 + 4 = 25

25. (a) Let us first find the number of ways in which the letters of the word W can be arranged so that all vowel's come together

MSSSSPP IIII

$\therefore$ Number of ways $= \dfrac{8!}{4!\,2!}$

Now, total number of ways in which the letters of the word W can be arrange

$$= \frac{11!}{4!\,4!\,2!}$$

Thus, required number of ways $= \dfrac{11!}{4!\,4!\,2!} - \dfrac{8!}{4!\,2!}$

$$= \frac{8!}{4!\,2!}\left(\frac{11 \times 10 \times 9}{4!} - 1\right)$$

$$= \frac{8!}{4!\,2!}\left(\frac{11 \times 10 \times 9}{4 \times 3 \times 2 \times 1} - 1\right)$$

$$= \frac{8!}{4!\,2!}\left(\frac{165 - 4}{4}\right)$$

$$= \frac{8!\,(161)}{(4)\,4!\,2!}$$

26. (b) Separate S first

| I | I | I | I | I | I | I | I | I | I | I | P | P | M | [8 gaps]

| I | I | I | I | I | I | I | I | I | PP | M |

$\Rightarrow \dfrac{7!}{4!\,2!} \cdot {}^8C_4 - \dfrac{6!}{4!} \cdot {}^7C_4,$

S's separated but P's may or may not be separated simplified to

$\dfrac{6!\,7!}{4!\,4!} = \dfrac{10!}{4!\,4!} \Rightarrow k = 1$

27. (6) We have, 2 identical white balls (W), 3 identical red balls (R) and 4 green balls of different shades (G_1, G_2, G_3, G_4).

Let us first find the number of ways in which they can be arranged in a row, so that no ball is separated from the balls of the same colour.

i.e. WW RRR $G_1\,G_2\,G_3\,G_4$

$\therefore$ Number of ways $= 3! \times 4!$

Also, total number of ways of arranging these balls in a row $= \dfrac{9!}{2!\,3!}$

$\therefore$ Required number of ways $= \dfrac{9!}{2!\,3!} - 3! \times 4!$

$$= \frac{9 \times 8 \times 7 \times 6 \times 5 \times 4}{2} - 3! \times 4!$$

$$= 6\,(7 \times 6 \times 5 \times 4 \times 3 \times 2) - 6 \times 4!$$

$$= 6\,(7! - 4!)$$

$\therefore \quad k = 6$

28. (9) Number of dearrangements in such problems is given by

$$n!\left\{1 - \frac{1}{1!} + \frac{1}{2!} - \frac{1}{3!} + \frac{1}{4!} - \dots + (-1)^n \frac{1}{n!}\right\}$$

Hence, the required number of dearrangements is

$$4!\left\{\frac{1}{2!} - \frac{1}{3!} + \frac{1}{4!}\right\} = 12 - 4 + 1 = 9$$

29. (4) Let the number of green balls be x. Then, the number of red balls is $2x$. Let the number of blue balls be y.

Then, $\quad x + 2x + y = 10$

$\Rightarrow \quad 3x + y = 10$

$\Rightarrow \quad y = 10 - 3x$

Clearly, x can take value 0, 1, 2, 3. The corresponding values of y are 10, 7, 4 and 1. Thus, the possibilities are (0, 10, 0), (2, 7, 1), (4, 4, 2) and (6, 1, 3), where (r, b, g) denotes the number of red, blue and green balls.

30. (1) ***Case I*** When n peoples are sitting in a row.

Let x be the number of persons to the left of the first person chosen, y be the number of persons between the first and second, z be the number of persons between the second and the third and u be the number of persons to the right of third person.

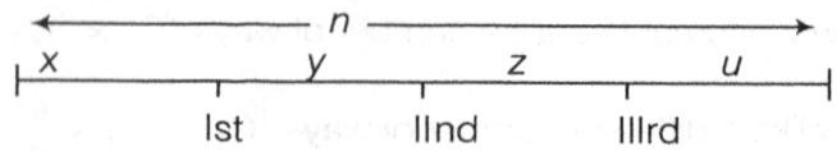

Then, $x, u \geq 0$; $y, z \geq 1$ and $x + y + z + u = n - 3$

Now, let $y_1 = y - 1$ and $z_1 = z - 1$. Then, $y_1 \geq 0$ and $z_1 \geq 0$ such that $x + y_1 + z_1 + u = n - 5$, where $x, y_1, z_1, u \geq 0$.

$\therefore$ Total number of ways of selecting 3 people out of n

= Number of non-negative integral solution of $x + y_1 + z_1 + u$

$= n - 5 = {}^{n-5+4-1}C_{4-1} = {}^{n-2}C_3$

i.e. $\quad P_n = {}^{n-2}C_3$

Case II When n peoples are sitting in a circle.

Let n persons be denoted by $a_1, a_2, \dots, a_n$.

Now, we have to select 3 persons so that no two of them are consecutive. For this we first find the number of ways in which 2 or 3 persons are consecutive. Now, number of ways in which 2 or 3 persons are consecutive is obtained as follows, with a_1 the number of such triples is $a_1 a_2 a_3, a_1 a_2 a_5, a_1 a_2 a_6, \dots, a_1 a_2 a_{n-1}$.

[Note that we have excluded $a_1 a_2 a_n$, since it will be repeated again for, if we start with a_n, then we shall get triples : $a_n a_1 a_2, a_n a_1 a_3$]

So, number of such triples when we start with a_1 is $(n - 3)$. Similarly, with $a_2, a_3, a_4, \dots$ we shall get the same number of triples that is $(n - 3)$.

But total number of triples is ${}^nC_3$.

Thus, required number of ways $= {}^nC_3 - n\,(n - 3)$

$$= \frac{n\,(n-1)\,(n-2)}{6} - n\,(n-3)$$

$$= \frac{n\,(n-4)\,(n-5)}{6}$$

$\because \quad P_n - Q_n = 6$

$\Rightarrow \quad {}^{n-2}C_3 - \dfrac{n\,(n-4)\,(n-5)}{6} = 6$

$\Rightarrow \quad \dfrac{(n-2)\,(n-3)\,(n-4)}{6} - \dfrac{n\,(n-4)\,(n-5)}{6} = 6$

$\Rightarrow \quad \dfrac{(n-4)}{6}\,[n^2 - 5n + 6 - n^2 + 5n] = 6$

$\Rightarrow \quad (n - 4) = 6$

$\therefore \quad n = 10 = 10^1$

$\therefore \quad x = 1$

31. (3) Let x be the number of books to the left of the first book chosen, y be the number of books between the first and second, z be the number of books between the second and third and u be the number of books to the right of the third book.

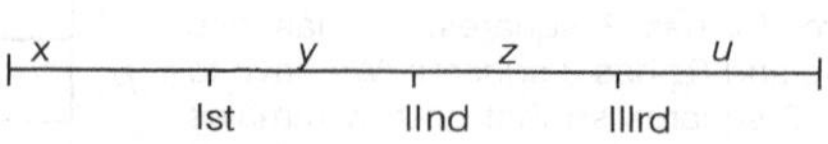

Then, $x, u \geq 0$, $y, z \geq 1$ and $x + y + z + u = 97$

Let $y_1 = y - 1$ and $z_1 = z - 1$. Then, $y_1 \geq 0$, $z_1 \geq 0$ such that

$$x + y_1 + z_1 + u = 95 \qquad \dots(i)$$

$\therefore$ Total number of ways in which 3 books can be selected = Number of non-negative integral solution of (i)

$$= {}^{95+4-1}C_{4-1} = {}^{98}C_3$$

$\therefore \quad k = 3$

Revisal Problems (JEE Main)

1. (c) ***Case I*** When only one flag is used

Number of signals made = $^3P_1 = 3$

Case II When only two flags are used

Number of signals made = $^3P_2 = 3 \times 2 = 6$

Case III When three flags are used

Number of signals made = $^3P_3 = 3! = 3 \times 2 \times 1 = 6$

Hence, required number of signals = $3 + 6 + 6 = 15$

2. (b) Number of numbers from 1000 to 9999 (both inclusive) using 0, 1, 2, 3, 4, 5, 6, 7, 8, 9 when

Case I Repeatition is allowed

We cannot place zero at thousand place. Thus, thousand place can be filled in 9 ways.

And unit's, ten's and hundreds place can be filled in 10 ways each.

•	•	•	•
9	10	10	10

$\therefore$ Number of numbers = $9 \times 10 \times 10 \times 10 = 9000$

Case II Repeatition is not allowed

We cannot place zero at thousand place. Thus, thousand place can be filled in 9 ways.

•	•	•	•
9	9	8	7

Since, one digit one of ten digits assign at thousand place, therefore hundreds place can be filled in 9 ways. Similarly, unit's and ten's place can be filled in 7 and 8 ways.

Number of numbers = $9 \times 9 \times 8 \times 7 = 4536$

$\therefore$ Required number of numbers that do not have all 4 different digit = $9000 - 4536 = 4464$

3. (d) Total number of words without any restriction is 6!.

Total number of words begning with *G* is 5!.

Total number of words ending with *S* is 5!.

Total number of words beginning with *G* and ending with *S* is 4!.

Thus, total number of required words = $6! - 2 \times 5! + 4!$.

$= 720 - 240 + 24 = 504$

4. (a) If all the girl sit together, then consider all the three girls as one group.

Now, we have to arrange $5 + 1 = 6$ persons

Number of ways arranging 6 persons is 6!.

And the three girls arranged themselves in 3! way.

$\therefore$ Number of ways of arranging all the girls together is $6! \times 3!$.

Total number of arrangement of 5 boys and 3 girls are 8!.

$\therefore$ Total number of ways in which all the girls are never together

= Total number of arrangement

– Total number of arrangement in which all the girls are always together

$= 8! - 6!3! = 40320 - 4320 = 36000$

5. (c) The total number of numbers that can be formed with the digits 1, 3, 5, and 7 taken all at a time is equal to the number of arrangements of 4 digit, taken all at a time

i.e. $^4P_4 = 4! = 24$

To find the sum of these 24 numbers, we have to find the sum of the digits at unit's, ten's, hundred's and thousand's places in all these numbers.

Consider the digits in the unit's places in all these numbers.

If 1 is the digit at unit's place, the remaining three place can be filled in 3! ways or we can say 1 occurs at unit's place

$$3! = (6)\text{times}.$$

Similarly, each digit occurs 6 times.

So, the total sum of the digits in the unit's place in all these numbers is

$$(1 + 3 + 5 + 7) \times 3! = 16 \times 6 = 96$$

Similarly, the sum of digit is 96 at ten's, hundred's and thousand places.

Hence, the sum of all the number is

$$16 \times 3! \times (10^0 + 10^1 + 10^2 + 10^3) = 16 \times 3! \times 1111$$

Aliterate Method

Trick Sum of all the numbers formed by taking all the given *n* digits.

= (Sum of all the *n* digits) $\times (n - 1)! \times$ (111...*n* times)

Hence, given digits are 1, 3, 5, and 7.

$\therefore$ Required sum = $(1 + 3 + 5 + 7) \times 3! \times (1111) = 16 \times 3! \times 1111$

6. (c) First of all, we arrange the letters N, I, N, T, N in following manner.

N	I	N	T	N

The above arrangement can be made is $\frac{5!}{3!}$ ways.

Now, there are 6 places and we have to choose only 3 place to arrange all the three *E*'s, this can be done in $^6C_3$ way.

Hence, total number of required permutation = $^6C_3 \times \frac{5!}{3!}$

7. (c) Out of 8 flowers, 4 particular flowers are always together. So, consider these four flowers as one group.

Now, we have to arrange 5 flowers in a circle.

Number of ways = $\frac{(5-1)!}{2} = \frac{4!}{2}$

And the 4 particular flowers which we have taken as one group can be arrange themselves in 4! ways.

$\therefore$ Required number of ways = $\frac{4!}{2} \times 4! = 288$ ways.

8. (c) According to the question, if all white roses come together, then we consider 3 white roses as one group.

Now, we have to arrange $6 + 1 = 7$ thing in a circle.

Then, number of arrangement = $\frac{(7-1)!}{2} = \frac{6!}{2}$

Now, 3 white roses can be arranged themselves in 3! ways.

$\therefore$ Required number of arrangement = $\frac{6!}{2} \times 3!$

$= \frac{6!}{2} \times 6 = 3 \times 6! = 3 \times 720 = 2160$

9. (a) First of all, we arrange the boys on the circle in 3! ways.

Now, arrangement of girls in between the boys in 4! ways.

$\therefore$ Required number of ways = $3! \times 4!$

10. (a) The number of ways in which 5 beads are choosen from 8 different beads is $^8C_5$.

Now, these 5 beads can be arranged in a ring into $\frac{(5-1)!}{2}$ ways.

$\therefore$ Required number of ways = $^8C_5 \times \frac{4!}{2} = \frac{8!}{5!\,3!} \times \frac{4!}{2!}$

$= \frac{1344}{2} = 672$

11. (a) Excluding two specified guest, 12 persons can be divided into two groups are containing 6 and other containing $(6-2)=4$ in $\left|\frac{(12-2)!}{6!\,4!}\right|$ ways i.e. $\frac{10!}{6!\,4!}$ ways.

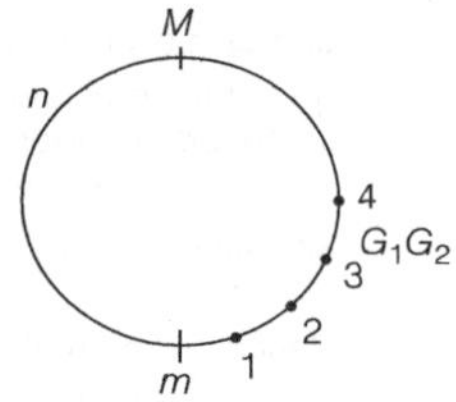

And can sit either side of master and mistress in 2! ways and can arrange themselves in $6!\times 4!$ ways.

Now, the two specified guest where 4 guest are seated have 5 gaps and can arrange themselves in 2! ways.

The number of ways, when G_1, G_2 will always be together is

$$\frac{10!}{6!\,4!}\times 2!\,6!\,4!\times 5\times 2!$$

$$=10!\times 4\times 5=20\times 10!$$

12. (d) Let the seat occupied by *A* be numbered as 1 and the remaining 5 seats be numbere as 2, 3, 4, 5, 6 in anticlockwise direction. Here, two cases arises.

***Case* I** *B* is on right of *A* i.e. at number 2. Then, seat number 3 can be occupied by *C* or *D* in ${}^2C_1$ ways and remaining 3 person can have remaining three seats in 3! ways.

Hence, the number of arrangements in this case is $2\times 6=12$.

***Case* II** *C* is at the right of *A* i.e. at number 2. Then, *B* can occupy any seat from number 3 or 4 or 5. Then, *D* must be on the right of *B*. So, we are left with 2 persons and 2 seats, which can be occupied in 2! ways.

Hence, the number of arrangement in this case is ${}^3C_1\times 2!=6$.

These cases are exclusive. So, by the same rule

Total number of arrangement is $12+6=18$

13. (c) There are 6 letters S, S, E, E, R, I

The following cases arise.

***Case* I** All lettrs different.

The required number of arranging 3 different letters from 4 different (S, E, R, I) types of letters $={}^4P_3=\frac{4!}{1!}=4!=24$

***Case* II** Two alike and one different

Required ways $={}^2C_1\times {}^3C_1\times \frac{3!}{2!}=2\times 3\times 3=18$

$\therefore$ Required number of permutation $=24+18=42$

14. (d) Let the city be represented by a rectangle whose sides are of length *a* and *b* North-South and West-East, respectively. Man has to go from *P* to *Q*. For this, he will have to travel a distance '*a*' vertically downward and a distance '*b*' horizontally from left to right.

Let $a_1, a_2 \ldots a_{m-1}$ denote the distances between consecutive street drawn horizontally beginning with the street passing through *P* and $b_1, b_2, \ldots b_{n-1}$ denote the distance between consecutive streets drawn vertically beginning with the street passing through *P*.

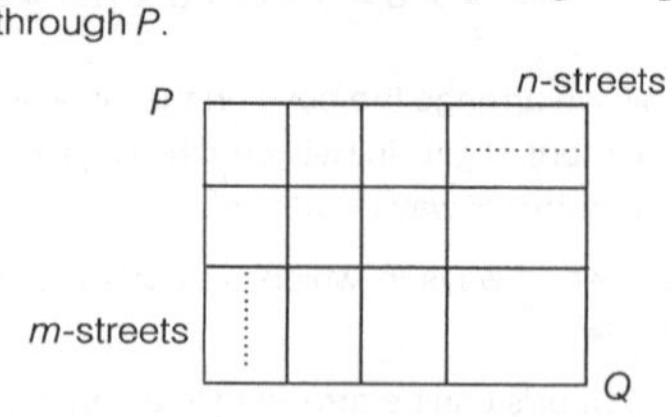

Each arrangement of $(m+n-2)$ things $a_1, a_2 \ldots a_{m-1}$, $b_1, b_2 \ldots b_{n-1}$ in a row, so that order of a_i's does not change and order of b_i's does not change corresponds to one path go from *P* to *Q*.

Therefore, the required number is equal to the number of arrangements of $(m+n-2)$ things such that order of a_i's does not change and order of b_i's does not change, which is equal to $\frac{(m+n-2)!}{(m-1)!\,(n-1)!}$.

Alternate Method

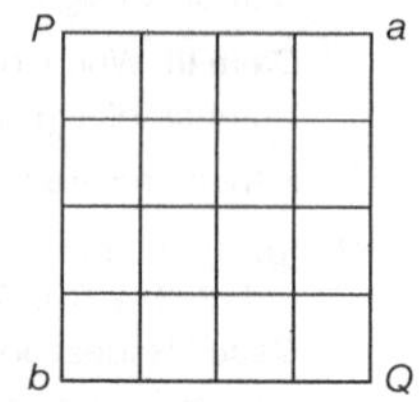

Let the city be represented by a rectangle whose sides are of length *a* and *b*.

For each path total distance covered in horizontal direction is *a* and that in vertical direction is *b*. *a* is the sum of length of $(m-1)$ horizontal line segments and *b* is the sum of length of $(n-1)$ vertical line segments.

Each path to go from *P* to *Q* will be a arrangement of $(m+n-2)$ lines segment of which $(m-1)$ are horizontal and $(n-1)$ are vertical.

Therefore, the required number is equal to number of arrangement of $(m+n-2)$ different things of which $(m-1)$ are one kind and $(n-1)$ are of another kind which is given by $\frac{(m+n-2)!}{(m-1)!\,(n-1)!}$

15. (c) We first select two men out of 9 in ${}^9C_2$ ways. Now, we exclude wives of these two selected men and so select 2 ladies from remaining 7 ladies in ${}^7C_2$ ways. Let *A*, *B* be two men and *X*, *Y* be the ladies playing in one set. Then, we have

(i) *A* and *X* playing against *B* and *Y*

(ii) *A* and *Y* playing against *B* and *X*

Then, the total number of ways is ${}^9C_2\times {}^7C_2\times 2$

$$=\frac{9!}{2!\,7!}\times\frac{7!}{2!\,5!}\times 2=\frac{9!}{2!\,2!\,5!}\times 2$$

$$=\frac{9!}{2!\,5!}=\frac{9\times 8\times 7\times 6}{2}$$

$$=9\times 4\times 7\times 6=1512$$

16. (c) In dictionary, the words at each stage are arranged in alphabetical order.

Starting with the letter E and arranging the other four letters Q, U, U, E in $\frac{4!}{2!}$ ways *i.e.* 12 ways.

Thus, there are 12 word start with E.

Now, the words starting with QE is $\frac{3!}{2!}$ i.e. 3

Thus, there are 3 words starting with QE.

Now, we have so far constructed 15 words.

The 16th word is QUEEU and hence 17th word is QUEUE.

17. (d) According to the question, the order of speakers S_1, S_2 and S_3 can be (not necessarily consecutive) $S_1S_2S_3$ or $S_3S_1S_2$.

For each order, we can select 3 slot out of 10 in ${}^{10}C_3$ ways.

After selecting these three slots in which speakers S_1, S_2, S_3 have only one way of arrangement as said, the remaining seven speakers can be arranged in seven slots in 7! ways.

Hence, total number of arrangements $=2\cdot{}^{10}C_3\times 7!$

$$=2\times\frac{10!}{7!\,3!}\times 7!$$

$$=2\times\frac{10!}{3!}=\frac{10!}{3}$$

18. (d) The number of ways to selecting 8 books from Mathematics, Physics, Chemistry and English is

$$x+y+z+w=8 \Rightarrow {}^{8+4-1}C_{4-1}={}^{11}C_3$$

19. (b) We can think of 4 groups. First consisting of 2 Indian, second consisting of 3 Americans, third consisting of 3 Italian and last one consisting of 4 Frenchmen.

These 4 groups can be arranged on a circle in 3! ways.

And contents of these packets can be arranged in 2!, 3!, 3! and 4! ways, respectively.

Hence, total number of ways $= 2! \times 3! \times 3! \times 4! \times 3!$

$= 2!\,(3!)^3\,4! = 2\,(3!)^3\,4!$

20. (d) We have to determine the total number of even numbers formed by using the given condition.

So, at unit's place we can use one of the digit 0, 2, 4, 6, 8. If 5 is at ten's place then, as per given condition, 7 should be at unit's place. In such a case, the number will not be an even number. So, 5 cannot be at ten's and one's places. Hence, 5 can be only at hundred's place. Now, have two cases arises.

Case I When 5 is at hundred's place. If 5 is at hundred's place, then 7 will be at ten's place. So, unit's place can be filled in 5 ways by using the digit 0, 2, 4, 6, 8.

$\therefore$ Total number of even numbers $= 1 \times 1 \times 5 = 5$

Case II When 5 is not at hundred's place. Now, hundred's place can be filled in 8 ways (0 and 5 can not be used at hundred's place).

In ten's place, we can use any one of the ten digits except 5.

So, ten's place can be filled in 9 ways.

At unit's place, we have to use one of the even digits 0, 2, 4, 6, 8.

So, unit's place can be filled in 5 ways.

So, total number of even numbers $= 8 \times 9 \times 5 = 360$

Hence, total number of required even numbers $= 360 + 5 = 365$

21. (d) According to the question, if terminal digits are even in six-digit number, then unit's place and lack's place are filled by the digits 2, 4 and 6 in 3 and 2 ways, respectively.

·					·
2					3

Remaining four places can be filled by five digits in 5! ways.

$\therefore$ Total number of required numbers $= 2 \times 5! \times 3$

$= 6 \times 5! = 6 \times 120 = 720$

22. (a) Numbers starting from 1,

1					

$\Rightarrow$ 5! = 120

121 st number, is 213456

122 nd, is 213465

123 rd is, 213546

124th, is 213564

23. (a) Total number of ways of answering 5 question out of 12 questions is ${}^{12}C_5$.

And the number of ways in which he does not select any questions from anyone section is $3 \times {}^8C_5$.

$\therefore$ Required number of ways $= {}^{12}C_5 - 3 \times {}^8C_5$

$$= \frac{12!}{5!\,7!} - 3 \times \frac{8!}{5!\,3!} = \frac{12 \times 11 \times 10 \times 9 \times 8}{5 \times 4 \times 3 \times 2 \times 1} - \frac{3 \times 8 \times 7 \times 6}{3 \times 2}$$

$= 9 \times 88 - 168 = 792 - 168 = 624$

24. (b) Number of ways teacher can select examples from arithmetic progression

$= {}^4C_1 + {}^4C_2 = 4 + 6 = 10$

Number of ways teacher can select examples from P and C

$= {}^5C_1 + {}^5C_2 = 5 + 10 = 15$

Number of ways teacher can select examples from Binomial theorem

$= {}^6C_1 + {}^6C_2 = 6 + 15 = 21$

Hence, total number of ways $= 10 \times 15 \times 21 = 3150$

25. (c) Number of kids = 25

Number of kids to be taken at a time = 5

Out of 25 kids 5 can be selected in ${}^{25}C_5$ ways.

Hence, teacher will have to go to the zoological garden ${}^{25}C_5$ times.

Number of selection of 5 kids out of 25 including a particular kid $= {}^1C_1 \times {}^{24}C_4 = {}^{24}C_4$

Hence, a particular kid will go to the zoological garden ${}^{24}C_4$ times.

26. (b) A number of the five digit will be of the form

$a\,x_1\,x_2\,x_3 0$, $a\,x_1\,x_2\,x_3 1$, $a\,x_1, x_2, x_3\ 2$, $a\,x_1, x_2, x_3, 3, \ldots a\,x_1, x_2, x_3\ 9$ where $a \in \{1, 2, 3, \ldots 9\}$ and $x_1, x_2, x_3 \in \{0, 1, 2, 3 \ldots 9\}$

Since, sum of the digits should be even therefore, if $a + x_1 + x_2 + x_3$ is an even number, then the digit at unit's place must be 0, 2, 4, 6, 8 and if $a + x_1 + x_2 + x_3$ is an odd number, then the digits at unit's place will be 1, 3, 5, 7, 9.

$\therefore$ Required number $= 9 \times 10 \times 10 \times 10 \times 5 = 45000$

Alternate Method

A five digit number is of the form a_1, a_2, a_3, a_4, a_5, where $1 \le a_1 \le 9$ and $0 \le a_i \le 9$ for $i \ge 2$.

Let $(x + x^2 + x^2 + \cdots + x^9)(1 + x + x^2 + \cdots + x^9)^4$

$= \alpha_1 x + \alpha_2 x^2 + \alpha_3 x^3 + \cdots + \alpha_{45} x^{45}$ ...(i)

Then, number of 5-digit numbers such that the sum of their digits is even

$= \alpha_2 + \alpha_4 + \cdots + \alpha_{44}$ ...(ii)

Putting $x = 1$ and -1 in Eq. (i), we get

$9 \times 10^4 = \alpha_1 + \alpha_2 + \alpha_3 + \cdots + \alpha_{45}$ ...(iii)

and $0 = -\alpha_1 + \alpha_2 - \alpha_3 + \cdots + \alpha_{45}$ ...(iv)

On adding Eqs. (iii) and (iv)

$9 \times 10^4 = 2(\alpha_2 + \alpha_4 + \alpha_6 + \cdots + \alpha_{44})$

$\Rightarrow \alpha_2 + \alpha_4 + \alpha_6 + \cdots + \alpha_{44} = \dfrac{9 \times 10^4}{2} = 4.5 \times 10^4 = 45000$

27. (d) According to the question, apples received by boys are (1, 3, 4) (2, 3, 3) or (2, 2, 4).

$\therefore$ Required number of ways

$$= 3!\left[\frac{8!}{1!\,3!\,4!} + \frac{8!}{2!\,3!\,3!\,2!} + \frac{8!}{2!\,2!\,4!\,2!}\right] = 6 \times 770 = 4620$$

But, it is given that the number of ways is $K \cdot {}^7P_3$

$\therefore$ $k \cdot {}^7P_3 = 4620$

$\Rightarrow$ $k \cdot \dfrac{7!}{4!} = 4620$

$\Rightarrow$ $k \cdot 7 \times 6 \times 5 = 4620$

$\Rightarrow$ $k \cdot 210 = 4620$

$\therefore$ $k = \dfrac{4620}{210} = 22$

28. (d) The number of ways in which 4 shoes can be chosen, so that there will be no complete pair $= {}^5C_4 \times 2^4 = 80$

29. (a) Required number of ways $= {}^3C_1{}^9C_1 + {}^3C_2{}^9C_2 + {}^3C_3{}^9C_3$

$= 3 \times 9 + 3 \times 36 + 1 \times 84$

$= 27 + 108 + 84 = 219$

30. (c) Let us consider the boxes to be different. Each object can kept in 2 ways.

Total number of ways $= 2^n$

This includes when all the things are in B_1 or B_2.

$\therefore$ Number of ways, no box remains empty $= 2^n - 2$

Since, the boxes are identical.

$\therefore$ Required number of ways $= \dfrac{2^n - 2}{2} = 2^{n-1} - 1$

31. (a) We know that, in a rectangle of $n \times p$ $(n < p)$

Number of rectangles of any size in $\frac{np}{4}(n+1)(p+1)$

and number of square of any size is $\sum_{r=1}^{n}(n+1-r)\times(p+1-r)$.

$\therefore$ Number of rectangles excluding square

$$= \frac{np}{4}(n+1)(p+1) - \sum_{r=1}^{n}(n+1-r)(r+1-r)$$

Put $n = 4$, $p = 6$ in above, we get

$$= \frac{6\times 4}{4}(4+1)(6+1) - \sum_{r=1}^{4}(5-r)(7-r)$$

$$= 6\times 7\times 5 - \sum_{r=1}^{4}(35 - 12r + r^2)$$

$$= 210 - \left(35\times 4 - 12\times\frac{4\times 5}{2} + \frac{4\times 5\times 9}{6}\right)$$

$$= 210 - (140 - 120 + 30) = 210 - 140 + 120 - 30 = 160$$

32. (c) $A^8, B_1, B_2, B_3, B_4, B_5, B_6, B_7, B_8$.

The number of ways taken 8 at a time, is

$$= {}^8C_0 + {}^8C_1 + \cdots + {}^8C_8 = 2^8$$

33. (a) Here, total factors in $a^2b^4c^5$ is 11, out of which 2 are alike of one kind, 4 are alike of second kind and 5 are alike of third kind.

$\therefore$ Required arrangement $= \frac{11!}{2!\,4!\,5!} = \frac{11\times 10\times 9\times 8\times 7\times 6}{2\times 4\times 3\times 2\times 1}$

$= 6930$

34. (c) ***Case I*** When the two identical digits are both unity as shown.

| 1 | x | y | 1 | anyone place out of 3 block for unity can be taken in 3 ways and the remaining two blocks can be filled in 9.8 ways.

$\therefore$ Total number of ways $= 3\cdot 9\cdot 8 = 216$.

Case II When the two identical digit are other than unit.

| 1 | x | x | y | | 1 | x | y | x | | 1 | y | x | x |

two x's can be taken in 9 ways and filled in three ways and y can be taken in 8 ways.

Total ways $= 9\cdot 8\cdot 3 = 216$

$\therefore$ Total ways $= 432$

35. (d) **Statement I**

There are 11 letters in the word ASSOCIATION in which 2 S's, 2 A's, 2 O's and 2 I's occur.

According to the question, 2 S come together, so consider 2S as one group.

Now, the arrangement of 2 A, 2 O, T, N, C and 2 S can be done in $\frac{8!}{2!2!}$ ways.

| A | A | O | O | T | N | C | 2S |

Now, there are 9 place in which we can arrange 2 I's

$\therefore$ Arrangement of 2 I's at 9 places in ${}^9C_2$.

$\therefore$ Total number of arrangement of the letters of the word ASSOCIATION in which 2 S come together and 2 I's are not together is $\frac{8!}{2!2!}\times {}^9C_2$.

Hence, Statement I is incorrect.
Clearly, Statement II is correct.

36. (d) Therefore, A occurs two times, L occurs three times and there are three other different letters P,R,E. We have, to select 4 letters.

$\therefore$ Required number

$=$ Coefficient of x^4 in $(1+x+x^2+x^3)(1+x+x^2)(1+x)^3$

$=$ Coefficient of x^4 in $(1-x)^{-5}(1-x^2)^3(1-x^3)(1-x^4)$

$=$ Coefficient of x^4 in $(1-x)^{-5}(1-x^3-x^4+x^7)(1-3x^2+3x^4-x^6)$

$=$ Coefficient of x^4 in $(1-x)^{-5}$ $-$ Coefficient of x in $(1-x)^{-5} + 2\cdot$ Coefficient of x^6 in $(1-x)^{-5} - 3\cdot$ Coefficient of x^2 in $(1-x)^{-5}$

$$= \frac{5\times 6\times 7\times 8}{4!} - 5 + 2 - 3\,\frac{6\cdot 5}{2!} = 22$$

Hence, Statement I is incorrect.

Clearly, Statement II is correct.

37. (b) **Statement I**

Two circles intersect maximum at two distinct points. How two circles can be selcted in ${}^8C_2$ ways.

Again, each selection of two circles gives two point of intersection.

Therefore, total number of point of intersection $= {}^8C_2 \times 2$

$= 28\times 2 = 56$

Statement II

Two cirlces intersect at two distinct points. Two straight line intersect at one point. One circle and one straight line intersect at two distinct point. Then, total number of point of intersection are as follows.

Number of ways of selection	Points of Intersection
Two staright lines : ${}^4C_2$	${}^4C_2\times 1 = 6$
Two circles : ${}^4C_2$	${}^4C_2\times 2 = 12$
One line and are circel : ${}^4C_1\times{}^4C_1$	${}^4C_1\times{}^4C_1\times 2 = 32$
Total	50

Statement II is correct, but not the correct explanation of Statement I.

38. (b) **Statement I** $\frac{(n+1)!}{(n-1)!} = (n+1)n$, is divisible by 6 for some $n \in N$ like, 2, 3 etc.

Statement II Product of three consecutive integrals is always divisible by 3!. Both the statements are correct but Statement II is not the correct explanation of Statement I.

39. (d) **Statement I** $50 = 2\times 5^2$

To find the exponent of 50 in 100!, we find exponent of 2 and 5.

Exponent of 2 is

$$\left[\frac{100}{2}\right] + \left[\frac{100}{2^2}\right] + \left[\frac{100}{2^3}\right] + \left[\frac{100}{2^4}\right] + \left[\frac{100}{2^5}\right] + \left[\frac{100}{2^6}\right]$$

$$= 50 + 25 + 12 + 6 + 3 + 1 = 97$$

Exponent of 5 is $\left[\frac{100}{5}\right] + \left[\frac{100}{5^2}\right] = 20 + 4 = 24$

Now, exponent of 25 in 100! is $\left[\frac{24}{2}\right] = 12$

Hence, exponent of 50 is 12.

Hence, Statement I is incorrect.

Clearly, Statement II is correct.

40. (a) It is given that, f is onto and $f(x) \neq x$

It means, we have to find dearrangement of $\{a, b, c, d, e\}$ onto $\{a, b, c, d, e\}$

$$= 5!\left[1 - \frac{1}{1!} + \frac{1}{2!} - \frac{1}{3!} + \frac{1}{4!} - \frac{1}{5!}\right]$$

$$= 5!\left[1 - 1 + \frac{1}{2} - \frac{1}{6} + \frac{1}{24} - \frac{1}{120}\right] = 5!\left[\frac{1}{2} - \frac{1}{6} + \frac{1}{24} - \frac{1}{120}\right]$$

$$= 5!\left[\frac{60 - 25 + 5 - 1}{120}\right] = 5!\times\frac{44}{120}$$

$$= 120\times\frac{44}{120} = 44$$

Hence, Statement I is correct. Statement II is correct and is the correct explanation of Statement I.

Revisal Problems (JEE Advanced)

1. (b) As, $\sum\sum\sum\sum_{0\le i<j<k<l\le n} n = n\cdot\sum\sum\sum\sum_{0\le i<j<k<l\le n} 1$

represents selection of four terms from $\{0, 1, 2 \ldots n\}$ i.e. $(n+1)$ terms

$\Rightarrow \quad {}^{n+1}C_4$

Thus, $\sum\sum\sum\sum_{0\le i<j<k<l\le n} n = n\cdot{}^{n+1}C_4$

Hence, option (b) is correct.

2. (c) Given, $n_1 = x_1x_2x_3$

and $\quad n_2 = y_1y_2y_3$

n_1 and n_2 can be subtracted without borrowing at any stage, if $x_i \le y_i$

Value of x_3	Value of y_3
9	0, 1, 2,...9
8	0, 1, 2,...8
7	0, 1, 2,...7
6	0, 1, 2,...6
5	0, 1, 2,...5
4	0, 1, 2,...4
3	0, 1, 2, 3
2	0, 1, 2
1	0, 1
0	0

Thus, x_3 and y_3 can be selected collectively by $10 + 9 + 8 + \cdots + 1 = 55$ ways. Similarly, (x_2, y_2) can be selected in 55 ways. But pair (x_1y_1) can be selected in $1 + 2 + 3 + \cdots + 9 = 45$ ways as in this pair we cannot have zero. Thus, total number of ways in $45\cdot(55)^2$.

3. (c) All numbers $r!$ $(r \ge 7)$ will be multiple of 7.

So, for remainder consider

$$S_6 = 1 + 2 + 6 + 24 + 120 + 720$$
$$= 873$$

which leaves the remainder 5, when divided by 7.

4. (d) Sum of 7 digits is a multiple of 9. Sum of numbers 1, 2, 3, 4, 5, 6, 7, 8, 9 is 45. So, two left digits should also have sum of 9. The pairs of left numbers are (1, 8), (2, 7), (3, 6), (4, 5) with each pair left numbers of 7-digit number is 7!

So, with all 4 pairs, total number is $4 \times 7!$

5. (c) Since, among 9! arrangements of digits 1, 2, 3, 4, 5, 6, 7, 8, 9 i.e. the number formed is of the form $x_1\,x_2\,x_3\,x_4\,x_5\,x_6\,x_7\,x_8\,x_9$. Taking any five consecutive positions product is divisible by 7 only, if $x_5 = 7$.

So, the required arrangement = 8!

Hence, option (c) is correct.

6. (c) First we take one number as 2 and one as $(n-2)$ and put them in $m\,(m-1)$ ways.

Now, remaining $(m-2)$ numbers can be anyone form 3, 4,... $(n-4)$, $(n-3)$ which we can do in $(n-5)^{(m-2)}$.

$\therefore$ Total number of ways $= m\,(m-1)\,(n-5)^{m-2}$

7. (b) A rational number of the desired category is of the form

$$2014\ x_1x_2\ldots x_k,\text{ where } 1 \le k \le 9$$

and $9 \ge x_1 \ge x_2 \ge \cdots \ge x_k \ge 1$. We can choose k digits out of 9 in ${}^9C_k$ ways and arrange them in decreasing order in just one way.

Thus, the desired number of rational number is

$${}^9C_1 + {}^9C_2 + \cdots + {}^9C_9 = 2^9 - 1$$

8. (a) The number of ways of choosing first couple is $({}^{15}C_1)\,({}^{15}C_1) = (15)^2$

The, number of ways of choosing second couple is $({}^{14}C_1)\,({}^{14}C_1) = (14)^2$ and so on.

Thus, the number of ways of choosing the couples is

$$(15)^2 + (14)^2 + (13)^2 + \cdots + 2^2 + 1^2$$
$$= \frac{15 \times 16 \times (15 \times 2 + 1)}{6} = \frac{15 \times 16 \times 31}{6}$$
$$\left[\because 1^2 + 2^2 + \cdots + n^2 = \frac{n\,(n+1)\,(2n+1)}{6}\right]$$
$$= 1240$$

9. (b) If n is odd, then $3^n = 4\lambda_1 - 1$, $5^n = 4\lambda_2 + 1$

$\Rightarrow \quad 2^n + 3^n + 5^n$ is divisible by 4, if $m \ge 2$.

Thus, $n = 3, 5, 7, 9, \ldots, 99$

i.e. n can take 49 different values.

If n is even, then $3^n = 4\lambda_1 + 1$, $5^n = 4\lambda_2 + 1$

$\Rightarrow 2^n + 3^n + 5^n$ is not divisible by 4.

as $2^n + 3^n + 5^n$ will be in the form of $4\lambda + 2$.

Then, total number of ways of selecting 'n' = 49.

10. (c) Number of ways of making 30 in 7 shots is coefficient of x^{30} in $(x^0 + x^1 + x^2 + \cdots + x^5)^7$

$\Rightarrow$ Coefficient of x^{30} in $\{(1 + n^2 + n^3) + x^4(x+1)\}^7$

$\Rightarrow$ Coefficient of x^{30} in $\{x^{28}(x+1)^7 + {}^7C_1x^{24}(x+1)^6\,(1 + x^2 + x^3) + {}^7C_2x^{20}\,(x+1)^5(x^3 + x + 1)^2 + \cdots\}$

$\Rightarrow \quad {}^7C_5 + {}^7C_1\,({}^6C_3 + {}^6C_2 + {}^6C_0) + {}^7C_2\,({}^5C_1 + 2)$

$\Rightarrow \quad 21 + 252 + 147 = 420$

11. (b) Here, $8! = 2^7\cdot 3^2\cdot 5^2\cdot 7^1$

Obviously, the factors are not multiple of either 2 or 3. So, the factors may be 1, 5, 7, 35 of which 5 and 35 are of the form $3m + 2$.

So, the sum = 40

12. (b) Any number will be of the form $4t + 1$, if it does not contain any power of 2, contain even number of odds of the form $4k + 3$ and any number of odds of the form $4k + 1$.

So, total factors $= 1 \times 3 \times 4 = 12$

But, one will be 1 also,

So, required number $= 12 - 1 = 11$

Alternate Method

Odd numbers of the type $4t + 1$ or $4t + 3$ and they are symmetrical.

$\Rightarrow$ Number of odd numbers $= (1 + 5)\,(1 + 3) = 24$

(any number of 5's and 3's and no 2's)

$\Rightarrow$ Required number = 12, but 1 is included

$\Rightarrow$ Required number of numbers $= 12 - 1 = 11$

13. (d) Triplet $x = y < z$, $x < y < z$, $y < z < x$ can be choosen in ${}^{n+1}C_2$, ${}^{n+1}C_3$, ${}^{n+1}C_3$ ways.

$\therefore \quad {}^{n+1}C_2 + 2\cdot{}^{n+1}C_3 = {}^{n+2}C_3 + {}^{n+1}C_3$

$$[\because {}^nC_r + {}^nC_{r-1} = {}^{n+1}C_r]$$

14. (d) In first round all the integers, which leaves the remainder 1, when divided by 15 will be marked.

Last number of this category is 991.

Next number to be marked in $(991 + 15 - 1000) = 6$

Again, second round of integers, which leaves the remainder 6, when divided by 15, will be marked.

Last number of this category is 996.

Next number to be marked is (996 + 15 − 1000) = 11

Thus, thrid round of integers which leave the remainder 11 when divided by 15, will be marked. Least number of this category is 986.

Next number to be marked is (986 + 15 − 1000) = 1 which is already been marked.

Hence, the number of the form $15\lambda + 1$, $15\lambda + 6$, $15\lambda + 11$, are marked. All these numbers can also be put in the form of $5\lambda_1 + 1\ (\lambda_1 \geq 0)$

Hence, total number of marked numbers $= 1 + \left[\frac{999}{5}\right] = 200$

$\therefore$ Unmarked numbers = 800

15. (b) Either, we have to choose

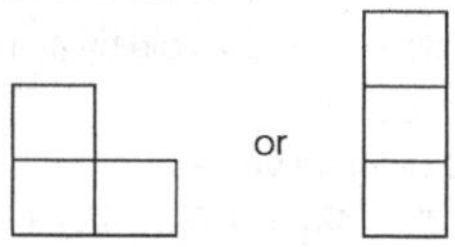

(i) Every square of (2 by 2) 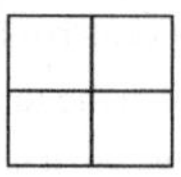

will contribute four 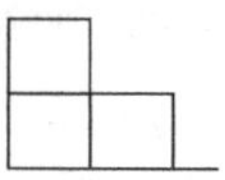 shaped

figure by removing anyone square out of it.

Number of ways to choose these square $= 7 \times 7 = 49$

So, total 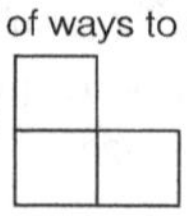 shaped figure $= 49 \times 4 = 196$

(ii) In every line, it is possible to have 6 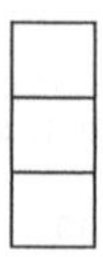 shaped figure.

So, total number of such figure $= 6 \times 8 \times 2 = 96$

$\therefore$ Total number of cases = 196 + 96 = 292

16. (c) First, we have to find $f(6)$

$$\therefore \quad 6 = 0(2) + 6(1) = 1(2) + 4(1)$$
$$= 2(2) + 2(1) = 3(2) + 0(1)$$

Number of 2's	Number of 1's	Number of permutation
0	6	1
1	4	$\frac{5!}{4!} = 5$
2	2	$\frac{4!}{2!2!} = 6$
3	0	$\frac{3!}{3!} = 1$
Total		13

$$\therefore \quad f(6) = 13$$

Now $\quad f(f(6)) = f(13)$

17. (b) Given digits 8, 7, 6, 4, 2, x, y.

We know that, any number is divisible by 3, if sum of digit is divisible by 3.

i.e. $8 + 7 + 6 + 4 + 2 + x + y$

$= (x + y + 27)$ is divisible by 3

So, x and y can take the values from 0, 1, 3, 5, 9.

Possible pairs are (5, 1) (3, 0) (9, 0) (9, 3) and (1, 5) (0, 3) (0, 9) (3, 9).

Hence, total number of ordered pair (x, y) is 8.

18. (b) Let A, B, C want to remain together and D, E do not want to remain together.

In	Out	Number of Selection
A, B, C, D	E	${}^5C_1 = 5$
A, B, C, E	D	${}^5C_1 = 5$
D	A, B, C, E	${}^5C_4 = 5$
E	A, B, C, D	${}^5C_4 = 5$
Total		20

19. (c) We have, $1000 = 2^3 \cdot 5^3$

$$2000 = 2^4 \cdot 5^3$$

So, $\quad a = 2^A \cdot 5^R,\ b = 2^B \cdot 5^S,\ c = 2^C \cdot 5^T$

$$\max(A, B) = 3,\ \max(A, C) = \max(B, C) = 4$$
$$\max(R, S) = \max(R, T) = \max(S, T) = 3$$

There are 10 ways of choosing R, S, T.

1 with all 3, 3 with one 2, 3 with one 1, 3 with one 0, C must be 4.

There are 7 ways of choosing A, B.

1 with both 3, 3 with $A = 3$, B not, 3 with $B = 3$ and A not.

Thus, 7 ways of choosing A, B, C have $7 \cdot 10 = 70$ ways.

Alternate Method

(c) LCM of $(a, b) = 1000 = 2^3 \cdot 5^3 = 2^{\alpha_1} \cdot 5^{\alpha_2}$

LCM of $(b, c) = 2000 = 2^4 \cdot 5^3 = 2^{\beta_1} \cdot 5^{\beta_2}$

LCM of $(a, c) = 2000 = 2^4 \cdot 5^3 = 2^{\gamma_1} \cdot 5^{\gamma_2}$

Max $(\alpha_1, \beta_1) = 4$, max$(\beta_1, \gamma_1) = 4$, max $(\gamma_1, \alpha_1) = 4$

Max $(\alpha_2, \beta_2) = 3$, max $(\beta_2, \gamma_2) = 3$, max $(\gamma_2, \alpha_2) = 3$

If $\alpha_1 > \beta_1 \Rightarrow \alpha_1 = 3, \beta_1 < 3$ and $\gamma_1 = 4$

$\therefore \alpha_1 = 3, \gamma_1 = 4$ and $\beta_1 = 0, 1, 2$ [so, 3 possibilites]

If $\alpha_1 < \beta_1 \Rightarrow \beta_1 = 3, \alpha_1 < 3$ and $\gamma_1 = 4$

$\therefore \alpha_1 = 0, 1, 2, \beta_1 = 3, \gamma_1 = 4$ [3 possibilities]

If $\alpha_1 = \beta_1 = 3$ and $\gamma_1 = 4$ [1 possibilities]

$\therefore$ Total of 7 possibilities for $\alpha_1, \beta_1, \gamma_1$.

If $\alpha_2 > \beta_2 \Rightarrow \alpha_2 = 3$

$\beta_2 < 3$ and $\gamma_2 = 3$ [3 possibilities]

If $\alpha_2 < \beta_2 \Rightarrow \beta_2 = 3, \alpha_2 < 3$ and $\gamma_2 = 3$ [3 possibilities]

If $\alpha_2 = \beta_2 = 3 \Rightarrow \gamma_2 \leq 3$ [4 possibilities]

$\therefore$ Total 10 possibilities for $\alpha_2, \beta_2, \gamma_2$.

Hence, total triplets $= 7 \times 10 = 70$

20. (c) Total number of ways to place 11 identical balls $= {}^{13}C_2$.

Total number of ways to place the balls so that the first box will have more balls than the second and third boxes combined $= {}^7C_2$

$\therefore$ Required cases $= {}^{13}C_3 - 3 \cdot {}^7C_2$

21. (c, d) If a, b and c are in AP, then a and c both are odd or both are even.

Case I When n is even

The number of ways of selection of two even numbers a and c is ${}^{n/2}C_2$. Number of ways of selection of two odd numbers is ${}^{n/2}C_2$.

Hence, the number of A.P is

$$2 \cdot {}^{n/2}C_2 = 2 \cdot \frac{\frac{n}{2}\left(\frac{n}{2} - 1\right)}{2} = \frac{n(n-2)}{4}$$

Case II When n is odd

The number of ways of selection of two odd numbers a and c is ${}^{\frac{n+1}{2}}C_2$. The number of ways of selection of two even numbers a and c is ${}^{\frac{n-1}{2}}C_2$.

Hence, the number of A.P's is

$$^{\frac{n+1}{2}}C_2 + {}^{\frac{n-1}{2}}C_2 = \frac{\left(\frac{n+1}{2}\right)\left(\frac{n+1}{2}-1\right)}{2} + \frac{\left(\frac{n-1}{2}\right)\left(\frac{n-1}{2}-1\right)}{2}$$

$$= \frac{1}{8}(n-1)(n+1) + (n-3) = \frac{(n-1)^2}{4}$$

22. (b, d) The combinatorial coefficient $^{n-1}C_p$ denotes the number of ways in which p different things can be selected out of n different things, if a particular thing is always excluded.

(d) First, we arranged $(n-2)$ white balls in a line. Since, balls are identical, then only one such arrangement is possible.

Now, there are $(n-2+1)$ i.e. $(n-1)$ places, where we can arrange p black balls.

$\therefore$ Number of arrangement of p identical black balls in $(n-1)$ places $= {}^{n-1}C_p$

Thus, required number of ways $= 1 \cdot {}^{n-1}C_p = {}^{n-1}C_p$

23. (b, c, d) We know that, distribution of n distinct objects in to r different boxes, if empty boxes are not allowed or in each box atleast one object is put $(n > r)$.

The number of ways

$$= r^n - {}^rC_1(r-1)^n + {}^rC_2(r-2)^n + \dots (-1)^{r-1}\,{}^rC_{r-1}\,1 \quad \dots(i)$$

Hence, $n = 5, r = 3$

$\therefore$ Number of ways in which the different books be distributed in 3 persons, so each gets atleast one is

$$(3)^5 - {}^3C_1(2)^5 + {}^3C_2 = 243 - 3 \times 32 + 3 = 150$$

(b) Number of parallelogram formed by one set of 6 parallel lines and other set of 5 parllel lines

$$= {}^6C_2 \times {}^5C_2 = 15 \times 10 = 150$$

(c) 5 different toys to be distributed among 3 children, so each gets atleast one

$$= (3)^5 - {}^3C_1(2)^5 + {}^3C_2$$

$$= 243 - 3 \times 32 + 3 = 150$$

(d) 3 Mathematics professors are assigned five different lectures, so that each professor get at least one lecture.

$$= (3)^3 - {}^3C_1(2)^5 + {}^3C_2 - 1$$

$$= 243 - 3 \times 32 + 3 = 150$$

24. (c)

(i) Required number of ways $= {}^9C_3 = \frac{{}^9C_0}{3!\,6!} = 84$

$\therefore$ **(i) → (t)**

(ii) Total number of triangles $= {}^9C_3 = \frac{9!}{3!\,6!} = 84$

There are 4 set of 3 collinear points which do not make triangles.

$\therefore \quad T = 84 - 4 = 80$

Total number of straight lines $= {}^9C_2 = \frac{9!}{2!\,7!} = 36$

$\therefore$ There are 8 straight lines which are sides of regular hexagon.

$\therefore \quad S = 36 - 8 = 28$

Now $\quad T - S = 80 - 28 = 52$

$\therefore$ **(ii) → (r)**

(iii) All are get their own mobile $= {}^5C_5 = 1$ ways

One of them do not get their own mobile $= {}^5C_4 = 5$

Two of them do not get their own

$$= {}^5C_3 \cdot \left(1 - \frac{1}{1!} + \frac{1}{2!}\right) = 10 \times \frac{1}{2} = 5$$

$\therefore$ Required number of ways $= 1 + 5 + 5 = 11$

$\therefore$ **(iii) → (p)**

(vi) Here, $12 = 2^2 \times 3$

Hence, groups of 4 that work out

$1, 1, 6, 2 = \frac{4!}{2!} = 12$

and $\quad 1, 1, 3, 4 = \frac{4!}{2!} = 12$

$2, 2, 3, 1 = \frac{4!}{2!} = 12$

Total = 36

$\therefore$ **(iv) → (q)**

(v) First we select 2 men out of 75 in ${}^5C_2$ ways. Now, we exude the wines of these two selected men and so select 2 ladies from remaining 3 ladies in ${}^3C_2$ ways.

Let A and B be two men and C, D are two ladies playing in one set then, we hare

(a) A and C playing against B, D

(b) A and D playing against B, C

Total number of ways $= 2 \times {}^5C_2 \times {}^3C_2 = 2\frac{5!}{2!3!} \times \frac{3!}{2!1!} = 60$

$\therefore$ **(v) → (s)**

25. (b) $a_n + b_n + c_n$

$b_n = a_{n-1}, \; c_n = a_{n-2}$

$\Rightarrow \quad a_n = a_{n-1} + a_{n-2}$

As, $\quad a_1 = 1, a_2 = 2, a_3 = 3, a_4 = 5, a_5 = 8$

$\Rightarrow \quad b_6 = 8$

26. (a) As, $a_n = a_{n-1} + a_{n-2}$, for $n = 17$

$\Rightarrow \quad a_{17} = a_{16} + a_{15}$

Explanations for Q. Nos. (27-29)

Let $N = x_1, x_2, x_3, \cdots, x_{10}$ be one of the such numbers.

$\because \Sigma x_i = 45$, hence N is divisible by 9, but 11111 is not divisible by 9. Hence, N should be divisible by $11111 \times 9 = 99999$.

Now, $x_1\, x_2\, x_3 \cdots x_9\, x_{10} = x_1x_2x_3x_4x_5 \times 10^5 + x_6x_7x_8x_9x_{10}$

$$= x_1x_2x_3x_4x_5 \times (99999) + x_1x_2x_3x_4x_5 + x_6x_7x_8x_9x_{10}$$

Now, $\quad x_1x_2x_3x_4x_5 < 9999$

and $\quad x_6x_7x_8x_9x_{10} < 99999$

$\therefore \quad x_1x_2x_3x_4x_5 + x_6x_7x_8x_9x_{10} < 2(99999)$

$\Rightarrow \quad x_1x_2x_3x_4x_5 + x_6x_7x_8x_9x_{10} = 9999$

$\Rightarrow \quad x_1 + x_6 = x_2 + x_7 = x_3 + x_8 = x_4 + x_9 = x_5 + x_{10} = 9$

27. (d) Smallest number = 1023489765.

28. (a) Biggest number = 9876501234.

29. (c) Total ways $= 9 \times 8 \times 6 \times 4 \times 2 \times 1 \times 1 \times 1 \times 1 \times 1 = 3456$

Explanations for Q. Nos. (30-32)

30. (b) $x_R + x_G + x_W + x_Y + x_B = 12, x_R, x_G, x_W, x_Y, x_B \geq 0$.

Total number of non-negative integral solutions of above equation $= {}^{12+5-1}C_{5-1} = {}^{16}C_4$

31. (b) $x_R + x_G + x_W + x_Y + x_B = 12$, where

$x_R, x_G, x_W, x_Y, x_B \geq 1$

$\therefore$ Number of solutions $= {}^{7+5-1}C_{5-1} = {}^{11}C_4$

32. (c) $x_R + x_G + x_W + x_Y + x_B = 16$

Where, $\quad x_R, x_G, x_W, x_Y, x_B \geq 2$

$\therefore$ Number of solutions $= {}^{6+5-1}C_{5-1} = {}^{10}C_4$

Explanations for Q. Nos. (33-35)

33. (a) $x_1 < x_2 < x_3 < x_4 < x_5 < x_6 \Rightarrow {}^9C_6$ ways

34. (b) $x_1 < x_2 < x_3 = x_4 < x_5 < x_6 \Rightarrow {}^9C_4$ ways

35. (c) $x_1 < x_2 < x_3 \le x_4 < x_5 < x_6$

$\Rightarrow \quad {}^9C_6 + {}^9C_4 = {}^9C_3 + {}^9C_4 = {}^{10}C_4$ ways

36. (d) Required number of ways $= (3)^5 - {}^3C_1(3-1)^5 - {}^3C_2(1)^5$

$= 243 - 3 \times 32 + 3 = 150$

37. (b) When balls are identical, but boxes are different.

Box I	Box II	Box III	Number of ways
1	2	2	1
1	1	3	1
1	3	1	1
2	1	2	1
3	1	1	1
2	2	1	1
		Total	6 ways

Alternate Method $x_1 + x_2 + x_3 = 5,\ x_1 \ge 1,\ x_2 \ge 1,\ x_3 \ge 1.$

$\Rightarrow \quad k + l + m = 2,\ k \ge 0,\ l \ge 0,\ m \ge 0$

$\therefore$ Number of solutions $= {}^{2+3-1}C_{3-1} = {}^4C_2 = 6.$

38. (a) Here, two cases arises.

Case I (2, 2, 1) $\Rightarrow \dfrac{{}^5C_2 \cdot {}^3C_2 \cdot {}^1C_1}{2!} = 15$

Case II (1, 1, 3) $\Rightarrow \dfrac{{}^5C_1 \cdot {}^4C_1 \cdot {}^3C_3}{2!} = 10$

$\therefore$ Total number of ways $= 15 + 10 = 25$

39. (d) Let's count the Δ's right away

FS	*SS*	*TS*
1	1	1
2	2	1, 2, 3
3	3	1, 2, 3, 4
4	4	1, 2, 3, 4

$FS \rightarrow$ First side, $\quad SS \rightarrow$ Second side

$TS \rightarrow$ Third side

$\therefore$ Total number of ways $= 1 + 3 + 4 + 4 = 12$

40. (c) The value of equal sides can vary from 1 to $2P$.

Hence, third side $< 2, < 4, < 6, \ldots, < 4P$.

FS	*SS*	*TS*
1	1	1
2	2	1, 2, 3
3	3	1, 2, 3, 4, 5
4	4	1, 2, 3, 4, 5, 6, 7
:	:	:
p	p	$1, 2, \ldots, (2p-1)$
$p+1$	$p+1$	$1, 2, 3, \ldots, 2p$
:	:	:
$2p$	$2p$	$1, 2, 3, \ldots, 2p$

$\therefore$ Number of Δ's $= \{1 + 3 + 5 + 7 + \cdots + (2p-1)\}$
$+ \{2p + 2p + \cdots + p \text{ times}\}$
$= p^2 + 2p^2 = 3p^2$

41. (d) If $c = 3 \Rightarrow$ triangle having 3 as largest sides and which are isosceles or equilateral are 3, 3, 1; 3, 3, 2; 3, 3, 3; 3, 2, 2

$\therefore$ Answer is (d).

42. (c) The number of routes through P, is

$= \dfrac{5!}{3! \cdot 2!} \times \dfrac{8!}{5! \cdot 3!} = 10 \times 56 = 560$

43. (b) The number of shortest routes through PQ, is

$= \dfrac{5!}{3! \times 2!} \times \dfrac{7!}{4! \times 3!} = 10 \times 35 = 350$

44. (b) The number of shortest routes through P and R, is

$= \dfrac{5!}{3! \times 2!} \times \dfrac{4!}{3! \times 1!} \times \dfrac{4!}{2! \times 2!} = 240$

45. (a) $\quad \theta(1672) = \theta(8 \times 209)$

$= \theta(8)\,\theta(209) \quad [\because \text{HCF}(8, 209) = 1]$

$= \theta(2^3)\,\theta(11 \times 19)$

$= \theta(2^3)\,\theta(11)\,\theta(19) \quad [\because \text{HCF}(11, 19) = 1]$

$= (2^2 \cdot 1)\,11^{1-1}(11-1)\,19^{1-1}(19-1)$

$[\because \theta(r^n) = r^{n-1}(r-1),\ r \text{ is prime}]$

$= 4 \times 10 \times 18 = 720$

46. (b) Given $\theta(1) = 1$

Now, $\quad \theta(2^0) = \theta(2^1) = 2^{1-1} \cdot (2-1) = 1$

$\Rightarrow \quad \theta(1) = \theta(2) = 1$

If r is an odd prime, then

$\theta(r^n) = r^{n-1}(r-1) = r^n - r^{n-1}$

which is even.

$\therefore$ The number of natural number is 2.

47. (c) $\theta(61^n) = 223260$

$\Rightarrow \quad 61^{n-1} \times 60 = 223260$

$\Rightarrow \quad 61^{n-1} = \dfrac{223260}{60}$

$\Rightarrow \quad 61^{n-1} = 3721$

$\Rightarrow \quad (61)^{n-1} = (61)^2$

$\Rightarrow \quad n - 1 = 2$

$\Rightarrow \quad n = 3$

48. (8) Here, $P_n = {}^{n-2}C_3$ and $P_{n+1} = {}^{n-1}C_3$

Hence, $\quad {}^{n-1}C_3 - {}^{n-2}C_3 = 15$

or ${}^{n-2}C_3 + {}^{n-2}C_2 - {}^{n-2}C_3 = 15$

or $\quad {}^{n-2}C_2 = 15$

$\Rightarrow \quad \dfrac{(n-2)!}{2!\,(n-4)!} = 15$

$\Rightarrow \quad \dfrac{(n-2)(n-3)}{2} = 15$

$\Rightarrow \quad (n-2)(n-3) = 30$

$\Rightarrow \quad (n-2)(n-3) = 6 \times 5$

$\Rightarrow \quad (n-2)(n-3) = (8-2)(8-3)$

$\therefore \quad n = 8$

49. (1) ${}^{2002}C_{1001} = \dfrac{(2002)!}{(1001)!\,(1001)!}$

Number of zero's in (2002) ! are,

$= 400 + 80 + 16 + 3 = 499$

Number of zero's in $(1001!)^2 = 2\,(200 + 40 + 8 + 1) = 498$

Hence, number of zero's in $\dfrac{(2002)!}{(1001!)^2} = 1$

50. (3)

11 Points

5 (one circle) | 6 (out of which no 4 lies on the same circle)

$\Rightarrow$ maximum number of circles

$= {}^5C_2 \cdot {}^6C_1 + {}^6C_2 \cdot {}^5C_1 + {}^6C_3 + 1 = 156 = 52(3)$

$\therefore \quad k = 3$

Alternate Method Maximum number of circles

$= {}^{11}C_3 - {}^5C_3 + 1.$

JEE Main & AIEEE Archive

1. (c) Given, $n(A) = 2$ and $n(B) = 4$

$\therefore \quad n(A \times B) = 8$

The number of subsets of $A \times B$ having 3 or more elements

$= {}^8C_3 + {}^8C_4 + \cdots + {}^8C_8 = 2^8 - {}^8C_0 - {}^8C_1 - {}^8C_2$

$= 256 - 1 - 8 - 28 = 219 \quad [\because 2^n = {}^nC_0 + {}^nC_1 + \cdots + {}^nC_n]$

2. (b) Given, $T_n = {}^nC_3$

$T_{n+1} = {}^{n+1}C_3$

$\therefore \quad T_{n+1} - T_n = {}^{n+1}C_3 - {}^nC_3 = 10 \quad$ [given]

$\Rightarrow \quad {}^nC_2 + {}^nC_3 - {}^nC_3 = 10 \Rightarrow {}^nC_2 = 10$

$\therefore \quad n = 5$

3. (d) **Given** 10 identical white balls, 9 identical green balls and 7 identical black balls.

To find The number of ways for selecting atleast one ball.

Number of ways to choose zero or more white balls

$= (10 + 1)$ [since, all white balls are mutually identical]

Number of ways to choose zero or more green balls

$= (9 + 1)$ [since, all green balls are mutually identical]

Number of ways to choose zero or more black balls $= (7 + 1)$ [since, all black balls are mutually identical]

Hence, number of ways to choose zero or more balls of any colour $= (10 + 1)(9 + 1)(7 + 1)$

Also, number of ways to choose a total of zero balls = 1

Hence, the number, if ways to choose atleast one ball (irrespective of any colour) $= (10 + 1)(9 + 1)(7 + 1) - 1 = 879$

[$10 \times 9 \times 7 - 1$ = number of ways to select atleast one ball of each colour (in case there is some confusion)]

4. (a) Let the number of ways of distributing n identical objects among r persons such that each person gets atleast one object is same as the number of ways of selecting $(r - 1)$ places out of $(n - 1)$ different places, i.e. ${}^{n-1}C_{r-1}$.

Statement I Here, $n = 10$ and $r = 4$

$\therefore \quad$ Required number of ways $= {}^{10-1}C_{4-1} = {}^9C_3$

Statement II Required number of ways $= {}^9C_3$

Hence, both the statements are correct but Statement II is not a correct explanation of Statement I.

5. (b) If out of n points, m are collinear, then

Number of triangles $= {}^nC_3 - {}^mC_3$

$\therefore \quad$ Number of triangles $= {}^{10}C_3 - {}^6C_3 = 120 - 20 \Rightarrow N = 100$

6. (c) The number of ways in which 4 novels can be selected

$= {}^6C_4 = 15$

The number of ways in which 1 dictionary can be selected

$= {}^3C_1 = 3$

Now, we have 5 places in which middle place is fixed.

$\therefore$ 4 novels can be arranged in 4! ways.

$\therefore$ The total number of ways $= 15 \times 4! \times 3 = 15 \times 24 \times 3 = 1080$

7. (a) Since, the number of ways that child can buy the six ice-creams is equal to the number of different ways of arranging 6 A's and 4 B's in a row.

So, number of ways to arrange 6 A's and 4 B's in a row $= \dfrac{10!}{6!\,4!}$

$= {}^{10}C_4$

And number of integral solutions of the equation

$x_1 + x_2 + x_3 + x_4 + x_5 = 6 = {}^{6+5-1}C_{5-1} = {}^{10}C_4 \neq {}^{10}C_5$

Hence, Statement I is incorrect and Statement II is correct.

8. (a) Given word is MISSISSIPPI.

Here, I = 4 times, S = 4 times,
P = 2 times M = 1 time

_M_I_I_I_I_P_P_

$\therefore$ Required number of words

$= {}^8C_4 \times \dfrac{7!}{4!\,2!} = {}^8C_4 \times \dfrac{7 \times 6!}{4!\,2!} = 7 \cdot {}^8C_4 \cdot {}^6C_4$

9. (c) Required number of ways $= {}^{12}C_4 \times {}^8C_4 \times {}^4C_4$

$= \dfrac{12!}{8! \times 4!} \times \dfrac{8!}{4! \times 4!} \times 1 = \dfrac{12!}{(4!)^3}$

10. (b) Total number of ways $= {}^{10}C_1 + {}^{10}C_2 + {}^{10}C_3 + {}^{10}C_4$

$= 10 + 45 + 120 + 210 = 385$

11. (d) In the word 'SACHIN', order of alphabets is A, C, H, I, N and S. Number of words start with $A = 5!$, so with C, H, I, N.

Now, words start with S and after that ACHIN are in ascending order of position, so $5 \cdot 5! = 600$ words are in dictionary before words with S start and position of this word is 601.

12. (a) Given that, $f(x) = {}^{7-x}P_{x-3}$. The above function is defined, if $7 - x \geq 0$ and $x - 3 \geq 0$ and $7 - x \geq x - 3$.

$\Rightarrow \quad x \leq 7, x \geq 3$ and $x \leq 5$

$\therefore \quad D_f = \{3, 4, 5\}$

Now, $f(3) = {}^4P_0 = 1$

$f(4) = {}^3P_1 = 3$ and $f(5) = {}^2P_2 = 2$

$\therefore \quad R_f = \{1, 2, 3\}$

13. (c) Total number of ways in which all letters can be arranged in alphabetical order = 6!.

There are two vowels (A, E) in the word 'GARDEN'. Total number of ways in which these two vowels can be arranged = 2!

$\therefore$ Total number of required ways $= \dfrac{6!}{2!} = 360$

14. (b) The required number of ways $= {}^{8-1}C_{3-1} = {}^7C_2$

$= \dfrac{7!}{2!\,5!} = \dfrac{7 \cdot 6}{2 \cdot 1} = 21$

15. (b) The number of choices available to him

$= {}^5C_4 \times {}^8C_6 + {}^5C_5 \times {}^8C_5$

$= \dfrac{5!}{4!\,1!} \times \dfrac{8!}{6!\,2!} + \dfrac{5!}{5!\,0!} \times \dfrac{8!}{5!\,3!} = 5 \times \dfrac{8 \times 7}{2} + 1 \times \dfrac{8 \times 7 \times 6}{3 \times 2}$

$= 5 \times 4 \times 7 + 8 \times 7$

$= 140 + 56 = 196$

16. (a) First, we fix the position of men, the number of ways to sit men $= 5!$ and the number of ways to sit women $= {}^6P_5$

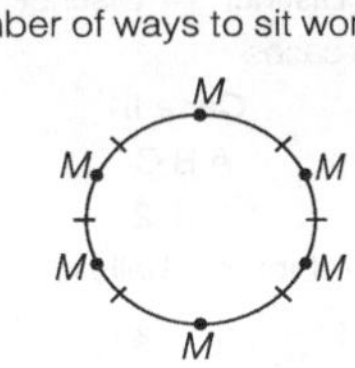

$\therefore$ Total number of ways $= 5! \times {}^6P_5 = 5! \times 6!$

17. (b) Here, we use ${}^nC_{r-1} + {}^nC_r = {}^{n+1}C_r \quad \ldots$(i)

Now, ${}^nC_{r+1} + {}^nC_{r-1} + 2 \cdot {}^nC_r = {}^nC_{r+1} + {}^nC_{r-1} + {}^nC_r + {}^nC_r$

$= {}^nC_{r+1} + {}^{n+1}C_r + {}^nC_r \quad$ [using Eq. (i)]

$= {}^{n+1}C_{r+1} + {}^{n+1}C_r \quad$ [using Eq. (i)]

$= {}^{n+2}C_{r+1} \quad$ [using Eq. (i)]

JEE Advanced & IIT JEE Archive

1. (7) **Plan** Reducing the equation to a newer equation, where sum of variables is less. Thus, finding the number of arrangements becomes easier.

As, $n_1 \geq 1$, $n_2 \geq 2$, $n_3 \geq 3$, $n_4 \geq 4$, $n_5 \geq 5$

Let $n_1 - 1 = x_1 \geq 0$

$$n_2 - 2 = x_2 \geq 0, \cdots, n_5 - 5 = x_5 \geq 0$$

$\Rightarrow$ New equation will be

$$x_1 + 1 + x_2 + 2 + \cdots + x_5 + 5 = 20$$

$\Rightarrow$ $$x_1 + x_2 + x_3 + x_4 + x_5 = 20 - 15 = 5$$

Now, $$x_1 \leq x_2 \leq x_3 \leq x_4 \leq x_5$$

x_1	x_2	x_3	x_4	x_5
0	0	0	0	5
0	0	0	1	4
0	0	0	2	3
0	0	1	1	3
0	0	1	2	2
0	1	1	1	2
1	1	1	1	1

So, 7 possible cases will be there.

2. (5) **Plan** Number of line segment joining pair of adjacent point

$$= n$$

Number of line segment obtained joining n points on a circle

$$= {}^nC_2$$

Number of red line segments $= {}^nC_2 - n$

Number of blue line segments $= n$

$\therefore$ $${}^nC_2 - n = n$$

$\Rightarrow$ $$\frac{n(n-1)}{2} = 2n$$

$\Rightarrow$ $$n = 5$$

3. (b) **Concept Involved** Distribution of objects into groups.

Here, the student should be particular about objects and groups.

i.e.

Objects/Groups	Groups	Objects
Distinct Identical	Distinct	Identical
Distinct Distinct	Identical	Identical

Description of Situation

Here, 5 distinct balls are distributed amongst 3 persons so that each gets atleast one ball.

i.e. Distinct $\rightarrow$ Distinct

So, we should make cases

Case I	*Case II*
A B C	A B C
1 2 3	1 2 2

Number of ways to distribute 5 balls is

$$\left({}^5C_1 \cdot {}^4C_1 \cdot {}^3C_1 \times \frac{3!}{2!}\right)\left({}^5C_2 \cdot {}^4C_2 \cdot {}^2C_2 \times \frac{3!}{2!}\right) = 60 + 90$$

$$= 150$$

4. (c) There are two possible cases

Case **I** Five 1's, one 2's, one 3's

$$\text{Number of numbers} = \frac{7!}{5!}$$

$$= 42$$

Case **II** Four 1's, three 2's

$$\text{Number of numbers} = \frac{7!}{4!\,3!}$$

$$= 35$$

$$\text{Total number of numbers} = 42 + 35$$

$$= 77$$

5. (b)

(i) If ENDEA is fixed word, then assume this as a single letter.

Total number of letters = 5

Total number of arrangements = 5!.

(i) $\rightarrow$ (p)

(ii) If E is at first and last places, then total number of permutations $= 7!/2! = 21 \times 5!$

(ii) $\rightarrow$ (s)

(iii) If D, L, N are not in last five positions

$$\leftarrow \text{D, L, N, N} \rightarrow \leftarrow \text{E, E, E, A, O} \rightarrow$$

$$\text{Total number of permutations} = \frac{4!}{2!} \times \frac{5!}{3!} = 2 \times 5!$$

(iii) $\rightarrow$ (q)

(iv) Total number of odd positions = 5

Permutations of AEEEO are $\frac{5!}{3!}$

Total number of even positions = 4

Number of permutations of N, N, D, L $= \frac{4!}{2!}$

$$\text{Hence, total number of permutations} = \frac{5!}{3!} \times \frac{4!}{2!}$$

$$= 2 \times 5!$$

(iv) $\rightarrow$ (q)

6. (c) Arrange the letters of the word COCHIN as in the order of dictionary CCHINO.

Consider the words starting from C.

There are 5! such words. Number of words with the two C's occupying first and second place = 4!.

Number of words starting with CH, CI, CN is 4! each.

Similarly, number of words before the first word starting with CO = 4! + 4! + 4! + 4! = 96.

The word starting with CO found first in the dictionary is COCHIN. There are 96 words before COCHIN.

7. (c) Since, r, s, t are prime numbers.

$\therefore$ Selection of p and q are as under

p	q	Number of ways
r^0	r^2	1 way
r^1	r^2	1 way
r^2	r^0, r^1, r^2	3 ways

$\therefore$ Total number of ways to select $r = 5$

Selection of s as under

s^0	s^4	1 way
s^1	s^4	1 way
s^2	s^4	1 way
s^3	s^4	1 way
s^4		5 ways

$\therefore$ Total number of ways to select $s = 9$

Similarly, the number of ways to select $t = 5$

$\therefore$ Total number of ways $= 5 \times 9 \times 5 = 225$

3. Probability

DPP-1 Equally Likely, Mutually Exclusive and Exhaustive Events, Classical Definition of Probability

1. Let S be the sample space of the experiment.

(i) $S = \{1, 2, 3, 4, 5, 6\}$

$A = \{1, 3, 5\}$

$B = \{4, 6\}$

Clearly, A and B are not equally likely as

$$P(A) = \frac{3}{6} = \frac{1}{2},\ P(B) = \frac{2}{6} = \frac{1}{3}$$

and $P(A) \neq P(B)$

i.e. chance of occurence of event A is not same as chance of occurence of event B.

$\because$ $A \cap B = \phi$

$\therefore$ A and B are mutually exclusive.

$\because$ $A \cup B \neq S$

$\therefore$ A and B are not exhaustive events.

(ii) Total number of ways of drawing a ball, $n(S) = {}^9C_1$

$n(E_1) = {}^2C_1$

$n(E_2) = {}^3C_1$

$n(E_3) = {}^4C_1$

$n(E_1 \cap E_2) = n,\ (E_2 \cap E_3) = n,\ (E_1 \cap E_3) = 0$

$\therefore$ $P(E_1) = \frac{2}{9},\ P(E_2) = \frac{3}{9},\ P(E_3) = \frac{4}{9}$

Since, chance of occurence of each of the event is not same, therefore the events are not equally likely.

$\because$ $E_1 \cap E_2 = E_2 \cap E_3 = E_1 \cap E_3 = \phi$

Therefore, the events are mutually exclusive.

$\because$ $E_1 \cup E_2 \cup E_3 = S$

Therefore the events are mutually exhaustive.

(iii) $$S = \begin{Bmatrix} (1,1), (1,2), \ldots, (1,6) \\ (2,1), (2,2), \ldots, (2,6) \\ \vdots \qquad \vdots \\ (6,1), (6,2), \ldots, (6,6) \end{Bmatrix}$$

$A = \{(1, 1), (2, 2), (3, 3), (4, 4), (5, 5), (6, 6)\}$,

$B = \{(4, 6), (6, 4), (5, 5), (6, 5), (5, 6), (6, 6)\}$

$\because$ $P(A) = \frac{6}{36} = \frac{1}{6}$ and $P(B) = \frac{6}{36} = \frac{1}{6}$ are same.

i.e. chance of occurence of event A and B are same, therefore the events are equally likely.

$\because$ $A \cap B = \{(5, 5), (6, 6)\} \neq \phi$,

Therefore, the events are not mutually exclusive.

$\because$ $A \cup B \neq S$,

Therefore, the events are not mutually exhaustive.

(iv) $n(S) = {}^{52}C_1 = 52$

$n(E_1) = {}^{13}C_1 = 13$

$n(E_2) = {}^{13}C_1 = 13$

$n(E_3) = {}^{13}C_1 = 13$

$n(E_4) = {}^{13}C_1 = 13$

$n(E_1 \cap E_2) = n,\ (E_2 \cap E_3) = n,$

$(E_3 \cap E_4) = n,\ (E_1 \cap E_4) = 0$

$\Rightarrow$ $P(E_1) = P(E_2) = P(E_3) = P(E_4)$

$= \frac{13}{52} = \frac{1}{4}$

Since, chance of occurence of each of the event is same, therefore the given events are mutually exclusive.

$\because$ $E_1 \cap E_2 = E_2 \cap E_3 = E_3 \cap E_4 = E_1 \cap E_4 = \phi$

Thus, the given events are mutually exclusive.

$\because$ $E_1 \cup E_2 \cup E_3 \cup E_4 = S$

Hence, the given events are mutually exhaustive.

(v) $n(S) = {}^{52}C_1 = 52$

$n(A) = {}^{13}C_1 = 13$

$n(B) = {}^{12}C_1 = 12$

$n(A \cap B) = {}^3C_1 = 3$

$\Rightarrow$ $P(A) = \frac{13}{52} = \frac{1}{4},$

$P(B) = \frac{12}{52} = \frac{3}{13}$

$P(A \cup B) = \frac{3}{52}$

Since, chance of occurence of events A and B are not same, therefore the events are not equally likely.

$\because$ $A \cap B$ = {jack of heart, queen of heart, king of heart} $\neq \phi$, therefore the events are not mutually exclusive.

Since, $A \cup B \neq S$, therefore the events are not mutually exhaustive.

2. Let S be the sample space of the experiment.

$$\therefore \quad S = \begin{Bmatrix} (1,1), (1,2), (1,3), (1,4), (1,5), (1,6) \\ (2,1), (2,2), (2,3), (2,4), (2,5), (2,6) \\ \vdots \\ (6,1), (6,2), (6,3), (6,4), (6,5), (6,6) \end{Bmatrix}$$

Clearly, $n(S) = 6 \times 6 = 36$

Now, $A = \{(2, 1), (2, 2), (2, 3), (2, 4), (2, 5), (2, 6), (4, 1), (4, 2),$
$(4, 3), (4, 4), (4, 5)\ (4, 6), (6, 1), (6, 2)$
$(6, 3), (6, 4), (6, 5), (6, 6)\}$

$B = \{(1, 1), (1, 2)\ (1, 3), (1, 4), (1, 5), (1, 6),$
$(3, 1), (3, 2), (3, 3), (3, 4), (3, 5), (3, 6),$
$(5, 1), (5, 2), (5, 3), (5, 4), (5, 5)\ (5, 6)\}$

$C = \{(1, 1)\}$

(i) False, as A contains 18 sample point.

(ii) True, as C contains only single sample point

(iii) True, as $A \cap B = \phi$

(iv) True, as A' = event of getting an odd number on the first die $= B$ and B' = event of getting an even number on the first die A, also $A' \cap B' = A \cap B = \phi$

(v) True [using part (iv)]

(vi) True, as $A \cup B = S$

3. Let S be the sample space of the experiment.

$\therefore$ $S = \{HHH, HTH, HHT, THH, TTT, TTH, THT, HTT\}$

(i) Let A be the event of getting exactly two heads, i.e.

$A = \{HTH, HHT, THH\}$

$\therefore$ $P(A) = \frac{n(A)}{n(S)} = \frac{3}{8}$

(ii) Let B be the event of getting atleast two heads, i.e.

$B = \{HTH, HHT, THH, HHH\}$

$\therefore$ $P(B) = \frac{n(B)}{n(S)} = \frac{4}{8} = \frac{1}{2}$

(iii) Let C be the event of getting atleast one head and one tail, i.e.

$$C = \{HTH, HHT, THH, TTH, THT, HTT\}$$

$$\therefore \quad P(C) = \frac{n(C)}{n(S)} = \frac{6}{8} = \frac{3}{4}$$

4. (a) We know that, in an ordinary or a non-leap year, there are 52 weeks and 2 days.

These last 2 days can be (Sunday, Monday), (Monday, Tuesday), (Tuesday, Wednesday), (Wednesday, Thursday), (Thursday, Friday), (Friday, Saturday), (Saturday, Sunday).

$\therefore$ P (that an ordinary year has 53 Sunday) $= \frac{2}{7}$

5. (c) Let A be the event of getting 10 on first throw, B be the event of getting 11 on second throw and C be the event of getting 12 on third throw.

$$\therefore \quad P(A) = \frac{3}{36}, \quad P(B) = \frac{2}{36}, \quad P(C) = \frac{1}{36}$$

Since, the trials are independent, therefore

$$P(A \cap B \cap C) = P(A) \cdot P(B) \cdot P(C)$$

$$= \frac{3}{36} \times \frac{2}{36} \times \frac{1}{36} = \frac{1}{36} \times \frac{1}{36} \times \frac{1}{6} = \frac{1}{7776}$$

6. (d) Let S be the sample space of the experiment, then $n(S) = 52$.

Let A and B be the events of getting a red card and a queen respectively, then

$$n(A) = 26, \quad n(B) = 4$$

$$n(A \cap B) = 2$$

$$\therefore \quad P(A \cup B) = P \text{ (getting red card or a queen)}$$

$$= P(A) + P(B) - P(A \cap B)$$

$$= \frac{26}{52} + \frac{4}{52} - \frac{2}{52} = \frac{28}{52} = \frac{7}{13}$$

7. (b) Total number of ways of arranging n persons around a table i.e. $n(S) = (n-1)!$

Let E be the event of arranging n persons around a table, so that two specified individuals sitting next to each other, i.e.

$$n(E) = (n-2)! \times 2!$$

$$\therefore \quad P(E) = \frac{(n-2)! \times 2!}{(n-1)!} = \frac{2}{(n-1)}$$

$$\Rightarrow \quad P(\bar{E}) = 1 - P(E) = 1 - \frac{2}{(n-1)}$$

$$= \frac{(n-1)-2}{(n-1)} = \frac{n-3}{(n-1)} = \frac{(n-3)}{(n-1)}$$

$$\therefore \text{ Odd against of an event } E = \frac{P(\bar{E})}{P(E)} = \frac{\frac{(n-3)}{(n-1)}}{\frac{2}{(n-1)}} = \frac{(n-3)}{2}$$

$$= (n-3) : 2$$

8. (d) $\because$ Total number of two-digit numbers that can be formed by digits 1, 2, 3, 4, 5 $= 5^2 = 25$

Now, as we know that a number is divisible by 4, if last two-digits of number divisible by 4.

So, the possible two-digit numbers divisible by 4 and formed by digits 1, 2, 3, 4, 5 are 12, 24, 32, 44, 52.

$\therefore$ Required probability $= \frac{5}{25} = \frac{1}{5}$

9. (d) We have a bag containing 5 W, 7 B and 4 R balls, where W, B and R denote white, black and red balls, respectively.

Total number of ways of selecting 3 balls $= {}^{16}C_3$

Number of ways of selecting 3 white balls $= {}^{5}C_3$

$\therefore$ Required probability $= \frac{{}^5C_3}{{}^{16}C_3} = \frac{5 \times 4 \times 3}{16 \times 15 \times 14} = \frac{1}{56}$

10. (b) Total number of cards = 52

Let A be the event of getting a king and B be the event of getting a queen, then

$$n(A) = 4, \; n(B) = 4 \text{ and } n(A \cap B) = 0$$

$\therefore$ P (either a king or a queen) $= P(A \cup B)$

$$= P(A) + P(B)$$

$$= \frac{4}{52} + \frac{4}{52} = \frac{8}{52} = \frac{2}{13}$$

11. (b) The total number of ways in which 8 persons can speak is ${}^8P_8 = 8!$. The number of ways in which A, B and C can be arranged in the specified speaking order is ${}^8C_3$. There are 5! ways in which the other five can speak. So, favourable number of ways is ${}^8C_3 \times 5!$.

$$\therefore \quad \text{Required probability} = \frac{{}^8C_3 \times 5!}{8!}$$

$$= \frac{8!}{3! \times 5!} \times 5! \times \frac{1}{8!}$$

$$= \frac{1}{3!} = \frac{1}{6}$$

12. (c) We have, a bag containing 3R and 3W balls, where R and W denote red and white balls, respectively.

Total number of ways of selecting two balls $= {}^6C_2$

Number of ways of selecting one red and one white balls

$$= {}^3C_1 \times {}^3C_1 = 9$$

$$\therefore \quad \text{Required probability} = \frac{9}{{}^6C_2} = \frac{9}{\frac{6 \times 5}{2 \times 1}}$$

$$= \frac{9}{15} = \frac{3}{5}$$

13. (a) Let A be the event of getting first two balls are red and B be the event of getting second two balls are blue.

$$\therefore \quad P(AB) = P(A) \cdot P(B) = \frac{{}^4C_2}{{}^7C_2} \times \frac{{}^3C_2}{{}^7C_2}$$

$$= \frac{4 \times 3}{7 \times 6} \times \frac{3 \times 2}{7 \times 6} = \frac{2}{49}$$

14. (a) Total number of ways of placing 3 letters in 3 envelopes

$$= {}^3P_3 = 3! = 6$$

Number of ways of placing 3 letters in 3 envelopes such that only one letter goes in right envelope $= {}^3C_1 \times 1 = 3$

$$\therefore \quad \text{Required probability} = \frac{3}{6} = \frac{1}{2}$$

15. (b) Total number of switches = 10

Number of ways of choosing any three switches $= {}^{10}C_3$

i.e. $\quad n(S) = {}^{10}C_3$

Let E be event of favourable cases.

$\therefore$ $n(E)$ = Number of ways of choosing three switches, so that room will be lighted

= Number of ways of choosing 3 switches containing atleast one switch of non-fused bulb

Clearly, number of ways of choosing 3 switches of fused bulbs only $= {}^4C_3 = 4$

$$\therefore \quad n(E) = {}^{10}C_3 - {}^4C_3$$

$$\therefore \quad P(E) = \frac{n(E)}{n(S)} = \frac{{}^{10}C_3 - {}^4C_3}{{}^{10}C_3} = 1 - \frac{{}^4C_3}{{}^{10}C_3}$$

$$= 1 - \frac{4 \times 3 \times 2}{10 \times 9 \times 8} = 1 - \frac{1}{30} = \frac{29}{30}$$

16. (c) We have, 4 red, 3 pink, 5 yellow and 8 white roses in a lawn.

Let A be the event that a blind man touch a red rose and B be the event that a blind man touch a white rose.

$\therefore$ $n(A) = 4$

and $n(B) = 8$

$\therefore$ $P(A \cup B) = P(A) + P(B)$ $\quad [\because A \cap B = \phi]$

$$= \frac{4}{20} + \frac{8}{20}$$

$$= \frac{12}{20} = \frac{3}{5}$$

17. (b) Number of ways of choosing two numbers from numbers 1, 2, ..., 40 = $^{40}C_2$.

As, we know that sum of two numbers is odd, if one is even and other is odd.

$\therefore$ Number of favourable ways = $^{20}C_1 \times {^{20}C_1}$

$= 20 \times 20 = 400$

$\therefore$ Required probability = $\frac{400}{^{40}C_2} = \frac{400 \times 2}{40 \times 39} = \frac{20}{39}$

18. (c) Number of ways of selecting three integers from 20 integers $= {^{20}C_3}$

As we know that, product of three integers is even, if atleast one of the number is even.

Now, let us first find the number of ways of selecting 3 odd integers = $^{10}C_3$

$\therefore$ Number of favourable ways = $^{20}C_3 - {^{10}C_3}$

$\therefore$ Required probability = $\frac{^{20}C_3 - {^{10}C_3}}{^{20}C_3}$

$$= 1 - \frac{10 \times 9 \times 8}{20 \times 19 \times 18}$$

$$= 1 - \frac{9 \times 4}{19 \times 18}$$

$$= 1 - \frac{2}{19} = \frac{17}{19}$$

19. (c) Total number of ways of selecting 5 people out of 9 = $^9C_5$

Let A be the event that selection of 5 people consist a certain married couple and B be the event that selection of 5 people never consist a certain married couple.

$\therefore$ $n(A) = {^7C_3}$

$n(B) = {^7C_5}$

$n(A \cap B) = 0$

$\therefore$ $P(A \cup B) = P(A) + P(B)$

$$= \frac{^7C_3 + {^7C_2}}{^9C_5}$$

$$= \frac{^8C_3}{^9C_5} = \frac{^8C_5}{^9C_5}$$

$$= \frac{8 \times 7 \times 6 \times 5 \times 4}{9 \times 8 \times 7 \times 6 \times 5}$$

$$= \frac{4}{9}$$

20. (a) Let S be the sample space of the experiment

$$S = \begin{Bmatrix} (1, 1), (1, 2), \ldots, (1, 6) \\ \vdots \quad \vdots \quad \vdots \\ (6, 1), (6, 2), \ldots, (6, 6) \end{Bmatrix}$$

Let A be the event of getting 4 atleast once, i.e.

$A = \{(4, 1)\ (4, 2), (4, 3), (4, 4), (4, 5), (4, 6), (1, 4),$
$(2, 4)\ (3, 4), (5, 4)\ (6, 4)\}$

$\therefore$ Required probability = $\frac{11}{36}$

21. (a) Let the side of the small cube = x

$\Rightarrow$ $64x^3 = a^3$

$\therefore$ $x = \frac{a}{4}$

There are 12 edges and from each edge we get 2 cubes with only two coloured faces.

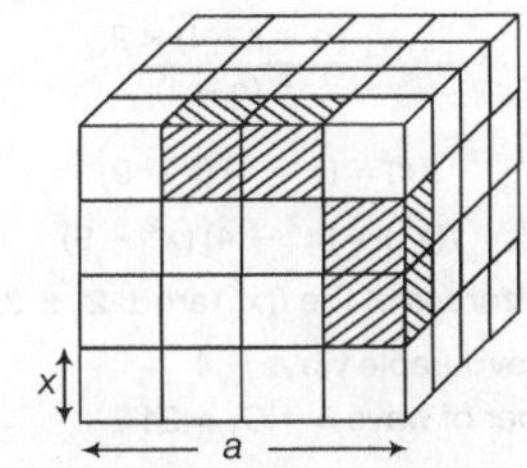

Faces = $12 \times 2 = 24$

$n(S) = {^{64}C_2}, n(A) = {^{24}C_2}$

$\therefore$ Required probability = $\frac{n(A)}{n(S)} = \frac{^{24}C_2}{^{64}C_2} = \frac{24 \times 23}{64 \times 63} = \frac{23}{168}$

22. (a) Total number of ways of selecting 4 people, $n(S) = {^{12}C_4}$

Let E be the event of getting exactly one married couple in selecting of 4 people. Clearly, 4 people may consist 3 men and 1 woman (one of the wives of selected men) or 2 men and 2 women (it include the wife of exactly one of the selected men) or 1 man and 3 women (include the wife of selected man).

$\therefore$ $n(E) = {^6C_3} \times {^3C_1} + {^6C_2} \times {^1C_1} \times {^4C_1} + {^6C_2} \times {^1C_1} \times {^4C_1} + {^6C_1} \times {^1C_1} \times {^5C_2}$

$= 60 + 60 + 60 + 60$

$= 240$

$\therefore$ $P(E) = \frac{n(E)}{n(S)} = \frac{240}{^{12}C_4}$

$$= \frac{240 \times 4 \times 3 \times 2 \times 1}{12 \times 11 \times 10 \times 9}$$

$$= \frac{24 \times 2}{11 \times 9} = \frac{16}{33}$$

23. (a) Let A be the event of dialing the correct telephone number.

$\therefore$ $n(A) = 1$

Now, as the last 4 digit involved 1, 7 and 9 with one of these number appearing twice, therefore

Required number of ways = Number of ways of forming 4-digit number using digit 1, 7, 9 and having two digits identical

$= 3 \times \frac{4!}{2!} = 36$

$\therefore$ $P(A) = \frac{n(A)}{n(S)} = \frac{1}{36}$

$\therefore$ Odd in favour of event $A = \frac{P(A)}{P(\bar{A})} = \frac{\frac{1}{36}}{1 - \frac{1}{36}} = \frac{1}{35}$

24. (b) We have, n different books that can be arrange in a shelf in $n!$ ways.

Now, select r books from $(n-2)$ books that can be done in $^{n-2}C_r$ ways.

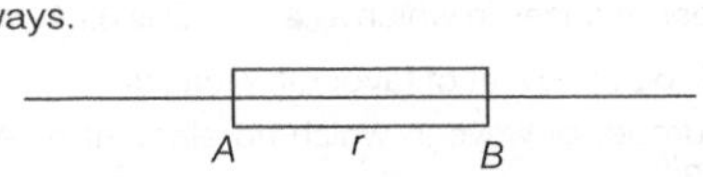

Now, we left with $(n - 2 - r)$ books and the number of arrangement of $(n - 2 - r)$ books and there are exactly books between A and B is $(n - 2 - r + 1)! \times r! \times 2!$.

$\therefore$ Required probability $= \dfrac{^{n-2}C_r\,(n-r-1)!\times r!\times 2!}{n!}$

$= \dfrac{(n-2)!\,(n-r-1)!\times r!\ 2!}{r!\,(n-r-2)!\times n!}$

$= \dfrac{(n-r-1)\,(n-r-2)!\times 2!}{(n-r-2)!\times n\,(n-1)}$

$= \dfrac{(n-r-1)\times 2}{n\,(n-1)}$

25. (a) Let $f(x) = (x-4)(x-9)$

$\Rightarrow f(x^2) = (x^2-4)(x^2-9)$

Thus, roots or zeroes of the $f(x^2)$ are $\pm 2, \pm 3$.

$\therefore$ Number of favourable ways = 4

and total number of ways $= {^{21}C_1} = 21$

$\therefore$ Required probability $= \dfrac{4}{21}$

26. (b) Here, $n(S) = \dfrac{9!}{4!\,5!} = 126$

and $n(A) = 0$ to F and F to P

$= \dfrac{5!}{2\times 3!}\times\dfrac{4!}{2!\times 2!} = 60$

$\therefore P(A) = \dfrac{n(A)}{n(S)} = \dfrac{60}{126} = \dfrac{10}{21}$

Explanations for Q. Nos. (27-29)

Let $A = \{a_1, a_2, \ldots, a_{10}\}$

For each $a_i \in A$ $(1 \le i \le 10)$, we have the following four choices :

(i) $a_i \in P$ and $a_i \in Q$

(ii) $a_i \in P$ and $a_i \notin Q$

(iii) $a_i \notin P$ and $a_i \in Q$

(iv) $a_i \notin P$ and $a_i \notin Q$

For one element a_i of A, total number of cases is 4.

Let S be the sample space.

$\therefore n(S) = 4^{10}$

27. (c) Number of cases in which $a_i \in P \cup Q$ is 3.

Now, let E be the event of favourable cases, $n(E)$ = Number of ways in which all element of A will belong to $P \cup Q = 3^{10}$.

$\therefore$ Required probability $= \dfrac{3^{10}}{4^{10}} = \left(\dfrac{3}{4}\right)^{10}$

28. (d) Let F be the event of favourable cases.

$\therefore n(F)$ = Number of ways in which exactly 4 elements of A will belong to $P \cup Q = {^{10}C_4}\,(3)^4$

$\therefore$ Required probability $= \dfrac{n(F)}{n(S)} = \dfrac{^{10}C_4\,(3)^4}{4^{10}}$

$= \dfrac{10\times 9\times 8\times 7\times 3^4}{2^{19}\times 2\times 4\times 3\times 2\times 1}$

$= \dfrac{5\times 3\times 7\times 3^4}{2^{19}}$

$= \dfrac{8505}{2^{19}}$

29. (b) Number of cases in which $a_i \notin P \cap Q$ is 3.

Now, let G be the event of favourable cases.

$n(G)$ = Number of ways in which no element of A will belong to $P \cap Q = 3^{10}$

$\therefore$ Required probability $= \dfrac{n(G)}{n(S)} = \dfrac{3^{10}}{4^{10}}$

$= \left(\dfrac{3}{4}\right)^{10}$

30. (d) Let p_1 be the probability of being an answer correct of question attempted from Section I, then $p_1 = \dfrac{1}{5}$. Again, let p_2 be the probability of being an answer correct of question attempted from Section II, then $p_2 = \dfrac{1}{15}$

$\therefore$ Required probability $= \dfrac{1}{5}\times\dfrac{1}{15} = \dfrac{1}{75}$

31. (a) Scoring 10 marks from four questions can be obtained as 3 marks + 3 marks + 3 marks + 1 mark = 10 i.e. to answer 3 questions from Section II and 1 question from Section I correctly.

$\therefore$ Required probability $= \dfrac{1}{5}\left(\dfrac{1}{15}\right)^3$

32. (b) To get 40 marks, candidate has to answer all questions correctly and its probability is $\left(\dfrac{1}{5}\right)^{10}\left(\dfrac{1}{15}\right)^{10}$.

Hence, probability of getting a score less than 40

$= 1-\left(\dfrac{1}{5}\right)^{10}\left(\dfrac{1}{15}\right)^{10} = 1-\left(\dfrac{1}{75}\right)^{10}$

33. (c) Total number of cases $= {^{20}C_4}$

Let the numbers, which are in AP are $a, a+d, a+2d, a+3d$.

$\because a \ge 1, a+3d \le 20$

$\Rightarrow 3d \le 19 \Rightarrow d \le \dfrac{19}{3}$

$\therefore d = \pm 1, \pm 2, \pm 3, \pm 4, \pm 5$ and ± 6

Hence, Statement II is incorrect.

If $d = 1$, then $a + 3d \le 20 \Rightarrow a \le 17$, so 17 cases will be there.

Similarly for $d = -1$, 17 cases will be there.

Total case for $d = \pm 1$ is 34.

Similarly,

d	± 1	± 2	± 3	± 4	± 5	± 6
Cases	34	28	22	16	10	4

So, total favourable cases $= \dfrac{114}{2!} = 57$

[since, combination line 1, 2, 3, 4 treated same as 4, 3, 2, 1]

$\therefore$ Required probability $= \dfrac{57}{^{20}C_4} = \dfrac{1}{85}$

Hence, Statement I is correct.

34. (b) We have, $ax + by = 0$ and $cx + dy = 0$

where $a, b, c, d \in \{0, 1\}$

Clearly, $x = 0, y = 0$ is always a solution of given linear equations.

$\therefore P$ (that the system of equations has a solution) = 1

Thus, Statement II is correct.

As, we know that a system of homogeneous equation $AX = 0$ has a unique solution, if $|A| \ne 0$, so let us find the number of such determinants.

Clearly, the number of all possible determinants of the form $\begin{vmatrix} a & b \\ c & d \end{vmatrix} = 2^4 = 16$

Out of which only 10 determinants given by

$\begin{vmatrix} 1 & 1 \\ 1 & 1 \end{vmatrix}, \begin{vmatrix} 0 & 0 \\ 0 & 0 \end{vmatrix}, \begin{vmatrix} 0 & 0 \\ 1 & 1 \end{vmatrix}, \begin{vmatrix} 1 & 1 \\ 0 & 0 \end{vmatrix}, \begin{vmatrix} 0 & 1 \\ 0 & 1 \end{vmatrix}, \begin{vmatrix} 1 & 0 \\ 1 & 0 \end{vmatrix}, \begin{vmatrix} 1 & 0 \\ 0 & 0 \end{vmatrix}, \begin{vmatrix} 0 & 1 \\ 0 & 0 \end{vmatrix},$

$\begin{vmatrix} 0 & 0 \\ 1 & 0 \end{vmatrix}, \begin{vmatrix} 0 & 0 \\ 0 & 1 \end{vmatrix}$ vanish and remaining six determinants have non-zero values.

$\therefore$ Required probability $= \dfrac{6}{16} = \dfrac{3}{8}$

Hence, Statement I is correct.

35. (1) Let α and β be the roots of the quadratic equation $ax^2 + bx + c = 0$, where $a, b, c \in R$ and $a \neq 0$, then the quadratic equation whose roots are α^2, β^2 is

$$a^2x^2 - (b^2 - 2ac)\,x + c^2 = 0.$$

As, these two equations are same.

We have,

$$a^2 = a\,,\ b^2 - 2ac = -\,b, c = c^2$$

$\Rightarrow$ $a = 1, c = 0, 1,\ b^2 + b = 2ac$

$\Rightarrow$ $a = 1, c = 0, 1,\ b\,(b + 1) = 2c$

If $c = 0$, then

$$b\,(b + 1) = 0$$

$\Rightarrow$ $b = 0, b = -\,1$

If $c = 1$, then

$$b\,(b + 1) = 2$$

$\Rightarrow$ $b^2 + b - 2 = 0$

$\Rightarrow$ $b^2 + 2b - b - 2 = 0$

$\Rightarrow$ $b\,(b + 2) - 1\,(b + 2) = 0$

$\Rightarrow$ $(b + 2)\,(b - 1) = 0$

$\Rightarrow$ $b = -\,2, 1$

$\therefore$ The possible quadratic equations are

$$x^2 = 0$$

$$x^2 - x = 0$$

$$x^2 - 2x + 1 = 0$$

$$x^2 + x + 1 = 0$$

$\therefore$ Number of ways of choosing a quadratic equation $= {}^4C_1$

and number of ways of choosing a quadratic equation having equal roots $= {}^2C_1$

$\therefore$ Required probability $= \dfrac{2}{4} = \dfrac{1}{2}$

Now, $n - m = 2 - 1 = 1$

36. (3) Here, $n(S) = 9 \times 10 = 90$

and $n(A) = 1 + 2 + 3 + \ldots + 7$

$$= \frac{7 \times 8}{2} = 28$$

x (tens)	y (units)
3	0
4	{0, 1}
5	{0, 1, 2}
⋮	
9	{0, 1, 2, 3, 4, 5, 6}

Now, $p = \dfrac{28}{90} = \dfrac{14}{45}$

$\therefore$ $(n - 3m) = 45 - 42 = 3$

37. (2) Total number of 5-digit numbers formed by 0, 1, 2, 3, 4 and 5

$$= {}^6P_5 - {}^5P_5 = 6\,! - 5\,! = 5 \times 5\,! = 600$$

Number of 5-digit numbers which are divisible by 6

$$= 24 + 24 + 18 + 24 + 18 = 108$$

1	2	4	5	0

$4\,! = 24$

1	3	4	5	2

or $4\,! = 24$

0	1	4	5	2

$3 \times 3 \times 2 \times 1 = 18$

1	2	3	5	4

or $4\,! = 24$

0	1	2	5	4

$3 \times 3 \times 2 \times 1 = 18$

$\therefore$ Required probability $= \dfrac{108}{600} = \dfrac{18}{100}$, i.e. 18%

$\therefore$ $m = 18$

i.e. number of digits in m is 2.

38. (8) There are four subsets of {1, 2, 3, ..., 9} that adds to greater than 21.

i.e. 24 {7, 8, 9}, {6, 9, 8} 23, {5, 8, 9} 22, {6, 7, 9} 22

The number of 3×3 array having 7, 8, 9 as a row is 3(3!) (6!);

Which is true for each of the four sets.

Hence, the number of 3×3 array having a row that sums > 21, is 4(3) (3!) (6!) also total ways = 9!

$\therefore$ Required probability $= \dfrac{4(3)\,(3!)\,(6!)}{9!}$

$$= \frac{1}{7}$$

$\therefore$ $m + n = 8$

39. (2) Let x denotes the number on red die and y denotes the number on blue die.

Then, $x + y \leq 6$ (as the number on green has to be less than or equal to 6).

$\therefore$ $x \geq 1$

and $y \geq 1$

So, $x + y \leq 4$

$\therefore$ $x + y + t = 4$

Now, $n(A) = {}^6C_2 = 15$

and $n(S) = 216$

$\therefore$ Required probability $= \dfrac{15}{216} = \dfrac{5}{72}$

Now, $a + b = 77$

i.e. number of digits is 2.

40. (2) The number of ways of choosing the first square is 64 and that for the second square is 63. Therefore, the number of ways of choosing the first and second square is $64 \times 63 = 4032$.

Now, we proceed to find the number of favourable ways. If the first happens to be any of the four squares in the corner, the second square can be chosen in two ways.

If the first square happens to be any of the 24 square on either side of the chess board, the second square can be chosen in 3 ways. If the first squares happens to be any of the 36 remaining squares, the second square can be chosen in 4 ways.

So, number of favourable ways

$$= 4(2) + 24(3) + 36(4) = 224$$

Hence, required probability $= \dfrac{224}{4032}$

$$= \frac{1}{18}$$

$\therefore$ $m + n = 1 + 18 = 19$

i.e. number of disits is 2.

DPP-2 Addition and Subtraction Theorems of Probability

1. We have, $P(A \cup B) = \frac{3}{4}$

$P(A \cap B) = \frac{1}{4}$

and $P(\bar{A}) = \frac{2}{3}$

(i) $P(A) = 1 - P(\bar{A}) = 1 - \frac{2}{3} = \frac{1}{3}$

(ii) $P(A \cup B) = P(A) + P(B) - P(A \cap B)$

$\Rightarrow P(B) = P(A \cup B) + P(A \cap B) - P(A)$

$= \frac{3}{4} + \frac{1}{4} - \frac{1}{3}$

$= 1 - \frac{1}{3} = \frac{2}{3}$

(iii) $P(A \cap \bar{B}) = P(A) - P(A \cap B)$

$= \frac{1}{3} - \frac{1}{4} = \frac{1}{12}$

(iv) $P(\bar{A} \cap B) = P(B) - P(A \cap B)$

$= \frac{2}{3} - \frac{1}{4} = \frac{5}{12}$

2. (i) Venn diagram of $(\bar{A} \cap \bar{B}) \cup (A \cap B)$,

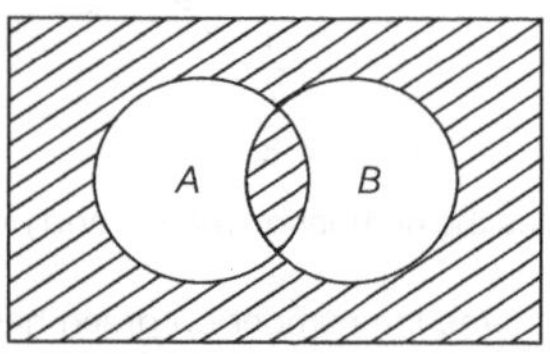

(ii) Venn diagram of $\bar{B} \cup (\bar{A} \cap B)$,

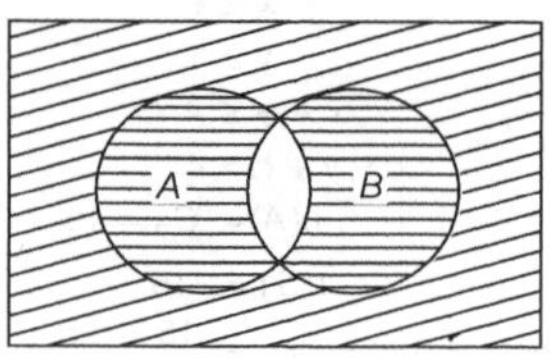

3. We have, $n(S) = 125$

$n(E) = 70$

$n(M) = 55$

$n(E \cap M) = 30$

$P(E) = \frac{70}{125}$

$P(M) = \frac{55}{125}$

$P(E \cap M) = \frac{30}{125}$

(i) Atleast passed in one subject,

$P(E \cup M) = P(E) + P(M) - P(E \cap M)$

$= \frac{70}{125} + \frac{55}{125} - \frac{30}{125}$

$= \frac{19}{25}$

(ii) Passed in only one subject,

$= P(A \cap \bar{B}) + P(\bar{A} \cap B)$

$= P(A) - P(A \cap B) + P(B) - P(A \cap B)$

$= \frac{70}{125} - \frac{30}{125} + \frac{55}{125} - \frac{30}{125}$

$= \frac{13}{125}$

4. (a) We have, A and B are mutually exclusive events.

$\therefore \quad P(A \cup B) = P(A) + P(B)$

and $P(A \cap B) = 0$

(a) $\because \quad P(A) + P(B) \le 1 \Rightarrow P(A) \le 1 - P(B)$

$\Rightarrow \quad P(A) \le P(\bar{B})$

(b) $P(\bar{A} \cap \bar{B}) = 1 - P(A \cup B) = 1 - P(A) - P(B) = P(\bar{A}) - P(B)$

(c) $P(\bar{A} \cup \bar{B}) = 1 - P(A \cap B) = 1$

(d) $P(\bar{A} \cap B) = P(B) - P(A \cap B) = P(B)$

Hence, option (a) is correct.

5. (a) We have, $S = \{1, 2, 3, \ldots, 10\}$

$A = $ (Minimum number is 4)

$= \{4, 5, 6, 7, 8, 9, 10\}$

$B = $ {Maximum number is 8}

$= \{1, 2, 3, 4, 5, 6, 7, 8\}$

$A \cap B = \{5, 6, 7\}$

$P(A) = \frac{{}^6C_2}{{}^{10}C_3}, P(B) = \frac{{}^7C_2}{{}^{10}C_3}, P(A \cap B) = \frac{{}^3C_2}{{}^{10}C_3}$

$\therefore \quad P(A \cup B) = \frac{{}^6C_2}{{}^{10}C_3} + \frac{{}^7C_2}{{}^{10}C_3} - \frac{{}^3C_2}{{}^{10}C_3}$

$= \frac{15}{120} + \frac{21}{120} - \frac{3}{120} = \frac{33}{120} = \frac{11}{40}$

Hence, option (a) is correct.

6. (a) We have, 6 white, 5 black and 4 red balls.

Total number of balls = 6 + 5 + 4 = 15

Total number of white or black balls = 6 + 5 = 11

$\therefore$ Probability of getting either a white or black ball $= \frac{11}{15}$

7. (a) We have, odd against A is 2 : 1.

$\therefore \quad P(A) = \frac{1}{3}$

Now, in favour of $A \cup B = 3 : 1$

$\therefore \quad P(A \cup B) = \frac{3}{4}$

For events A and B,

$P(A \cup B) \ge P(A)$ or $P(B)$

$\therefore \quad P(B) \le P(A \cup B) = \frac{3}{4}$

Now, $\quad P(A \cup B) = P(A) + P(B) - P(A \cap B)$

$\Rightarrow \quad \frac{3}{4} = \frac{1}{3} + P(B) - P(A \cap B)$

$\Rightarrow \quad P(B) - P(A \cap B) = \frac{3}{4} - \frac{1}{3} = \frac{5}{12}$

$\Rightarrow \quad P(B) \ge \frac{5}{12} \quad [\because P(A \cap B) \ge 0]$

Since, $\quad \frac{5}{12} \le P(B) \le \frac{3}{4}$

Hence, option (a) is correct.

8. (c) We have, $S = \{1, 2, 3, 4, 5, 6\}$

A (number greater than 3) $= \{4, 5, 6\}$

B {number less than 5} $= \{1, 2, 3, 4\}$

$A \cap B = \{4\}$

$P(A) = \frac{3}{6}, P(B) = \frac{4}{6}, P(A \cap B) = \frac{1}{6}$

$P(A \cup B) = P(A) + P(B) - P(A \cap B)$

$= \frac{3}{6} + \frac{4}{6} - \frac{1}{6} = 1$

Hence, option (c) is correct.

9. (d) We have, $S = \{1, 2, 3, 4, \ldots, 100\}$

Since, three distinct numbers are to be selected from 100 natural numbers $= {}^{100}C_3$

A = All the three of them are divisible by 2 and 3

$\Rightarrow$ divisible by 6 i.e. $\{6, 12, 18, \ldots, 96\}$

$$n(A) = {}^{16}C_3$$

Required probability $= \dfrac{{}^{16}C_3}{{}^{100}C_3} = \dfrac{4}{1155}$

Hence, option (d) is correct.

10. (a) We have, 52 cards in which face cards and that are removed.

$\therefore$ Total number of card remaining $= (52 - 16) = 36$

A = The card drawn is an ace

H = The card drawn is a heart

S = The card drawn is a spade

Now, $\quad P(A) = \dfrac{4}{36}, P(H) = \dfrac{9}{36}, P(S) = \dfrac{9}{36}$

$\Rightarrow \quad 9P(A) = 4P(H)$

$$P(A \cap H) = \frac{1}{36}, P(A \cap S) = \frac{1}{36}$$

$$P(A \cup S) = P(A) + P(S) - P(A \cap S)$$

$$= \frac{4}{36} + \frac{9}{36} - \frac{1}{36} = \frac{12}{36}$$

Hence, option (a) is correct.

11. (a) Since, $\quad P(A) = P(B) = 2P(C)$

Now, $\quad P(A) + P(B) + P(C) = 1$

$\therefore \quad P(C) = \dfrac{1}{5}$

$$P(A) = P(B) = \frac{2}{5}$$

Probability of B or C wins $= \dfrac{2}{5} + \dfrac{1}{5} = \dfrac{3}{5}$

Hence, option (a) is correct.

12. (d) We have, $P(E) \le P(F)$

and $\quad P(E \cap F) > 0$

Here, $\quad E \cap F \ne \phi$

$\therefore$ E is may or may not be subset of F.

$\therefore$ None of the given option is hold.

Hence, option (d) is correct.

13. (a) We have, $n(S) = 100$

$$n(P \cap C') = 25$$
$$n(P' \cap C) = 20$$
$$n(P \cap C) = 15$$

$\therefore$ Required probability,

$$P(P \cap C') + P(P' \cap C) = \frac{25}{100} + \frac{20}{100}$$
$$= \frac{45}{100} = \frac{9}{20}$$

Hence, option (a) is correct.

14. (b) We have, $\quad P(A \cup B) = 0.6$

and $\quad P(A \cap B) = 0.2$

$\therefore \quad P(A) + P(B) = P(A \cup B) + P(A \cap B)$

$= 0.6 + 0.2$

$= 0.8$

Now, $\quad P(\bar{A}) + P(\bar{B}) = 2 - [P(A) + P(B)]$

$= 2 - 0.8$

$= 1.2$

Hence, option (b) is correct.

15. (a) Here, $\quad 0 \le \dfrac{1 + 3p}{3} \le 1$

$\Rightarrow \quad -\dfrac{1}{3} \le p \le \dfrac{2}{3} \quad \ldots$(i)

$$0 \le \frac{1-p}{4} \le 1$$

$\Rightarrow \quad -3 \le p \le 1 \quad \ldots$(ii)

$$0 \le \frac{1-2p}{2} \le 1$$

$\Rightarrow \quad -\dfrac{1}{2} \le p \le \dfrac{1}{2} \quad \ldots$(iii)

From Eqs. (i), (ii) and (iii), we get

$$-\frac{1}{3} \le p \le \frac{1}{2}$$

16. (a) From Venn diagram, we can see that

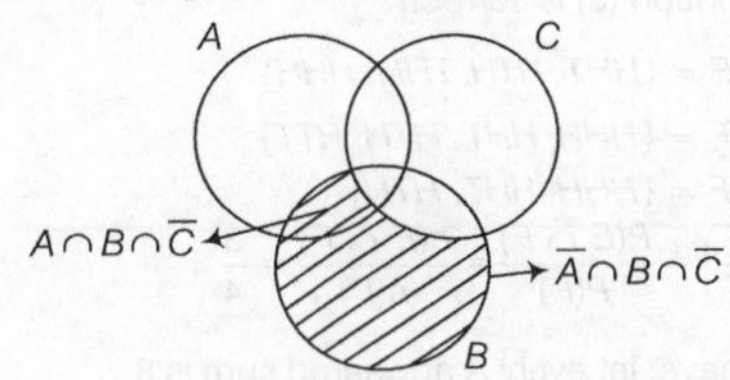

$$P(B \cap C) = P(B) - P(A \cap B \cap \bar{C}) - P(\bar{A} \cap B \cap \bar{C})$$

$$= \frac{3}{4} - \frac{1}{3} - \frac{1}{3} = \frac{1}{12}$$

Hence, option (a) is correct.

17. (a) We have, $\quad \sum_{i=1}^{6} P_i = \sum_{i=1}^{6} \lambda \quad$ [$\because P_i$ = Probability of tail i]

$\Rightarrow \quad 1 = \lambda(1 + 2 + 3 + 4 + 5 + 6) \quad [\because \Sigma P_i = 1]$

$\Rightarrow \quad \lambda = \dfrac{1}{21}$

Probability of an even number $= \dfrac{2}{21} + \dfrac{4}{21} + \dfrac{6}{21} = \dfrac{12}{21} = \dfrac{4}{7}$

Hence, option (a) is correct.

18. (c) Let $X = \{a_1, a_2, a_3, \ldots, a_n\}$ and $P(X) = A = \{\{x_1\}, \{x_2\}, \ldots, \{x_n\}\}$

For $x_i \in A (1 \le i \le n)$, $\quad [\because \{x_i\} = B$ subset of $X]$

We have, following 7 choices,

(i) $x_i \in P, Q, R$

(ii) $x_1 \in P$ but $x_i \notin Q$ and R

(iii) $x_i \in P$ and Q but $x_i \notin R$

(iv) $x_i \in P$ and R but not $x_i \notin Q$

(v) $x_i \in R$ but $x_i \notin P$ and Q

(vi) $x_i \notin R$ and Q but $x_i \notin P$

(vii) $x_i \in Q$ but $x_i \notin Q$ and R

$\therefore$ Total number of choices picking P, Q and R is 7^n.

For x_i does not belong to $P \cap R, Q \cap R, R \cap P$ are three choice.

$\therefore \quad P \cap Q \cap R = \phi = 3^n$ ways

Thus, probability of $P \cup Q \cup R = \left(\dfrac{3}{7}\right)^n$

Hence, option (c) is correct.

19. (d) $(P \cap Q \cap R)$ contains just 1 element is 5^n.

Thus, probability of $P \cap Q \cap R$ contains just 1 element $= \left(\dfrac{5}{7}\right)^n$.

20. (c) $P \cap Q \cap R$ contain no elements, then the number of ways $Q = R$ is 3^n.

Thus, $\quad$ probability $= \left(\dfrac{3}{7}\right)^n$

DPP-3 Conditional Probability and Multiplication Theorem of Probability

1. (a) $P(B/A) = \dfrac{P(A \cap B)}{P(A)} = \dfrac{1/10}{1/4} = \dfrac{2}{5}$

Hence, option (a) is correct.

2. (c) $P(\bar{A}/\bar{B}) = \dfrac{P(\bar{A} \cap \bar{B})}{P(\bar{B})} = \dfrac{1 - P(A \cup B)}{P(\bar{B})}$

Hence, option (c) is correct.

3. (c) We have, $P(A \cup B) = P(A \cap B)$

Now, $P(A \cup B) = P(A) + P(B) - P(A \cap B)$

$\Rightarrow P(A) + P(B) = 2P(A \cap B)$

$\Rightarrow P(A) + P(B) = 2P(A) \cdot P(B/A) \left[\because P(B/A) = \dfrac{P(A \cap B)}{P(A)}\right]$

Hence, option (c) is correct.

4. (a) $\because$ $E = \{HHT, HTH, THH, HHH\}$

$F = \{HHH, HHT, HTH, HTT\}$

$E \cap F = \{HHH, HHT, HTH\}$

$\therefore P\left(\dfrac{E}{F}\right) = \dfrac{P(E \cap F)}{P(F)} = \dfrac{n(E \cap F)}{n(F)} = \dfrac{3}{4}$

5. (d) We have, let event A appeared sum is 8.

$A = \{(2, 6), (3, 5) (4, 4), (5, 3) (6, 2)\}$

Let event B be the second die always exhibits 4.

$B = \{(1, 4), (2, 4), (3, 4), (4, 4), (5, 4), (6, 4)\}$

$A \cap B = \{(4, 4)\}$

$P(A) = 5/36, P(B) = 6/36, P(A \cap B) = 1/36$

To find $P(A/B) = \dfrac{P(A \cap B)}{P(B)} = \dfrac{1/36}{6/36} = \dfrac{1}{6}$

Hence, option (d) is correct.

6. (a) Total number of balls in a bag = (3 + 6 + 7) = 16

Let A be the event first balls is white and B be the event second ball is blue.

$\therefore \quad P(A) = \dfrac{6}{16}, P(B/A) = \dfrac{7}{15}$

$P(A \cap B) = P(A) \times P(B/A) = \dfrac{6 \times 7}{16 \times 15} = \dfrac{7}{40}$

Hence, option (a) is correct.

7. (c) Here, 4 red and 4 blue balls and four balls are drawn one-by-one.

There are two ways in which balls are alternate in colour *RBRB, BRBR*.

$\therefore$ Required probability = $P(RBRB) + P(BRBR)$

$= \dfrac{4}{8} \times \dfrac{4}{7} \times \dfrac{3}{6} \times \dfrac{3}{5} + \dfrac{4}{8} \times \dfrac{4}{7} \times \dfrac{3}{6} \times \dfrac{3}{5} = \dfrac{2 \times 4 \times 4 \times 3 \times 3}{8 \times 7 \times 6 \times 5} = \dfrac{6}{35}$

Hence, option (c) is correct.

8. (c) We have $P(\bar{A}) = 0.3, P(B) = 0.4, P(A \cap \bar{B}) = 0.5$

$\therefore \quad P(A) = 0.7, P(\bar{B}) = 0.6$

$\Rightarrow P(A \cup \bar{B}) = P(A) + P(\bar{B}) - P(A \cap \bar{B}) = 0.7 + 0.6 - 0.5 = 0.8$

$\Rightarrow P[B \cup (A \cup \bar{B})] = P(B) + P(A \cup \bar{B}) - P[B \cap (A \cup \bar{B})]$

$\Rightarrow 1 = 0.4 + 0.8 - P[B \cap (A \cup \bar{B})] \quad [\because P \cdot (B \cup (A \cup \bar{B})) = 1]$

$\Rightarrow P[B \cap (A \cup \bar{B})] = 0.2$

Now, $P(B / A \cup \bar{B}) = \dfrac{P[B \cap (A \cup \bar{B})]}{P(A \cup \bar{B})} = \dfrac{0.2}{0.8} = \dfrac{1}{4}$

Hence, option (c) is correct.

9. (a) Here, in 52 cards there are 12 faces cards and four suits, in each suits have 3 face cards.

$\therefore$ Required probability $= \left(\dfrac{3}{52} \times \dfrac{2}{51} \times \dfrac{1}{50}\right) \times 4 = \dfrac{1}{5525}$

Hence, option (a) is correct.

10. (a) Total number of balls (2 + 4 + 3) = 9 out of which 2 black, 4 white and 3 red balls, here ballls are in sequence

BBWWWWRRR

$\therefore$ Required probability

$= \dfrac{2}{9} \times \dfrac{1}{8} \times \dfrac{4}{7} \times \dfrac{3}{6} \times \dfrac{2}{5} \times \dfrac{1}{4} \times \dfrac{3}{2} \times \dfrac{2}{2} \times \dfrac{1}{1} = \dfrac{1}{1260}$

Hence, option (a) is correct.

11. (c) We have, $S = \{1, 2, 3, 5, 7, 11, 12, 13, 17, 19\}$

Let A be the event one number is 11 in pair of number, then

$P(A) = \dfrac{{}^9C_1}{{}^{10}C_2} = \dfrac{9 \times 2}{10 \times 9} = \dfrac{2}{10} = \dfrac{1}{5}$

B be the event whose sum of the number is even sum of two number, B even iff we select two number from any odd or even number.

$\therefore \quad P(B) = \dfrac{{}^8C_2}{{}^{10}C_2} + \dfrac{{}^2C_2}{{}^{10}C_2} = \dfrac{28}{45} + \dfrac{1}{45} = \dfrac{29}{45}$

$A \cap B = \{(1, 11) (3, 11) (5, 11) (7, 11) (13, 11) (17, 11) (19, 11)\}$

$P(A \cap B) = \dfrac{7}{45}$

Required probability $= P(A/B) = \dfrac{P(A \cap B)}{P(B)} = \dfrac{7/45}{29/45} = \dfrac{7}{29} = 0.24$

Hence, option (c) is correct.

12. (d) We have, let A be the event of face one and B be the event of either face one or face two has turn up.

$P(A) = 0.10, P(B) = 0.10 + 0.32 = 0.42$

Required probability,

$P(A/B) = \dfrac{P(A \cap B)}{P(B)} = \dfrac{0.10}{0.42} = \dfrac{5}{21} \quad [\because A \cup B = A]$

Hence, option (d) is correct.

13. (b) Let A be the event automobile will be stolen and found with in one week.

$\therefore \quad P(A \cap B) = 0.0006$

and B be the event automobile will be stolen, then

$P(B) = 0.0015 \Rightarrow P(A \cap B) = 0.0006$

Required probability $= P(A/B) = \dfrac{P(A \cap B)}{P(B)} = \dfrac{0.0006}{0.0015} = 0.4$

Hence, option (b) is correct.

14. (a) Let A be the event the person will get an electric contract and B be the event the person will get plumbing contract we have,

$P(A) = \dfrac{2}{5}, P(\bar{B}) = \dfrac{4}{7}, P(B) = 3/7$

$P(A \cup B) = \dfrac{2}{3}$

To find $P(A \cup B)$

$\therefore \quad P(A \cap B) = P(A) + P(B) - P(A \cup B)$

$= \dfrac{2}{5} + \dfrac{3}{7} - \dfrac{2}{3} = \dfrac{42 + 45 - 70}{105} = \dfrac{17}{105}$

Hence, option (a) is correct.

15. (c)

Red box		Green box	
5 *R*	0 *G*	1 *R*	6 *G*
4 *R*	1 *G*	2 *R*	5 *G*
3 *R*	2 *G*	3 *R*	4 *G*
2 *R*	3 *G*	4 *R*	3 *G*
1 *R*	4 *G*	5 *R*	2 *G*
0 *R*	5 *R*	6 *R*	1 *G*

Let E be event in desired, then $P(E) = \dfrac{{}^6C_0 \cdot {}^6C_5 + {}^6C_4 \cdot {}^6C_1}{{}^{12}C_5} = \dfrac{4}{33}$

Hence, $\quad p + q = 4 + 33 = 37$

DPP-4 Independent Events

1. (c) Given, $P(A) = 0.4$, $P(B) = p$, $P(A \cup B) = 0.7$

To find, value of p

We know that, if A and B are independent, then

$$P(A \cup B) = P(A) + P(B) - P(A) \times P(B)$$

$$\Rightarrow \quad 0.7 = 0.4 + p - 0.4 \times p$$

$$\Rightarrow \quad 0.3 = 0.6p$$

$$\therefore \quad p = \frac{0.3}{0.6} = \frac{1}{2}$$

2. (a) Number of three letters words from English alphabets $= (26)^3$

[$\because$ repeating of letters allowed]

Total number of three-digit numbers $= (10)^3$ [$\because$repetition allowed]

If letters palindrome, favourable number of ways $= 1 \times {}^{26}C_1 \times {}^{26}C_1$

Probability that a letter palindrome,

$$P(A) = \frac{{}^{26}C_1 \times {}^{26}C_1}{(26)^3} = \frac{26 \times 26}{(26)^3} = \frac{1}{26}$$

Probability that a digit palindrome,

$$P(B) = \frac{{}^{10}C_1 \times {}^{10}C_1 \times 1}{(10)^3} = \frac{1}{10}$$

$$P(\overline{A}) = 1 - \frac{1}{26} = \frac{25}{26}$$

$$P(\overline{B}) = 1 - \frac{1}{10} = \frac{9}{10}$$

$$P(A \cup B) = 1 - P(\overline{A}) \cdot P(\overline{B})$$

$$= 1 - \frac{25}{26} \cdot \frac{9}{10}$$

$$= \frac{260 - 225}{260}$$

$$= \frac{35}{260} = \frac{7}{52}$$

3. (a) Given, $P(A) = \frac{1}{5}$, $P(B) = \frac{1}{6}$

$$P(A \cap C) = \frac{1}{20}$$

$$P(B \cup C) = \frac{3}{8}$$

Now, since A and C are independent.

So, $$P(A \cap C) = P(A) \cdot P(C) = \frac{1}{20}$$

$$\Rightarrow \quad \frac{1}{5} \cdot P(C) = \frac{1}{20}$$

$$\Rightarrow \quad P(C) = \frac{1}{4}$$

$$P(B \cup C) = P(B) + P(C) - P(B \cap C)$$

$$\Rightarrow \quad \frac{3}{8} = \frac{1}{6} + \frac{1}{4} - P(B \cap C)$$

$$\Rightarrow \quad P(B \cap C) = \frac{5}{12} - \frac{3}{8} = \frac{1}{24}$$

Now, $$P(B) \cdot P(C) = \frac{1}{6} \times \frac{1}{4} = \frac{1}{24} = P(B \cap C)$$

$\Rightarrow B$ and C are independent.

4. (d) Given, $$P(A) = P(A/B) = \frac{1}{4}$$

$\Rightarrow A$ and B are independent.

$$\therefore \quad P(B/A) = \frac{1}{2} = P(B)$$

$$P\left(\frac{\overline{A}}{B}\right) = 1 - P(A) = 1 - \frac{1}{4} = \frac{3}{4}$$

and $$P\left(\frac{\overline{B}}{A}\right) = 1 - P(B) = 1 - \frac{1}{2} = \frac{1}{2}$$

5. (d) Let A, B and C be three events such that,

$P(A) = p_1$, $P(B) = p_2$, $P(C) = p_3$

Now, $$P(\overline{A}) = 1 - P(A) = (1 - p_1)$$

$$P(\overline{B}) = 1 - P(B) = (1 - p_2)$$

$$P(\overline{C}) = 1 - P(C) = (1 - p_3)$$

Probability of that happening of A, B and C,

$$P(\overline{A} \cap \overline{B} \cap \overline{C}) = P(\overline{A}) \cdot P(\overline{B}) \cdot P(\overline{C})$$

$$= (1 - p_1)(1 - p_2)(1 - p_3)$$

$\therefore$ Probability of happening of atleast one of them

$= 1 -$ (Probability of not happening of all of them)

$= 1 - (1 - p_1)(1 - p_2)(1 - p_3)$

6. (b) Given, $P(\overline{A \cup B}) = \frac{1}{6}$

$$P(A \cap B) = \frac{1}{4}$$

and $$P(\overline{A}) = \frac{1}{4}$$

Now, $$P(\overline{A \cup B}) = 1 - P(A \cup B)$$

$$\Rightarrow \quad \frac{1}{6} = 1 - P(A \cup B)$$

$$\Rightarrow \quad P(A \cup B) = 1 - \frac{1}{6} = \frac{5}{6}$$

and $$P(\overline{A}) = 1 - P(A)$$

$$\Rightarrow \quad \frac{1}{4} = 1 - P(A)$$

$$\Rightarrow \quad P(A) = 1 - \frac{1}{4} = \frac{3}{4}$$

Now, $$P(A \cup B) = P(A) + P(B) - P(A \cap B)$$

$$\Rightarrow \quad \frac{5}{6} = \frac{3}{4} + P(B) - \frac{1}{4}$$

$$\Rightarrow \quad \frac{5}{6} = \frac{1}{2} + P(B)$$

$$\Rightarrow \quad P(B) = \frac{5}{6} - \frac{1}{2} = \frac{5 - 3}{6} = \frac{2}{6} = \frac{1}{3}$$

$$\therefore \quad P(A/B) = \frac{P(A \cap B)}{P(B)} = \frac{1/4}{1/3} = \frac{3}{4} = P(A)$$

$\Rightarrow A$ and B are independent.

$$\because \quad P(A) \neq P(B)$$

$\Rightarrow A$ and B are not equally likely.

7. (c) All the three persons has three options to apply a house.

$\therefore$ Total number of cases $= (3)^3$

Now, favourable cases $= 3$ (with all has applied for house 1 or 2 or 3)

$\therefore$ Required probability $= \frac{3}{3^3} = \frac{1}{9}$

8. (d) As, $$P\left(\frac{C}{D}\right) = \frac{P(C \cap D)}{P(D)} \quad \left[\begin{array}{l}\because \quad C \subset D \\ \therefore P(C \cap D) = P(C)\end{array}\right]$$

$$= \frac{P(C)}{P(D)} \quad \ldots(i)$$

As, $$P(D) \leq 1$$

$$\therefore \quad \frac{1}{P(D)} \geq 1 \quad \ldots(ii)$$

and $$\frac{P(C)}{P(D)} \geq P(C)$$

From Eqs. (i) and (ii), we have

$$P\left(\frac{C}{D}\right) = \frac{P(C)}{P(D)} \geq P(C)$$

9. (a) Given, A, B and C are pairwise independent events and $P(C) > 0, P(A \cap B \cap C) = 0$.

To find, $P\left(\frac{\bar{A} \cap \bar{B}}{C}\right)$

We have,

$$P\left(\frac{\bar{A} \cap \bar{B}}{C}\right) = \frac{P(\bar{A} \cap \bar{B} \cap C)}{P(C)}$$

$$P\left(\frac{\bar{A} \cap \bar{B}}{C}\right) = \frac{P(C) - P(A \cap C) - P(B \cap C) + P(A \cap B \cap C)}{P(C)} \quad \ldots(i)$$

Now, $P(A \cap C) = P(A) \cdot P(C)$ [$\because A$ and C are independent]

and $P(B \cap C) = P(B) \cdot P(C)$

$$\therefore \quad P\left(\frac{\bar{A} \cap \bar{B}}{C}\right) = \frac{P(C) - P(A) \cdot P(C) - P(B) \cdot P(C) + 0}{P(C)}$$

$$= 1 - P(A) - P(B)$$

$$= P(\bar{A}) - P(B)$$

10. (d) Since, $P(A) = \frac{4}{10} = \frac{2}{5}$

For independent events,

$$P(A \cap B) = P(A) \cdot P(B)$$

$$\Rightarrow \quad P(A \cap B) \le \frac{2}{5}$$

$$\Rightarrow \quad P(A \cap B) = \frac{1}{10}, \frac{2}{10}, \frac{3}{10}, \frac{4}{10}$$

[maximum 4 outcomes may be in $A \cap B$]

Case I When $P(A \cap B) = \frac{1}{10}$

$$P(A) \cdot P(B) = \frac{1}{10}$$

$$\Rightarrow \quad P(B) = \frac{1}{10} \times \frac{5}{2} = \frac{1}{4} \quad \text{[not possible]}$$

Case II When $P(A \cap B) = \frac{2}{10}$, then

$$P(A) \cdot P(B) = \frac{2}{10}$$

$$\Rightarrow \quad \frac{2}{5} \cdot P(B) = \frac{2}{10}$$

$$\Rightarrow \quad P(B) = \frac{1}{2} = \frac{5}{10}$$

$\Rightarrow B$ may have 5 outcomes.

Case III When $P(A \cap B) = \frac{3}{10}$, then

$$P(A) \cdot P(B) = \frac{3}{10}$$

$$\Rightarrow \quad \frac{2}{5} \times P(B) = \frac{3}{10}$$

$$\Rightarrow \quad P(B) = \frac{3}{10} \times \frac{5}{2} = \frac{3}{4}, \text{ not possible}$$

Case IV When $P(A \cap B) = \frac{4}{10}$, then

$$P(A) \cdot P(B) = \frac{4}{10}$$

$$\Rightarrow \quad \frac{2}{5} \cdot P(B) = \frac{4}{10}$$

$$\Rightarrow \quad P(B) = 1 = \frac{10}{10}$$

$\therefore B$ may have 10 outcomes.

Hence, B must have 5 or 10 outcomes.

11. (b) Even chance means probability is half. Suppose, n cards are drawn, then

$$P(E) = P(S \text{ or } FS \text{ or } FFS \ldots n \text{ terms})$$

$$= \frac{P(S) \cdot \{1 - (PF)^n\}}{1 - P(F)} = 1 - \left(\frac{3}{4}\right)^n \ge \frac{1}{2}$$

$$\Rightarrow \quad \left(\frac{3}{4}\right)^n \le \frac{1}{2}$$

$$\therefore \quad n_{\min} = 3$$

12. (a) Given, $P(a) = 0.3, P(b) = 0.5$ and $P(c) = 0.2$

Hence, a, b and c are exhaustive.

P (same horse wins all the three races)

$$= P(aaa \text{ or } bbb \text{ or } ccc)$$

$$= (0.3)^3 + (0.5)^3 + (0.2)^3$$

$$= \frac{27 + 125 + 8}{1000} = \frac{160}{1000} = \frac{4}{25}$$

P (each horse wins exactly one race)

$$= P(abc \text{ or } acb \text{ or } bca \text{ or } bac \text{ or } cab \text{ or } cba)$$

$$= 0.3 \times 0.5 \times 0.2 \times 6$$

$$= 0.18 = \frac{9}{50}$$

13. (c) First cube has five red faces and 1 blue face.

$$\therefore \quad P(R) = \frac{5}{6}, P(B) = \frac{1}{6}$$

Let second cube has n red faces and $(6 - n)$ blue faces.

$$\therefore \quad P(R) = \frac{n}{6}, P(B) = \frac{6 - n}{6}$$

Probability that the two top faces shows same colour $= \frac{1}{2}$

$$\Rightarrow \quad \frac{5}{6} \times \frac{n}{6} + \frac{1}{6} \times \frac{6 - n}{6} = \frac{1}{2}$$

$$\Rightarrow \quad \frac{5n}{36} + \frac{6 - n}{36} = \frac{1}{2}$$

$$\Rightarrow \quad 4n + 6 = \frac{36}{2} = 18$$

$$\Rightarrow \quad 4n + 6 = 18$$

$$\Rightarrow \quad 4n = 12$$

$$\therefore \quad n = 3$$

14. (a) Let A, B, C and D are the persons and their respective probabilities to solve the problem is $\frac{1}{2}, \frac{3}{4}, \frac{1}{4}, \frac{1}{8}$.

i.e. $P(A) = \frac{1}{2}, P(B) = \frac{3}{4}, P(C) = \frac{1}{4}$ and $P(D) = \frac{1}{8}$

Probability that A is fail to solve the problem,

$$P(\bar{A}) = 1 - P(A) = 1 - \frac{1}{2} = \frac{1}{2}$$

Similarly, B, C and D are fail to solve the problem in

$$P(\bar{B}) = 1 - P(B) = 1 - \frac{3}{4} = \frac{1}{4}$$

$$P(\bar{C}) = 1 - P(C) = 1 - \frac{1}{4} = \frac{3}{4}$$

$$P(\bar{D}) = 1 - P(D) = 1 - \frac{1}{8} = \frac{7}{8}$$

Now, probability that all of them are fail to solve the problem,

$$P(\bar{A} \cap \bar{B} \cap \bar{C} \cap \bar{D}) = P(\bar{A}) \cdot P(\bar{B}) \cdot P(\bar{C}) \cdot P(\bar{D})$$

$$= \frac{1}{2} \cdot \frac{1}{4} \cdot \frac{3}{4} \cdot \frac{7}{8} = \frac{21}{256}$$

Probability that, the problem is solved correctly by atleast one of them

$= 1 -$ (Problem is not solved by all of them)

$$= 1 - \frac{21}{256} = \frac{235}{256}$$

15. (a) Since, the committee of 3 persons have to be selected from 3 men and 2 women in such a way that, committee have exactly 2 women and one man and chairperson is woman.

Favourable number of ways $= {}^2C_2 \times {}^3C_1 \times {}^2C_1 = 6$

Total number of ways $= {}^5C_3 \times {}^3C_1 = 30$

Required probability $= \frac{6}{30} = \frac{1}{5}$

16. (a) Let A, B and C be three persons.

Then, $P(A) = \frac{75}{100} = \frac{3}{4} = P(B) = P(C)$

$\therefore$ Required probability $= P(A \cap B \cap C)$

$$= P(A) \cdot P(B) \cdot P(C)$$
$$= \frac{3}{4} \times \frac{3}{4} \times \frac{3}{4} = \frac{27}{64}$$

17. (b) Probability that bird is killed in a shot $= \frac{1}{3}$

Probability that bird is not killed in a shot $= 1 - \frac{1}{3} = \frac{2}{3}$

Now, probability that the bird is not killed in three shots

$$= \frac{2}{3} \times \frac{2}{3} \times \frac{2}{3} = \frac{8}{27}$$

Thus, probability that bird is killed $= 1 - \frac{8}{27} = \frac{19}{27}$

18. (c) Given, odds against solving a problem by A are 4 : 3.

$\therefore$ $P(\overline{A}) = \frac{4}{4+3} = \frac{4}{7}$

and $P(A) = \frac{3}{4+3} = \frac{3}{7}$

and odds in favour of B are 7 : 5.

$\therefore$ $P(B) = \frac{7}{7+5} = \frac{7}{12}$

and $P(\overline{B}) = \frac{5}{7+5} = \frac{5}{12}$

P (Only one will solve the problem) $= P(A) \cdot P(\overline{B}) + P(\overline{A}) \cdot P(B)$

$$= \frac{3}{7} \times \frac{5}{12} + \frac{4}{7} \times \frac{7}{12} = \frac{5}{28} + \frac{1}{3}$$
$$= \frac{15+28}{84} = \frac{43}{84}$$

19. (a) The probability of getting a head in single toss of a coin,

$$p = \frac{1}{2} \quad \text{[say]}$$

The probability of getting 5 or 6 in a single throw of a die,

$$q = \frac{2}{6} \quad \text{[say]}$$

$\therefore$ Required probability $= p + (1-p)(1-q)p + (1-p)(1-q)$
$(1-p)(1-q)p + \cdots$

$$= p + (1-p)(1-q)p + (1-p)^2(1-q)^2 p + \cdots$$
$$= \frac{p}{1-(1-p)(1-q)}$$
$$= \frac{\frac{1}{2}}{1 - \frac{1}{2} \times \frac{2}{3}} = \frac{3}{4}$$

20. (b) Let A denote the event that a sum of 5 occurs, B the event that a sum of 7 occurs and C the event that neither a sum of 5 nor a sum of 7 occurs.

We have, $P(A) = \frac{4}{36} = \frac{1}{9}$

$P(B) = \frac{6}{36} = \frac{1}{6}$

and $P(C) = \frac{26}{36} = \frac{13}{18}$

Thus, $P(A$ occurs before $B)$

$$= \frac{1}{9} + \frac{13}{18} \times \frac{1}{9} + \left(\frac{13}{18}\right)^2 \times \frac{1}{9} + \cdots$$
$$= \frac{\frac{1}{9}}{1 - \frac{13}{18}} = \frac{2}{5} \qquad \left[\because S = \frac{a}{1-r}\right]$$

21. (c)

3 W 2 B	2 W 3 B
I	II

When a ball is drawn from second bag and placed in the first bag.

$\therefore$ Probability that drawn ball is white $= \frac{2}{5}$

Now, there are six balls in the Bag I, we draw a ball from Bag I and put it in Bag II, i.e.

4 W 2 B	1 W 3 B
I	II

$\therefore$ Probability that drawn ball is black $= \frac{2}{6}$

Draw a ball from Bag II and put it in to Bag I, i.e.

4 W 1 B	1 W 4 B
I	II

$\therefore$ Probability of drawing a white ball $= \frac{1}{5}$

Now, draw a ball from Bag I and put it in Bag II,

i.e.

5 W 1 B	4 B
I	II

$\therefore$ Probability of drawing a black balls $= \frac{1}{6}$

Required probability $= \frac{2}{5} \times \frac{2}{6} \times \frac{1}{5} \times \frac{1}{6}$

$$= \frac{4}{25 \times 36} = \frac{1}{25 \times 9} = \frac{1}{225}$$

22. (c) Let A_1 be the event that India win in first test match.

A_2 be the event that India win in second test match.

A_3 be the event that India win in third test match.

It is given that, probability that India win against West Indies,

$\frac{1}{2} = P(A)$ [say]

$\Rightarrow$ $P(\overline{A}) = \frac{1}{2}$

Required probability $= P(\overline{A}_1 \cap A_2 \cap A_3) + P(A_1 \cap \overline{A}_2 \cap A_3)$

$$= P(\overline{A}_1) \cdot P(A_2) \cdot P(A_3) + P(A_1) \cdot P(\overline{A}_2) \cdot P(A_3)$$
$$= \frac{1}{2} \times \frac{1}{2} \times \frac{1}{2} + \frac{1}{2} \times \frac{1}{2} \times \frac{1}{2}$$
$$= \frac{1}{8} + \frac{1}{8} = \frac{2}{8} = \frac{1}{4}$$

23. (b) Die marked with 1, 2, 2, 3, 3, 3 is thrown 3 times,

$$P(1) = \frac{1}{6}, P(2) = \frac{2}{6}, P(3) = \frac{3}{6}$$

P (4 or 6) = P [112(3 times) or 123 (6 case) or 222]

$$= 3 \times \frac{1}{6} \times \frac{1}{6} \times \frac{2}{6} + 6 \times \frac{1}{6} \times \frac{2}{6} \times \frac{3}{6} + \frac{2}{6} \times \frac{2}{6} \times \frac{2}{6}$$

$$= \frac{6 + 36 + 8}{216} = \frac{50}{216}$$

24. (c) Let $q = 1 - p$

Since, head appears first time in an even throw i.e. 2 or 4 or 6 etc.

$$\therefore \quad \frac{2}{5} = qp + q^3p + q^5p + \cdots$$

$$\Rightarrow \quad \frac{2}{5} = \frac{qp}{1-q^2}$$

$$\Rightarrow \quad \frac{2}{5} = \frac{(1-p)p}{1-(1-p)^2}$$

$$\Rightarrow \quad \frac{2}{5} = \frac{1-p}{2-p}$$

$$\Rightarrow \quad 2(2-p) = 5(1-p)$$

$$\Rightarrow \quad 4 - 2p = 5 - 5p$$

$$\Rightarrow \quad 3p = 1$$

$$\therefore \quad p = \frac{1}{3}$$

25. (a) Let r and s be the remainders, when a and b are divided by 5.

So, $a = 5p + r, 0 \le r \le 4, p \in N$

$b = 5q + s, 0 \le s \le 4, q \in N$

$$\therefore \quad a^2 + b^2 = 25(p^2 + q^2) + 10(pr + qs) + r^2 + s^2$$

$$= 5k + (r^2 + s^2), k \in N$$

Thus, $a^2 + b^2$ will be divisible by 5, if $r^2 + s^2$ is divisible by 5.

Therefore, total number of ways is equal to the number of ways of selecting r and s, which is $s^2 = 25$.

For favourable ways, we should have $r^2 + s^2$ is divisible by 5 and $0 \le r^2 + s^2 \le 32$.

So, $r^2 + s^2 = 0, 5, 10, 20, 25$ or 30

Thus,

r	0	1	2	3	4
s	0	2, 3	1, 4	1, 4	2, 3

So, number of favourable ways is 9.

Hence, the required probability is $\frac{9}{25}$.

26. (a) Probability (ending on R in the second step)

= Probability (landing on B, then move to R)

+ Probability (landing on W, then move to R)

$$= \frac{1}{6} \cdot (1) + \frac{1}{6} \cdot \left(\frac{1}{2}\right) = \frac{1}{4}$$

27. (a) Let $L \to$ denote loss and $W \to$ denote win.

Then, L and W can be filled at 14 places in 2^{14} ways.

$$\therefore \quad n(S) = 2^{14}$$

Now, 13 L's and 1W's can be arranged at 14 places in 14 ways.

Hence, $n(A) = 14$

$$\therefore \text{ Required probability} = \frac{n(A)}{n(S)} = \frac{14}{2^{14}} = \frac{7}{2^{13}}$$

28. (b) Set A have 4 elements and set B have 5 elements.

$\therefore$ Total number of functions from A to $B = (5)^4 = 625$

Number of monotonic function (either increasing or decreasing)

$$= 2 \times {}^5C_4 = 2 \times 5 = 10$$

$$\text{Required probability} = \frac{10}{625} = \frac{2}{125}$$

29. (a, b) Given, $P(X/Y) = \frac{1}{2}$ and $P(Y/X) = \frac{1}{3}$

and $P(X \cap Y) = \frac{1}{6}$

Now $P(X) = \frac{P(X \cap Y)}{p(Y/X)}$

$$\Rightarrow \quad P(X) = \frac{1/6}{1/3} = \frac{1}{6} \times \frac{3}{1} = \frac{1}{2}$$

$$\text{and} \quad P(Y) = \frac{P(X \cap Y)}{P(X/Y)} = \frac{1/6}{1/2} = \frac{1}{6} \times \frac{2}{1} = \frac{1}{3}$$

$$\because \quad P(X/Y) = P(X) = \frac{1}{2}$$

$$\text{and} \quad P(Y/X) = P(Y) = \frac{1}{3}$$

$\therefore$ X and Y are independent events.

Now, $P(X \cup Y) = P(X) + P(Y) - P(X \cap Y)$

$$= P(X) + P(Y) - P(X) \cdot P(Y)$$

$$= \frac{1}{2} + \frac{1}{3} - \frac{1}{2} \times \frac{1}{3}$$

$$= \frac{5}{6} - \frac{1}{6} = \frac{4}{6} = \frac{2}{3}$$

$$\text{and} \quad P(\overline{X} \cap Y) = P(Y) - P(X \cap Y) = \frac{1}{3} - \frac{1}{6} = \frac{1}{6}$$

30. (a, d) Let $P(E) = e$ and $P(F) = f$

$$\therefore \quad P(\overline{E}) = (1-e) \text{ and } P(\overline{F}) = (1-f)$$

$$\text{Now,} \quad P(E \cup F) - P(E \cap F) = \frac{11}{25}$$

$$\Rightarrow \quad e + f - 2ef = \frac{11}{25} \quad \ldots\text{(i)}$$

$$\text{and} \quad P(\overline{E} \cap \overline{F}) = \frac{2}{25}$$

$$\Rightarrow \quad (1-e)(1-f) = \frac{2}{25}$$

$$\Rightarrow \quad 1 - e - f + ef = \frac{2}{25} \quad \ldots\text{(ii)}$$

[$\because$ E and F are independent, $\overline{E}$ and $\overline{F}$ are also independent, i.e. $P(\overline{E} \cap \overline{F}) = P(\overline{E}) \cdot P(\overline{F})$]

From Eqs. (i) and (ii), we have

$$ef = \frac{12}{25} \quad \ldots\text{(iii)}$$

$$\text{and} \quad e + f = \frac{7}{5} \quad \ldots\text{(iv)}$$

On solving Eqs. (iii) and (iv), we get

$$e\left(\frac{7}{5} - e\right) = \frac{12}{25}$$

$$\Rightarrow \quad e^2 - \frac{7}{5}e + \frac{12}{25} = 0$$

$$\Rightarrow \quad \left(e - \frac{3}{5}\right)\left(e - \frac{4}{5}\right) = 0$$

$$\therefore \quad e = \frac{3}{5} \text{ or } \frac{4}{5}$$

$$\Rightarrow \quad f = \frac{4}{5} \text{ or } \frac{3}{5}$$

31. (b) If p_1, p_2, p_3, p_4 are the probabilities of success in a single throw for A, B, C and D.

Then, $$P(A) = p_1 = \frac{1}{2}$$

$$\Rightarrow \quad P(B) = p_2 = \frac{1}{8} + {}^3C_1 \cdot \frac{1}{8}$$
$$= \frac{1}{2}$$

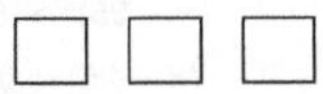

(all even + exactly one even)

$$P(C) = p_3 = \frac{1}{2^5} + \frac{{}^5C_1}{2^5} + \frac{{}^5C_2}{2^5}$$
$$= \frac{16}{2^5} = \frac{1}{2}$$

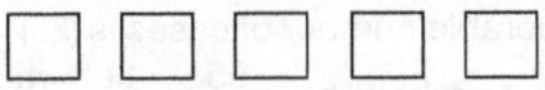

(i) all even (ii) one even and 4 odd

(iii) 3 even and 2 odd

$$P(D) = p_4 = \frac{1}{2^7} + \frac{{}^7C_1}{2^7} + \frac{{}^7C_3}{2^7} + \frac{{}^7C_5}{2^7}$$
$$= \frac{1 + 7 + 35 + 21}{2^7}$$
$$= \frac{64}{2^7} = \frac{1}{2}$$

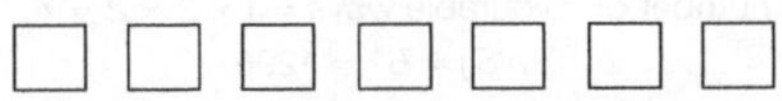

So, the probability of success is same for all in the single throw.

Hence, all equiprobable to win.

If they throw in succession, i.e. A, B, C and D. Then,

$$P(A) = P(S \text{ or } FFFFS \text{ or } \ldots)$$
$$= \frac{P(S)}{1 - [P(F)]^4} = \frac{1}{2} \cdot \frac{1}{1 - \frac{1}{16}} = \frac{8}{15}$$

Similarly, $$P(B) = \frac{4}{15}$$
$$P(C) = \frac{2}{15}$$
$$P(D) = \frac{1}{15}$$

Hence, both the statements are correct and Statement II is not the correct explanation.

32. (d) If T comes to start, then Australia can win only in one case that is TT.

$\therefore$ Probability of Australia wins $= \frac{1}{4}$

and probability of India wins $= \frac{3}{4}$

Hence, Statement II is correct, but Statement I is incorrect.

33. (d) We know that,

$$P(H_i/E) = \frac{P(H_i \cap E)}{P(E)} = \frac{P(E/H_i)P(H_i)}{P(E)}$$

or $$P(H_i/E)\,P(E) = P(E/H_i)P(H_i)$$

or $$P(E) = \frac{P(E/H_i)\,P(H_i)}{P(H_i/E)}$$

Now, $$0 < P(E) < 1 \qquad \text{[given]}$$

$$\Rightarrow \quad 0 < \frac{P(E/H_i)P(H_i)}{P(H_i/E)} < 1$$

$$\Rightarrow \quad P(E/H_i)\,P(H_i) < P(H_i/E)$$

But, if $$P(H_i \cap E) = 0$$

then $$P(H_i/E) = P(E/H_i) = 0$$

$\Rightarrow$ $P(E/H_i)\,P(H_i) < P(H_i/E)$, which is not true.

Hence, Statement I is not correct.

Also, $H_1, H_2, \ldots, H_n$ are mutually exclusive and exhaustive events.

Therefore, $$\sum_{i=1}^{n} P(H_i) = 1$$

Hence, Statement II is correct.

Explanations for Q. Nos. (34-35)

For the event that X is not hit, either the game ends with X hits Y in Ist round or both fails in Ist and X hits Y in IInd or both fails in first two rounds and X hits Y in IIIrd and so on.

34. (d) $P(X \text{ is not hit}) = p_1 \cdot \bar{p}_2 + \bar{p}_1\bar{p}_2 p_1 \cdot \bar{p}_2 + (\bar{p}_1)^2 \cdot (\bar{p}_2)^2 \cdot p_1 \cdot \bar{p}_2 + \ldots$

$$= \frac{2}{5} \cdot \frac{4}{5} + \frac{3}{5} \cdot \frac{4}{5} \cdot \frac{2}{5} \cdot \frac{4}{5} + \ldots \infty$$

$$= \frac{\frac{2}{5} \cdot \frac{4}{5}}{1 - \frac{3}{5} \cdot \frac{4}{5}} = \frac{8}{13}$$

35. (a) P (both the duelists are hit)

$$= p_1 \cdot p_2 + \bar{p}_1\bar{p}_2 p_1 \cdot p_2 + (\bar{p}_1)^2\,(\bar{p}_2)^2\,p_1 \cdot p_2 + \ldots \infty$$

$$= \frac{p_1 \cdot p_2}{1 - \bar{p}_1 \cdot \bar{p}_2} = \frac{\frac{2}{5} \cdot \frac{1}{5}}{1 - \frac{3}{5} \cdot \frac{4}{5}} = \frac{2}{13}$$

36. (a) For the event that duel ends after the nth round of shots, both duelists should fail till $(n-1)$th round and in nth round atleast one of them should be hitted.

$\therefore$ P (duel ends after the nth round)

$$= (\bar{p}_1 \cdot \bar{p}_2)^{n-1} \cdot (p_1 \cdot p_2 + p_1 \cdot \bar{p}_2 + \bar{p}_1 \cdot p_2)$$

$$= \left(\frac{3}{5} \cdot \frac{4}{5}\right)^{n-1} \cdot \left(\frac{2}{5} \cdot \frac{1}{5} + \frac{2}{5} \cdot \frac{4}{5} + \frac{3}{5} \cdot \frac{1}{5}\right)$$

$$= \left(\frac{12}{5}\right)^{n-1} \cdot \frac{13}{5^{n+1}}$$

37. (5) In a throwing 3 die, $n(S) = (6)^3$

Number of ways that atleast 2 of the three numbers obtained are same $= 6 \times 6$

Thus, required probability $= \frac{6^2}{6^3} = \frac{1}{6}$

But $$\frac{m}{n} = \frac{1}{6}$$

Now, $$n - m = 6 - 1 = 5$$

38. (9)

$$10 \begin{cases} 4\,D \\ 6\,G \end{cases} \xrightarrow{\;5\;} \begin{cases} 2\,D \\ 3\,G \end{cases}$$

$\therefore$ Required probability $= \frac{{}^4C_2 \cdot {}^6C_3}{{}^{10}C_5} = \frac{10}{21}$

Now, $$3m - n = 9$$

39. (3) Total number of ways of distribution of 7 prizes $= {}^{10}C_7$

Favourable number of ways of distributing 7 prizes out of 10 prizes, such that the remaining prize are one A, one B and one C

$$= {}^5C_4 \times {}^3C_2 \times {}^2C_1$$

Required probability $= \frac{{}^5C_4 \times {}^3C_2 \times {}^2C_1}{{}^{10}C_7} = \frac{5 \times 3 \times 2}{120} = \frac{1}{4}$

Now, $$\frac{m}{n} = \frac{1}{4}$$

$\therefore$ $$|m - n| = |1 - 4| = 3$$

40. (5) Total number of ways of selecting 3 tickets, one at a time (without replacement) $= {}^7C_1 \times {}^6C_1 \times {}^5C_1$

There are 4 odd digits and 3 even digits. We have to select 3 digits, one at a time, in sequence odd-even-odd and even-odd-even.

Favourable number of ways in the sequence odd-even-odd

$$= {}^4C_1 \times {}^3C_1 \times {}^3C_1$$

and favourable number of ways in the sequence even-odd-even

$$= {}^3C_1 \times {}^4C_1 \times {}^2C_1$$

$$\text{Required probability} = \frac{{}^4C_1 \times {}^3C_1 \times {}^3C_1 + {}^3C_1 \times {}^4C_1 \times {}^2C_1}{{}^7C_1 \times {}^6C_1 \times {}^5C_1}$$

$$= \frac{4 \times 3 \times 3 + 3 \times 4 \times 2}{7 \times 6 \times 5} = \frac{2}{7}$$

But, $\frac{m}{n} = \frac{2}{7}$

$\therefore$ $|m - n| = 5$

41. (2) Total arrangement of 2 red, 3 green and 4 blue counters in a row

$$= \frac{(2+3+4)!}{2!\,3!\,4!} = \frac{9!}{2!\,3!\,4!} = 1260$$

First, we arrange 2 red counter and 3 green counters.

Number of ways $= \frac{5!}{3!2!} = \frac{120}{6 \times 2} = 10$

we have, $R\ R\ G_1\ G_1\ G_1$

Now, there are 6 places at we place 4 blue counters in ${}^6C_4$ ways.

$\therefore$ Required probability $= \frac{15 \times 10}{1260}$

$$= \frac{15}{126} = \frac{5}{42}$$

But, $\frac{m}{n} = \frac{5}{42}$

$\therefore$ $|8m - n| = |5 \times 8 - 42| = 2$

42. (3) Here, $n(S) = 216$

x_1 = Number appearing on first die

x_2 = Number appearing on second die

x_3 = Number appearing on third die

$x_1 + x_2 + x_3 \le 10$ $[\because x_1, x_2, x_3 \in [1, 6]]$

$\Rightarrow$ $x_1 + x_2 + x_3 \le 7$ [after giving 1 each to x_1, x_2, x_3]

$\therefore$ $x_1 + x_2 + x_3 + X = 7$ [adding X as a false beggar]

Total number of solution $= {}^{7+3}C_3$

$$= {}^{10}C_3 = 120$$

Now, number of solutions when any one of x_1, x_2, x_3 takes the value 7 is

$x_1 + x_2 + x_3 + X = 1$

$\Rightarrow$ ${}^{1+3}C_3 = 4$

$\therefore$ Total number of ways $= {}^{10}C_3 - {}^4C_3 \times {}^3C_1$

$$= 108$$

$\therefore$ Required probability $= \frac{108}{216} = \frac{1}{2}$

Now, $m + n = 3$

43. (9) We have,

$$x + \frac{100}{x} > 50$$

or $x^2 + 100 > 50x$

$\Rightarrow$ $x^2 - 50x + 100 > 0$

$\Rightarrow$ $x^2 - 50x + (25)^2 + 100 - (25)^2 > 0$

$\Rightarrow$ $(x - 25)^2 - 525 > 0$

$\Rightarrow$ $(x - 25)^2 > 525$

$\Rightarrow$ $(x - 25) < \sqrt{525}$

or $x - 25 > \sqrt{525}$

$\Rightarrow$ $x < 25 - \sqrt{525}$

or $25 + \sqrt{525}$

As, x is a positive integer and $\sqrt{525} \approx 22.91$

We must have, $x \le 2$ or $x \ge 48$

Thus, the favourable number of cases is $(2 + 53) = 55$

$\therefore$ Required probability $= \frac{55}{100} = \frac{11}{20} = \frac{m}{n}$

$\Rightarrow$ $m = 11, n = 20$

Now, $|m - n| = |11 - 20| = 9$

44. (1) Here, 'a' can take only one value i.e. 2

'b' can take 1 or 3 i.e. two values

'c' can take 2 or 4 i.e. two values

and 'd' can take only one value i.e. 5

$\therefore$ Total number of favourable ways $= 1 \times 2 \times 2 = 4$

$$n(S) = 6^4 = 1296$$

$\therefore$ $P(E) = \frac{4}{1296}$

$$= \frac{1}{324} = \frac{1}{abc}$$

$\therefore$ $a + b - c = 3 + 2 - 4 = 1$

45. (1) We have, $n(S) = (5)^5 = 3125$

For computing favourable outcomes, 2 boxes which are to remain empty, can be selected in ${}^5C_2$ ways and 5 marbles can be placed in the remaining 3 boxes in groups of 221 or 311 in

$$3!\left(\frac{5!}{2!2!2!} + \frac{5!}{3!2!}\right) = 150 \text{ ways}$$

Now, $n(E) = {}^5C_2 \times 150$

Required probability, $P(E) = \frac{n(E)}{n(S)}$

$$= \frac{{}^5C_2 \times 150}{3125}$$

$$= \frac{10 \times 150}{3125}$$

$$= \frac{12}{25} = \frac{m}{n}$$

$\therefore$ $|2m - n| = |2 \times 12 - 25|$

$= |24 - 25|$

$= 1$

DPP-5 Theorem of Total Probability

1. (d) Consider, the events ε_1, ε_2 and A.

ε_1 = selection of 1st purse

ε_2 = selection of 2nd purse

A = drawn coin is copper

Now, $P(\varepsilon_1) = \frac{1}{2}$

$P(\varepsilon_2) = \frac{1}{2}$

$P(A/\varepsilon_1) = \frac{4}{7}$

$P(A/\varepsilon_2) = \frac{6}{8}$

$\therefore$ Required probability $P(A) = P(\varepsilon_1) \times P(A/\varepsilon_1) + P(\varepsilon_2) \times P(A/\varepsilon_2)$

$$= \frac{1}{2} \times \frac{4}{7} + \frac{1}{2} \times \frac{6}{8} = \frac{37}{56}$$

Hence, option (d) is correct.

2. (a) Consider, the events ε_1, ε_2, ε_3 and A.

ε_1 = selection of group A

ε_2 = selection of group B

ε_2 = selection of group C

A = introducing a new product

$P(\varepsilon_1) = 0.5,\ P(\varepsilon_2) = 0.3,\ P(\varepsilon_3) = 0.2$

$P(A/\varepsilon_1) = 0.7,\ P(A/\varepsilon_2) = 0.6,\ P(A/\varepsilon_3) = 0.5$

$\therefore$ Required probability

$$P(A) = P(\varepsilon_1) \times P(A/\varepsilon_1) + P(\varepsilon_2) \times P(A/\varepsilon_2) + P(\varepsilon_s) \times P(A/\varepsilon_3)$$

$= 0.5 \times 0.7 + 0.3 \times 0.6 + 0.2 \times 0.5$

$= 0.35 + 0.18 + 0.10$

$= 0.63$

Hence, option (a) is correct.

3. (c) We have,

$$\sum_{i=1}^{6} P_i = \sum_{i=1}^{6} \lambda i$$

[$\because P_i$ = probability of number shows on a face of die]

$\Rightarrow \quad 1 = \lambda\,(1 + 2 + 3 + 4 + 5 + 6) \quad [\because \Sigma p_i = 1]$

$\Rightarrow \quad \lambda = \frac{1}{21}$

$\therefore$ Required probability $= 3\lambda = 3 \times \frac{1}{21} = \frac{1}{7}$

Hence, option (c) is correct.

4. (c) Consider the $\varepsilon_1, \varepsilon_2, \varepsilon_3, \cdots, \varepsilon_n$ and A event.

ε_1 = 1st bag be selected

ε_2 = 2nd bag be selected

$\vdots \quad \vdots$

ε_n = nth bag be selected

A = both balls draw are white

$P(\varepsilon_1) = P(\varepsilon_2) = P(\varepsilon_3) = \cdots = P(\varepsilon_n) = \frac{1}{n}$

$P(A/\varepsilon_1) = \frac{1}{^3C_2}$

$P(A/\varepsilon_2) = \frac{1}{^4C_2}$

$\vdots \quad \vdots \quad \vdots$

$P(A/\varepsilon_n) = \frac{1}{^{n+2}C_2}$

$\therefore$ Required probability,

$$P(A) = P(\varepsilon_1) \times P(A/\varepsilon_1) + P(\varepsilon_2) \times P(A/\varepsilon_2) \cdots P(\varepsilon_n) \times P(A/\varepsilon_n)$$

$$= \frac{1}{n} \times \frac{1}{^3C_2} + \frac{1}{n} \times \frac{1}{^4C_2} + \cdots + \frac{1}{n} \times \frac{1}{^{n+2}C_2}$$

$$= \frac{1}{n} \sum_{i=1}^{n} \frac{1}{^{i+2}C_2}$$

$$= \frac{2}{n} \sum_{i=1}^{n} \frac{1}{(i+2)(i+1)}$$

$$= \frac{2}{n} \times \frac{n}{2 \times n + 2} \qquad \left[\because \Sigma \frac{1}{(i+2)(i+1)} = \frac{n}{2(n+2)}\right]$$

$$= \frac{1}{n+2}$$

Hence, option (c) is correct.

5. (d) Here, $P(6) = (51, 15, 24, 42)$

or $\{11 \text{ and } (22 \text{ or } 13 \text{ or } 31) \text{ or } (22 \text{ and } 11)\}$

$\therefore \quad P(6) = \frac{148}{1296}$

6. (c) We have, 3 red and 3 green balls.

There are four possible, if three balls are drawn.

ε_1 = all are red balls

ε_2 = all are green balls

ε_3 = 1 red and 2 green balls

ε_4 = 2 red and 1 green balls

3 blue balls are put into bag.

Now, there are four possible

1 = 3 green, 3 blue

2 = 3 red, 3 blue

3 = 2 red, 1 green, 3 blue

4 = 1 red, 2 green, 3 blue

$\therefore$ Required probability to find all different colour,

$$P(A) = P(\varepsilon_3) \times P(A/\varepsilon_3) + P(\varepsilon_4) \times P(A/\varepsilon_4)$$

$$= \frac{^3C_1 \times {^3C_2}}{^6C_3} \times \frac{^2C_1 \times {^1C_1} \times {^3C_1}}{^6C_3} + \frac{^3C_2 \times {^3C_1}}{^6C_3} \times \frac{^1C_1 \times {^2C_1} \times {^3C_1}}{^6C_3}$$

$$= \frac{3 \times 3 \times 2 \times 3}{20 \times 20} + \frac{3 \times 3 \times 2 \times 3}{20 \times 20}$$

$$= \frac{2 \times 3 \times 3 \times 2 \times 3}{20 \times 20} = \frac{27}{100} = 27\%$$

Hence, option (c) is correct.

7. (b) Consider, the events ε_1, ε_2 and A.

ε_1 = intercepted the missile

ε_2 = not intercepted the missile

A = hit the targets

$P(A) = P(\varepsilon_1) \times P(A/\varepsilon_1) + P(\varepsilon_2) \times P(A/\varepsilon_2)$

$P(A) = 1/3 \times 0 + \frac{2}{3} \times \frac{3}{4}$

$$\left[\because P(\varepsilon_1) = \frac{1}{3},\ P(\varepsilon_2) = 1 - \frac{1}{3} = \frac{2}{3}\right]$$

$= \frac{1}{2}$

$P(A/\varepsilon_1) = 0$

[since, intercepted the missile does not hit the target.]

$\therefore$ Required probability $= \frac{1}{2} \times \frac{1}{2} \times \frac{1}{2} = \frac{1}{8}$

Hence, option (b) is correct.

8. (c) Consider, the event ε_1, ε_2 and A.

ε_1 = odd number appears on the top face

ε_2 = even number appears on the top face

A = 2 number appears on the second roll

$$P(\varepsilon_1) = \frac{2}{6}, P(\varepsilon_2) = \frac{4}{6}$$

$$P(A/\varepsilon_1) = \frac{2}{6}, P(A/\varepsilon_2) = \frac{1}{6}$$

∴ Required probability,

$$P(A) = P(\varepsilon_1) \times P(A/\varepsilon_1) + P(\varepsilon_2) \times P(A/\varepsilon_2)$$
$$= \frac{2}{6} \times \frac{2}{6} + \frac{4}{6} \times \frac{1}{6}$$
$$= \frac{8}{36} = \frac{2}{9}$$

Hence, option (c) is correct.

9. (b) Consider, the events ε_1, ε_2 and A.

ε_1 = jockey B rides the horse A

ε_2 = jockey C rides the horse B

A = house A wins

$$P(\varepsilon_1) = 2/3, P(\varepsilon_2) = 1/3$$
$$P(A/\varepsilon_1) = 1/6, P(A/\varepsilon_2) = 3/6$$

∴ Required probability

$$P(A) = P(\varepsilon_1) \times P(A/\varepsilon_1) + P(\varepsilon_2) \times P(A/\varepsilon_2)$$
$$= \frac{2}{3} \times \frac{1}{6} + \frac{1}{3} \times \frac{3}{6}$$
$$= \frac{2+3}{18} = \frac{5}{18}$$

Hence, option (b) is correct.

10. (c) We have, 3 goods and 2 defective articles in A and 4 goods and 1 defective articles in B.

A new to C is formed by taking 3 articles from A and 2 from B.

Good	defective	good	defective	good	defective
3	2	4	1	3 + 2	0 + 0
3	2	4	1	3 + 2	1 + 0
3	2	4	1	3 + 2	2 + 0
3	2	4	1	3 + 2	1 + 0
3	2	4	1	3 + 2	1 + 1
3	2	4	1	3 + 2	2 + 1

Required probability

$$= \frac{{}^3C_3 \times {}^4C_2 \times 0}{{}^5C_3 \times {}^5C_3} + \frac{{}^3C_2 \times {}^2C_1 \times {}^4C_2}{{}^5C_3 \times {}^5C_3} \times \frac{1}{5}$$
$$+ \frac{{}^3C_1 \times {}^2C_2 \times {}^4C_2}{{}^5C_3 \times {}^5C_3} \times \frac{2}{5} + \frac{{}^3C_3 \times {}^1C_1 \times {}^4C_1}{{}^5C_3 \times {}^5C_3} \times \frac{1}{5}$$
$$+ \frac{{}^3C_2 \times {}^2C_2 \times {}^1C_1 \times {}^4C_{12}}{{}^5C_3 \times {}^5C_3} \times \frac{2}{5} + \frac{{}^3C_1 \times {}^2C_2 \times {}^1C_1 \times {}^4C_1}{{}^5C_3 \times {}^5C_3} \times \frac{3}{5}$$
$$= \frac{36}{500} + \frac{36}{500} + \frac{4}{500} + \frac{48}{500} + \frac{36}{500}$$
$$= \frac{160}{500} = \frac{8}{25}$$

Hence, option (c) is correct.

11. (a) Consider, the events ε_1, ε_2 and A.

ε_1 = fair dice, ε_2 = unfair dice

A = dice show face 3

$$P(\varepsilon_1) = 25\% = \frac{1}{4}, P(\varepsilon_2) = 75\% = \frac{3}{4}$$

$$P(A/\varepsilon_1) = \frac{1}{6}, P(\varepsilon_2) = \frac{2}{6}$$

∴ Required probability,

$$P(\varepsilon_1/A) = \frac{P(\varepsilon_1) \times P(A/\varepsilon_1)}{P(\varepsilon_1) \times P(A/\varepsilon_1) + P(\varepsilon_2) \times P(A/\varepsilon_2)}$$

[using Baye's theorem]

$$= \frac{\frac{1}{4} \times \frac{1}{6}}{\frac{1}{4} \times \frac{1}{6} \times \frac{3}{4} \times \frac{2}{6}} = \frac{1}{7}$$

Hence, option (a) is correct.

12. (b) Consider, the event ε_1, ε_2 and A.

ε_1 = specific home is unlocked, ε_2 = specific home is locked

A = master key open houses

$$P(\varepsilon_1) = 40\% = \frac{40}{100}, P(\varepsilon_2) = 60\% = \frac{60}{100}$$

$$P(A/\varepsilon_1) = 1, P(A/\varepsilon_2) = 3/8$$

∴ Required probability,

$$P(A) = P(\varepsilon_1) \times P(A/\varepsilon_1) + P(\varepsilon_2) \times P(A/\varepsilon_2)$$
$$= \frac{40}{100} \times 1 + \frac{60}{100} \times \frac{3}{8} = \frac{320+180}{800} = \frac{500}{800} = \frac{5}{8}$$

Hence, option (b) is correct.

13. (a) Consider, the events, ε_1, ε_2 and A.

ε_1 = red marbles in first draw

ε_2 = green marbles in first draw

A = red marble in second draw without replacement

$$P(\varepsilon_1) = \frac{6}{9}, P(\varepsilon_2) = \frac{3}{9}$$

$$P(A/\varepsilon_1) = \frac{5}{8} P,(A/\varepsilon_2) = \frac{6}{8}$$

∴ Required probability

$$P(A) = P(\varepsilon_1) \times P(A/\varepsilon_1) + P(\varepsilon_2) \times P(A/\varepsilon_2)$$
$$= \frac{6}{9} \times \frac{5}{8} + \frac{3}{9} \times \frac{6}{8} = \frac{30+18}{72} = \frac{48}{72} = \frac{2}{3}$$

Hence, option (a) is correct.

14. (d) We have, the probability that a radar will detect on object is on cycle is p.

So, the probability that a radar will not detect an object is an cycle is $(1-p)$.

∴ Required probability $= 1 - (1-p)^n$

Hence, option (d) is correct.

15. (b) $P(E) = P_{11} \cdot P_{12} + P_{12} \cdot P_{22} + P_{13} \cdot P_{32}$

$P_{11} \cdot P_{12}$ = party 1 in power and party 1 wins in the Ist and party 1 in power and party 2 wins.

or $P_{12} \cdot P_{22}$ = party 1 in power and party 2 wins in the Ist and party 2 wins.

or $P_{13} \cdot P_{32}$ = party 1 in power and party 3 wins.

∴ $P(E) = P_{11} \cdot P_{12} + P_{12} \cdot P_{22} + P_{13} \cdot P_{32} = 0.24$

DPP-6 **Baye's Theorem**

1. (d) Let E_1 be the event that chosen bag is A,

E_2 be the event that chosen bag is B

and A be the event that chosen ball is red.

Now, $P(E_1) = P(E_2) = \frac{1}{2}$

$$P\left(\frac{A}{E_1}\right) = \frac{3}{5}, \quad P\left(\frac{A}{E_2}\right) = \frac{5}{9}$$

To find $P\left(\frac{E_1}{A}\right)$

By Baye's theorem, we have

$$P\left(\frac{E_1}{A}\right) = \frac{P(E_1)\,P\left(\frac{A}{E_1}\right)}{P(E_1)\,P\left(\frac{A}{E_1}\right) + P(E_2)\,P\left(\frac{A}{E_2}\right)}$$

$$= \frac{\frac{1}{2} \times \frac{3}{5}}{\frac{1}{2} \times \frac{3}{5} + \frac{1}{2} \times \frac{5}{9}} = \frac{\frac{3}{5}}{\frac{3}{5} + \frac{5}{9}} = \frac{\frac{3}{5}}{\frac{52}{45}} = \frac{3}{5} \times \frac{45}{52} = \frac{27}{52}$$

$\therefore \quad P(E_2 / A) = \frac{25}{52}$

2. (a)

$$10 \begin{cases} 4 \text{ rotten} \\ 6 \text{ good} \end{cases} \xrightarrow{2} \text{P (other is good/one of them is good)}$$

$$\therefore \quad P(A/B) = \frac{\frac{^6C_2}{^{18}C_2}}{\frac{^6C_1 \cdot {}^4C_1}{^{10}C_2} + \frac{^6C_2}{^{10}C_2}} = \frac{15}{24 + 15} = \frac{15}{39} = \frac{5}{13}$$

3. (b) Let E_1 = coin shows head, E_2 = Coin shows tail

A = Drawn ball is blue

$P(E_1) = \frac{1}{2} = P(E_2)$

$P(A/E_1)$ = Probability of drawing a blue ball from bag I $= \frac{2}{7}$

$P\left(\frac{A}{E_2}\right)$ = Probability of drawing a blue ball from bag II $= \frac{6}{8}$

To find, $P\left(\frac{E_1}{A}\right)$

By Baye's theorem, we have

$$P\left(\frac{E_1}{A}\right) = \frac{P(E_1)\,P\left(\frac{A}{E_1}\right)}{P(E_1)\,P\left(\frac{A}{E_1}\right) + P(E_2)\,P\left(\frac{A}{E_2}\right)}$$

$$= \frac{\frac{1}{2} \times \frac{2}{7}}{\frac{1}{2} \times \frac{2}{7} + \frac{1}{2} \times \frac{6}{8}} = \frac{\frac{2}{7}}{\frac{2}{7} + \frac{6}{8}} = \frac{2}{7} \times \frac{56}{58} = \frac{8}{29}$$

4. (c) Let E_1 be the event that the letter came from 'DHARMAPURI' and E_2 be the event that the letter came from 'KRISHNAGIRI'.

Let A be the event that two consecutive letters RI are visible as post mark.

Since, letters have come either from KRISHNAGIRI or 'DHARMAPURI'.

$\therefore \quad P(E_1) = P(E_2) = \frac{1}{2}$

If E_1 has occurred, then it means that letters came from 'DHARAMPURI'.

In the word 'DHARMAPURI' there are 10 letters in which RI occurs in the end.

Considering RI as one letter there are 9 letters out of which one can be in 9 ways.

$$\therefore \quad P\left(\frac{A}{E_1}\right) = \frac{1}{9}$$

If E_2 has occurred, then the letters came from 'KRISHNAGIRI'.

In the word 'KRISHNAGIRI' there are 11 letters in which RI occurs twice. Considering one of the two RI as one letter there are 10 letters.

$$\therefore \quad P\left(\frac{A}{E_2}\right) = \frac{2}{10}$$

By Baye's theorem, we have

$$P\left(\frac{E_2}{A}\right) = \frac{P(E_2)\,P\left(\frac{A}{E_2}\right)}{P(E_1)\,P\left(\frac{A}{E_1}\right) + P(E_2)\,P\left(\frac{A}{E_2}\right)}$$

$$= \frac{\frac{1}{2} \times \frac{2}{10}}{\frac{1}{2} \times \frac{1}{9} + \frac{1}{2} \times \frac{2}{10}} = \frac{\frac{2}{10}}{\frac{1}{9} + \frac{2}{10}} = \frac{2}{10} \times \frac{90}{28} = \frac{9}{14}$$

5. (a) Let $E_1, E_2, E_3, \cdots, E_{100}$ and A be the events defined as follow :

E_1 : There are one coin of 1 rupee rest are different.

E_2 : There are two coin of 1 rupee rest are different.

E_{100} : All coins are 1 rupee coin

A : Drawing a coin from the purse

$\therefore$ Required probability

$$= \frac{\frac{1}{100} \times \frac{1}{100}}{\frac{1}{100} \times \frac{1}{100} + \frac{1}{100} \times \frac{2}{100} + \frac{1}{100} \times \frac{3}{100} + \cdots + \frac{1}{100} \times \frac{100}{100}}$$

$$= \frac{\frac{1}{100}}{\frac{1}{100} + \frac{2}{100} + \frac{3}{100} + \cdots + \frac{100}{100}} = \frac{\frac{1}{100}}{\frac{1}{100}(1 + 2 + 3 + 4 + \cdots + 100)}$$

$$= \frac{1}{\frac{100 \times 101}{2}} = \frac{1}{50 \times 101} = \frac{1}{5050}$$

6. (c) Let E_1, E_2 and A be the event defined as follow :

E_1 : The driver are not under the influence

E_2 : The driver are under the influence of alcohol

A : Car met with an accident

Now, $P(E_1) = \frac{80}{100} = \frac{4}{5}$

$P(E_2) = \frac{20}{100} = \frac{1}{5}$

$P\left(\frac{A}{E_1}\right) = 0.0001$

$P\left(\frac{A}{E_2}\right) = 0.001$

To find, $P\left(\frac{E_2}{A}\right)$

By Baye's theorem, we have

$$P\left(\frac{E_2}{A}\right) = \frac{P(E_2)\,P\left(\frac{A}{E_2}\right)}{P(E_1)\,P\left(\frac{A}{E_1}\right) + P(E_2)\,P\left(\frac{A}{E_2}\right)} = \frac{\frac{1}{5} \times 0.001}{\frac{4}{5} \times 0.0001 + \frac{1}{5} \times 0.001}$$

$$= \frac{0.001}{0.0004 + 0.001} = \frac{0.001}{0.0014} = \frac{10}{14} = \frac{5}{7}$$

7. (c) Let E_1 : The event that fair die is chosen.

E_2 : The event that rigged die is chosen.

A : The event that tossing die three times it shows 5 on all the faces.

$$P(E_1) = \frac{3}{4},\ P(E_2) = \frac{1}{4}$$

$$P\left(\frac{A}{E_1}\right) = {}^3C_3\left(\frac{1}{6}\right)^3 = \frac{1}{6^3}$$

$$P\left(\frac{A}{E_2}\right) = 1$$

Now, $$P\left(\frac{E_2}{A}\right) = \frac{P(E_2)\,P\left(\frac{A}{E_2}\right)}{P(E_1)\,P\left(\frac{A}{E_1}\right) + P(E_2)\,P\left(\frac{A}{E_2}\right)}$$

$$= \frac{\frac{1}{4}\times 1}{\frac{3}{4}\times\frac{1}{6^3} + \frac{1}{4}\times 1} = \frac{\frac{1}{4}}{\frac{3}{216}+1} = \frac{216}{219}$$

8. (a) R : They obtained the same result.

$$B_1 : A \cap \bar{B};\ P(B_1) = \frac{1}{8}\cdot\frac{11}{12}$$

$$B_2 : \bar{A} \cap B;\ P(B_2) = \frac{7}{8}\cdot\frac{1}{12}$$

$$B_3 : A \cap B;\ P(B_3) = \frac{1}{8}\cdot\frac{1}{12}$$

$$B_4 : \bar{A} \cap \bar{B};\ P(B_4) = \frac{7}{8}\cdot\frac{11}{12}$$

Now, $P(R/B_1) = 0$, $P(R/B_2) = 0$, $P(R/B_3) = 1$,

$$P(R/B_4) = \frac{7}{8}\cdot\frac{11}{12}\cdot\frac{1}{1001}$$

$$P(B_3/4) = \frac{\frac{1}{8}\cdot\frac{1}{12}}{\frac{1}{8}\cdot\frac{1}{12} + \frac{7}{8}\cdot\frac{11}{12}\cdot\frac{1}{1001}} = \frac{13}{14}$$

9. (c) Let E_1, E_2, E_3 and A be the events defined as follow :

E_1 : The bolt is manufactured by machine A.

E_2 : The bolt is manufactured by machine B.

E_3 : The bolt is manufactured by machine C.

A : The bolt is defective.

Now, $P(E_1) = \frac{35}{100} = \frac{7}{20}$, $\quad P(E_2) = \frac{25}{100} = \frac{1}{4}$

$$P(E_3) = \frac{40}{100} = \frac{4}{10} = \frac{2}{5}$$

$P\left(\frac{A}{E_1}\right)$: Probability that the bolt drawn in defective given that it is manufactured by machine $A = \frac{2}{100} = \frac{1}{50}$

Similarly, $$P\left(\frac{A}{E_2}\right) = \frac{1}{100}$$

$$P\left(\frac{A}{E_3}\right) = \frac{3}{100}$$

By Baye's theorem, we have

$$P\left(\frac{E_3}{A}\right) = \frac{P(E_3)\,P\left(\frac{A}{E_3}\right)}{P(E_1)\,P\left(\frac{A}{E_1}\right) + P(E_2)\,P\left(\frac{A}{E_2}\right) + P(E_3)\,P\left(\frac{A}{E_3}\right)}$$

$$= \frac{\frac{2}{5}\times\frac{3}{100}}{\frac{7}{20}\times\frac{1}{50} + \frac{1}{4}\times\frac{1}{100} + \frac{2}{5}\times\frac{3}{100}}$$

$$= \frac{\frac{6}{500}}{\frac{7}{1000} + \frac{1}{400} + \frac{6}{500}} = \frac{\frac{6}{500}}{\frac{14+5+24}{2000}} = \frac{\frac{6}{500}}{\frac{43}{2000}}$$

$$= \frac{6}{500}\times\frac{2000}{43} = \frac{24}{43}$$

10. (c) Given, $P(A) = \frac{1}{5}$, $\quad P(B) = \frac{7}{25}$

$$P(B/A) = \frac{9}{10}$$

To find $P(A/B)$

Now, $$P\left(\frac{A}{B}\right) = \frac{P(A \cap B)}{P(B)} = \frac{P(A)\,P\left(\frac{B}{A}\right)}{P(B)}$$

$$= \frac{\frac{1}{5}\times\frac{9}{10}}{\frac{7}{25}} = \frac{9}{50}\times\frac{25}{7} = \frac{9}{14} = \frac{18}{28}$$

11. (c) Let us define the following events

E_1 : Person goes by car, $\qquad E_2$: Person goes by scooter

E_3 : Person goes by bus, $\qquad E_4$: Person goes by train

A : Person reaches late

Then, in question gives that,

$$P(E_1) = \frac{1}{7},\ P(E_2) = \frac{3}{7},\ P(E_3) = \frac{2}{7},\ P(E_4) = \frac{1}{7}$$

$$P\left(\frac{A}{E_1}\right) = \frac{2}{9},\qquad P(A/E_2) = \frac{1}{9}$$

$$P\left(\frac{A}{E_3}\right) = \frac{4}{9},\qquad P\left(\frac{A}{E_4}\right) = \frac{1}{9}$$

To find $P\left(\frac{E_1}{\bar{A}}\right)$ [since, reaches in time not late]

By Baye's theorem, we have

$$P(E_1/\bar{A}) = \frac{P(E_1)P(\bar{A}/E_1)}{P(E_1)\,P(\bar{A}/E_1) + P(E_2)\,P(\bar{A}/E_2) + P(E_3)\,P(\bar{A}/E_3) + P(E_4)\,P(\bar{A}/E_4)} \quad \ldots\text{(i)}$$

Now, $$P\left(\frac{\bar{A}}{E_1}\right) = 1 - \frac{2}{9} = \frac{7}{9},\ P\left(\frac{\bar{A}}{E_3}\right) = 1 - \frac{4}{9} = \frac{5}{9}$$

$$P(\bar{A}/E_2) = 1 - \frac{1}{9} = \frac{8}{9},\ P(\bar{A}/E_4) = 1 - \frac{1}{9} = \frac{8}{9}$$

On substituting these values in Eq. (i), we get

$$P\left(\frac{E_1}{\bar{A}}\right) = \frac{\frac{7}{9}\times\frac{1}{7}}{\frac{7}{9}\times\frac{1}{7} + \frac{8}{9}\times\frac{3}{7} + \frac{5}{9}\times\frac{2}{7} + \frac{8}{9}\times\frac{1}{7}}$$

$$= \frac{7}{7+24+10+8} = \frac{7}{49} = \frac{1}{7}$$

12. (b) Given, $P(u_i) \propto i$, where $i = 1, 2, 3, \cdots, n$

$\Rightarrow P(u_i) = k\cdot i$, where k constant of proportionality

But, $\qquad \Sigma\, P(u_i) = 1$

$$\Rightarrow \quad \Sigma ki = 1 \Rightarrow k\cdot\Sigma i = 1 \Rightarrow k = \frac{2}{n(n+1)}$$

$$\Rightarrow \quad P(u_i) = \frac{2i}{n(n+1)}$$

By total probability theorem,

$$P(w) = \sum_{i=1}^{n} P(u_i)\,P\left(\frac{w}{u_i}\right) = \sum_{i=1}^{n}\frac{2i}{n(n+1)}\times\frac{i}{n+1}$$

$$= \frac{2}{n(n+1)^2}\cdot\frac{n(n+1)(2n+1)}{6} = \frac{2n+1}{3n+3}$$

$$\therefore \lim_{n\to\infty} P(w) = \lim_{n\to\infty}\left(\frac{2n+1}{3n+3}\right) = \frac{2}{3}$$

13. (a) If $P(u_i) = C$ (constant), then

By using Baye's theorem,

$$P\left(\frac{u_n}{w}\right) = \frac{P(u_n)\,P(w/u_n)}{\sum_{i=1}^{n} P(u_i)\,P\left(\frac{w}{u_i}\right)} = \frac{C \times \frac{n}{n+1}}{C\left[\frac{1}{n+1} + \frac{2}{n+1} + \cdots + \frac{n}{n+1}\right]}$$

$$= \frac{n}{n+1} \times \frac{n+1}{\frac{n(n+1)}{2}} = \frac{2}{n+1}$$

14. (b) $P\left(\frac{W}{E}\right) = \frac{P(W \cap E)}{P(E)}$

$$= \frac{\frac{1}{n} \times \frac{2}{n+1} + \frac{1}{n} \times \frac{4}{n+1} + \frac{1}{n} \times \frac{6}{n+1} + \cdots + \frac{1}{n} \times \frac{n}{n+1}}{\frac{1}{n} + \frac{1}{n} + \frac{1}{n} + \cdots + \frac{1}{n}\left(\frac{n}{2}\text{ times}\right)}$$

$$= \frac{\frac{2}{n(n+1)}\left(1 + 2 + 3 + \cdots + \frac{n}{2}\right)}{\frac{1}{n} \times \frac{n}{2}} \quad \text{[here, } n \text{ being even]}$$

$$= \frac{4}{n(n+1)}\left[\frac{\frac{n}{2}\left(\frac{n}{2} + 1\right)}{2}\right] = \frac{4 \cdot n(n+2)}{8 \cdot n(n+1)} = \frac{n+2}{2(n+1)}$$

15. (a) Let p be the probability that 6 occurs, then

$$p = \frac{1}{6}$$

and $$q = 1 - p = \frac{5}{6}$$

$$p(X = 3) = q \cdot q \cdot p = q^2 p = \left(\frac{5}{6}\right)^2 \times \frac{1}{6} = \frac{25}{216}$$

16. (b) $P(X \geq 3) = 1 - P(X \leq 2)$

$$= 1 - \left(\frac{1}{6} + \frac{5}{6} \times \frac{1}{6}\right) = 1 - \left(\frac{1}{6} + \frac{5}{36}\right)$$

$$= 1 - \frac{11}{36} = \frac{25}{36}$$

17. (d) For $X \geq 6$,

$$\text{Required probability} = \left(\frac{5}{6}\right)^5 \cdot \frac{1}{6} + \left(\frac{5}{6}\right)^6 \cdot \frac{1}{6} + \left(\frac{5}{6}\right)^7 \cdot \frac{1}{6} + \cdots \infty$$

$$= \frac{1}{6}\left(\frac{5}{6}\right)^5 \left[1 + \frac{5}{6} + \left(\frac{5}{6}\right)^2 + \cdots \infty\right]$$

$$= \frac{(5)^5}{(6)^6} \times \frac{1}{1 - \frac{5}{6}} \quad \left[\because S_\infty = \frac{a}{1-r}\right]$$

$$= \frac{(5)^5}{(6)^6} \times 6 = \left(\frac{5}{6}\right)^5$$

For $X > 3$,

$$\text{Required probability} = \frac{5^3}{6^4} + \frac{5^4}{6^5} + \frac{5^5}{6^6} + \cdots \infty$$

$$= \frac{1}{6} \times \frac{5^3}{6^3}\left[1 + \frac{5}{6} + \left(\frac{5}{6}\right)^2 + \cdots \infty\right]$$

$$= \frac{1}{6} \times \frac{5^3}{6^3} \times \frac{1}{1 - \frac{5}{6}} = \frac{5^3}{6^4} \times 6 = \frac{5^3}{6^3}$$

Hence, the conditional probability that $X \geq 6$, given $X > 3$

$$= \frac{\left(\frac{5}{6}\right)^5}{\left(\frac{5}{6}\right)^3} = \left(\frac{5}{6}\right)^2 = \frac{25}{36}$$

18. (b) Let E_1, E_2 and A be the event defined as follow :

E_1 : Head appears on the coin.

E_2 : Tail appears on the coin.

A : Drawn ball from U_2 is white.

Now, $P(E_1) = \frac{1}{2}, P(E_2) = \frac{1}{2}$

$P(A/E_1)$ = Probability that drawn ball from U_2 is white when 1 ball is transferred from U_1 to U_2.

$$= \left[\left(\frac{3}{5} \times 1\right) + \left(\frac{2}{5} \times \frac{1}{2}\right)\right] = \left(\frac{3}{5} + \frac{1}{5}\right) = \frac{4}{5}$$

$P\left(\frac{A}{E_2}\right)$ = Probability that drawn ball from U_2 is white when 2 balls are transferred from U_1 to U_2

$$= \left[\frac{{}^3C_2}{{}^5C_2} \times 1 + \frac{{}^2C_2}{{}^5C_2} \times \frac{1}{3} + \frac{{}^3C_1 \times {}^2C_1}{{}^5C_2} \times \frac{2}{3}\right]$$

$$= \left[\frac{3}{10} \times 1 + \frac{1}{10} \times \frac{1}{3} + \frac{6}{10} \times \frac{2}{3}\right]$$

$$= \left[\frac{3}{10} + \frac{1}{30} + \frac{12}{30}\right] = \frac{22}{30}$$

P(drawing a white ball from U_2)

$$= P(E_1)\,P\left(\frac{A}{E_1}\right) + P(E_2)\,P\left(\frac{A}{E_2}\right)$$

$$= \frac{1}{2} \times \frac{4}{5} + \frac{1}{2} \times \frac{22}{30} = \frac{4}{10} + \frac{11}{30}$$

$$= \frac{12 + 11}{30} = \frac{23}{30}$$

19. (d) P (head appeared/white ball from U_2)

$$= \frac{1}{2} \times \frac{\left[\frac{{}^3C_1}{{}^5C_1} \times \frac{{}^2C_1}{{}^2C_1} \times \frac{{}^2C_1}{{}^5C_1} \times \frac{{}^1C_1}{2C_1}\right]}{\frac{23}{30}} = \frac{1}{2} \times \frac{\frac{3}{5} \times 1 + \frac{2}{5} \times \frac{1}{2}}{\frac{23}{30}} = \frac{12}{23}$$

20. (d) Let E_1, E_2, E_3 and A be the event defined as follow :

E_1 : Both balls are from B_1

E_2 : Both balls are from B_2

E_3 : Both balls are from B_3

A : One ball is white and other is red.

Now, $$P(E_1) = P(E_2) = P(E_3) = \frac{1}{3}$$

$$P\left(\frac{A}{E_1}\right) = \frac{{}^1C_1 \times {}^3C_1}{{}^6C_2} = \frac{3}{15}$$

$$P\left(\frac{A}{E_2}\right) = \frac{{}^2C_1 \times {}^3C_1}{{}^9C_2} = \frac{2 \times 3}{36} = \frac{1}{6}$$

$$P\left(\frac{A}{E_3}\right) = \frac{{}^3C_1 \times {}^4C_1}{{}^{12}C_2} = \frac{3 \times 4}{66} = \frac{2}{11}$$

To find $P\left(\frac{E_2}{A}\right)$

$$P\left(\frac{E_2}{A}\right) = \frac{P(E_2)\,P\left(\frac{A}{E_2}\right)}{P(E_1)\,P\left(\frac{A}{E_1}\right) + P(E_2)\,P\left(\frac{A}{E_2}\right) + P(E_3)\,P\left(\frac{A}{E_3}\right)}$$

$$= \frac{\frac{1}{3} \times \frac{1}{6}}{\frac{1}{3} \times \frac{3}{15} + \frac{1}{3} \times \frac{1}{6} + \frac{1}{3} \times \frac{2}{11}}$$

$$= \frac{\frac{1}{6}}{\frac{3}{15} + \frac{1}{6} + \frac{2}{11}} = \frac{1}{6} \times \frac{330}{181} = \frac{55}{181}$$

21. (a) P (all three drawn balls are of same colour)

$= P(\text{all are white}) + P(\text{all are red}) + P$ (all are black)

$= \frac{1}{6} \times \frac{2}{9} \times \frac{3}{12} + \frac{3}{6} \times \frac{3}{9} \times \frac{4}{12} + \frac{3}{6} \times \frac{4}{9} \times \frac{5}{12}$

$= \frac{6}{648} + \frac{36}{648} + \frac{40}{648} = \frac{82}{648}$

Explanations for Q. Nos. (22-24)

Let A : She gets a success, T : She studies 10 h

S : She studies 7 h, F : She studies 4 h

$\therefore \quad P(T) = 0.1, P(S) = 0.2, P(F) = 0.7$

$P\left(\frac{A}{T}\right) = 0.8, P\left(\frac{A}{S}\right) = 0.6, P\left(\frac{A}{F}\right) = 0.4$

22. (c) $P(A) = P(A \cap T) + P(A \cap S) + P(A \cap F)$

$= P(T)\,P(A/T) + P(S)\,P(A/S) + P(F)\,P(A/F)$

$= (0.1)(0.8) + (0.2)(0.6) + (0.7)(0.4)$

$= 0.08 + 0.12 + 0.28 = 0.48$

23. (b) $P\left(\frac{F}{A}\right) = \frac{P(F \cap A)}{P(A)} = \frac{(0.7)(0.4)}{0.48} = \frac{0.28}{0.48} = \frac{7}{12}$

24. (d) $P\left(\frac{F}{\bar{A}}\right) = \frac{P(F \cap \bar{A})}{P(\bar{A})} = \frac{P(F) - P(F \cap A)}{0.52}$

$= \frac{(0.7) - 0.28}{0.52} = \frac{21}{26}$

Explanations for Q. Nos. (25-27)

Let E_1: The event of getting an even number on the die.

E_2 : Getting an odd number on the die.

A : First two draw shows blue balls.

$P(E_1) = \frac{3}{6} = \frac{1}{2} = P(E_2)$

$P\left(\frac{A}{E_2}\right) = \frac{4}{9}, P\left(\frac{A}{E_1}\right) = \frac{1}{36}$

25. (b) Conditional probability, $P\left(\frac{A}{E_2}\right) = \frac{4}{9}$

26. (a) Let $p = P\left(\frac{E_1}{A}\right) = \frac{P(E_1)\,P(A/E_1)}{P(E_1)\,P(A/E_1) + P(E_2)\,P(A/E_2)}$

$= \frac{\frac{1}{2} \times \frac{1}{36}}{\frac{1}{2} \times \frac{1}{36} + \frac{1}{2} \times \frac{16}{36}} = \frac{1}{17}$

Again, let $\quad q = P\left(\frac{E_2}{A}\right) = \frac{\frac{1}{2} \times \frac{16}{36}}{\frac{1}{72} + \frac{16}{72}} = \frac{\frac{16}{72}}{\frac{17}{12}} = \frac{16}{17}$

$q = \frac{16}{17} = \frac{1}{17} \times 16 = 16p \qquad \left[\because p = \frac{1}{17}\right]$

27. (c) E : Third ball drawn is red

$P(E) = P(E \cap E_1) + P(E \cap E_2)$

$= \frac{1}{17} \cdot \frac{5}{6} + \frac{16}{17} \cdot \frac{2}{6} = \frac{5 + 32}{102} = \frac{37}{102}$

28. (1) Let E_1, E_2, E_3 and A be the events defined as follows

E_1 : Bolt is manufactured by machine A.

E_2 : Bolt is manufactured by machine B.

E_3 : Bolt is manufactured by machine C.

A : Bolt is defection.

Then, $\quad P(E_1) = \frac{25}{100}, P(E_2) = \frac{35}{100}, P(E_3) = \frac{40}{100}$

$P\left(\frac{A}{E_1}\right) = \frac{5}{100}, P\left(\frac{A}{E_2}\right) = \frac{4}{100}, P\left(\frac{A}{E_3}\right) = \frac{2}{100}$

$\therefore$ Required probability $= P(E_2/A) = \frac{P(E_2)P\left(\frac{A}{E_2}\right)}{\sum_{i=1}^{n} P(E_i)\,P\left(\frac{A}{E_i}\right)}$

$= \frac{\frac{35}{100} \times \frac{4}{100}}{\frac{25}{100} \times \frac{5}{100} + \frac{35}{100} \times \frac{4}{100} + \frac{40}{100} \times \frac{2}{100}}$

$= \frac{140}{125 + 140 + 80} = \frac{140}{345} = \frac{28}{69}$

But given that, $\quad \frac{m}{n} = \frac{28}{69}$

$\Rightarrow \quad \left|\frac{2m - n}{13}\right| = \left|\frac{2 \times 28 - 69}{13}\right| = 1$

29. (2) Let E_1, E_2 and A be the events defined as follow :

E_1 : Bag A chosen.

E_2 : Bag B chosen.

A : Ball is red.

Now, $\quad P(E_1) = P(E_2) = \frac{1}{2}$

$P\left(\frac{A}{E_1}\right) = \frac{3}{5}, P\left(\frac{A}{E_2}\right) = \frac{5}{9}$

By Baye's theorem, we have

$P(E_2/A) = \frac{P(E_2)\,P(A/E_2)}{P(E_1)\,P(A/E_1) + P(E_2)\,P(A/E_2)}$

$= \frac{\frac{1}{2} \times \frac{5}{9}}{\frac{1}{2} \times \frac{3}{5} + \frac{1}{2} \times \frac{5}{9}} = \frac{\frac{5}{9}}{\frac{3}{5} + \frac{5}{9}} = \frac{5}{9} \times \frac{45}{52} = \frac{25}{52}$

But it is given that, $\frac{m}{n} = \frac{25}{52}$

$\Rightarrow \quad |2m - n| = |50 - 52| = 2$

30. (5) Let E_1, E_2, E_3 and A be the event defined as follow :

E_1 : Examinee guesses the answer.

E_2 : Examinee copies the answer.

E_3 : Examinee knows the answer.

A : Examinee answer correctly.

We have, $P(E_1) = \frac{1}{3}\ P(E_2) = \frac{1}{6}$

Since, E_1, E_2 and E_3 are mutually exclusive and exhaustive events.

$\therefore \quad P(E_1) + P(E_2) + P(E_3) = 1$

$\Rightarrow \quad P(E_3) = 1 - [P(E_2) + P(E_1)]$

$= 1 - \left(\frac{1}{3} - \frac{1}{6}\right) = \frac{1}{2}$

If E_1 has already occurred, then the examinee guesses. Since, there are four choices out of which only one is correct, therefore the probability that he answers correctly given that he has made a guess is $\frac{1}{4}$ i.e. $P\left(\frac{A}{E_1}\right) = \frac{1}{4}$. It is given that, $P\left(\frac{A}{E_2}\right) = \frac{1}{8}$ and $P\left(\frac{A}{E_3}\right) = 1$.

By Baye's theorem, we have

$P\left(\frac{E_3}{A}\right) = \frac{P(E_3)\,P(A/E_3)}{P(E_1)\,P(A/E_1) + P(E_2)\,P(A/E_2) + P(E_3)\,P(A/E_3)}$

$= \frac{\frac{1}{2} \times 1}{\frac{1}{3} \times \frac{1}{4} + \frac{1}{6} \times \frac{1}{8} + \frac{1}{2} \times 1} = \frac{24}{29}$

But $\quad \frac{m}{n} = \frac{24}{29}$

$\therefore \quad |m - n| = |24 - 29| = 5$

DPP-7 Binomial Distribution and Probability Distribution

1. Let p = Probability of getting head on tossing a coin

n = Number of times coin tossed = 10

x = Number of times getting head

Clearly, $X \sim B\left(10, \frac{1}{2}\right)$

$\therefore \quad P(X = x) = {}^{10}C_x\left(\frac{1}{2}\right)^x\left(\frac{1}{2}\right)^{10-x}, x = 0, 1, 2, \cdots, 10$

$\Rightarrow \quad P(X = x) = {}^{10}C_x\left(\frac{1}{2}\right)^{10}, x = 0, 1, 2, \cdots, 10$

(i) P (exactly six heads) $= P(X = 6)$

$= {}^{10}C_6\left(\frac{1}{2}\right)^{10} = {}^{10}C_4\left(\frac{1}{2}\right)^{10}$

$= \frac{10 \times 9 \times 8 \times 7}{4 \times 3 \times 2 \times 1} \times \frac{1}{2^{10}}$

$= \frac{105}{2^9} = \frac{105}{512}$

(ii) P (atleast six heads) $= P(X = 6) + P(X = 7) + P(X = 8) + P(X = 9) + P(X = 10)$

$= {}^{10}C_6\left(\frac{1}{2}\right)^{10} + {}^{10}C_7\left(\frac{1}{2}\right)^{10} + {}^{10}C_8\left(\frac{1}{2}\right)^{10} + {}^{10}C_9\left(\frac{1}{2}\right)^{10} + {}^{10}C_{10}\left(\frac{1}{2}\right)^{10}$

$= \left(\frac{1}{2}\right)^{10}({}^{10}C_4 + {}^{10}C_3 + {}^{10}C_2 + {}^{10}C_1 + {}^{10}C_0)$

$= \frac{1}{2^{10}} \times (210 + 120 + 45 + 10 + 1)$

$= \frac{386}{2^{10}} = \frac{193}{512}$

(iii) P (atmost six heads)

$= P(X = 0) + P(X = 1) + P(X = 2) + P(X = 3) + P(X = 4) + P(X = 5) + P(X = 6)$

$= {}^{10}C_0\left(\frac{1}{2}\right)^{10} + {}^{10}C_1\left(\frac{1}{2}\right)^{10} + {}^{10}C_2\left(\frac{1}{2}\right)^{10} + {}^{10}C_3\left(\frac{1}{2}\right)^{10} + {}^{10}C_4\left(\frac{1}{2}\right)^{10} + {}^{10}C_5\left(\frac{1}{2}\right)^{10} + {}^{10}C_6\left(\frac{1}{2}\right)^{10}$

$= \left(\frac{1}{2^{10}}\right)({}^{10}C_0 + {}^{10}C_1 + {}^{10}C_2 + \cdots + {}^{10}C_6)$

$= \frac{1}{2^{10}}[2^{10} - ({}^{10}C_7 + {}^{10}C_8 + {}^{10}C_9 + {}^{10}C_{10})]$

$= \frac{1}{2^{10}}[2^{10} - ({}^{10}C_3 + {}^{10}C_2 + {}^{10}C_1 + {}^{10}C_0)]$

$= \frac{1}{2^{10}}[2^{10} - (120 + 45 + 10 + 1)]$

$= 1 - \frac{176}{2^{10}} = 1 - \frac{2^4 \times 11}{2^{10}}$

$= 1 - \frac{11}{64} = \frac{53}{64}$

2. (a) Let p = Probability of winning a match by India $= \frac{1}{2}$

n = Nnumber of matches = 5

x = Number of matches win by India

Clearly, $X \sim B\left(5, \frac{1}{2}\right)$

$\therefore \quad P(X = x) = {}^5C_x\left(\frac{1}{2}\right)^x\left(\frac{1}{2}\right)^{5-x}, x = 0, 1, 2, \cdots, 5$

$= {}^5C_x\left(\frac{1}{2}\right)^5, x = 0, 1, 2, \cdots, 5$

$\therefore$ P (that India wins atleast three matches)

$= P(X = 3) + P(X = 4) + P(X = 5)$

$= {}^5C_3\left(\frac{1}{2}\right)^5 + {}^5C_4\left(\frac{1}{2}\right)^5 + {}^5C_5\left(\frac{1}{2}\right)^5$

$= \left(\frac{1}{2}\right)^5({}^5C_3 + {}^5C_4 + {}^5C_5)$

$= \left(\frac{1}{2}\right)^5(10 + 5 + 1)$

$= \frac{2^4}{2^5} = \frac{1}{2}$

3. (c) Let P = Probability of getting head on tossing a coin $= \frac{1}{2}$

n = Number of times coin tossed = 7

x = Number of times man get head

Clearly, $X \sim B\left(7, \frac{1}{2}\right)$

$\therefore \quad P(X = x) = {}^7C_x\left(\frac{1}{2}\right)^x\left(\frac{1}{2}\right)^{7-x}, x = 0, 1, 2 \cdots 7$

$= {}^7C_x\left(\frac{1}{2}\right)^7$

P (that man wins the toss on more than three occassions)

$= 1 - P$ (that man wins the toss atmost three occassion)

$= 1 - [P(X = 0) + P(X = 1) + P(X = 2) + P(X = 3)]$

$= 1 - \left[{}^7C_0\left(\frac{1}{2}\right)^7 + {}^7C_1\left(\frac{1}{2}\right)^7 + {}^7C_2\left(\frac{1}{2}\right)^7 + {}^7C_3\left(\frac{1}{2}\right)^7\right]$

$= 1 - \left(\frac{1}{2}\right)^7({}^7C_0 + {}^7C_1 + {}^7C_2 + {}^7C_3)$

$= 1 - \left(\frac{1}{2}\right)^7(1 + 7 + + 21 + 35)$

$= 1 - \frac{64}{2^7} = 1 - \frac{1}{2} = \frac{1}{2}$

4. (c) The man is one step away from starting point in two cases :

(i) He takes 6 steps forward and 5 steps backward.

(ii) He takes 5 steps forward and 6 steps backward.

$\therefore$ Required probability $= {}^{11}C_6(0.4)^6(0.6)^5 + {}^{11}C_5(0.4)^5(0.6)^6 = 0.37$

5. (d) Let P = Probability of getting head on tossing a coin $= \frac{1}{2}$

n = Number of coins tossed = 8

x = Number of times getting head

Clearly, $X \sim B\left(8, \frac{1}{2}\right)$

$\therefore \quad P(X = x) = {}^8C_x\left(\frac{1}{2}\right)^x\left(\frac{1}{2}\right)^{8-x}, x = 0, 1, 2, \cdots, 8$

$= {}^8C_x\left(\frac{1}{2}\right)^8, x = 0, 1, 2, \cdots, 8$

Now, P (getting atleast 6 heads)

$= P(X = 6) + P(X = 7) + P(X = 8)$

$= {}^8C_6\left(\frac{1}{2}\right)^8 + {}^8C_7\left(\frac{1}{2}\right)^8 + {}^8C_8\left(\frac{1}{2}\right)^8$

$= \left(\frac{1}{2}\right)^8({}^8C_6 + {}^8C_7 + {}^8C_8)$

$= \frac{1}{2^8}(28 + 8 + 1) = \frac{37}{256}$

6. (d) Let P be the probability of hitting a target.

$\therefore \quad p = \frac{3}{4}$ and $q = \frac{1}{4}$

n = Number of trials = 5

x = Number of times man hit the target

Clearly, $X \sim B(n, p) \Rightarrow X \sim B\left(5, \frac{3}{4}\right)$

$\therefore \quad P(X = x) = {}^5C_x \left(\frac{3}{4}\right)^x \left(\frac{1}{4}\right)^{5-x}$

$x = 0, 1, 2, 3, 4, 5$

Required probability $= P(X = 3) + P(X = 4) + P(X = 5)$

$$= {}^5C_3\left(\frac{3}{4}\right)^3\left(\frac{1}{4}\right)^2 + {}^5C_4\left(\frac{3}{4}\right)^4\left(\frac{1}{4}\right) + {}^5C_5\left(\frac{3}{4}\right)^5$$

$$= \frac{10 \times 27}{4^5} + \frac{5 \times 81}{4^5} + \frac{243}{4^5}$$

$$= \frac{270 + 405 + 243}{4^5}$$

$$= \frac{459}{519}$$

7. (a) Let p be the probability of getting head $= \frac{1}{2}$

q be the probability of getting tail $= \frac{1}{2}$

Let getting a head is success and getting a tail is failure.

Let n = Number of trials = 10, x = Number of successes

Clearly, $X \sim B\left(10, \frac{1}{2}\right)$

$\therefore \quad P(X = x) = {}^{10}C_x\left(\frac{1}{2}\right)^x\left(\frac{1}{2}\right)^{10-x}, \quad x = 0, 1, 2, \cdots, 10$

$$= {}^{10}C_x\left(\frac{1}{2}\right)^{10}$$

Required probability = Probability of getting 5 heads and 5 tails

$$= P(X = 5) = {}^{10}C_5\left(\frac{1}{2}\right)^{10}$$

$$= \frac{10 \times 9 \times 8 \times 7 \times 6}{5 \times 4 \times 3 \times 2 \times 1} \times \frac{1}{2^{10}}$$

$$= \frac{2 \times 9 \times 2 \times 7}{2^{10}} = \frac{63}{2^8} = \frac{63}{256}$$

8. (d) Given, probability distribution is

$X = x$	0	1	2	3
$P(X = x)$	k	$3k$	$3k$	k

As we know that, total probability of distribution is 1.

$\therefore$ We have, $k + 3k + 3k + k = 1$

$\Rightarrow \quad 8k = 1 \Rightarrow k = \frac{1}{8}$

Now, $\text{Var}(X) = E[X - E(X)]^2 = E(X^2) - [E(X)]^2 \quad \ldots(i)$

Consider, $\quad E(X^2) = \Sigma x^2 \cdot P(X = x)$

$$= 0^2 \cdot P(X = 0) + 1^2 P(X = 1) + 2^2 P(X = 2) + 3^2 P(X = 3)$$

$$= 1(3k) + 4 \cdot (3k) + 9 \cdot (k) = 24k = 24 \times \frac{1}{8} = 3$$

and $E(X) = \Sigma xP(X = x) = 0 \cdot P(X = 0)$

$$+ 1 \cdot P(X = 1) + 2 \cdot P(X = 2) + 3 \cdot P(X = 3)$$

$$= 3 \cdot k + 2 \cdot (3k) + 3 \cdot (k)$$

$$= 12k = 12 \times \frac{1}{8} = \frac{3}{2}$$

$\therefore \text{Var}(X) = 3 - \frac{9}{4} = \frac{3}{4}$

9. (c) Let p be the probability that the person kill a bird $= \frac{3}{4}$

n = Number of trials = 5

x = Number of times he kill the bird

Clearly, $\quad X \sim B\left(5, \frac{3}{4}\right)$

$\therefore \quad P(X = x) = {}^5C_x\left(\frac{3}{4}\right)^x\left(\frac{1}{4}\right)^{5-x}, \quad$ where $x = 0, 1, 2, 3, 4, 5$

Now, required probability = Probability that he will not get the success

$$P(X = 0) = {}^5C_0\left(\frac{3}{4}\right)^0\left(\frac{1}{4}\right)^{5-0}$$

$$= \frac{1}{4^5} = \frac{1}{2^{10}} = \frac{1}{1024}$$

10. (a) Let p = Probability that a student selected at random is a swimmer

$$= 1 - \frac{1}{5} = \frac{4}{5}$$

n = Number of students selected = 5

x = Number of swimmers

Clearly, $X \sim B(n, p)$, where $n = 5$ and $p = \frac{4}{5}$

$\therefore \quad P(X = x) = {}^nC_x p^x q^{n-x}, \; x = 0, 1, 2, 3, 4, 5$

Now, probability that exactly 4 students are swimmer $= P(X = 4)$

$${}^5C_4 p^4 q = {}^5C_4\left(\frac{4}{5}\right)^4\left(\frac{1}{5}\right)$$

11. (d) Let X follows binomial distribution with parameters n and p i.e. $X \sim B(n, p)$.

We have, $\quad E(X) = np = 2 \quad \ldots(i)$

and $\quad \text{Var}(X) = npq = 1 \quad \ldots(ii)$

Using Eqs. (i) and (ii), we get $q = \frac{1}{2} \Rightarrow p = \frac{1}{2}$ and $n = 4$

Thus, $\quad X \sim B\left(4, \frac{1}{2}\right)$

$\therefore \quad P(X = x) = {}^4C_x\left(\frac{1}{2}\right)^x\left(\frac{1}{2}\right)^{4-x}$

$$= {}^4C_x\left(\frac{1}{2}\right)^4, \; x = 0, 1, 2, \cdots, 4$$

Now, $\quad$ required probability $= P(X \geq 1)$

$$1 - P(X = 0) = 1 - {}^4C_0\left(\frac{1}{2}\right)^4$$

$$= 1 - \frac{1}{16} = \frac{15}{16}$$

12. (a) Let $\quad p$ = Probability of getting an even number on the die

$$= \frac{3}{6} = \frac{1}{2}$$

n = Number of times die tossed = 5

x = Number of times getting an even number

Clearly, $\quad X \sim B\left(5, \frac{1}{2}\right)$

$\therefore \quad P(X = x) = {}^5C_x\left(\frac{1}{2}\right)^x\left(\frac{1}{2}\right)^{5-x}, \; x = 0, 1, 2, 3, 4, 5$

$$= {}^5C_x\left(\frac{1}{2}\right)^5, \; x = 0, 1, 2, 3, 4, 5$$

Required probability $= P(X = 3) = {}^5C_3\left(\frac{1}{2}\right)^5$

$$= 10 \times \frac{1}{2^5} = \frac{5}{16}$$

13. (b) Let X be the number of times head occurs.

Then, X follows a binomial distribution with parameters n and $p = \frac{1}{2}$

$\therefore \quad P(X = x) = {}^nC_x\left(\frac{1}{2}\right)^x\left(\frac{1}{2}\right)^{n-x}, x = 0, 1, 2, \cdots, n$

We have, $\quad P(X = 6) = P(X = 8)$

$\Rightarrow \quad {}^nC_6\left(\frac{1}{2}\right)^6\left(\frac{1}{2}\right)^{n-6} = {}^nC_8\left(\frac{1}{2}\right)^8\left(\frac{1}{2}\right)^{n-8}$

$\Rightarrow \quad {}^nC_6\left(\frac{1}{2}\right)^n = {}^nC_8\left(\frac{1}{2}\right)^n$

$\Rightarrow \quad {}^nC_6 = {}^nC_8$

$[\because {}^nC_r = {}^nC_s \Rightarrow \text{either } r = s \text{ or } r + s = n]$

$\Rightarrow \quad n = 6 + 8$

$\therefore \quad n = 14$

14. (c) Let p be the probability of success and q be the probability of failure.

$\therefore$ We have, $\quad p = 2q$

$\because \quad p + q = 1$

$\Rightarrow \quad q = \frac{1}{3}$

$\Rightarrow \quad p = \frac{2}{3}$

Let $\quad n$ = Number of trials = 4

x = Number of successes

Clearly, $X \sim B\left(4, \frac{2}{3}\right)$

$\therefore \quad P(X = x) = {}^4C_x\left(\frac{2}{3}\right)^x\left(\frac{1}{3}\right)^{4-x}, \quad x = 0, 1, 2, 3, 4$

Required probability $= P(X = 3) + P(X = 4)$

$= {}^4C_3\left(\frac{2}{3}\right)^3\left(\frac{1}{3}\right) + {}^4C_4\left(\frac{2}{3}\right)^4\left(\frac{1}{3}\right)^0$

$= 4 \times \frac{2^3}{3^4} + \frac{2^4}{3^4}$

$= 2 \times \frac{2^4}{5^4} + \frac{2^4}{3^4}$

$= \frac{2^4}{3^4} \times 3 = \frac{16}{27}$

15. (d) Let p be the probability of getting 5 on a die $= \frac{1}{6}$

n = Number of times die thrown = 7

x = Number of times getting 5

Clearly, $\quad X \sim B\left(7, \frac{1}{6}\right)$

$\therefore \quad P(X = x) = {}^7C_x\left(\frac{1}{6}\right)^x\left(\frac{5}{6}\right)^{7-x}, x = 0, 1, 2, \cdots, 7$

Required probability $= P(X = 4) = {}^7C_4\left(\frac{1}{6}\right)^4\left(\frac{5}{6}\right)^3$

16. (d) Let p = Probability of getting head on tossing a coin

n = Number of times coin tossed = 10

x = Number of times getting head

Clearly, $\quad X \sim B\left(10, \frac{1}{2}\right)$

$\therefore \quad P(X = x) = {}^{10}C_x\left(\frac{1}{2}\right)^x\left(\frac{1}{2}\right)^{10-x}, x = 0, 1, 2, \cdots, 10$

$= {}^{10}C_x\left(\frac{1}{2}\right)^{10}, x = 0, 1, 2, \cdots, 10$

Now, P (getting an odd number of heads)

$= P(X = 1) + P(X = 3) + P(X = 5) + P(X = 7) + P(X = 9)$

$= {}^{10}C_1\left(\frac{1}{2}\right)^{10} + {}^{10}C_3\left(\frac{1}{2}\right)^{10} + {}^{10}C_5\left(\frac{1}{2}\right)^{10} + {}^{10}C_7\left(\frac{1}{2}\right)^{10} + {}^{10}C_9\left(\frac{1}{2}\right)^{10}$

$= \left(\frac{1}{2}\right)^{10}({}^{10}C_1 + {}^{10}C_3 + {}^{10}C_5 + {}^{10}C_3 + {}^{10}C_1)$

$= \frac{1}{2^{10}}(2 \cdot {}^{10}C_1 + 2 \cdot {}^{10}C_3 + {}^{10}C_5)$

$= \frac{1}{2^{10}}(20 + 240 + 252) = \frac{512}{2^{10}} = \frac{1}{2}$

17. (c) Let p = Probability of hitting a bird $= \frac{1}{4}$

q = Probability of not hitting a bird $= \frac{3}{4}$

Now, suppose man fires n times and let E be the event that target hit atleast once.

$\therefore \quad P(E) = 1 - P(\text{no success}) = 1 - {}^nC_0\left(\frac{1}{4}\right)^0\left(\frac{3}{4}\right)^n = 1 - \left(\frac{3}{4}\right)^n$

$\because \quad P(E) > \frac{2}{3}$

$\Rightarrow \quad 1 - \left(\frac{3}{4}\right)^n > \frac{2}{3} \Rightarrow \left(\frac{3}{4}\right)^n < \frac{1}{3} \Rightarrow 3^{n+1} < 4^n$

$\therefore \quad n = 4$

18. (a) Let p = Probability of getting head $= \frac{1}{2}$

q = Probability of getting tail $= \frac{1}{2}$

n = Number of times coin tossed

x = Number of times getting head

Clearly, $\quad X \sim B\left(n, \frac{1}{2}\right)$

$\therefore P(X = x) = {}^nC_x\left(\frac{1}{2}\right)^x\left(\frac{1}{2}\right)^{n-x}, \quad x = 0, 1, 2, \cdots, n$

$= {}^nC_x\left(\frac{1}{2}\right)^n$

Case I If n is even.

Required probability

$= P(X = 1) + P(X = 3) + P(X = 5) + \cdots + P(X = n - 1)$

$= {}^nC_1\left(\frac{1}{2}\right)^n + {}^nC_3\left(\frac{1}{2}\right)^n + \cdots + {}^nC_{n-1}\left(\frac{1}{2}\right)^n$

$= \left(\frac{1}{2}\right)^n({}^nC_1 + {}^nC_3 + \cdots + {}^nC_{n-1}) = \frac{1}{2^n} \times 2^{n-1} = \frac{1}{2}$

Case II If n is odd.

Required probability $= P(X = 1) + P(X = 3) + \cdots + P(X = n)$

$= {}^nC_1\left(\frac{1}{2}\right)^n + {}^nC_3\left(\frac{1}{2}\right)^n + \cdots + {}^nC_n\left(\frac{1}{2}\right)^n$

$= \left(\frac{1}{2}\right)^n({}^nC_1 + {}^nC_3 + \cdots + {}^nC_n) = \left(\frac{1}{2}\right)^n(2^{n-1}) = \frac{1}{2}$

Clearly, in both the cases, we get the probability $\frac{1}{2}$.

19. (b) Given, $E = \{x \text{ is a prime number}\} = \{2, 3, 5, 7\}$

$\therefore \quad P(E) = P(X = 2) + P(X = 3) + P(X = 5) + P(X = 7)$

$= 0.23 + 0.12 + 0.20 + 0.07 = 0.62$

and $\quad F = \{x < 4\} = \{1, 2, 3\}$

$\Rightarrow \quad P(F) = P(X = 1) + P(X = 2) + P(X = 3)$

$= 0.15 + 0.23 + 0.12 = 0.5$

and $\quad E \cap F = \{2, 3\}$

$\therefore \quad P(E \cap F) = P(X = 2) + P(X = 3) = 0.23 + 0.12 = 0.35$

Now, $\quad P(E \cup F) = P(E) + P(E) - P(E \cap F)$

$= 0.62 + 0.5 - 0.35 = 0.77$

20. (b) Let p = Probability of getting a score of 9 on throwing a pair of dice

$$= \frac{4}{36} = \frac{1}{9}$$

n = Number of times a pair of dice thrown = 3

x = Number of times getting a score of 9

Clearly, $X \sim B\left(3, \frac{1}{9}\right)$

$\therefore \quad P(X = x) = {}^3C_x\left(\frac{1}{9}\right)^x\left(\frac{8}{9}\right)^{3-x}, x = 0, 1, 2, 3$

Required probability

$$= P(X = 2) = {}^3C_2\left(\frac{1}{9}\right)^2\left(\frac{8}{9}\right)^1 = 3 \times \frac{1}{81} \times \frac{8}{9} = \frac{8}{243}$$

21. (b) We have, p = Probability of success

Let n = Number of trials = 5

q = Probability of failure

x = Number of times getting failure

Clearly, $X \sim B(5, q)$

$\therefore \quad P(X = x) = {}^5C_x q^x p^{5-x}, \; x = 0, 1, 2, \cdots, 5$

$\because \quad P(X \geq 1) \geq \frac{31}{32} \Rightarrow 1 - P(X = 0) \geq \frac{31}{32}$

$\Rightarrow \quad P(X = 0) \leq \frac{1}{32} \Rightarrow {}^5C_0 q^0 p^5 \leq \frac{1}{32}$

$\Rightarrow \quad p^5 \leq \frac{1}{32}$

$\therefore \quad p \leq \frac{1}{2}$ and $p \geq 0$

$\Rightarrow \quad p \in \left[0, \frac{1}{2}\right]$

22. (c) Let $X \sim B\left(n, p = \frac{1}{4}\right)$

$\because \quad P(x \geq 1) \geq \frac{9}{10} \Rightarrow 1 - P(X = 0) \geq \frac{9}{10}$

$\Rightarrow \quad P(X = 0) \leq \frac{1}{10} \Rightarrow {}^nC_0 p^0 \cdot q^{n-0} \leq \frac{1}{10}$

$\Rightarrow \quad \left(\frac{3}{4}\right)^n \leq \frac{1}{10}$

$\Rightarrow \quad \left(\frac{4}{3}\right)^n \geq 10$

$\Rightarrow \quad n[\log_{10} 4 - \log_{10} 3] \geq \log_{10} 10 = 1$

$\therefore \quad n \geq \frac{1}{\log_{10} 4 - \log_{10} 3}$

23. (a) Clearly, in a single throw of a dice, probability of getting 1 is $\frac{1}{6}$ and probability of not getting 1 is $\frac{5}{6}$.

Then, getting 1 in even number of trials is getting 1 i.e. in 2nd trial or in 4th or in 6th trial and so on.

Required probability $= \frac{5}{6} \times \frac{1}{6} + \left(\frac{5}{6}\right)^3 \times \frac{1}{6} + \left(\frac{5}{6}\right)^5 \times \frac{1}{6} + \cdots \infty$

$$= \frac{5}{36}\left[1 + \frac{25}{36} + \left(\frac{25}{36}\right)^2 + \cdots\right] = \frac{5}{36}\left[\frac{1}{1 - \frac{25}{36}}\right] = \frac{5}{36} \times \frac{36}{11} = \frac{5}{11}$$

24. (b)

(i) $P(\text{success}) = \frac{1}{2}; P(\text{failure}) = \frac{1}{2}$

Now, suppose n bombs are to dropped. Let E be the event that the bridge is destroyed.

$$P(E) = 1 - P(0 \text{ or } 1 \text{ success}) = 1 - \left[{}^nC_0\left(\frac{1}{2}\right)^n + {}^nC_1\left(\frac{1}{2}\right)\left(\frac{1}{2}\right)^{n-1}\right]$$

$$= 1 - \left[\left(\frac{1}{2}\right)^n + \frac{n}{2^n}\right]$$

$\because \quad P(E) > 0.9$

$\Rightarrow \quad 1 - \left[\left(\frac{1}{2}\right)^n + \frac{n}{2^n}\right] > \frac{9}{10}$

$\Rightarrow \quad \frac{1}{10} > \frac{n+1}{2^n} \Rightarrow \frac{2^n}{n+1} > 10$

$\therefore \quad n = 7$

(i)→ (s)

(ii) We have a bag containing 2 red, 3 white, and 5 black balls. Let n number of times a ball is drawn and getting a red ball is success.

$\therefore \quad P(\text{success}) = \frac{2}{10} = \frac{1}{5}; P(\text{failure}) = \frac{4}{5}$

Let E be the event of getting a red ball for first time.

$$P(E) = P(S \text{ or } FS \text{ or } FFS \text{ or } \cdots) \geq \frac{1}{2}$$

$\Rightarrow P(S) + P(F)P(S) + P(F)P(F)P(S) + \cdots \geq \frac{1}{2}$

$\Rightarrow \quad P(S)\left(\frac{1 - P(F)^n}{1 - P(F)}\right) \geq \frac{1}{2}$

$\Rightarrow \quad 1 - [P(F)]^n \geq \frac{1}{2}$

$\Rightarrow \quad [P(F)]^n \leq \frac{1}{2}$

$\Rightarrow \quad \left(\frac{4}{5}\right)^n \leq \frac{1}{2}$

$\therefore \quad n = 4$

(ii)→ (p)

(iii) B_1 : Animal hides in Bush I

B_2 : Animal hides in Bush II

$$P(B_1) = \frac{4}{5}; P(B_2) = \frac{1}{5}; P(H) = \frac{1}{2}$$

Let x bullets → Bush I

and $(10 - x)$ bullets → Bush II

To maximise probability

$$= \frac{4}{5}\left\{1 - \left(\frac{1}{2}\right)^x\right\} + \frac{1}{5}\left\{1 - \left(\frac{1}{2}\right)^{10-x}\right\}$$

$$\frac{dP}{dx} = -\frac{4}{5}\left(\frac{1}{2}\right)^x \log\left(\frac{1}{2}\right) + \frac{1}{5}\left(\frac{1}{2}\right)^{10-x} \log\left(\frac{1}{2}\right) = 0$$

$\Rightarrow \quad x = 6$

(iii) → (r)

25. (b)

(i) Given S_2 reaches the semifinal, hence all other players $1 + (2^n - 1)$ are equally likely to wins the final with probability p.

$\therefore \quad (2^n - 1)p + \frac{1}{4} = 1$

$(2^n - 1)p = \frac{3}{4}$

$\Rightarrow \quad p = \frac{3}{4(2^n - 1)}$

If $p = \frac{1}{20}$

$\Rightarrow \quad 60 = 4(2^n - 1)$

$\Rightarrow \quad 2^n = 16 \Rightarrow n = 4$

Alternate Method

Number of ways in which S_2 and 3 other players out of $2^n - 1 = \lambda$ say can be taken for semifinals $= {}^{\lambda}C_3$

Number of ways in which S_1, S_2 and 2 others out of $(\lambda - 1)$ can be taken for semifinal $= {}^{\lambda-1}C_2$

$\therefore$ $P(S_1, S_2$ and two others reach semifinal)

$$= \frac{{}^{\lambda-1}C_2}{{}^{\lambda}C_3} = \frac{3}{\lambda} = \frac{3}{2^n - 1}$$

Now, S_1, S_2 and two others reach the semifinal with probability $= \dfrac{3}{2^n - 1}$

Probability (S_1 wins the tournament)

$$= \frac{3}{2^n - 1} \cdot \frac{1}{2} \cdot \frac{1}{2} = \frac{3}{4(2^n - 1)} = \frac{1}{20}$$

$$\Rightarrow \quad n = 4$$

(i) → (q)

(ii) Let n number of chances be allowed. The total number of ways of ticking one or more alternatives out of 4 is

$${}^4C_1 + {}^4C_2 + {}^4C_3 + {}^4C_4 = 15$$

Out of these 15 combinations, only one is correct. The probability of ticking the alternatives correct at the first trial $= \dfrac{1}{15}$, that at the second trial is $\dfrac{14}{15} \cdot \dfrac{1}{14} = \dfrac{1}{15}$ and that at the third trial $= \dfrac{14}{15} \cdot \dfrac{13}{14} \cdot \dfrac{1}{13} = \dfrac{1}{15}$ and so on.

Thus, the probability that the candidate will get makrs on the questions, if he is allowed n trials is

$$\underbrace{\frac{1}{15} + \frac{1}{15} + \frac{1}{15} + \cdots + \frac{1}{15}}_{n \text{ times}} = \frac{1}{15} \times n = \frac{n}{15} \geq \frac{1}{3}$$

$$\Rightarrow \quad n \geq 5$$

$$\therefore \quad n = 5$$

(ii) → (r)

(iii) We know that, in a pack of 52 cards, there are 12 face cards and 4 aces.

Now, number of ways of selecting 20 cards from 40 cards $= {}^{40}C_{20}$

and number of ways of selecting 2 cards from selected 20 cards $= {}^{20}C_2$

$\therefore$ Total number of selecting two cards $= {}^{40}C_{20} \times {}^{20}C_2$

Now, number of ways of selecting 20 cards and two aces

$$= {}^4C_2 \cdot {}^{38}C_{20}$$

$$\text{Required probability} = \frac{{}^4C_2 \cdot {}^{38}C_{20}}{{}^{40}C_{20} \times {}^{20}C_2}$$

$$\therefore \quad p = {}^4C_2 = 6$$

(iii) → (s)

(iv) Let p = Probability of getting head on tossing a coin $= \dfrac{1}{2}$

n = Number of tosses

x = Number of times getting head

Clearly, $\quad X \sim B\left(n, \dfrac{1}{2}\right)$

$$\therefore \quad P(X = x) = {}^nC_x\left(\frac{1}{2}\right)^x\left(\frac{1}{2}\right)^{n-x}, \; x = 0, 1, 2, \cdots, n$$

Now, $P(E_1) = P$ (that both heads and tails are present in n tosses)

$$= P(X = 1) + P(X = 2) + \cdots + P(X = n - 1)$$

$$= 1 - [P(X = 0) + P(X = n)]$$

$$= 1 - \left[{}^nC_0\left(\frac{1}{2}\right)^n + {}^nC_n\left(\frac{1}{2}\right)^n\right] = 1 - 2 \cdot \left(\frac{1}{2}\right)^n$$

$P(E_2) = P$ (that the coin shows up heads atmost once)

$$= P(X = 0) + P(X = 1) = {}^nC_0\left(\frac{1}{2}\right)^n + {}^nC_1\left(\frac{1}{2}\right)^n$$

$$= \left(\frac{1}{2}\right)^n + n\left(\frac{1}{2}\right)^n = \left(\frac{1}{2}\right)^n (n + 1)$$

$P(E_1 \cap E_2) = P$ (both heads and tails are present in n tosses and heads present atmost once)

$$= P(X = 1) = {}^nC_1\left(\frac{1}{2}\right)^n = n\left(\frac{1}{2}\right)^n$$

$\therefore$ E_1 and E_2 are independent.

Now, we have

$$P(E_1 \cap E_2) = P(E_1) \cdot P(E_2)$$

$$\Rightarrow \quad n\left(\frac{1}{2}\right)^n = \left[1 - 2 \cdot \left(\frac{1}{2}\right)^n\right]\left(\frac{1}{2}\right)^n (n + 1)$$

$$\Rightarrow \quad \frac{n}{n + 1} = 1 - \frac{1}{2^{n-1}}$$

$$\Rightarrow \quad \frac{1}{2^{n-1}} = 1 - \frac{n}{n + 1}$$

$$\therefore \quad n = 3$$

(iv) → (p)

26. (b) X can win after the $(n + 1)$th game in the following two mutually exclusive ways.

(i) X wins exactly one of the first n games, draw $(n - 1)$ games and wins the $(n + 1)$ th game.

(ii) X losses exactly one of the first n games, wins exactly one of the first n games and draws $(n - 2)$ games and wins the $(n + 1)$ th game.

For (i), the probability is $({}^nP_1 ab^{n-1})a$ and for (ii), the probability is $({}^nP_2 acb^{n-2})a$.

$\therefore$ Required probability $(P_n) = {}^nP_1 a^2b^{n-1} + {}^nP_2 a^2cb^{n-2}$

$$= na^2b^{n-1} + n(n - 1)a^2cb^{n-2}$$

$$= a^2[nb^{n-1} + n(n - 1)b^{n-2}c]$$

27. (a) Clearly, on substituting $n = 3$ and interchanging a and c, we can get the required probability.

Required probability $= c^2(3b^{3-1} + 3(3 - 1)b^{3-2}a) = 3bc^2(b + 2a)$

28. (b) X will win the match with probability

$$\sum_{n=1}^{\infty} P_n = \sum_{n=1}^{\infty} a^2(nb^{n-1} + n(n - 1)b^{n-2}c)$$

$$= a^2 \sum_{n=1}^{\infty} nb^{n-1} + a^2c \sum_{n=1}^{\infty} n(n - 1)b^{n-2}$$

$$= a^2(1 + 2b + 3b^2 + \cdots) + a^2c \, (2 \cdot 1 + 3 \cdot 2b + 4 \cdot 3b^2 + \cdots)$$

$$= a^2(1 - b)^{-2} + 2a^2c(1 + 3b + 6b^2 + \cdots)$$

$$= a^2(1 - b)^{-2} + 2a^2c \, (1 - b)^{-3}$$

$$= \frac{a^2}{(1 - b)^2} + \frac{2a^2c}{(1 - b)^3} = \frac{a^2(1 - b) + 2a^2c}{(1 - b)^3}$$

$$= \frac{a^2(a + c) + 2a^2c}{(a + c)^3} \qquad [\because a + b + c = 1]$$

$$= \frac{a^3 + 3a^2c}{(a + c)^3}$$

Explanations for Q. Nos. (29 - 31)

We have, p = Probability of getting head on tossing a coin

X = Number of tosses before the appearence of the first head

Clearly, $X \geq 0$ and

$P(X = n)$ = Probability of the first head appearing at $(n + 1)$th trial

$$= (1 - p)^n p \qquad [\because \text{trials are independent}]$$

29. (a) $P(X = 5) = (1 - p)^5 \cdot p$

30. (d) $P(3 \le X \le 6) = P(X = 3) + P(X = 4) + P(X = 5) + P(X = 6)$

$$= (1-p)^3 p + (1-p)^4 p + (1-p)^5 p + (1-p)^6 p$$
$$= (1-p)^3 p[1 + (1-p) + (1-p)^2 + (1-p)^3)]$$
$$= (1-p)^3 p\left[\frac{1-(1-p)^4}{1-(1-p)}\right] \qquad [\because 1-p \ne 1]$$
$$= (1-p)^3 p\left[\frac{1-(1-p)^4}{p}\right] = (1-p)^3 - (1-p)^7$$

31. (a) $P(X \ge 4) = 1 - P(X < 4)$

$$= 1 - [P(X=0) + P(X=1) + P(X=2) + P(X=3)]$$
$$= 1 - [p + (1-p)p + (1-p)^2 p + (1-p)^3 p]$$
$$= 1 - p[1 + (1-p) + (1-p)^2 + (1-p)^3]$$
$$= 1 - p\left[\frac{1-(1-p)^4}{1-(1-p)}\right] \qquad [\because 1-p \ne 1]$$
$$= 1 - p\left[\frac{1-(1-p)^4}{p}\right] = 1 - [1-(1-p)^4] = (1-p)^4$$

32. (6) We have,

p = Probability of getting head on flipping a coin

$(1-p)$ = Probability of getting tail on flipping a coin

Let n = Number of times coin flipped = 8

x = Number of times getting head

Clearly $X \sim B(8, p)$

$\therefore \quad P(X = x) = {}^8C_x\, p^x (1-p)^{8-x},\ x = 0, 1, 2, \cdots, 8$

Now, we have

$P(\text{getting 3 head and five tails}) = \frac{1}{25} P(\text{getting 5 heads and 3 tails})$

$$\Rightarrow \quad P(X=3) = \frac{1}{25} P(X=5)$$
$$\Rightarrow \quad {}^8C_3\, p^3 (1-p)^5 = \frac{1}{25}\, {}^8C_5 p^5 (1-p)^{8-5}$$
$$\Rightarrow \quad {}^8C_3\, p^3 (1-p)^5 = \frac{1}{25}\, {}^8C_5\, p^5 (1-p)^3$$
$$\Rightarrow \quad (1-p)^2 = \frac{1}{25} p^2$$
$$\Rightarrow \quad \left(\frac{p}{1-p}\right)^2 = (5)^2 \quad \Rightarrow \quad \frac{p}{1-p} = 5$$
$$\Rightarrow \quad p = \frac{5}{6} = \frac{m}{n}$$
$$\therefore \quad m + n - 5 = 5 + 6 - 5 = 6$$

33. (7) Let p = Probability of hitting a target = $\frac{1}{5}$

Now, suppose man fire n times. Let E be the event that target is hit.

$\therefore \quad P(E) = 1 - P\text{ (no success)} = 1 - {}^nC_0\, p^0 q^n = 1 - \left(\frac{4}{5}\right)^n$

$$\because \quad P(E) > \frac{3}{4} \quad \Rightarrow \quad 1 - \left(\frac{4}{5}\right)^n > \frac{3}{4}$$
$$\Rightarrow \quad 1 - \frac{3}{4} > \left(\frac{4}{5}\right)^n \quad \Rightarrow \quad \left(\frac{4}{5}\right)^n < \frac{1}{4}$$
$$\Rightarrow \quad 4^{n+1} < 5^n$$
$$\therefore \quad n = 7$$

34. (7) Clearly, X is a binomial variate with parameters n and $p = \frac{1}{2}$ such that

$$P(X = x) = {}^nC_x\, p^x q^{n-x}$$
$$= {}^nC_x \left(\frac{1}{2}\right)^x \left(\frac{1}{2}\right)^{n-x} = {}^nC_x \left(\frac{1}{2}\right)^n$$

Now, $P(x=4), P(x=5)$ and $P(x=6)$ are in an AP.

$$\Rightarrow \quad 2P(x=5) = P(x=4) + P(x=6)$$
$$\Rightarrow \quad 2 \cdot {}^nC_5 \left(\frac{1}{2}\right)^n = {}^nC_4 \left(\frac{1}{2}\right)^n + {}^nC_6 \left(\frac{1}{2}\right)^n$$
$$\Rightarrow \quad 2 \cdot {}^nC_5 = {}^nC_4 + {}^nC_6$$
$$\Rightarrow \quad 2 \cdot \frac{n!}{5!\,(n-5)!} = \frac{n!}{4!\,(n-4)!} + \frac{n!}{6!\,(n-6)!}$$
$$\Rightarrow \quad \frac{2}{5(n-5)} = \frac{1}{(n-4)(n-5)} + \frac{1}{6 \cdot 5}$$
$$\Rightarrow \quad n^2 - 21n + 98 = 0$$
$$\Rightarrow \quad (n-7)(n-14) = 0$$
$$\Rightarrow \quad n = 7 \text{ or } 14$$
$$\therefore \quad n = 7$$

35. (9) Let E_1 be the event that coin A is selected,

E_2 be the event that coin B is selected,

E_3 be the event that coin C is selected,

and F be the event of getting 2 heads and one tail on tossing coin three times.

$$\therefore \quad P(E_1) = P(E_2) = P(E_3) = \frac{1}{3}$$
$$P(F/E_1) = {}^3C_2 \left(\frac{1}{2}\right)^2 \cdot \frac{1}{2} = 3 \cdot \left(\frac{1}{2}\right)^3 = \frac{3}{8}$$
$$P(F/E_2) = {}^3C_2 \left(\frac{2}{3}\right)^2 \left(\frac{1}{3}\right) = 3 \cdot \frac{4}{27} = \frac{4}{9}$$
$$P(F/E_3) = {}^3C_2 \left(\frac{1}{3}\right)^2 \left(\frac{2}{3}\right) = 3 \cdot \frac{2}{27} = \frac{2}{9}$$

Now, it is given that $P(E_1/F) = \frac{k}{25}$

$$\Rightarrow \quad \frac{P(E_1) \cdot P(F/E_1)}{\left[\begin{array}{c}P(E_1) \cdot P(F/E_1) + P(E_2) \cdot P(F/E_2) \\ + P(E_3) P(F/E_3)\end{array}\right]} = \frac{k}{25}$$
$$\Rightarrow \quad \frac{\frac{1}{3} \cdot \frac{3}{8}}{\frac{1}{3} \cdot \frac{3}{8} + \frac{1}{3} \cdot \frac{4}{9} + \frac{1}{3} \cdot \frac{2}{9}} = \frac{k}{25}$$
$$\Rightarrow \quad \frac{\frac{3}{8}}{\frac{3}{8} + \frac{4}{9} + \frac{2}{9}} = \frac{k}{25}$$
$$\Rightarrow \quad \frac{\frac{3}{8}}{\frac{27 + 48}{72}} = \frac{k}{25}$$
$$\Rightarrow \quad \frac{3}{8} \times \frac{72}{75} = \frac{k}{25}$$
$$\Rightarrow \quad \frac{9}{25} = \frac{k}{25}$$
$$\therefore \quad k = 9$$

DPP-8 **Probability Through Statistical** (Stochatic) **Tree Diagram**

1.

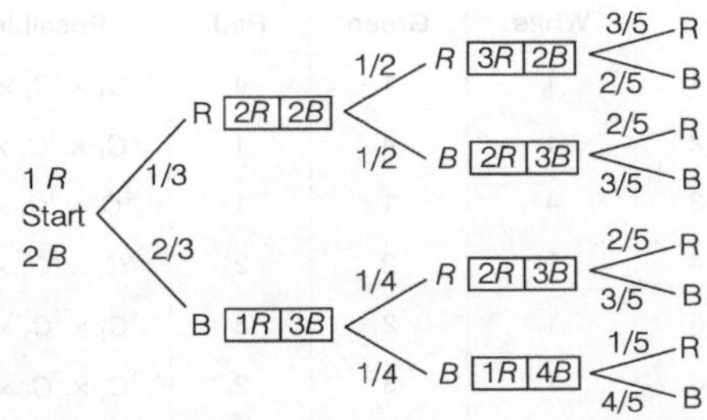

(i) $P(A) = 1 - P(RRR) = 1 - \frac{1}{3}\cdot\frac{1}{2}\cdot\frac{3}{5} = 1 - \frac{1}{10} = \frac{9}{10}$

(ii) $P(B) = \frac{1}{3}\cdot\frac{1}{2}\cdot\frac{2}{5} + \frac{1}{3}\cdot\frac{1}{2}\cdot\frac{2}{5} + \frac{2}{3}\cdot\frac{1}{4}\cdot\frac{2}{5} = \frac{3}{15} = \frac{1}{5}$

(iii) $P(C) = P\left(\frac{RRR}{(RRR \cup BBB)}\right) = \frac{P(RRR)}{P(RRR) + P(BBB)}$

$$= \frac{\frac{1}{3}\cdot\frac{1}{2}\cdot\frac{3}{5}}{\frac{1}{3}\cdot\frac{1}{2}\cdot\frac{3}{5} + \frac{2}{3}\cdot\frac{3}{4}\cdot\frac{4}{5}} = \frac{\frac{1}{10}}{\frac{1}{10} + \frac{4}{10}} = \frac{1}{5}$$

2. We have, three coins in which two of them are fair and one is biased.

$\therefore \quad P(H_1) = \frac{1}{2}, P(H_2) = \frac{1}{2}$ and $P(H_3) = \frac{3}{4}$

$P(T_1) = \frac{1}{2}, P(T_2) = \frac{1}{2}$ and $P(T_3) = \frac{1}{4}$

Required probability $= P(H_1 H_2 T_3) + P(H_1 T_2 H_3) + P(T_1 H_2 H_3)$

$$= \frac{1}{2}\times\frac{1}{2}\times\frac{1}{4} + \frac{1}{2}\times\frac{1}{2}\times\frac{3}{4} + \frac{1}{2}\times\frac{1}{2}\times\frac{3}{4} = \frac{7}{16}$$

3. Let E_1, E_2 and E_3 be the right answers of the given three questions, respectively.

$\therefore \quad P(E_1) = \frac{1}{5}, \; P(\bar{E}_1) = \frac{4}{5}$

$P(E_2) = \frac{1}{5}, \; P(\bar{E}_2) = \frac{4}{5}$

$P(E_3) = \frac{1}{4}, \; P(\bar{E}_3) = \frac{3}{4}$

(i) Required probability

$P(E_1 \bar{E}_2 \bar{E}_3) + P(\bar{E}_1 E_2 \bar{E}_3) + P(\bar{E}_1 \bar{E}_2 E_3)$

$$= \frac{1}{5}\times\frac{4}{5}\times\frac{3}{4} + \frac{4}{5}\times\frac{1}{5}\times\frac{3}{4} + \frac{4}{5}\times\frac{4}{5}\times\frac{1}{4}$$

$$= \frac{12 + 12 + 16}{100} = \frac{40}{100} = \frac{2}{5}$$

(ii) Required probability $= 1 - P(\bar{E}_1 \bar{E}_2 \bar{E}_3)$

$$= 1 - \frac{4}{5}\times\frac{4}{5}\times\frac{3}{4} = 1 - \frac{48}{100} = \frac{52}{100} = \frac{13}{25}$$

4. (a) Given, families have one, two or three children.

A family have four grand children is possible, if they have two or three children.

Now, two children have two possible cases (2, 2), (1, 3) and (3, 1) and three children have three possible cases (1, 1, 2), (1, 2, 1) and (2, 1, 1).

Required probability $= \frac{1}{2}\times\frac{1}{2}\times\frac{1}{2} + \frac{1}{2}\times\frac{1}{4}\times\frac{1}{4} + \frac{1}{2}\times\frac{1}{4}\times\frac{1}{4} +$

$\frac{1}{4}\times\frac{1}{4}\times\frac{1}{4}\times\frac{1}{2} + \frac{1}{4}\times\frac{1}{4}\times\frac{1}{2}\times\frac{1}{4} + \frac{1}{4}\times\frac{1}{2}\times\frac{1}{4}\times\frac{1}{4}$

$$= \frac{1}{8} + \frac{1}{32} + \frac{1}{32} + \frac{1}{128} + \frac{1}{128} + \frac{1}{128}$$

$$= \frac{16 + 4 + 4 + 1 + 1 + 1}{128} = \frac{27}{128}$$

5. (b)

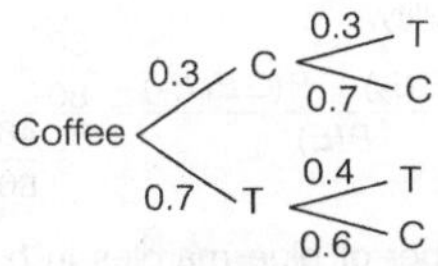

Monday , Tuesday, Wednesday

$P(E) = CCT$ or CTT

$= (0.3)(0.7) + (0.7)(0.4)$

$= 0.21 + 0.28$

$= 0.49$

6. (c) If S_1 and S_2 are in the same pair, then exactly onewins. The number of ways 16 players are divided into 8 pairs is

$n(S) = \frac{16!}{(2!)^8 \times 8!}$

The number of ways in which S_1 and S_2 are in same pairs,

$n(E) = \frac{14!}{(2!)^7 \times 7!}$

$\therefore$ The probability of S_1 and S_2 in the same pairs $= \frac{\frac{14!}{(2!)^7 \times 7!}}{\frac{16!}{(2!)^8 \times 8}} = \frac{1}{15}$

The probability of anyone winning in the pair of S_1 S_2 is P (certain event) = 1

Hence, the probability that the pair of S_1, S_2 being in two pairs separately and anyone of S_1, S_2 wins is given by the probability S_1, S_2 being in two pairs separately and S_1 wins, S_2 loses + the probability of S_1, S_2 being in two pairs separately and S_1 loses, S_2 wins, it is given by

$$\left(1 - \frac{1}{15}\right)\times\frac{1}{2}\times\frac{1}{2} + \left(1 - \frac{1}{15}\right)\times\frac{1}{2}\times\frac{1}{2} = \frac{7}{15}$$

Required probability $= \frac{1}{15} + \frac{7}{15} = \frac{8}{15}$

7. (d)

3^a **ends in** → 7^b **ends in** ↓	1	3	7	9
1			8	
3				
7	8			
9				8

Out of 16 cases, 3 are favourable.

$\Rightarrow \quad P = \frac{3}{16}$

8. (c) Consider the events E_1, E_2, E_i and G.

E_1 = A receives the signal correct

E_2 = B receives the signal correct

E_i = Signal received by B is green

G = Original signal is green

$P(E_1) = \frac{3}{4}, P(E_2) = \frac{3}{4}$

$P(\bar{E}_1) = \frac{1}{4}, \; P(\bar{E}_2) = \frac{1}{4}$

$P(G) = \frac{4}{5}, \; P(\bar{G}) = \frac{1}{5}$

$P(E) = P(G E_1 E_2) + P(G \bar{E}_1 \bar{E}_2) + P(\bar{G} E_1 \bar{E}_2) + P(\bar{G} \bar{E}_1 E_2)$

$$P(E) = \frac{4}{5} \times \frac{3}{4} \times \frac{3}{4} + \frac{4}{5} \times \frac{1}{4} \times \frac{1}{4} + \frac{1}{5} \times \frac{3}{4} \times \frac{1}{4} + \frac{1}{5} \times \frac{1}{4} \times \frac{3}{4}$$
$$= \frac{36 + 4 + 3 + 3}{80} = \frac{46}{80}$$

Required probability,

$$P(G/E) = \frac{P(GE_1E_2) + P(G\bar{E}_1\bar{E}_2)}{P(E)} = \frac{\frac{36}{80} + \frac{4}{80}}{\frac{46}{80}} = \frac{20}{23}$$

9. (b,c,d) Let number of blue marbles in b and number of green marbles is g.

Hence $\dfrac{bg}{{}^{b+g}C_2} = \dfrac{1}{2}$

$\Rightarrow (b + g)(g + b - 1) = 4bg$

$\Rightarrow (b + g)^2 - (b + g) = 4bg$

$\Rightarrow b^2 + g^2 + 2bg - b - g = 4bg$

$\Rightarrow g^2 - 2bg - g + b^2 - b = 0$

$\Rightarrow g^2 - (2b + 1)g + (b^2 - b) = 0$

$D = (2b + 1)^2 - 4(b^2 - b)$

$= (8b + 1)$ must be a perfect square

Hence, possible values of b are 3,6,10.

10. (b) (i) We have, $\{D, 1, 2, 3, 4, \cdots, 10\}$

Possible cases when sum and positive difference, both are multiples of 4 are (2, 6) (2, 10) (4, 8) (6, 10)

$\therefore$ Required probability $= \dfrac{6}{{}^{11}C_2} = \dfrac{6}{55}$

Hence, $x = 6$

$\therefore$ **(i) → (q)**

(ii) Let the number of green socks be n.

$\therefore$ Total number of socks are

$(2 + 2 + 2 + n) = n + 6$

Probability of same colours when two socks are taken

$\Rightarrow \dfrac{{}^2C_2}{{}^{(n+6)}C_2} + \dfrac{{}^2C_2}{{}^{(n+6)}C_2} + \dfrac{{}^2C_2}{{}^{(n+6)}C_2} + \dfrac{{}^nC_2}{{}^{(n+6)}C_2} = \dfrac{1}{5}$

$\Rightarrow \dfrac{3 \times 2}{(n + 6)(n + 5)} + \dfrac{n(n - 1)}{(n + 6)(n + 5)} = \dfrac{1}{5}$

$\Rightarrow \dfrac{6 + n^2 - n}{n^2 + 11n + 30} = \dfrac{1}{5}$

$\Rightarrow 30 + 5n^2 - 5n = n^2 + 11n + 30$

$\Rightarrow 4n^2 = 16n \Rightarrow n = 4$

Hence, The number of green socks = 4

$\therefore$ **(ii) → (p)**

(iii) Let there be x red socks and y blue socks and $x > y$, then

$\dfrac{{}^xC_2 + {}^yC_2}{{}^{x+y}C_2} = \dfrac{1}{2}$

$\Rightarrow \dfrac{x(x - 1) + y(y - 1)}{(x + y)(x + y - 1)} = \dfrac{1}{2}$

$\Rightarrow 2x^2 - 2x + 2y^2 - 2y = x^2 + 2xy + y^2 - x - y$

$\Rightarrow x^2 - 2xy + y^2 = x + y$

$\Rightarrow (x - y)^2 = x + y$

$\Rightarrow |x - y| = x + y$

Now, $x + y \le 17$

$\Rightarrow x - y \le 17$

As, $(x - y)$ must be integer.

So, $x - y = 4$

$\Rightarrow x + y = 16$

Hence, Maximum number of red socks = 10

$\therefore$ **(iii) → (s)**

11. (c) We have, 6 balls of 3 different colours, white, green and red balls in which atleast one ball of each different colour. There are following possible cases :

	White	Green	Red	Possible way
1	1	1	4	${}^1C_1 \times {}^1C_1 \times {}^4C_1 = 4$
2	1	4	1	${}^1C_1 \times {}^4C_1 \times {}^1C_1 = 4$
3	4	1	1	${}^4C_1 \times {}^1C_1 \times {}^1C_1 = 4$
4	2	2	2	${}^2C_1 \times {}^2C_1 \times {}^2C_1 = 8$
5	1	2	3	${}^1C_1 \times {}^2C_1 \times {}^3C_1 = 6$
6	1	3	2	${}^1C_1 \times {}^3C_1 \times {}^2C_1 = 6$
7	3	1	2	${}^3C_1 \times {}^1C_1 \times {}^2C_1 = 6$
8	2	1	3	${}^2C_1 \times {}^1C_1 \times {}^3C_1 = 6$
9	2	3	1	${}^2C_1 \times {}^3C_1 \times {}^1C_1 = 6$
10	3	2	1	${}^3C_1 \times {}^2C_1 \times {}^1C_1 = 6$

The probability that the bag contains 2 balls of each colour = 1/10

12. (a) Required probability

$$= \frac{{}^1C_1 \times {}^1C_1 \times {}^4C_1}{4 + 4 + 4 + 8 + 6 + 6 + 6 + 6 + 6 + 6}$$
$$= \frac{4}{56} = \frac{1}{14}$$

13. (b) Required probability $= \dfrac{{}^1C_1 \times {}^1C_1 \times {}^4C_1 + {}^2C_1 \times {}^2C_1 \times {}^2C_1}{56}$

$$= \frac{4 + 8}{56} = \frac{12}{56} = \frac{3}{14}$$

14. (a) Let X be the sum of the numbers appearing on the faces of dice and $P(X)$ be the probability.

X	2	3	4	5	6	7	8	9	10	11	12
$P(X)$	$\frac{1}{36}$	$\frac{2}{36}$	$\frac{3}{36}$	$\frac{4}{36}$	$\frac{5}{36}$	$\frac{6}{36}$	$\frac{5}{36}$	$\frac{4}{36}$	$\frac{3}{36}$	$\frac{2}{36}$	$\frac{1}{36}$

(a) $P(A_3) = P(X = 3) + P(X = 6) + P(X = 9) + P(X = 12) = \dfrac{12}{36}$

(b) $P(A_4) = P(X = 4) + P(X = 8) + P(X = 12) = \dfrac{9}{36}$

(c) $P(A_5) = P(X = 5) + P(X = 10) = \dfrac{7}{36}$

(d) $P(A_6) = P(X = 6) + P(X = 12) = \dfrac{6}{36}$

Therefore, most probable is $P(A_3)$.

15. (c)

(a) $P(A_3) = \dfrac{12}{36}, P(A_4) = \dfrac{9}{36}, P(A_3 \cap A_4) = \dfrac{1}{36}$

$P(A_3 \cap A_4) \ne P(A_3) \times P(A_4)$

(b) $P(A_4) = \dfrac{9}{36}, P(A_6) = \dfrac{6}{36}, P(A_4 \cap A_6) = \dfrac{1}{36}$

$P(A_4 \cap A_6) \ne P(A_4) \times P(A_6)$

(c) $P(A_2) = \dfrac{18}{36}, P(A_3) = \dfrac{12}{36}, P(A_2 \cap A_3) = \dfrac{6}{36}$

$P(A_2 \cap A_3) = P(A_2) \times P(A_3)$

(d) $P(A_4) = \dfrac{9}{36}, P(A_2) = \dfrac{18}{36}, P(A_4 \cap A_2) = \dfrac{9}{36}$

$P(A_4 \cap A_2) \ne P(A_4) \times P(A_2)$

Hence, (c) is independent event.

16. (d) Number of possible pairs are 25.

Revisal Problems (JEE Main)

1. (c) Clearly, probability of chosen any month out of 12 month $= \frac{1}{12}$

There are 7 possible ways in which the month can start and it will be a Friday on 13th day, if the first day of the month is Sunday.

So, its probability $= \frac{1}{7}$

Thus, required probability $= \frac{1}{12} \times \frac{1}{7} = \frac{1}{84}$

Hence, the probability that the 13th day of a randomly selected month is Friday $= \frac{1}{84}$

2. (c) Total number of ways of putting 4 letters in 4 envelopes = 4!

Number of ways of putting in right envelopes = 1

Now, required probability = 1 − P (all of them are in correct envelopes)

$$= 1 - \frac{1}{4!} = 1 - \frac{1}{24} = \frac{23}{24}$$

3. (d) We have,

$$P(A) = P\left(\frac{A}{B}\right) = \frac{1}{4} \text{ and } P\left(\frac{B}{A}\right) = \frac{1}{2}$$

$\because \quad P(A) = P\left(\frac{A}{B}\right) = \frac{P(A \cap B)}{P(B)}$

$\Rightarrow \quad P(A \cap B) = P(A) \cdot P(B)$

$\Rightarrow$ A and B are independent events.

Consider

$$P\cdot\left(\frac{A'}{B}\right) = \frac{P(A' \cap B)}{P(B)} = \frac{P(B) - P(A \cap B)}{P(B)}$$

$$= 1 - \frac{P(A \cap B)}{P(B)}$$

$$= 1 - P\left(\frac{A}{B}\right) = 1 - \frac{1}{4} = \frac{3}{4}$$

and $\quad P\left(\frac{B'}{A'}\right) = \frac{P(B' \cap A')}{P(A')} = \frac{P(A \cup B)'}{P(A')} = \frac{1 - P(A \cup B)}{1 - P(A)} \quad \ldots(i)$

Clearly, $\quad P(A) = \frac{1}{4}, P(B) = P\left(\frac{B}{A}\right) = \frac{1}{2}$

and $\quad P(A \cap B) = P(A) \cdot P(B) = \frac{1}{4} \times \frac{1}{2} = \frac{1}{8}$

From Eq. (i), we have

$$P\left(\frac{B'}{A'}\right) = \frac{1 - [P(A) + P(B) - P(A \cap B)]}{1 - P(A)}$$

$$= \frac{1 - \left(\frac{1}{4} + \frac{1}{2} - \frac{1}{8}\right)}{1 - \frac{1}{4}} = \frac{1 - \left(\frac{5}{8}\right)}{1 - \frac{1}{4}} = \frac{\frac{3}{8}}{\frac{3}{4}} = \frac{1}{2}$$

Alternate Method

$\because \quad P(A) = P\left(\frac{A}{B}\right) = \frac{P(A \cap B)}{P(B)}$

$\Rightarrow \quad P(A \cap B) = P(A) \cdot P(B)$

$\Rightarrow$ A and B are independent events.

$\Rightarrow$ A' and B are independent events.

$\Rightarrow$ A' and B' are independent events

$\Rightarrow \quad P\left(\frac{A'}{B}\right) = P(A') = 1 - P(A)$

$$= 1 - \frac{1}{4} = \frac{3}{4}$$

and $\quad P\left(\frac{B'}{A'}\right) = P(B') = 1 - P(B) = 1 - P\left(\frac{B}{A}\right)$

$$= 1 - \frac{1}{2} = \frac{1}{2}$$

4. (c) There are 366 days in a leap year, in which 52 weeks and two days present. The possibilities for last two days are (Sunday, Monday), (Monday, Tuesday), (Tuesday, Wednesday), (Wednesday, Thursday), (Thursday, Friday), (Friday, Saturday), (Saturday, Sunday).

$$P\text{ (53 Sunday)} = \frac{2}{7}$$

$$P\text{ (53 Monday)} = \frac{2}{7}$$

$$P\text{ (53 Sunday and 53 Monday)} = \frac{1}{7}$$

$\therefore \quad P\text{ (53 Sunday or 53 Monday)} = \frac{2}{7} + \frac{2}{7} - \frac{1}{7} = \frac{3}{7}$

5. (b) Let A be the event that component A fail during one year, B be the event that component B fail during one year. and C be the event that component C fail during one year.

We have, $P(A) = 0.15$, $P(B) = 0.05$ and $P(C) = 0.10$

Now, let E be the event that equipment fail before the end of the year.

$\therefore \quad P(E) = 1 - P(\bar{E})$

= 1 − P (equipment will not fail during one year)

= 1 − P (all components working during the whole year)

$= 1 - P(\bar{A}\bar{B}\bar{C})$

$= 1 - P(\bar{A})\,P(\bar{B})\,P(\bar{C}) \quad$ [$\because A, B, C$ are independent]

$= 1 - (0.85)\,(0.95)\,(0.90)$

$= 1 - 0.72675 = 0.27325$

6. (d) Let A be the event of getting a leap year and B be event of getting 53 Sunday in a year.

Now, as we know that a leap year is fallen within 4 years.

$\therefore \quad P(A) = \frac{25}{100} = \frac{1}{4}$

and $\quad P(\bar{A}) = \frac{3}{4}$

$\therefore$ Required probability $= P(AB) + P(\bar{A}B)$

$$= P(A) \cdot P\left(\frac{B}{A}\right) + P(\bar{A}) \cdot P\left(\frac{B}{\bar{A}}\right)$$

$$= \frac{1}{4} \cdot \frac{2}{7} + \frac{3}{4} \cdot \frac{1}{7} = \frac{5}{28}$$

7. (b) There are 10 digits 0, 1, 2, 3, 4, 5, 6, 7, 8, 9.

The last two digits can be dialled in $^{10}P_2$

= 90 ways, out of which only one way is favourable.

$\therefore$ Required probability $= \frac{1}{90}$

8. (d) Let p = Probability of getting 2 $= \frac{1}{2}$

q = Probability of getting 3 $= \frac{1}{2}$

Clearly, a total of 12 is obtained only if we get 2 on 3 coins and 3 on rest 2 coins.

$\therefore$ Required probability $= {}^5C_3\left(\frac{1}{2}\right)^3\left(\frac{1}{2}\right)^2 = {}^5C_2\left(\frac{1}{2}\right)^5$

$$= 10 \times \frac{1}{32} = \frac{5}{16}$$

9. (b) Let p = Probability of throwing 9 with two dice $= \frac{4}{36} = \frac{1}{9}$

and q = Probability of not throwing 9 with two dice $= \frac{8}{9}$

Let A be the event that A win the game and B be the event that B win the game.

Now, as A start the game, therefore B win, if he throw 9 in 2nd or 4th or 6th and so on attempt.

∴ Probability that B win the game

$$= P(\bar{A}B \text{ or } \bar{A}\bar{B}\bar{A}B \text{ or } \bar{A}\bar{B}\bar{A}\bar{B}\bar{A}B \text{ or} \cdots)$$
$$= P(\bar{A}B) + P(\bar{A}\bar{B}\bar{A}B) + P(\bar{A}\bar{B}\bar{A}\bar{B}\bar{A}B) + \cdots$$
$$= P(\bar{A})P(B) + P(\bar{A})P(\bar{B})P(\bar{A})P(B) + \cdots$$
$$= qp + q^3 p + \cdots$$
$$= \frac{qp}{1-q^2} = \frac{\frac{1}{9} \times \frac{8}{9}}{1 - \frac{64}{81}} = \frac{8}{17}$$

10. (b) Clearly, the sample space S of the given experiment is

$$S = \begin{cases} (1, 1), (1, 2), \cdots (1, 6) \\ (2, 1), (2, 2), \cdots, (2, 6) \\ \vdots \quad \vdots \quad \vdots \\ (6, 1), (6, 2), \cdots, (6, 6) \end{cases}$$

Let A be the event that sum of the digits be 6 and B be the event that atleast one of them show 4.

∴ $A = \{(2, 4), (4, 2), (1, 5), (5, 1), (3, 3)\}$

$B = \{(1, 4), (2, 4), (3, 4), (4, 4), (5, 4), (6, 4), (4, 1), (4, 2), (4, 3), (4, 5), (4, 6)\}$

$A \cap B = \{(2, 4), (4, 2)\}$

Required probability $= P\left(\frac{B}{A}\right) = \frac{P(B \cap A)}{P(A)} = \frac{\frac{2}{36}}{\frac{5}{36}} = \frac{2}{5}$

11. (b) Clearly, matches played by India are four and maximum points in any match are 2.

∴ Maximum point in four matches can be 8 only.

Required probability = Probability of India getting 7 points + Probability of India getting 8 points ...(i)

Note that, 7 points can be get in 4 matches, if India get 1 point in one of the 4 matches and 2 points in each of rest 3 matches and 8 points can be get only if India get 2 point in each of the 4 matches.

From Eq. (i), we have

Required probability $= {}^4C_1 (0.05)(0.5)^3 + {}^4C_4 (0.5)^4$

$= 4(0.05)(0.5)^3 + (0.5)^4$

$= 0.0250 + 0.0625$

$= 0.0875$

12. (c) Let E_1 denote an event that a coin with two heads is selected and E_2 be an event that a fair coin is selected. Let A be the event that the toss results in heads.

∴ $P(E_1) = \frac{1}{n+1}, P(E_2) = \frac{n}{n+1}, P\left(\frac{A}{E_1}\right) = 1$

$P\left(\frac{A}{E_2}\right) = \frac{1}{2}$

Now, $P(A) = P(E_1) \cdot P\left(\frac{A}{E_1}\right) + P(E_2) \cdot P\left(\frac{A}{E_2}\right)$

⇒ $\frac{7}{12} = \frac{1}{n+1} \cdot 1 + \frac{n}{n+1} \cdot \frac{1}{2}$

⇒ $\frac{7}{12} = \frac{2+n}{2(n+1)}$

⇒ $7n + 7 = 12 + 6n$

∴ $n = 5$

13. (a) The possible pair which give the sum 100 are (1, 99), (2, 98) ... (99, 1).

Thus, number of such pairs = 99

Now, required pairs are (12, 88), (13, 87), ..., (88, 12) and these are 77 in numbers.

∴ Required probability $= \frac{77}{99} = \frac{7}{9}$

14. (b) We have, $P(A \cap B) = \frac{1}{6}$ and $P(A^c \cap B^c) = \frac{1}{3}$

∵ $P(A^c \cap B^c) = \frac{1}{3}$

⇒ $P[(A \cup B)^c] = \frac{1}{3}$

⇒ $1 - P(A \cup B) = \frac{1}{3}$

⇒ $P(A \cup B) = \frac{2}{3}$

⇒ $P(A) + P(B) - P(A \cap B) = \frac{2}{3}$

⇒ $P(A) + P(B) = \frac{2}{3} + \frac{1}{6} = \frac{4+1}{6} = \frac{5}{6}$ ...(i)

∵ A and B are independent.

∴ We have, $P(A \cap B) = P(A) \cdot P(B)$

⇒ $P(A) \cdot P(B) = \frac{1}{6}$ ...(ii)

Now, consider

$[P(A) - P(B)]^2 = [P(A)]^2 + [P(B)]^2 - 2P(A)P(B)$

$= [P(A) + P(B)]^2 - 4P(A)P(B)$

$= \left(\frac{5}{6}\right)^2 - 4 \times \frac{1}{6} = \frac{25}{36} - \frac{4}{6} = \frac{1}{36}$

⇒ $P(A) - P(B) = \pm \frac{1}{6}$ ...(iii)

On solving Eqs. (ii) and (iii), we get $P(A) = \frac{1}{2}$ or $\frac{1}{3}$

15. (c) Let A be the event that student A solve the problem, B be the event that student B solve the problem and C be the event that student C solve the problem.

∴ $P(A) = \frac{1}{2} \Rightarrow P(\bar{A}) = \frac{1}{2}$

$P(B) = \frac{1}{3} \Rightarrow P(\bar{B}) = \frac{2}{3}$

$P(C) = \frac{1}{4} \Rightarrow P(\bar{C}) = \frac{3}{4}$

Required probability = 1 − P (None of them solved the problem)

$= 1 - P(\bar{A}\bar{B}\bar{C})$

$= 1 - P(\bar{A})P(\bar{B})P(\bar{C})$

$= 1 - \frac{1}{2} \times \frac{2}{3} \times \frac{3}{4}$

$= 1 - \frac{1}{4} = \frac{3}{4}$

16. (d) Let $X \sim B(n, p)$.

We have $E(X) = np = 4$ ...(i)

and $\text{Var}(X) = npq = 2$ ...(ii)

From Eqs. (i) and (ii), we get

$q = \frac{1}{2} \Rightarrow p = \frac{1}{2}$ and $n = 8$

Thus, $X \sim B\left(8, \frac{1}{2}\right)$

∴ $P(X = x) = {}^8C_x \left(\frac{1}{2}\right)^x \left(\frac{1}{2}\right)^{8-x}, x = 0, 1, 2, \cdots, 8$

$= {}^8C_x \left(\frac{1}{2}\right)^8, x = 0, 1, 2, \cdots, 8$

Now, required probability $= P(X = 2)$

$= {}^8C_2 \left(\frac{1}{2}\right)^8 = \frac{8 \times 7}{2 \times 1} \times \frac{1}{2^8}$

$= \frac{2^2 \times 7}{2^8} = \frac{28}{256}$

17. (a) Let A be the event that the product of digits is zero and B be the event that the sum of the digits is 8.

Clearly, $A = \{00, 01, 02, 03, 04, 05, 06, 07, 08, 09, 10, 20, 30, 40\}$

$B = \{08, 17, 26, 35, 44\}$

$A \cap B = \{08\}$

$\therefore \quad P(A) = \frac{14}{50}, \quad P(B) = \frac{5}{50}$

$P(A \cap B) = \frac{1}{50}$

$\therefore \quad P\left(\frac{B}{A}\right) = \frac{P(A \cap B)}{P(A)} = \frac{\frac{1}{50}}{\frac{14}{50}} = \frac{1}{14}$

18. (b) We have, an urn containing 9 balls of which 3 are red, 4 are blue and 2 are green.

Total number of ways of drawing 3 balls $= {}^9C_3$

Number of ways of drawing 3 balls of different colours $= {}^3C_1 \times {}^4C_1 \times {}^2C_1$

$\therefore$ Required probability $= \frac{{}^3C_1 \times {}^4C_1 \times {}^2C_1}{{}^9C_3}$

$= \frac{3 \times 4 \times 2 \times 3 \times 2}{9 \times 8 \times 7} = \frac{2}{7}$

19. (d) Total number of ways of arrangement $= \frac{8!}{2!\,4!}$

Now, let us first arrange 2 A's and 1 N in a row, that can be done in $\frac{4!}{2!}$ ways.

Now, we left with 5 places, in which 4's can be arranged in ${}^5P_4$ ways = 5

$\therefore$ Number of favourable ways $= 5 \times \frac{4!}{2!}$

Required Probability $= \frac{5! \times \frac{4!}{2!}}{\frac{8!}{4!\,2!}} = \frac{5! \times 4!}{2! \times 8!} \times 4! \times 2! = \frac{1}{14}$

20. (a) We have, 4 machines and out of which 2 are faulty.

Now, probability that only 2 tests are needed

= Probability that in first two tests we get the faulty machines or both non-faulty

$= 2\left[\frac{{}^2C_1}{{}^4C_1} \times \frac{{}^1C_1}{{}^3C_1}\right] = \left[\frac{2}{4} \times \frac{1}{3}\right] \times 2 = \frac{1}{3}$

21. (b) Let M, P and C be the events of passing in Mathematics, Physics and Chemistry, respectively.

$\therefore \quad P(M) = m, P(P) = p, P(C) = c$

$P(M \cup P \cup C) = \frac{75}{100} = \frac{3}{4}$ ...(i)

Now, $P(M \cap P) + P(P \cap C) + P(M \cap C)$
$- 2P(M \cap P \cap C) = \frac{50}{100} = \frac{1}{2}$ ...(ii)

and $P(M \cap P) + P(P \cap C) + P(M \cap C)$
$- 3P(P \cap M \cap C) = \frac{40}{100} = \frac{2}{5}$ ...(iii)

From Eqs. (ii) and (iii), we have

$\frac{1}{2} + 2P(M \cap P \cap C) - 3P(M \cap P \cap C) = \frac{2}{5}$

$\Rightarrow \quad P(M \cap P \cap C) = \frac{1}{2} - \frac{2}{5} = \frac{5-4}{10}$

$\Rightarrow \quad P(M)\,P(P)\,P(C) = \frac{1}{10}$

[$\because M, P, C$ are independent]

$\therefore \quad pmc = \frac{1}{10}$

22. (a) We have, E and F are two independent events, then

$P(E \cap \bar{F} \text{ or } \bar{E} \cap F) = \frac{11}{25}$ and $P(\bar{E} \cap \bar{F}) = \frac{2}{25}$

$\because \quad P(\bar{E} \cap \bar{F}) = \frac{2}{25}$

$\Rightarrow \quad P[\overline{(E \cup F)}] = \frac{2}{25}$

$\Rightarrow \quad 1 - P(E \cup F) = \frac{2}{25}$

$\Rightarrow \quad P(E \cup F) = \frac{23}{25}$

$\Rightarrow \quad P(E) + P(F) - P(E \cap F) = \frac{23}{25}$ ...(i)

Now, consider $P(E \cap \bar{F} \text{ or } \bar{E} \cap F) = \frac{11}{25}$

$\Rightarrow \quad P(E \cap \bar{F}) + P(\bar{E} \cap F) = \frac{11}{25}$

$\Rightarrow \quad P(E) - P(E \cap F) + P(F) - P(E \cap F) = \frac{11}{25}$

$\Rightarrow \quad P(E) + P(F) - 2P(E \cap F) = \frac{11}{25}$ ...(ii)

$\Rightarrow \quad \frac{23}{25} + P(E \cap F) - 2P(E \cap F) = \frac{11}{25}$ [using Eq. (i)]

$\Rightarrow \quad P(E \cap F) = \frac{12}{25}$ ...(iii)

Putting the value of $P(E \cap F)$ in Eq. (i), we get

$P(E) + P(F) = \frac{35}{25} = \frac{7}{5}$ ...(iv)

Now, consider

$[P(E) - P(E)]^2 = [P(E) + P(F)]^2 - 4P(E)\,P(F)$

$= \left(\frac{7}{5}\right)^2 - 4\,P(E \cap F)$

[$\because E$ and F are independent]

$= \frac{49}{25} - \frac{48}{25} = \frac{1}{25}$

$\therefore \quad P(E) - P(F) = \pm\frac{1}{5}$

$\Rightarrow \quad P(E) - P(F) = \frac{1}{5}$ ...(v)

or $\quad P(E) - P(F) = -\frac{1}{5}$ ...(vi)

On solving Eqs. (iv) and (v), we get

$P(E) = \frac{4}{5}$

and $\quad P(F) = \frac{3}{5}$

23. (d) Total number of ways of selected 7 coupons at random

$= {}^{15}C_1 \times {}^{15}C_1 \times {}^{15}C_1 \times {}^{15}C_1 \times {}^{15}C_1 \times {}^{15}C_1 \times {}^{15}C_1 = (15)^7$

Number of ways of selected 7 coupons are at a time with replacement from coupons 1 to 9

$= {}^9C_1 \times {}^9C_1 \times {}^9C_1 \times {}^9C_1 \times {}^9C_1 \times {}^9C_1 \times {}^9C_1 = (9)^7$

Number of ways of selecting 7 coupons from 1 to 8 $= (8)^7$

$\therefore$ Required probability $= \frac{9^7 - 8^7}{(15)^7}$

24. (c) Total number of triangles; which can be formed

$= {}^6C_3 = \frac{6 \times 5 \times 4}{3 \times 2 \times 1} = 20$

Number of equilateral triangles = 2

$\therefore$ Required probability $= \frac{2}{20} = \frac{1}{10}$

25. (a) We know that,

P (exactly one of A or B occurs) $= P(A) + P(B) - 2P(A \cap B)$

$\therefore \quad P(A) + P(B) - 2P(A \cap B) = p \quad \ldots(\text{i})$

Similarly, $\quad P(B) + P(C) - 2P(B \cap C) = P \quad \ldots(\text{ii})$

$P(C) + P(A) - 2P(C \cap A) = P \quad \ldots(\text{iii})$

Adding Eqs. (i), (ii) and (iii), we get

$$2P(A) + 2P(B) + 2P(C) - 2P(A \cap B) - 2P(B \cap C) - 2P(C \cap A) = 3p$$

$$\Rightarrow P(A) + P(B) + P(C) - P(A \cap B) - P(B \cap C) - P(C \cap A) = \frac{3p}{2} \quad \ldots(\text{iv})$$

We are also given that, $P(A \cap B \cap C) = p^2$

$\therefore P$ (atleast one of A, B, C) $= P(A \cup B \cup C)$

$$= P(A) + P(B) + P(C) - P(A \cap B) - P(B \cap C) - P(C \cap A) + P(A \cap B \cap C)$$

$$= \frac{3p}{2} + p^2 = \frac{3p + 2p^2}{2}$$

26. (b) The number of ways to arrange 7 white and 3 black balls in a row = 10!

Now, let us first arrange the 7 white balls in a row that can be done in 7!. ways.

Now, we have 8 places for 3 black balls that can be arrange in $^8P_3$ ways $= \frac{8!}{5!}$

Thus, number of favourable ways $= 7! \times \frac{8!}{5!} = 7 \times 6 \times 8!$

$$\therefore \text{ Required probability} = \frac{7 \times 6 \times 8!}{10!} = \frac{7 \times 6 \times 8!}{10 \times 9 \times 8!} = \frac{7 \times 6}{10 \times 9} = \frac{7}{15}$$

27. (a) Since, m and n are selected from 1 to 100, therefore total number of ways of selecting m and n = (100 × 100).

Note that $7^1 = 7, \quad 7^2 = 49,$

$7^3 = 343, \quad 7^4 = 2401$

and $\quad 7^5 = 16807$ etc.

i.e. 1, 3, 7 and 9 will be the last digits in the powers of 7. Now, let us find favourable cases.

Clearly, maximum possible choices for (n, m) are $\frac{n}{m} \rightarrow$

1, 1 1, 2 1, 3 ... 1, 100

2, 1, 2, 2 2, 3 ... 2,100

⋮

100, 1 100, 2 100, 3 ... 100, 100

Now, for $m = 1, n = 3, 7, 11, \ldots, 97$

$\therefore$ Favourable cases = 25

For $m = 2, n = 4, 8, 12, \cdots, 100$

$\therefore$ Favourable cases = 25

Similarly, for every m, favourable cases = 25

$\therefore$ Total favourable cases = 100 × 25

$$\therefore \text{ Required probability} = \frac{100 \times 25}{100 \times 100} = \frac{1}{4}$$

28. (c) Let A be the event that the Indian man is seated adjacent to his wife and B be the event that the American man is seated adjacent to his wife.

$$\therefore \quad P(A) = \frac{8! \times 2}{9!} = \frac{2}{9}$$

[since, considering Indian man and his wife as one unit]

$$P(B) = \frac{2^4 \times 5!}{9!}$$

[since, considering American man and his wife as one unit]

$$P(A \cap B) = \frac{2^5 \times 4!}{9!}$$ [since, the married couple as one unit]

$$\therefore \quad P\left(\frac{A}{B}\right) = \frac{P(A \cap B)}{P(B)} = \frac{\frac{2^5 \times 4!}{9!}}{\frac{2^4 \times 5!}{9!}} = \frac{2 \times 4!}{5!} = \frac{2}{5}$$

29. (c) Let p = probability of ticking correct answer $= \frac{1}{3}$

q = probability of ticking wrong answer $= \frac{2}{3}$

n = number of question or number of trials = 5

x = number of times getting success i.e. ticking correct answer

Clearly, $\quad X \sim B\left(5, \frac{1}{3}\right)$

$$\therefore P(X = x) = {}^5C_x \left(\frac{1}{3}\right)^x \left(\frac{2}{3}\right)^{5-x}, x = 0, 1, 2, 3, 4, 5$$

$\therefore$ Required probability

$$= P(X \geq 4) = P(X = 4) + P(X = 5)$$

$$= {}^5C_4 \left(\frac{1}{3}\right)^4 \left(\frac{2}{3}\right) + {}^5C_5 \left(\frac{1}{3}\right)^5 = 5 \cdot \frac{2}{3^5} + \frac{1}{3^5} = \frac{11}{3^5}$$

30. (a) Clearly, the number of outcomes in sample space of the given experiment is $6 \times 6 \times 6 \times 6 = 6^4$, i.e. $n(S) = 6^4$

Now, let us find favourable cases.

Favourable case D_4 shows a number, then

only 1 of $D_1\ D_2\ D_3$ shows same number or

only 2 of $D_1\ D_2\ D_3$ shows same number or

all 3 of $D_1\ D_2\ D_3$ shows same number.

$\therefore$ Number of favourable cases

$$= {}^6C_1({}^3C_1 \times 5 \times 5 + {}^3C_2 \times 5 + {}^3C_3)$$

$$\therefore \text{ Required probability} = \frac{{}^6C_1\ (75 + 15 + 1)}{6^4} = \frac{91}{6^3} = \frac{91}{216}$$

31. (a) Clearly, the sample space,

$$S = \begin{cases} (1, 1), (1, 2), \cdots, (1, 6) \\ (2, 1), (2, 2), \cdots, (2, 6) \\ \vdots \quad\quad \vdots \quad\quad\quad \vdots \\ (6, 1), (6, 2), \cdots, (6, 6) \end{cases}$$

Let A be the event of getting the sum of numbers on the dice atleast 10.

$\therefore \quad A = \{(5, 5), (6, 4), (4, 6), (6, 5), (5, 6), (6, 6)\}$

$$\therefore \quad P(A) = \frac{6}{36} = \frac{1}{6}$$

32. (a) The total number of ways of dividing $2n$ boys equally into a first and then a second subgroup $= {}^{2n}C_n \times {}^nC_n = {}^{2n}C_n = \frac{(2n)!}{n!\,n!}$

Now, there are two ways in which the two tallest boys lie in different groups and corresponding to each way the remaining $(2n - 2)$ boys can be divided into two groups in

${}^{2n-2}C_{n-1} \times {}^{n-1}C_{n-1}$ ways $= {}^{2n-2}C_{n-1}$

$\therefore$ The favourable number of ways $= 2 \cdot ({}^{2n-2}C_{n-1})$

$$= 2 \cdot \frac{(2n-2)!}{(n-1)!\,(n-1)!}$$

$$\text{Hence, the required probability} = \frac{\frac{2 \cdot (2n-2)!}{(n-1)!\,(n-1)!}}{\frac{(2n)!}{n!\,n!}}$$

$$= \frac{2 \cdot (2n-2)!}{(n-1)!\,(n-1)!} \times \frac{n!\,n!}{(2n)!}$$

$$= \frac{2(n)\,(n)}{(2n)\,(2n-1)} = \frac{n}{2n-1}$$

33. (a) Clearly, total number of ways of selecting 13 cards from a pack of 52 cards = $^{52}C_{13}$

Number of ways of selecting 13 cards, so that 9 cards are of same suit = Number of ways of selecting (9 cards from spade and 4 from remaining or 9 from club and 4 from remaining or 9 from heart and 4 from remaining or 9 from diamond and 4 from remaining)

$$= {}^{13}C_9 \cdot {}^{39}C_4 + {}^{13}C_9 \cdot {}^{39}C_4 + {}^{13}C_9 \cdot {}^{39}C_4 + {}^{39}C_9 \cdot {}^{39}C_4$$
$$= 4 \times {}^{13}C_9 \cdot {}^{39}C_4$$

$\therefore$ Required probability $= \dfrac{4 \times {}^{13}C_9 \times {}^{39}C_4}{^{52}C_{13}}$

34. (c) Clearly, the sample space,

$$S = \begin{Bmatrix} (1,1),(1,2),\cdots,(1,6) \\ (2,1),(2,2),\cdots,(2,6) \\ \vdots \\ (6,1),(6,2),\cdots,(6,6) \end{Bmatrix}$$

Let A be the event that 6 comes up on first die and B be the event that total of 7 is obtained.

$\therefore$ $A = \{(6,1),(6,2),\cdots,(6,6)\}$

$B = \{(3,4),(4,3),(5,2),(2,5),(1,6),(6,1)\}$

$A \cap B = \{(6,1)\}$

Now, $P(A \cap B) = \dfrac{1}{36}$

$P(A) = \dfrac{6}{36} = \dfrac{1}{6}$

$P(B) = \dfrac{6}{36} = \dfrac{1}{6}$

$\because$ $P(A \cap B) = P(A) \cdot P(B)$

$\Rightarrow$ A and B are independent events.

As, $A \cap B \neq \phi$, therefore A and B are not mutually exclusive.

Also, as $A \cup B \neq S$, therefore A and B are not mutually exhaustive.

35. (a) Total number of ways of selecting 2 person = $^{13}C_2$

Number of ways of selecting 2 person such that it consist atleast one woman

= Number of ways of selecting one man and one woman + Number of ways of selecting two women

$= {}^5C_1 \times {}^8C_1 + {}^5C_2$

$\therefore$ Required probability $= \dfrac{^5C_1 \times {}^8C_1 + {}^5C_2}{^{13}C_2}$

$$= \dfrac{40 + 10}{\dfrac{13 \times 12}{2 \times 1}} = \dfrac{50}{13 \times 6} = \dfrac{25}{39}$$

36. (c) Clearly, the sample space,

$$S = \begin{Bmatrix} (1,1),(1,2),\cdots,(1,6) \\ (2,1),(2,2),\cdots,(2,6) \\ \vdots \quad \vdots \quad \vdots \\ (6,1),(6,2),\cdots,(6,6) \end{Bmatrix}$$

Let A be the event that total of numbers turned up on both the dies is 8, and B be the event that number turn up on second die is 5.

$\therefore$ $A = \{(3,5),(5,3),(6,2),(2,6),(4,4)\}$

$B = \{(1,5),(2,5),(3,5),(4,5),(5,5),(6,5)\}$

$A \cap B = \{(3,5)\}$

Now, required probability $= P\left(\dfrac{B}{A}\right) = \dfrac{P(B \cap A)}{P(A)}$

$$= \dfrac{\dfrac{1}{36}}{\dfrac{5}{36}} = \dfrac{1}{5}$$

37. (a) Let p = probability of getting success

q = probability of getting failure

n = number of independent trials = 3

x = number of times getting success

Clearly, $X \sim B(3, p)$

$\therefore$ $P(X = x) = {}^3C_x\, p^x (q)^{3-x}$, $x = 0, 1, 2, 3$

$\because$ $P(X = 2) = 12\, P(X = 3)$

$\Rightarrow$ ${}^3C_2 p^2 q = 12\, {}^3C_3 p^3 q^0$

$\Rightarrow$ $3q = 12p$

$\Rightarrow$ $q = 4p$ ...(i)

$\because$ $p + q = 1$

$\Rightarrow$ $5p = 1$ [using Eq. (i)]

$\therefore$ $p = \dfrac{1}{5}$

38. (c) Let p = probability of getting a white ball $= \dfrac{2}{6} = \dfrac{1}{3}$

q = probability of getting a black ball $= \dfrac{4}{6} = \dfrac{2}{3}$

n = number of times ball drawn or number of trials = 5

x = number of times getting white ball

Clearly, $X \sim B\left(5, \dfrac{1}{3}\right)$

$\therefore$ $P(X = x) = {}^5C_x \left(\dfrac{1}{3}\right)^x \left(\dfrac{2}{3}\right)^{5-x}$, $x = 0, 1, 2, 3, 4, 5$

Required probability $= P(X = 4) + P(X = 5)$

$$= {}^5C_4 \left(\dfrac{1}{3}\right)^4 \left(\dfrac{2}{3}\right) + {}^5C_5 \left(\dfrac{1}{3}\right)^5$$
$$= 5 \cdot \dfrac{2}{3^5} + \dfrac{1}{3^5} = \dfrac{11}{3^5} = \dfrac{11}{243}$$

39. (d) Here, $p = 1/5 = 0.2$

$q = 0.8$

$P(E) = 1 - P(0 \text{ or } 1)$

$= 1 - [{}^8C_0 (0.2)^0 \cdot (0.8)^8 + {}^8C_1 (0.2)^1 \cdot (0.8)^7]$

$= 1 - [(0.8)^8 + 8(0.2)(0.8)^7]$

$= 1 - [(0.8)^8 + 2(0.8)^8]$

$= 1 - 3(0.8)^8$

40. (a) There are 6 different letters. Let us select 6 squares taking atleast one from each row and then arranging in each selection. Let us first select places in each row such that no row remains empty.

R_1	R_2	R_3	R_4	Number of selections
2	2	1	1	${}^4C_2 \times {}^3C_2 \times {}^2C_1 \times {}^1C_1 = 36$
3	1	1	1	${}^4C_3 \times {}^3C_1 \times {}^2C_1 \times {}^1C_1 = 24$
1	3	1	1	${}^4C_1 \times {}^3C_3 \times {}^2C_1 \times {}^1C_1 = 8$
2	1	2	1	${}^4C_2 \times {}^3C_1 \times {}^2C_2 \times {}^1C_1 = 18$
1	2	2	1	${}^4C_1 \times {}^3C_2 \times {}^2C_2 \times {}^1C_1 = 12$

$\therefore$ Total number of selection of 6 squares is $36 + 24 + 8 + 18 + 12 = 98$. For each selection of 6 squares, the number of arrangements of 6 letters is $6! = 720$, therefore the required number of ways $= 98 \times 720$.

Now, number of arrangement of 6 letters in 10 squares taking 6 at a time $= {}^{10}P_6 = \dfrac{10!}{4!}$.

$\therefore$ Required probability $= \dfrac{98 \times 720}{10!} \times 4!$

$$= \dfrac{98 \times 4!}{10 \times 9 \times 8 \times 7} = \dfrac{7}{15}$$

41. (a) Here, $n(S)$ = length of the interval [0, 5]

$= 5 - 0 = 5$

$n(E)$ = length of the interval ≤ 5 in which p belongs such that the given equation has real roots.

Now, $x^2 + px + \frac{1}{4}(p+2) = 0$ will have real roots,

if $p^2 - 4 \cdot 1 \cdot \frac{1}{4}(p+2) \geq 0$

$\Rightarrow p^2 - p - 2 \geq 0$

$\Rightarrow (p+1)(p-2) \geq 0$

$\Rightarrow p \leq -1$ or $p \geq 2$

$\because p \in [0, 5]$, so $E = [2, 5]$

$\therefore n(E)$ = length of the interval [2, 5] = 3

$\therefore$ Required probability $= \frac{3}{5}$

42. (c) Let terms of an AP $a, a+d, a+2d, a+3d$.

$\because a \geq 1, a + 3d \leq 20$

$\Rightarrow 3d \leq 19$

$\Rightarrow d \leq \frac{19}{3}$

So, $d = \pm 1, \pm 2, \pm 3, \pm 4, \pm 5$ and ± 6.

Hence, Statement II is incorrect.

If $d = 1$, then $a + 3d \leq 20 \Rightarrow a \leq 17$

So, 17 cases will be there. Similarly, for $d = -1$, we get 17 cases.

$\therefore$ Total cases for $d = \pm 1$ is 34.

Similarly,

d	±1	±2	±3	±4	±5	±6
Cases	34	28	22	16	10	4

So, total favourable cases $= \frac{114}{2}$

Total number of cases $= {}^{20}C_4$

$\therefore$ Required probability $= \frac{114}{{}^{20}C_4 \times 2} = \frac{1}{85}$

Hence, Statement I is correct.

43. (d) Clearly, the sample space S of the given experiment is

$S = \{(H, 1), (H, 2), (H, 3), (H, 4), (H, 5), (H, 6), T\}$

Note that, elementary events are not equally likely as the probabilities assigned to the 7 elementary events are

$\frac{1}{12}, \frac{1}{12}, \frac{1}{12}, \frac{1}{12}, \frac{1}{12}, \frac{1}{12}, \frac{1}{2}$

Now, let $A = \{(H, 1), (H, 2), (H, 5)\}$

$\therefore P(A) = P(H,1) + P(H,2) + P(H,5)$

[$\because$ sample point of event A are mutually exclusive]

$= \frac{1}{12} + \frac{1}{12} + \frac{1}{12} = \frac{3}{12} = \frac{1}{4}$

Hence, Statement I is incorrect and Statement II is correct by definition.

44. (a) For independent events A and B, we have

$P\left(\frac{A}{B}\right) = P(A) \Rightarrow \frac{P(A \cap B)}{P(B)} = P(A)$

$\Rightarrow P(A \cap B) = P(A) \cdot P(B)$

$\Rightarrow P(A \cap B) \neq 0$ [$\because P(A) \neq 0$ and $P(B) \neq 0$]

$\Rightarrow A \cap B \neq \phi$

$\therefore A$ and B cannot be mutually exclusive.

Thus, Statements I and II are correct and Statement II is a correct explanation of Statement I.

45. (b) We must have,

$0 \leq \frac{1+4p}{4} \leq 1, 0 \leq \frac{1-p}{4} \leq 1$ and $0 \leq \frac{1-2p}{4} \leq 1$

$\Rightarrow \frac{-1}{4} \leq p \leq \frac{3}{4}, -3 \leq p \leq 1, \frac{-3}{2} \leq p \leq \frac{1}{2}.$

Again, as these events are pairwise mutually exclusive, so we have

$0 \leq \frac{1+4p}{4} + \frac{1-p}{4} + \frac{1-2p}{4} \leq 1 \Rightarrow -3 \leq p \leq 1$

Taking intersection of all four intervals of p, we get

$\frac{-1}{4} \leq p \leq \frac{1}{2}.$

Hence, Statement I is correct and Statement II is corrcet by the definition but Statement II is not a correct explanation for Statement I.

Revisal Problems (JEE Advanced)

1. (a) We have, $10^{99} = 2^{99} \times 5^{99}$

Total number of divisor $= (99+1)(99+1) = (100)^2$

and $10^{88} = 2^{88} \times 5^{88}$

The multiple of 10^{88}, which is positive divisor of $10^{99} = 2^{11} \times 5^{11}$

$\therefore$ Total number of divisor of $2^{11} \times 5^{11} = (11+1)(11+1) = 12^2$

$\therefore$ Required probability $= \frac{(12)^2}{(100)^2} = \left(\frac{3}{25}\right)^2 = \frac{9}{625}$

where, $m = 9$ and $n = 625$

$\therefore m + n = 9 + 625 = 634$

Hence, option (a) is correct.

2. (d) Total number of determinant is formed of order 2 taken 0 and 1 element only $= 2^4 = 16$

There are three possible cases for determinant have positive value

i.e. $\begin{bmatrix} 1 & 0 \\ 1 & 1 \end{bmatrix}, \begin{bmatrix} 1 & 0 \\ 0 & 1 \end{bmatrix}, \begin{bmatrix} 1 & 1 \\ 0 & 1 \end{bmatrix}$

$\therefore$ Required probability $= 1 - \frac{3}{16} = \frac{13}{16}$

Hence, option (d) is correct.

3. (c) Let the x and y be the black and red colour balls, respectively.

Now, $x + y = 10$

and $\frac{{}^xC_1 \times {}^yC_1}{{}^{(x+y)}C_2} = \frac{8}{15} = \frac{xy}{{}^{10}C_2} = \frac{8}{15}$ [$\because x + y = 10$]

$= xy = \frac{8 \times 45}{15}$

$\Rightarrow xy = 24$ and $x + y = 10$

On solving these equations, we get

$x = 6, y = 4$ or $y = 6$ and $x = 4$

Now, here black ball is more than red balls.

$\therefore x = 6$ and $y = 4$

$\therefore$ Required probability $= \frac{{}^6C_1 \times {}^4C_1}{{}^{10}C_2} + \frac{{}^6C_2}{{}^{10}C_2}$

$= \frac{6 \times 4}{45} + \frac{6 \times 5}{10 \times 9}$

$= \frac{24 + 15}{45} = \frac{39}{45}$

Hence, option (c) is correct.

4. (b) Let $P(S) = P(1 \text{ or } 2) = \frac{1}{3}$

$$P(F) = P(3 \text{ or } 4 \text{ or } 5 \text{ or } 6) = \frac{2}{3}$$

$$P(A \text{ wins}) = P[SS \text{ or } SFSS \text{ or } SFSFSS \text{ or } \ldots) \text{ or } (FSS \text{ or } FSFSS \text{ or } \ldots)]$$

$$= \frac{\frac{1}{9}}{1 - \frac{2}{9}} + \frac{\frac{2}{27}}{1 - \frac{2}{9}}$$

$$= \frac{1}{9} \times \frac{9}{7} + \frac{2}{27} \times \frac{9}{7}$$

$$= \frac{5}{21}$$

5. (a) We have, $P(A) = P(7) = \frac{6}{36}$

and $P(B) = P(4) = \frac{3}{36}$

Since, equal throws are disregarded.

Hence, in each throw A is twice as likely to win as B.

Let $P(B) = p$

$P(A) = 2p$

$\therefore \quad 3p = 1$

$\Rightarrow \quad p = \frac{1}{3}$

6. (b) We have, {1, 2, 3, 4, 5}

Total number of selection a, b, c from {1, 2, 3, 4, 5,}, where a, b, c are different or may not be different is 5^3 i.e. 125.

Now, $ab + c$ is even. If ab and c both are odd or even.

a	b	c	Total number of ways
Odd	Odd	Odd	${}^3C_1 \times {}^3C_1 \times {}^3C_1 = 27$
Odd	Even	Even	${}^3C_1 \times {}^2C_1 \times {}^2C_1 = 12$
Even	Odd	Even	${}^2C_1 \times {}^3C_1 \times {}^2C_1 = 12$
Even	Even	Even	${}^2C_1 \times {}^2C_1 \times {}^2C_1 = 8$

$$\therefore \text{Required probability} = \frac{27 + 12 + 12 + 8}{125} = \frac{59}{125}$$

Hence, option (b) is correct.

7. (c) Let E_1 be the event Mr. A transfer one ball to Mr. B and E_2 be the event the same colour ball transfer to Mr. A.

Now, $P(E_1) = \frac{1}{5}$

$P(E_2) = \frac{2}{6}$

Now, this process is repeated five times.

$\therefore$ Required probability $= 5 \times P(E_1) \times P(E_2)$

$$= 5 \times \frac{1}{5} \times \frac{2}{6} = \frac{1}{3}$$

Hence, option (c) is correct.

8. (c) Here, there are n ropes lying on the floor.

Let two boys A and B anyone of n ropes hold.

$\therefore$ Total number of outcomes $= n + n = 2n$

Required probability $\frac{n}{{}^{2n}C_2} = \frac{1}{101}$

$$\Rightarrow \quad \frac{n \times 2}{2n(2n-1)} = \frac{1}{101}$$

$\therefore \quad n = 51$

Hence, option (c) is correct.

9. (c) We have, word 'STATISTICS' have S occur 3 times, T occur 3 times, I occur 2 times and A and C are different, and the word 'ASSISTANT' have S occurs 3 times, T occurs 2 times A occurs 2 times and I, and N different.

Proabability of S, T, I, A, C from word 'STATISTICS' are $\frac{3}{10}, \frac{3}{10}, \frac{2}{10}, \frac{1}{10}, \frac{1}{10}$ respectively and probability of S, T, A, I and N from word 'ASSISTANT' are $\frac{3}{9}, \frac{2}{9}, \frac{2}{9}, \frac{1}{9}, \frac{1}{9}$ respectively. In the given 'STATISTICS' and 'ASSISTANT' have same letters are S, T, and A.

$$\therefore \text{Required probability} = \frac{3}{10} \times \frac{3}{9} + \frac{3}{10} \times \frac{2}{9} + \frac{2}{10} \times \frac{2}{9} = \frac{9 + 6 + 4}{90} = \frac{19}{90}$$

Hence, option (c) is correct.

10. (b) If A fair coin is tossed $(m + n)$ times. The probability of atleast m times consecutive head occurs is $\frac{n+2}{2^{m+1}}$.

Here, $n = 4$ and $m = 3$

$$\therefore \text{Required probability} = \frac{4+2}{2^{3+1}} = \frac{6}{2^4} = \frac{6}{16} = \frac{3}{8}$$

Hence, option (b) is correct.

11. (c) Put $\tan 2x = t, -\infty < t < \infty$

$$[t^2] = t + a, a \in N$$

So, t should be an integer.

$\Rightarrow \quad (t^2) - t - a = 0$

$$\therefore \quad t = \frac{1 \pm \sqrt{1 + 4a}}{2}$$

Here, $1 + 4a$ is odd number and which is perfect square.

$$1 + 4a = 9, 25, \ldots, (63)^2$$

$\therefore$ Number of values that a can take = 31

$$\Rightarrow \quad p = \frac{31}{1000}$$

12. (c) Total number of ways of selecting three numbers from $\{1, 2, 3, \ldots, n\}$ with replacement $= n \times n \times n = n^3$

Now, let us find the favourable cases.

Let $x + y + z = 2n$, where $1 \le x \le n, 1 \le y \le n$ and $1 \le z \le n$.

Clearly, favourable number of cases is the number of solution of above equation

$= \text{Coefficient of } x^{2n} \text{ in } (x + x^2 + \cdots + x^n)^3$

$= \text{Coefficient of } x^{2n} \text{ in } x^3(1 + x + x^2 + \cdots + x^{n-1})^3$

$= \text{Coefficient of } x^{2n-3} \text{ in } (1 + x + x^2 + \cdots + x^{n-1})^3$

$= \text{Coefficient of } x^{2n-3} \text{ in } \left(\frac{1 - x^n}{1 - x}\right)^3$

$= \text{Coefficient of } x^{2n-3} \text{ in } (1 - x^n)^3(1 - x)^{-3}$

$= \text{Coefficient of } x^{2n-3} \text{ in } (1 - x^{3n} - 3x^n + 3x^{2n})(1 - x)^{-3}$

$= \text{Coefficient of } x^{2n-3} \text{ in } (1 - x)^{-3} + (-3)$

$= \text{Coefficient of } x^{n-3} \text{ in } \quad (1 - x)^{-3}$

$= {}^{2n-1}C_2 - 3\, {}^{n-1}C_2 = \frac{(n-1)(n-4)}{2}$

$$\therefore \text{Required probability} = \frac{(n-1)(n+4)}{2n^3}$$

13. (c, d) Here, $E_1 = \{5, 10, 15, 20, 25, \ldots\}$ and $E_2 = \{5, 15, 25, 35, \ldots\}$

(a) E_1 is not twice as likely to occur E_2.

Hence, statement is incorrect.

(b) $E_1 \cap E_2 = \{5, 15, 25, \ldots\}$

Hence, this statement is incorrect.

(c) Here, $P(E_1) = 1 - \left(\frac{8}{10}\right)^2 = \frac{36}{100}$,

and $P(E_2) = 1 - \left(\frac{9}{10}\right)^2 = \frac{9}{100}$

$\therefore \quad P(E_1 \cap E_2) = \frac{9}{100}$

$$P\left(\frac{E_2}{E_1}\right) = \frac{\frac{9}{100}}{\frac{36}{100}} = \frac{1}{4}$$

Hence, this statement is correct.

(d) $P\left(\frac{E_1}{E_2}\right) = \frac{\frac{9}{100}}{\frac{9}{100}} = 1$

Hence, this statement is correct.

14. (a, b, c, d)

(a) $P\left(\frac{A}{B^c}\right) = \frac{P(A \cap B^c)}{P(B^c)} = \frac{P(A) - P(A \cap B)}{1 - P(B)}$

(b) we have, $P(A \cup B) \le 1$

$\Rightarrow \quad P(A) + P(B) - P(A \cap B) \le 1$

$\Rightarrow \quad P(A \cap B) \ge P(A) + P(B) - 1$

(c) we have, $P\left(\frac{A}{B^c}\right) < P(A)$

$\Rightarrow \quad \frac{P(A \cap B^c)}{P(B^c)} < P(A)$

$\Rightarrow \quad \frac{P(A) - P(A \cap B)}{1 - P(B)} < P(A)$

$\Rightarrow \quad P(A) - P(A \cap B) < P(A) - P(A)P(B)$

$\Rightarrow \quad P(A)P(B) < P(A \cap B)$

$\Rightarrow \quad P(A) \times P(B) < P(B) \times P\left(\frac{A}{B}\right)$

$\Rightarrow \quad P(A) < P\left(\frac{A}{B}\right)$

(d) $P\left(\frac{A}{B^c}\right) + P\left(\frac{A^c}{B^c}\right) = \frac{P(A \cap B^c)}{P(B^c)} + \frac{P(A^c \cap B^c)}{P(B^c)}$

$$= \frac{P(A) - P(A \cap B) + 1 - P(A \cup B)}{P(B^c)}$$

$$= \frac{P(A) + 1 - [P(A \cup B) + P(A \cap B)]}{P(B^c)}$$

$$= \frac{P(A) + 1 - P(A) - P(B)}{P(B^c)}$$

$$= \frac{1 - P(B)}{P(B^c)} = \frac{P(B^c)}{P(B^c)} = 1$$

15. (b, c, d)

(a) $P(TTT \text{ or } HHH) = \frac{1}{8} + \frac{1}{8} = \frac{1}{4}$

Hecne, this statement is incorrect.

(b) $\frac{P(A \cap B)}{P(B)} = \frac{P(A \cap \bar{B})}{P(\bar{B})} = \frac{P(A) - P(A \cap B)}{1 - P(B)}$

$\Rightarrow \quad P(A \cap B)[1 - P(B)] = P(B)[P(A) - P(A \cap B)]$

$\Rightarrow P(A \cap B) - P(A \cap B)P(B) = P(A)P(B) - P(B)P(A \cap B)$

$\Rightarrow \quad P(A \cap B) = P(A) \cdot P(B)$

Since, A and B are independent events. Hence, this statement is correct.

(c) Let E_1 be the white ball is drawn in first drawn, E_2 be the event that black is drawn in second drawn and E be the event that white ball is drawn in second drawn.

$\therefore \quad P(E) = P(E_1) \times P\left(\frac{E}{E_1}\right) + P(E_2) \times \left(\frac{E}{E_2}\right)$

$$= \left(\frac{w}{w+b}\right)\left(\frac{d+w}{w+b+d}\right) + \frac{b}{(w+b)}\left(\frac{w}{w+b+d}\right)$$

$$= \frac{w(d+w+b)}{(w+b)(w+b+d)} = \frac{w}{w+b}$$

which is independent of d. Hence, this statement is correct.

(d) To prove that A, B, C are pairwise independent only.

Now, $P(A \cap B) = P[(A \cap B \cap \bar{C}) \cup P(A \cap B \cap C)]$

$= P(A \cap B \cap \bar{C}) + P(A \cap B \cap C)$

$= P(A)P(B)P(\bar{C}) + P(A)P(B)P(C)$

$= P(A) \times P(B)[P(\bar{C}) + P(C)]$

$= P(A) \times P(B)$

Hence, this statement is correct.

16. (b) Consider the events A, B, C and D.

A = good, good, good, defective items

B = good, good, good, good items

C = good, good, defective, good items

D = good, good, defective, defective items

$$P(A) = \frac{6 \times 5 \times 4 \times 3}{9 \times 8 \times 7 \times 6}$$

$$P(B) = \frac{6 \times 5 \times 4 \times 3}{9 \times 8 \times 7 \times 6}$$

$$P(C) = \frac{6 \times 5 \times 4 \times 3}{9 \times 8 \times 7 \times 6}$$

$$P(D) = \frac{6 \times 5 \times 3 \times 2}{9 \times 8 \times 7 \times 6}$$

$\therefore$ Required probability $= \frac{P(B)}{P(A) + P(B) + P(C) + P(D)}$

$$= \frac{\frac{6 \times 5 \times 4 \times 3}{9 \times 8 \times 7 \times 6}}{\frac{360 + 360 + 360 + 180}{9 \times 8 \times 7 \times 6}}$$

$$= \frac{360}{1260} = \frac{2}{7}$$

Hence, option (b) is correct.

17. (d) Required probability $= \frac{P(A) + P(B)}{P(A) + P(B) + P(C) + P(D)}$

$$= \frac{\frac{360}{9 \times 8 \times 7 \times 6} + \frac{360}{9 \times 8 \times 7 \times 6}}{\frac{360 + 360 + 360 + 180}{9 \times 8 \times 7 \times 6}} = \frac{4}{7}$$

Hence, option (d) is correct.

18. (b) We have, b red balls, $2b$ white balls and $3b$ blue balls.

$\therefore$ Total number of balls $= b + 2b + 3b = 6b$

A = The event have no two of the selected balls are same colour i.e. all are different

$$\therefore \quad P(A) = \frac{{}^bC_1 \times {}^{2b}C_1 \times {}^{3b}C_1}{{}^{6b}C_3}$$

$$= \frac{b \times 2b \times 3b \times 1 \times 2 \times 3}{6b(6b-1)(6b-2)}$$

$$= \frac{6b^2}{36b^2 - 18b + 2}$$

Now, $P(A) = 0.3$

$\therefore \quad \frac{6b^2}{36b^2 - 18b + 2} = 0.3 = \frac{3}{10}$

$\Rightarrow \quad 20b^2 = 36b^2 - 18b + 2$

$\Rightarrow \quad 16b^2 - 18b + 2 = 0$

$\Rightarrow \quad 8b^2 - 9b + 1 = 0$

$\Rightarrow \quad 8b^2 - 8b - b + 1 = 0$

$\Rightarrow \quad (8b - 1)(b - 1) = 0$

$\therefore \quad b = 1$

So, there is exactly one value of b for which $P(A) = 0.3$ this is less than 10.

Hence, option (b) is correct.

19. (d) Required probability

$$= \frac{{}^{3b}C_0\, {}^{3b}C_3}{{}^{6b}C_3} + \frac{{}^{3b}C_1\, {}^{3b}C_2}{{}^{6b}C_3}$$

$$= \frac{3b(3b-1)(3b-2)}{6b(6b-1)(6b-2)} + \frac{3b(3b)(3b-1)\times 3}{6b(6b-1)(6b-2)}$$

$$= \frac{3b(3b-1)(3b-2+9b)}{6b(6b-1)(6b-2)}$$

$$= \frac{3b(3b-1)(12b-2)}{6b(6b-1)(6b-2)} = \frac{1}{2}$$

Hence, option (d) is correct.

20. (a) $P\left(\frac{A}{B}\right) = \frac{P(A \cap B)}{P(B)} = \frac{0.3}{0.5} = \frac{3}{5}$ $\quad [\because A \cap B = A, A \subset B]$

Hence, option (a) is correct.

21. (c) We have, A_1 have one balls {1}

A_2 have two balls {1, 2}

A_3 have three balls {1, 2, 3}

A_4 have four balls {1, 2, 3, 4}

$P(A_1) = \frac{1}{10}, \quad P(A_2) = \frac{2}{10}, \quad P(A_3) = \frac{3}{10}, \quad P(A_4) = \frac{4}{10}$

$P\left(\frac{E_1}{A_1}\right) = 1, \quad P\left(\frac{E_2}{A_2}\right) = \frac{1}{2}, \quad P\left(\frac{E_3}{A_3}\right) = \frac{1}{3}, \quad P\left(\frac{E_4}{A_4}\right) = \frac{1}{4}$

$$\therefore \sum_{i=1}^{4} P(E_i \cap A_i) = P(E_1 \cap A_1) + P(E_2 \cap A_2) + P(E_3 \cap A_3) + P(E_4 \cap A_4)$$

$$= P(A_1) \times P\left(\frac{E_1}{A_1}\right) + P(A_2) \times P\left(\frac{E_2}{A_2}\right) + P(A_3) \times P\left(\frac{E_3}{A_3}\right) + P(A_4) \times P\left(\frac{E_4}{A_4}\right)$$

$$= \frac{1}{10} \times 1 + \frac{2}{10} \times \frac{1}{2} + \frac{3}{10} \times \frac{1}{3} + \frac{4}{10} \times \frac{1}{4} = \frac{2}{5}$$

Hence, option (c) is correct.

22. (b) $$P\left(\frac{A_3}{E_2}\right) = \frac{P(A_3) \times P\left(\frac{E_2}{A_3}\right)}{\left[P(A_1) \times P\left(\frac{E_2}{A_1}\right) + P(A_2) \times P\left(\frac{E_2}{A_2}\right) + P(A_3) \times P\left(\frac{E_2}{A_3}\right) + P(A_4) \times P\left(\frac{E_2}{A_4}\right)\right]}$$

$$= \frac{\frac{3}{10} \times \frac{1}{3}}{\frac{1}{10} \times 0 + \frac{2}{10} \times \frac{1}{2} + \frac{3}{10} \times \frac{1}{3} + \frac{4}{10} \times \frac{1}{4}} = \frac{1}{3}$$

Hence, option (b) is correct.

23. (d) Sum of the four number is less than 5.

i.e. only one possible case (1 + 1 + 1 + 1) = 4

$$\therefore \text{Required probability} = \frac{1}{{}^1C_1 \times {}^2C_1 \times {}^3C_1 \times {}^4C_1} = \frac{1}{24}$$

Hence, option (d) is correct.

24. (a) Here, $P_n = P_{n-1} + P_{n-2}$ and $P_1 = 1, P_2 = 1$

So, $\quad P_3 = 2, P_4 = 3$

25. (b) Using, $P_n = P_{n-1} + P_{n-2}$

$\therefore \quad P_8 = 21$

26. (a) Since, n = odd game always gets over with A so, probability that A wins = 1.

27. (b) Total number of balls = (4 white + 9 black) = 13 balls

Total number of selection of 13 balls with replacement = 13^4

∴ Required probability

$P(BBBB) + P(BBBW) + P(BBWB) + P(BWBB) + P(WBBB) + P(WBBW) + P(WBWB) + P(BWBW)$

$$= \left(\frac{9}{13}\right)^4 + \frac{9^3 \times 4}{13^4} + \frac{9^3 \times 4}{13^4} + \frac{9^3 \times 4}{13^4} + \frac{9^3 \times 4}{13^4} + \frac{9^2 \times 4^2}{13^4} + \frac{9^2 \times 4^2}{13^4} + \frac{9^2 \times 4^2}{13^4}$$

$$= \frac{9^2}{13^4}(81 + 36 + 36 + 36 + 36 + 16 + 16 + 16)$$

$$= \frac{9^2}{13^4}(273) = \frac{81 \times 273}{(169)^2}$$

Hence, option (b) is correct.

28. (b)

29. (c) Since, $$P_{(r)} = P_{(r-2)}\left(\frac{4}{13}\right)^2 + P_{r-1}\left(\frac{9}{13}\right)$$

[by Recursion formula]

$$P_{(r)} = \frac{16 \cdot 12^r - (-3)^r}{15 \cdot (13)^r}$$

30. (b) Total number of selection of set of students $\{a_1, a_2, a_3, \ldots, a_n\}$ plans to visit two institutes next month = $2^n \times 2^n = 4^n$

Suppose, r students visit 1st institute, where r varies from 0 to n.

∴ Total number of selection = ${}^nC_r$ ways.

So, $(n - r)$ student visit 2nd institute, because 1st and 2nd institute have no common student.

∴ Total number of selection for $n - r = 2^{n-r}$ ways.

The total number of student visit 1st and 2nd institute is chosen in ${}^nC_r 2^{n-r}$ ways.

But r can varies from 0 to n.

$$\therefore \sum_{r=0}^{n} {}^nC_r 2^{n-r} = (1+2)^n = 3^n.$$

Now, required probability = $\frac{3^n}{4^n} = \left(\frac{3}{4}\right)^n$

Hence, option (b) is correct.

31. (c) Suppose, r and $n - r$ students visit 1st and 2nd institute, respectively. Any number of student chosen from r student, so total number of student selecting from either in r and $n - r = {}^nC_r 2^r$

But r can varie from 0 to n.

$$\therefore \text{Total number of ways for selecting} = \sum_{r=0}^{n} {}^nC_r 2^r = (1+2)^n = 3^n$$

$$\therefore \text{Hence, required probability} = \frac{3^n}{4^n} = \left(\frac{3}{4}\right)^n$$

Hence, option (c) is correct.

32. (d) Some number of student visit both institute.
Let r number of student visit both institute.
$\therefore$ Number of ways of selection $= {}^nC_r \times {}^nC_r = ({}^nC_r)^2$
But r can varies from 0 to n.

$$\therefore \quad \sum_{r=0}^{n} ({}^nC_r)^2 = {}^{2n}C_n$$

$$\therefore \quad \text{Required probability} = \frac{{}^{2n}C_n}{4^n}$$

Hence, option (d) is correct.

33. (c) $$P(E) = \frac{{}^9C_3\left[{}^3C_1 \cdot {}^2C_2 \cdot \frac{5!}{3!} + {}^3C_2 \cdot {}^1C_1 \cdot \frac{5!}{2!2!}\right]}{9^5}$$

$$= \frac{{}^9C_3(60 + 90)}{9^5}$$

$$= \frac{{}^9C_3 \times 150}{9^5}$$

34. (b) $$\text{Favourable} = \begin{bmatrix} a_{11} & a_{12} & a_{13} \\ a_{21} & a_{22} & a_{23} \\ a_{31} & a_{32} & a_{33} \end{bmatrix}$$

Here, $a_{11} = a_{22} = a_{33} = 0$
a_{12}, a_{13}, a_{23} can be filled in 9 ways each a_{21}, a_{31}, a_{32} has one way.

$$\therefore \quad n(E) = 9^3$$

$$\text{and} \quad n(S) = 9^9$$

$$\therefore \quad P(E) = \frac{n(E)}{n(S)}$$

$$= \frac{9^3}{9^9} = \frac{1}{9^6}$$

35. (b) Total number of ways of selecting five number from $\{-4, -3, -2, -1, 0, 1, 2, 3, 4\}$ with replacement $= 9^5$

Now, $x_1 + x_2 + x_3 + x_4 + x_5 = 0, -4 \le x_i \le 4$

$\therefore$ Favourable cases

= Coefficient of x^0 in
$$(x^{-4} + x^{-3} + x^{-2} + x^{-1} + x^0 + x^1 + x^2 + x^3 + x^4)^5$$

= Coefficient of x^0 in
$$\left(\frac{1 + x + x^2 + x^3 + x^4 + x^5 + x^6 + x^7 + x^8}{x^4}\right)^5$$

= Coefficient of x^{20} in
$$(1 + x + x^2 + x^3 + x^4 + x^5 + x^6 + x^7 + x^8)^5$$

= Coefficient of x^{20} in $\left(\frac{1 - x^9}{1 - x}\right)^5$

= Coefficient of x^{20} in $(1 - x^9)^5 (1 - x)^{-5}$

= Coefficient of x^{20} in $(1 - {}^5C_1x^9 + {}^5C_2x^{18} \ldots)(1 - x^{-5})$

= Coefficient of x^{20} in $(1 - x)^{-5} - 5$

= Coefficient of $x^{11}(1 - x)^{-5}$ + 10 coefficient of $x^2 (1 - x)^{-5}$

$= {}^{24}C_4 - 5 \times {}^{15}C_4 + 10 \times {}^6C_4$

$$= \frac{24!}{4!20!} - \frac{5 \times 15!}{4!11!} + 10 \times \frac{6!}{4!2!}$$

$= 10626 - 6825 + 150$
$= 3951$

$$\therefore \text{Required probability} = \frac{3951}{9^5}$$

Hence, option (b) is correct.

36. (3) Let, A be the even of choosing 5-digits number and E be the event the number is a palindrome ($abcba$).

$$\therefore \quad n(A) = 9(10^4) = 90000$$

$$\text{and} \quad n(A \cap B) = 9 \times 10^2 = 900$$

$$\therefore \quad P(B/A) = \frac{n(B \cap A)}{n(A)} = \frac{1}{100} = p$$

$$\therefore \quad 300p = 3$$

37. (8) Let n pairs of socks in drawer.
$\therefore$ Total number of socks in drawer $= 2n$
Total number of out comes when two socks are drawn $= {}^{2n}C_2$
Favourable case $= {}^nC_1$

$$\therefore \text{Required probability} = \frac{{}^nC_1}{{}^{2n}C_2} = \frac{1}{(2n - 1)}$$

$$\text{Now,} \quad \frac{1}{2n - 1} = \frac{1}{15}$$

$$\Rightarrow \quad n = 8$$

Hence, 8 pairs of socks in drawer.

38. (6) Total number of ways in which papers of 4 students, can be checked by seven teachers $= 7^4$

Now, choosing two teachers out of $7 = {}^7C_2 = 21$

The number of ways in which 4 papers can be checked by exactly two teachers $= 2^4 - 2 = 14$

$\therefore$ Favourable ways $= 21 \times 14$

$$\text{Required probability} = \frac{21 \times 14}{7^4} = \frac{6}{49}$$

$$\text{Now,} \quad 49A = 49 \times \frac{6}{49} = 6$$

39. (4) The total number of mapping is n^n. The number of one-one mapping is $n!$.

$$\text{Required probability} = \frac{n!}{n^n} = \frac{3}{32} = \frac{6}{64} = \frac{24}{256}$$

$$\Rightarrow \quad \frac{n!}{n^n} = \frac{4!}{4^4}$$

On comparing, we get $n = 4$

40. (5) We have, n identical red balls, $2n$ identical black balls and $3n$ identical white balls.

Total number of selection of n balls from n, $2n$ and $3n$ balls,

$$x_1 + x_2 + x_3 = n$$

= Coefficient of x^n $(1 + x + x^2 + \cdots + x^n)$
$$(1 + x + x^2 + \cdots + x^{2n})(1 + x + \cdots + x^{3n})$$

= Coefficient of x^n in $(1 - x^{n+1})(1 - x^{2n+1})(1 - x^{3n+1})(1 - x)^{-3}$.

= Coefficient of x^n in $(1 - x)^{-3} = {}^{n+3-1}C_{3-1} = {}^{n+2}C_2$

Favourable case $= 1 + 1 + 1 = 3$

$$\therefore \text{Required probability} = \frac{3}{{}^{n+2}C_2}$$

$$\therefore \quad \frac{6}{(n + 2)(n + 1)} \ge \frac{1}{6}$$

$$\Rightarrow \quad n^2 + 3n - 34 \ge 0$$

$$n = \frac{-3 \pm \sqrt{9 + 136}}{2} = \frac{-3 \pm \sqrt{145}}{2}$$

$$= \frac{-3 \pm 12.04}{2} > 4.56$$

$$\therefore \quad n \ge 5$$

JEE Main & AIEEE Archive

1. (a) Given that, $P(A \cap B) = \frac{1}{4}, P(\bar{A}) = \frac{1}{4}$

and $\overline{P(A \cup B)} = \frac{1}{6}$

$\Rightarrow \quad 1 - P(A \cup B) = \frac{1}{6} \quad [\because P(A) + P(\bar{A}) = 1]$

$\Rightarrow 1 - P(A) - P(B) + P(A \cap B) = \frac{1}{6}$

$[\because P(A \cup B) = P(A) + P(B) - P(A \cap B)]$

$\Rightarrow \quad P(\bar{A}) - P(B) + \frac{1}{4} = \frac{1}{6}$

$\Rightarrow \quad P(B) = \frac{1}{4} + \frac{1}{4} - \frac{1}{6}$

$\Rightarrow \quad P(B) = \frac{1}{3}$

and $\quad P(A) = \frac{3}{4}$

Now, $\quad P(A \cap B) = P(A)P(B) = \frac{3}{4} \times \frac{1}{3} = \frac{1}{4}$

Hence, the events A and B are independent but not equally likely.

2. (c) Probability of guessing a correct answer, $p = \frac{1}{3}$

and probability of guessing a wrong answer, $q = \frac{2}{3}$

So, the probability of guessing 4 or more correct answers

$$= {}^5C_4\left(\frac{1}{3}\right)^4 \cdot \frac{2}{3} + {}^5C_5\left(\frac{1}{3}\right)^5 = 5 \cdot \frac{2}{3^5} + \frac{1}{3^5} = \frac{11}{3^5}$$

3. (b) Given, 3 numbers are drawn without replacement from the set {1, 2, 3, 4, 5, 6, 7, 8}.

To find The probability that their minimum is 3 given that their maximum is 6.

Let A denotes the event that the minimum of the 3 selected numbers is 3 and B denotes the event that the maximum of the 3 selected numbers is 6.

$$\text{Required probability} = P\left(\frac{A}{B}\right)$$

Now, $\quad P\left(\frac{A}{B}\right) = \frac{P(A \cap B)}{P(B)}$

$P(B)$ = Probability that the maximum number selected is 6.

If the maximum number is 6, then

(i) 6 should be one of the 3 selected numbers (1 way) and

(ii) the remaining two numbers should be less than 6 i.e. any 2 from 1 to 5 (${}^5C_2$ ways).

$\Rightarrow \quad P(B) = \frac{1 \times {}^5C_2}{{}^8C_3} = \frac{10}{{}^8C_3}$

Similarly, $P(A \cap B)$ = The probability that the minimum number is 3 and the maximum number is 6.

If the minimum number is 3 and the maximum number is 6, then

(i) 6 should be one of the 3 selected numbers (1 way).

(ii) 3 should be one of the 3 selected numbers (1 way) and

(iii) the remaining 1 number should lie between 3 and 6 i.e. any one of 4 and 5 (2 ways).

$\Rightarrow \quad P(A \cap B) = \frac{1 \times 1 \times {}^2C_1}{{}^8C_3} = \frac{2}{{}^8C_3}$

$\therefore \quad P\left(\frac{A}{B}\right) = \frac{P(A \cap B)}{P(B)} = \frac{2}{10} = \frac{1}{5}$

Hence, the required probability is $\frac{1}{5}$.

4. (b) Here, $n = 5$ and $r \geq 1$

$\therefore \quad P(X = r) = {}^nC_r\, p^{n-r} q^r$

$P(X \geq 1) = 1 - P(x = 0)$

$= 1 - {}^5C_0 \cdot p^5 \cdot q^0 \geq \frac{31}{32} \quad$ [given]

$\Rightarrow \quad p^5 \leq 1 - \frac{31}{32} = \frac{1}{32}$

$\therefore \quad p \leq \frac{1}{2}$ and $p \geq 0 \Rightarrow p \in \left[0, \frac{1}{2}\right]$

5. (a) As, $\quad P\left(\frac{C}{D}\right) = \frac{P(C \cap D)}{P(D)} \quad \left[\begin{array}{l}\because C \subset D \\ \therefore P(C \cap D) = P(C)\end{array}\right]$

$= \frac{P(C)}{P(D)} \quad \ldots$(i)

Also, $\quad P(D) \leq 1$

$\therefore \quad \frac{1}{P(D)} \geq 1$

and $\quad \frac{P(C)}{P(D)} \geq P(C) \quad \ldots$(ii)

From Eqs. (i) and (ii), we get

$$P\left(\frac{C}{D}\right) = \frac{P(C)}{P(D)} \geq P(C)$$

6. (a) $P\left(\frac{A^C \cap B^C}{C}\right) = \frac{P(A^C \cap B^C \cap C)}{P(C)}$

$= \frac{P(C) - P(A \cap C) - P(B \cap C) + P(A \cap B \cap C)}{P(C)} \quad \ldots$(i)

Given, $P(A \cap B \cap C) = 0$ and A, B, C are pairwise independent.

$\therefore \quad P(A \cap C) = P(A) \cdot P(C)$

and $\quad P(B \cap C) = P(B) \cdot P(C)$

$\therefore P\left(\frac{A^C \cap B^C}{C}\right) = \frac{P(C) - P(A) \cdot P(C) - P(B) \cdot P(C) + 0}{P(C)}$

$= 1 - P(A) - P(B)$

$= P(A^C) - P(B)$

7. (b) Total number of cases $= {}^9C_3 = 84$

Number of favourable cases $= {}^3C_1 \cdot {}^4C_1 \cdot {}^2C_1$

$= 24$

$\therefore \quad$ Required probability $= \frac{24}{84} = \frac{2}{7}$

8. (c) Here, $n(S) = {}^{20}C_4$

Statement I

Common difference is 1;

Total number of cases = 17

Common difference is 2;

Total number of cases = 14

Common difference is 3;

Total number of cases = 11

Common difference is 4;

Total number of cases = 8

Common difference is 5;

Total number of cases = 5

Common difference is 6;

Total number of cases = 2

$\therefore$ Required probability $= \frac{17 + 14 + 11 + 8 + 5 + 2}{{}^{20}C_4} = \frac{1}{85}$

Hence, Statement I is correct and Statement II is incorrect.

9. (a) According to the given condition,

$$1-\left(\frac{3}{4}\right)^n \geq \frac{9}{10}$$

$$\Rightarrow \quad \left(\frac{3}{4}\right)^n \leq 1-\frac{9}{10}=\frac{1}{10}$$

$$\Rightarrow \quad \left(\frac{4}{3}\right)^n \geq 10$$

Taking log on both sides, we get

$$n[\log 4-\log 3] \geq \log_{10} 10=1$$

$$\Rightarrow \quad n \geq \frac{1}{\log_{10} 4-\log_{10} 3}$$

10. (a) We have, $S=\{00, 01, 02, \ldots, 49\}$

Let A be the event that sum of the digits on the selected ticket is 8, then

$$A=\{08, 17, 26, 35, 44\}$$

Let B be the event that the product of the digits is zero, then

$$B=\{00, 01, 02, 03, \ldots, 09, 10, 20, 30, 40\}$$

$$\therefore \quad A \cap B=\{8\}$$

$\therefore$ Required probability $=P\left(\frac{A}{B}\right)=\frac{P(A \cap B)}{P(B)}$

$$=\frac{\frac{1}{50}}{\frac{14}{50}}=\frac{1}{14}$$

11. (c) Given that, $P(A)=\frac{1}{4}, P\left(\frac{A}{B}\right)=\frac{1}{2}$ and $P\left(\frac{B}{A}\right)=\frac{2}{3}$

We know that,

$$P\left(\frac{A}{B}\right)=\frac{P(A \cap B)}{P(B)} \quad \ldots(i)$$

and

$$P\left(\frac{B}{A}\right)=\frac{P(B \cap A)}{P(A)} \quad \ldots(ii)$$

$$\therefore \quad P(B)=\frac{P\left(\frac{B}{A}\right) \cdot P(A)}{P\left(\frac{A}{B}\right)}=\frac{\left(\frac{2}{3}\right)\left(\frac{1}{4}\right)}{\left(\frac{1}{2}\right)}=\frac{1}{3}$$

12. (d) Here, $A=\{4, 5, 6\}$ and $B=\{1, 2, 3, 4\}$

$\therefore \quad A \cap B=\{4\}$ [by addition theorem of probability]

Now, $\quad P(A \cup B)=P(A)+P(B)-P(A \cap B)$

$$\therefore \quad P(A \cup B)=\frac{3}{6}+\frac{4}{6}-\frac{1}{6}=1$$

13. (d) Probability of getting score 9 in a single throw $=\frac{4}{36}=\frac{1}{9}$

Probability of getting score 9 exactly in double throw

$$={}^3C_2 \times\left(\frac{1}{9}\right)^2 \times \frac{8}{9}=\frac{8}{243}$$

14. (c) Let the events,

$A=$ Ist aeroplane hit the target

$B=$ IInd aeroplane hit the target

and their corresponding probabilities are

$$P(A)=0.3 \text{ and } P(B)=0.2$$

$$\Rightarrow \quad P(\bar{A})=0.7 \text{ and } P(\bar{B})=0.8$$

$\therefore$ Required probability

$=P(\bar{A}) P(B)+P(\bar{A}) P(\bar{B}) P(\bar{A}) P(B)+\ldots$

$=(0.7)(0.2)+(0.7)(0.8)(0.7)(0.2)$

$+(0.7)(0.8)(0.7)(0.8)(0.7)(0.2)+\ldots$

$=0.14\left[1+(0.56)+(0.56)^2+\ldots\right]$

$$=0.14\left(\frac{1}{1-0.56}\right)=\frac{0.14}{0.44}=\frac{7}{22}=0.32$$

15. (a) Required probability $=P(X=0)+P(X=1)$

$$=\frac{e^{-5}}{0!} \cdot 5^0+\frac{e^{-5}}{1!} \cdot 5^1$$

$$=e^{-5}+5 e^{-5}=\frac{6}{e^5}$$

16. (b) Probability of getting success, $p=\frac{1}{2}$

and probability of failure, $q=\frac{1}{2}$

$\therefore \quad$ Required probability $={}^7C_2\left(\frac{1}{6}\right)^2\left(\frac{5}{6}\right)^5 \times \frac{1}{6}=\frac{{}^7C_2 \times 5^5}{6^8}$

17. (c) All the three persons has three options to apply a house.

$\therefore$ Total number of cases $=3^3$

Now, favourable cases $=3$ (as either all has applied for house 1 or 2 or 3)

$$\therefore \quad \text{Required probability}=\frac{3}{3^3}=\frac{1}{9}$$

18. (b) $P(X>1.5)=P(2)+P(3)+\cdots \infty$

$$=1-[P(0)+P(1)]$$

$$=1-\left(e^{-2}+\frac{e^{-2} \times 2}{1}\right)=1-\frac{3}{e^2}$$

19. (c) Given, probabilities of speaking truth are

$$P(A)=\frac{4}{5} \text{ and } P(B)=\frac{3}{4}$$

And their corresponding probabilities of not speaking truth are

$$P(\bar{A})=\frac{1}{5} \text{ and } P(\bar{B})=\frac{1}{4}$$

The probability that they contradict each other

$$=P(A) \times P(\bar{B})+P(\bar{A}) \times P(B)$$

$$=\frac{4}{5} \times \frac{1}{4}+\frac{1}{5} \times \frac{3}{4}$$

$$=\frac{1}{5}+\frac{3}{20}=\frac{7}{20}$$

20. (b) Given, $E=\{X$ is a prime number$\}=\{2, 3, 5, 7\}$

$$P(E)=P(X=2)+P(X=3)+P(X=5)+P(X=7)$$

$$=0.23+0.12+0.20+0.07=0.62$$

and $\quad F=\{X<4\}=\{1, 2, 3\}$

$$\Rightarrow \quad P(F)=P(X=1)+P(X=2)+P(X=3)$$

$$=0.15+0.23+0.12=0.5$$

and $\quad E \cap F=\{X$ is prime number as well as $<4\}=\{2, 3\}$

$$P(E \cap F)=P(X=2)+P(X=3)$$

$$=0.23+0.12=0.35$$

$\therefore$ Required probability,

$$P(E \cup F)=P(E)+P(F)-P(E \cap F)$$

$$=0.62+0.5-0.35=0.77$$

21. (d) Given that, mean $=4 \Rightarrow np=4$ and variance $=2$

$$\Rightarrow \quad npq=2 \Rightarrow 4q=2$$

$$\Rightarrow \quad q=\frac{1}{2}$$

$$\therefore \quad p=1-q=1-\frac{1}{2}=\frac{1}{2}$$

Also, $\quad n=8$

Probability of 2 successes $=P(X=2)={}^8C_2 p^2 q^6$

$$=\frac{8!}{2! \times 6!} \times\left(\frac{1}{2}\right)^2 \times\left(\frac{1}{2}\right)^6$$

$$=28 \times \frac{1}{2^8}=\frac{28}{256}$$

22. (d) The probability that Mr. A selected the loosing horse

$$= \frac{4}{5} \times \frac{3}{4} = \frac{3}{5}$$

The probability that Mr. A selected the winning horse $= 1 - \frac{3}{5}$

$$= \frac{2}{5}$$

23. (a) Since, $0 \le P(A) \le 1$, $0 \le P(B) \le 1$, $0 \le P(C) \le 1$

and $0 \le P(A) + P(B) + P(C) \le 1$

$\therefore \quad 0 \le \frac{3x+1}{3} \le 1$

$\Rightarrow \quad -\frac{1}{3} \le x \le \frac{2}{3}$ ...(i)

$0 \le \frac{1-x}{4} \le 1$

$\Rightarrow \quad -3 \le x \le 1$ ...(ii)

$0 \le \frac{1-2x}{2} \le 1$

$\Rightarrow \quad -\frac{1}{2} \le x \le \frac{1}{2}$ ...(iii)

and $\quad 0 \le \frac{3x+1}{3} + \frac{1-x}{4} + \frac{1-2x}{2} \le 1$

$\Rightarrow \quad 0 \le 13 - 3x \le 12$

$\Rightarrow \quad \frac{1}{3} \le x \le \frac{13}{3}$ ...(iv)

From Eqs. (i), (ii), (iii) and (iv), we get

$$\frac{1}{3} \le x \le \frac{1}{2}$$

24. (a) Given that, for Binomial distribution mean, $np = 4$ and variance, $npq = 2$

$\therefore \quad q = \frac{1}{2}$, but $p + q = 1$

$\Rightarrow \quad p = \frac{1}{2}$

and $\quad n \times \frac{1}{2} = 4 \Rightarrow n = 8$

We know that,

$$P(X = r) = {}^nC_r p^r q^{n-r}$$

$$\therefore \quad P(X = 1) = {}^8C_1 \left(\frac{1}{2}\right)^7 \left(\frac{1}{2}\right)^1$$

$$= 8 \times \frac{1}{2^8}$$

$$= \frac{1}{2^5} = \frac{1}{32}$$

25. (a) Since, the probabilities of solving the problem by A, B and C are $\frac{1}{2}, \frac{1}{3}$ and $\frac{1}{4}$, respectively.

$\therefore$ Probability that the problem is not solved $= P(\bar{A})\, P(\bar{B})\, P(\bar{C})$

$$= \left(1 - \frac{1}{2}\right)\left(1 - \frac{1}{3}\right)\left(1 - \frac{1}{4}\right) = \frac{1}{2} \times \frac{2}{3} \times \frac{3}{4} = \frac{1}{4}$$

Hence, the probability that the problem is solved $= 1 - \frac{1}{4} = \frac{3}{4}$

26. (b) The total number of ways in which numbers can be choosed $= 25 \times 25 = 625$

The number of ways in which either players can choose same numbers $= 25$

$\therefore$ Probability that they win a prize $= \frac{25}{625} = \frac{1}{25}$

Thus, the probability that they will not win a prize in a single trial

$$= 1 - \frac{1}{25} = \frac{24}{25}$$

27. (d) Since, A and B are two mutually exclusive events.

$\therefore \quad A \cap B = \phi \Rightarrow$ Either $A \subseteq \bar{B}$ or $B \subseteq \bar{A}$

$\Rightarrow \quad P(A) \le P(\bar{B})$ or $P(B) \le P(\bar{A})$

28. (b) Let A_1, A_2 and A_3 be the events of match winning in first, second and third matches respectively and whose probabilities are

$$P(A_1) = P(A_2) = P(A_3) = \frac{1}{2}$$

$\therefore$ Required probability $= P(A_1 A'_2 A_3) + P(A'_1 A_2 A_3)$

$$= P(A_1)P(A'_2)P(A_3) + P(A'_1)P(A_2)P(A_3)$$

$$= \left(\frac{1}{2}\right)^3 + \left(\frac{1}{2}\right)^3 = \frac{1}{8} + \frac{1}{8} = \frac{1}{4}$$

29. (a) Let the probability of getting a head be p and not getting a head be q.

Since, head appears first time in an even throw 2 or 4 or 6 ...

$\therefore \quad \frac{2}{5} = qp + q^3p + q^5p + \ldots$

$\Rightarrow \quad \frac{2}{5} = \frac{qp}{1-q^2}$

$\Rightarrow \quad \frac{2}{5} = \frac{(1-p)p}{1-(1-p)^2}$ $\quad [\because q = 1 - p]$

$\Rightarrow \quad \frac{2}{5} = \frac{1-p}{2-p}$

$\Rightarrow \quad 4 - 2p = 5 - 5p$

$\therefore \quad p = \frac{1}{3}$

JEE Advanced & IIT JEE Archive

1. (a) 3 boys and 2 girls

(1) B (2) B (3) B (4)

Girl can't occupy 4th position. Either girls can occupy 2 of 1,2,3 position or they can both be a position (1) or (2).

Hence, the total number of ways in which a girl can be seated is

$({}^3C_2 \times 2! \times 3!) + ({}^2C_1 \times 2! \times 3!) = 36 + 24 = 60$

$\therefore$ Required probability $= \frac{60}{120} = \frac{1}{2}$

2. (b) Probability $= \frac{\text{Number of favourable outcomes}}{\text{Number of total outcomes}}$

As, $x_1 + x_2 + x_3$ is odd.

So, all may be odd or one of them is odd and other two are even.

$\therefore$ Required probability

$$= \frac{{}^2C_1 \times {}^3C_1 \times {}^4C_1 + {}^1C_1 \times {}^2C_1 \times {}^4C_1 + {}^2C_1 \times {}^2C_1 \times {}^3C_1 + {}^1C_1 \times {}^3C_1 \times {}^3C_1}{{}^3C_1 \times {}^5C_1 \times {}^7C_1}$$

$$= \frac{24 + 8 + 12 + 9}{105} = \frac{53}{105}$$

3. (c) If x_1, x_2, x_3 are in AP.

Then, $\quad x_2 - x_1 = x_3 - x_2$

$\Rightarrow \quad 2x_2 = x_1 + x_3$

Since, x_1, x_2, x_3 are in AP, then

$$x_1 + x_3 = 2x_2$$

So, $x_1 + x_3$ should be even number.

Either both x_1 and x_3 are odd or both are even.

$\therefore$ Required probability $= \dfrac{{}^2C_1 \times {}^4C_1 + {}^1C_1 \times {}^3C_1}{{}^3C_1 \times {}^5C_1 \times {}^7C_1}$

$$= \frac{11}{105}$$

4. (a) It is simple application of independent event, us to solve a certain problem or any type of compitition each event in independent of other.

$P(A \cap B) = P(A) \cdot P(B)$, when A and B are independent events.

Probability that the problem is solved correctly by atleast one of them = 1 − (Problem is not solved by all)

$\therefore$ P (Problem is solved) = 1 − P (Problem is solved)

$$= 1 - P(\bar{A}) \cdot P(\bar{B}) \cdot P(\bar{C}) \cdot P(\bar{D})$$

$$= 1 - \left(\frac{1}{2} \cdot \frac{1}{4} \cdot \frac{3}{4} \cdot \frac{7}{8}\right)$$

$$= 1 - \frac{21}{256} = \frac{235}{256}$$

5. (a)As one of the dice shows a number appearing on one of P_1, P_2 and P_3.

Thus, three cases arises.

(i) All shows same number.

(ii) Number appearing on D_4 appears on any one of D_1, D_2, D_3.

(iii) Number appearing on D_4 appears on any two of D_1, D_2, D_3.

Sample space = $6 \times 6 \times 6 \times 6 = 6^4$ favourable events.

= Case I or Case II or Case III

Case I First we should select one number for D_4 which appears on all i.e. ${}^6C_1 \times 1$.

Case II For D_4 there are ${}^6C_1$ ways. Now, it appears on any one of D_1, D_2, D_3 i.e. ${}^3C_1 \times 1$.

For other two, there are 5×5 ways

$\Rightarrow$ $= {}^6C_1 \times {}^3C_1 \times 1 \times 5 \times 5$

Case III For D_4 there are ${}^6C_1$ ways now it appears on any two of D_1, D_2, D_3 $\Rightarrow$ ${}^3C_2 \times 1^2$

For other one, there are 5 ways

$$= {}^6C_1 \times {}^3C_2 \times 1^2 \times 5$$

Thus, probability $= \dfrac{{}^6C_1 + {}^6C_1 \times {}^3C_1 \times 5^2 + {}^6C_1 \times {}^3C_2 \times 5}{6^4}$

$$= \frac{6(1 + 75 + 15)}{6^4} = \frac{91}{216}$$

6. (b,d) It is based on law of total probability and Baye's Law.

It is given that ship would work, if atleast two of engines must work. If X be event that the ship works. Then, $X \Rightarrow$ either any two of E_1, E_2, E_3 works or all three engines E_1, E_2, E_3 works.

Given, $P(E_1) = \frac{1}{2}$, $P(E_2) = \frac{1}{4}$, $P(E_3) = \frac{1}{4}$

$$\therefore \quad P(X) = \left\{\begin{matrix} P(E_1 \cap E_2 \cap \bar{E}_3) + P(E_1 \cap \bar{E}_2 \cap E_3) \\ + P(\bar{E}_1 \cap E_2 \cap E_3) + P(E_1 \cap E_2 \cap E_3) \end{matrix}\right\}$$

$$= \left(\frac{1}{2} \cdot \frac{1}{4} \cdot \frac{3}{4} + \frac{1}{2} \cdot \frac{3}{4} \cdot \frac{1}{4} + \frac{1}{2} \cdot \frac{1}{4} \cdot \frac{1}{4}\right) + \left(\frac{1}{2} \cdot \frac{1}{4} \cdot \frac{1}{4}\right) = 1/4$$

(a) $P(X_1^c / X) = P\left(\dfrac{X_1^c \cap X}{P(X)}\right) = \dfrac{P(\bar{E}_1 \cap E_2 \cap E_3)}{P(X)}$

$$= \frac{\frac{1}{2} \cdot \frac{1}{4} \cdot \frac{1}{4}}{\frac{1}{4}} = \frac{1}{8}$$

(b) P (exactly two engines of the ship are functioning)

$$= \frac{P(E_1 \cap E_2 \cap \bar{E}_3) + P(E_1 \cap \bar{E}_2 \cap E_3) + P(\bar{E}_1 \cap E_2 \cap E_3)}{P(X)}$$

$$= \frac{\frac{1}{2} \cdot \frac{1}{4} \cdot \frac{3}{4} + \frac{1}{2} \cdot \frac{3}{4} \cdot \frac{1}{4} + \frac{1}{2} \cdot \frac{1}{4} \cdot \frac{1}{4}}{\frac{1}{4}} = \frac{7}{8}$$

(c) $P\left(\dfrac{X}{X_2}\right) = \dfrac{P(X \cap X_2)}{P(X_2)}$

$$= \frac{P \text{ (Ship is operating with } E_2 \text{ function)}}{P(X_2)}$$

$$= \frac{P(E_1 \cap E_2 \cap \bar{E}_3) + P(\bar{E}_1 \cap E_2 \cap E_3) + P(E_1 \cap E_2 \cap E_3)}{P(E_2)}$$

$$= \frac{\frac{1}{2} \cdot \frac{1}{4} \cdot \frac{3}{4} + \frac{1}{2} \cdot \frac{1}{4} \cdot \frac{1}{4} + \frac{1}{2} \cdot \frac{1}{4} \cdot \frac{1}{4}}{\frac{1}{4}} = \frac{5}{8}$$

(d) $P(X / X_1) = \dfrac{P(X \cap X_1)}{P(X_1)} = \dfrac{\frac{1}{2} \cdot \frac{1}{4} \cdot \frac{1}{4} + \frac{1}{2} \cdot \frac{3}{4} \cdot \frac{1}{4} + \frac{1}{2} \cdot \frac{1}{4} \cdot \frac{3}{4}}{1/2} = \dfrac{7}{16}$

7. (a, b)

(i) Conditional probability i.e. $P(A/B) = \dfrac{P(A \cap B)}{P(B)}$

(ii) $P(A \cup B) = P(A) + P(B) - P(A \cap B)$

(iii) Independent event, then

$$P(A \cap B) = P(A) \cdot P(B)$$

Here, $P(X/Y) = \frac{1}{2}$, $P\left(\frac{Y}{X}\right) = \frac{1}{3}$ and $P(X \cap Y) = 6$

$\therefore$ $P\left(\dfrac{X}{Y}\right) = \dfrac{P(X \cap Y)}{P(Y)}$

$\Rightarrow$ $\dfrac{1}{2} = \dfrac{1/6}{P(Y)}$

$\Rightarrow$ $P(Y) = \dfrac{1}{3}$ ...(i)

$P\left(\dfrac{Y}{X}\right) = \dfrac{1}{3}$

$\Rightarrow$ $\dfrac{P(X \cap Y)}{P(X)} = \dfrac{1}{3}$

$\Rightarrow$ $\dfrac{1}{6} = \dfrac{1}{3} P(X)$

$\therefore$ $P(X) = \dfrac{1}{2}$ ...(ii)

$P(X \cup Y) = P(X) + P(Y) - P(X \cap Y)$

$= \dfrac{1}{2} + \dfrac{1}{3} - \dfrac{1}{6} = \dfrac{2}{3}$ ...(iii)

$P(X \cap Y) = \dfrac{1}{6}$

and $P(X) \cdot P(Y) = \dfrac{1}{2} \cdot \dfrac{1}{3} = \dfrac{1}{6}$

$\Rightarrow$ $P(X \cap Y) = P(X) \cdot P(Y)$

for independent events, ...(iv)

$P(X^C \cap Y) = P(Y) - P(X \cap Y)$

$= \dfrac{1}{3} - \dfrac{1}{6} = \dfrac{1}{6}$ ...(v)

8. (a, d)

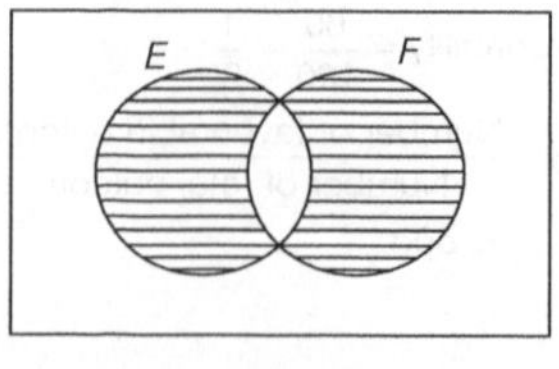

$$P(E \cup F) - P(E \cap F) = \frac{11}{25} \quad \dots\text{(i)}$$

(i.e. only E or only F)

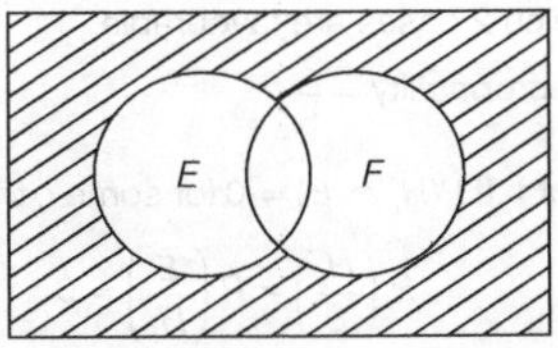

Neither of them occurs $= \frac{2}{25}$

$$\Rightarrow \quad P(\bar{E} \cap \bar{F}) = \frac{2}{25} \quad \dots\text{(ii)}$$

From Eq. (i),

$$P(E) + P(F) - 2P(E \cap F) = \frac{11}{25} \quad \dots\text{(iii)}$$

From Eq. (ii),

$$(1 - P(E))(1 - P(F)) = \frac{2}{25}$$

$$\Rightarrow \quad 1 - P(E) - P(F) + P(E) \cdot P(F) = \frac{2}{25} \quad \dots\text{(iv)}$$

From Eqs. (iii) and (iv), we get

$$P(E) + P(F) = \frac{7}{5} \text{ and } P(E) \cdot P(F) = \frac{12}{25}$$

$$\therefore \quad P(E) \cdot \left\{\frac{7}{5} - P(E)\right\} = \frac{12}{25}$$

$$\Rightarrow \quad [P(E)]^2 - \frac{7}{5}P(E) + \frac{12}{25} = 0$$

$$\Rightarrow \quad \left(P(E) - \frac{3}{5}\right)\left(P(E) - \frac{4}{5}\right) = 0$$

$$\therefore \quad P(E) = \frac{3}{5} \text{ or } \frac{4}{5}$$

$$\Rightarrow \quad P(F) = \frac{4}{5} \text{ or } \frac{3}{5}$$

Explanations for Q. Nos. (9-10)

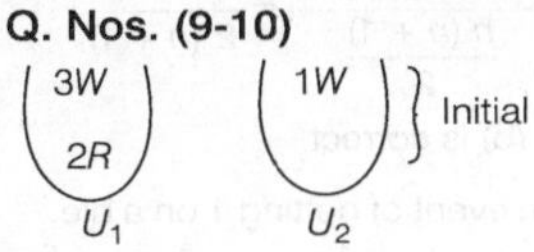

Head appears

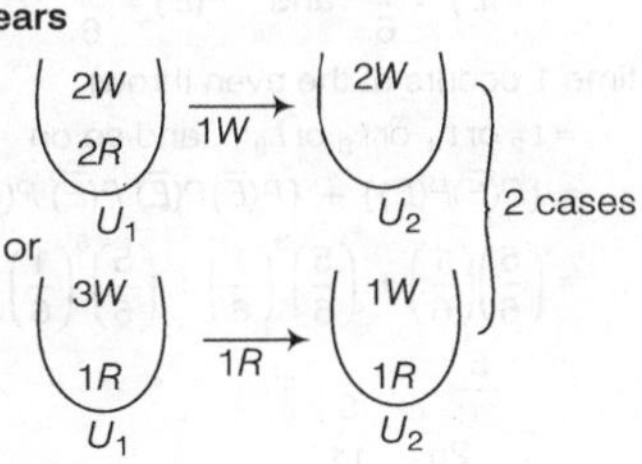

Tail appears

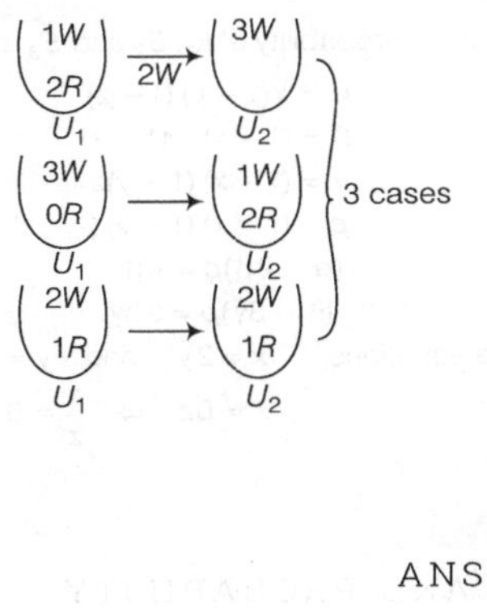

9. (b) Now, probability of the drawn ball from U_2 being white is

$$P(\text{white}/U_2) = P(H) \cdot \left\{\frac{{}^3C_1}{{}^5C_1} \times \frac{{}^2C_1}{{}^2C_1} + \frac{{}^2C_1}{{}^5C_1} \times \frac{{}^1C_1}{{}^2C_1}\right\}$$

$$+ P(H)\left\{\frac{{}^3C_2}{{}^5C_2} \times \frac{{}^3C_2}{{}^3C_2} + \frac{{}^2C_2}{{}^5C_2} \times \frac{{}^1C_1}{{}^3C_2} + \frac{{}^3C_1 \cdot {}^2C_1}{{}^5C_2} \times \frac{{}^2C_1}{{}^3C_2}\right\}$$

$$P(\text{white}/U_2) = \frac{1}{2}\left\{\frac{3}{5} \times 1 + \frac{2}{5} \times \frac{1}{2}\right\}$$

$$+ \frac{1}{2}\left\{\frac{3}{10} \times 1 + \frac{1}{10} \times \frac{1}{3} + \frac{6}{10} \times \frac{2}{3}\right\} = \frac{23}{30}$$

10. (d) P (Head appeared/white from V_2)

$$= P(H) \cdot \frac{\left\{\frac{{}^3C_1}{{}^5C_1} \times \frac{{}^2C_1}{{}^2C_1} + \frac{{}^2C_1}{{}^5C_1} \times \frac{{}^1C_1}{{}^2C_1}\right\}}{23/30}$$

$$= \frac{1}{2} \frac{\left\{\frac{3}{5} \times 1 + \frac{2}{5} \times \frac{1}{2}\right\}}{20/30}$$

$$= \frac{12}{23}$$

11. (c) From the tree-diagram it follows that

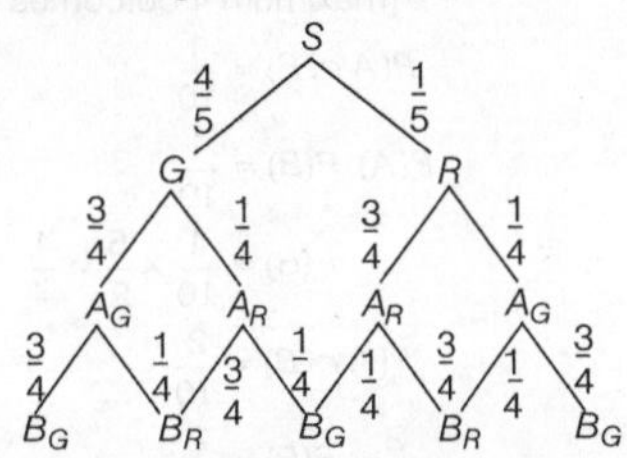

$$P(B_G) = \frac{46}{80}, \quad P(B_G / G) = \frac{10}{16} = \frac{5}{8}$$

$$\therefore \quad P(B_G \cap G) = \frac{5}{8} \times \frac{4}{5} = \frac{1}{2}$$

$$P(G / B_G) = \frac{\frac{1}{2}}{P(B_G)} = \frac{1}{2} \times \frac{80}{46} = \frac{20}{23}$$

12. (c) **Sample space** A die is thrown thrice, $n(S) = 6 \times 6 \times 6$.

Favorable events $\omega^{r_1} + \omega^{r_2} + \omega^{r_3} = 0$

i.e. (r_1, r_2, r_3) are ordered 3-triples which can take values,

$$\left.\begin{matrix}(1,2,3), & (1,5,3), & (4,2,3), & (4,5,3) \\ (1,2,6), & (1,5,6), & (4,2,6), & (4,5,6)\end{matrix}\right\} \text{ i.e. 8 ordered pairs and each}$$

can be arranged in $3!$ ways $= 6$

$$\therefore \quad n(E) = 8 \times 6$$

$$\Rightarrow \quad P(E) = \frac{8 \times 6}{6 \times 6 \times 6} = \frac{2}{9}$$

13. (a) $P(X = 3) = \frac{5}{6} \cdot \frac{5}{6} \cdot \frac{1}{6} = \frac{25}{216}$

14. (b) $P(X \ge 3) = \frac{5}{6} \cdot \frac{5}{6} \cdot 1 = \frac{25}{36}$

15. (d) $P[(X \ge 6)/(X > 3)] = \frac{P[(X > 3)/(X \ge 6)] \cdot P(X \ge 6)}{P(X > 3)}$

$$= \frac{1 \cdot \left[\left(\frac{5}{6}\right)^5 \cdot \left(\frac{1}{6}\right) + \left(\frac{5}{6}\right)^6 \cdot \left(\frac{1}{6}\right) + \cdots \infty\right]}{\left[\left(\frac{5}{6}\right)^3 \cdot \frac{1}{6} + \left(\frac{5}{6}\right)^4 \cdot \frac{1}{6} + \cdots \infty\right]}$$

$$= \frac{25}{36}$$

16. (b) The number of all possible determinants of the form

$$\begin{vmatrix} a & b \\ c & d \end{vmatrix} = 2^4 = 16$$

Out of which only 10 determinants given by

$$\begin{vmatrix} 1 & 1 \\ 1 & 1 \end{vmatrix}, \begin{vmatrix} 0 & 0 \\ 0 & 0 \end{vmatrix}, \begin{vmatrix} 1 & 1 \\ 0 & 0 \end{vmatrix}, \begin{vmatrix} 0 & 0 \\ 1 & 1 \end{vmatrix}, \begin{vmatrix} 0 & 1 \\ 0 & 1 \end{vmatrix}, \begin{vmatrix} 1 & 0 \\ 1 & 0 \end{vmatrix}, \begin{vmatrix} 1 & 0 \\ 0 & 0 \end{vmatrix}, \begin{vmatrix} 0 & 1 \\ 0 & 0 \end{vmatrix}, \begin{vmatrix} 0 & 0 \\ 1 & 0 \end{vmatrix}$$

Vanish and remaining six determinants have non-zero values.

$\therefore$ Required probability $= \frac{6}{16} = \frac{3}{8}$

So, Statement I is correct.

Hence, Statement II is also correct as the homogeneous equations have always a solution and Statement II does not explain the explanation of Statement I.

17. (d) Since, $P(A) = \frac{2}{5}$

For independent events,

$$P(A \cap B) = P(A) \cdot P(B) \Rightarrow P(A \cap B) \le \frac{2}{5}$$

$$\Rightarrow \quad P(A \cap B) = \frac{1}{10}, \frac{2}{10}, \frac{3}{10}, \frac{4}{10}$$

[maximum 4 outcomes may be in $A \cap B$]

1. Now, $P(A \cap B) = \frac{1}{10}$

$\Rightarrow \quad P(A) \cdot P(B) = \frac{1}{10}$

$\Rightarrow \quad P(B) = \frac{1}{10} \times \frac{5}{2} = \frac{1}{4}$ [not possible]

2. Now, $P(A \cap B) = \frac{2}{10}$

$\Rightarrow \quad \frac{2}{5} \times P(B) = \frac{2}{10}$

$\Rightarrow \quad P(B) = \frac{5}{10}$ [outcomes of $B = 5$]

3. Now, $P(A \cap B) = \frac{3}{10}$

$\Rightarrow \quad P(A)P(B) = \frac{3}{10}$

$\Rightarrow \quad \frac{2}{5} \times P(B) = \frac{3}{10}$

$\Rightarrow \quad P(B) = \frac{3}{4}$ [not possible]

4. Now, $P(A \cap B) = \frac{4}{10}$

$\Rightarrow \quad P(A) \cdot P(B) = \frac{4}{10}$

$\Rightarrow \quad P(B) = 1$ [outcomes of $B = 10$]

18. (c) $P\left(\frac{E^c \cap F^c}{G}\right) = \frac{P(E^c \cap F^c \cap G)}{P(G)}$

$$= \frac{P(G) - P(E \cap G) - P(G \cap F)}{P(G)}$$

$$= \frac{P(G)[1 - P(E) - P(F)]}{P(G)} \quad [\because P(G) \ne 0]$$

$$= 1 - P(E) - P(F) = P(E^c) - P(F)$$

19. (c) Let E = event when each American man is seated adjacent to his wife

and A = event when Indian man is seated adjacent to his wife

Now, $n(A \cap E) = (4!) \times (2!)^5$

Even when each American man is seated adjacent to his wife.

Again, $n(E) = (5!) \times (2!)^4$

$$\therefore \quad P\left(\frac{A}{E}\right) = \frac{n(A \cap E)}{n(E)} = \frac{(4!) \times (2!)^5}{(5!) \times (2!)^4} = \frac{2}{5}$$

Alternate Method

Fixing four American couples and one Indian man in between any two couples; we have 5 different ways in which his wife can be seated, of which 2 cases are favourable.

$\therefore$ Required probability $= \frac{2}{5}$

20. (d) **Statement I** If $P(H_i \cap E) = 0$ for some i, then

$$P\left(\frac{H_i}{E}\right) = P\left(\frac{E}{H_i}\right) = 0$$

If $P(H_i \cap E) \ne 0$ for $\forall \; = 1, 2, \dots, n$, then

$$P\left(\frac{H_i}{E}\right) = \frac{P(H_i \cap E)}{P(H_i)} \times \frac{P(H_i)}{P(E)}$$

$$= \frac{P\left(\frac{E}{H_i}\right) \times P(H_i)}{P(E)} > P\left(\frac{E}{H_i}\right) \cdot P(H_i) \quad [\because 0 < P(E) < 1]$$

Hence, Statement I may not always be true.

Statement II Clearly, $H_1 \cup H_2 \cup \dots \cup H_n = S$ (sample space)

$\Rightarrow \quad P(H_1) + P(H_2) + \dots + P(H_n) = 1$

21. (b) Here, $P(u_i) = ki, \Sigma P(u_i) = 1 \Rightarrow k = \frac{2}{n(n+1)}$

$$\therefore \quad \lim_{n \to \infty} P(W) = \lim_{n \to \infty} \sum_{i=1}^{n} \frac{2i^2}{n(n+1)^2}$$

$$= \lim_{n \to \infty} \frac{2n(n+1)(2n+1)}{6n(n+1)^2} = \frac{2}{3}$$

Hence, option (b) is correct.

22. (a) $P\left(\frac{u_n}{W}\right) = \frac{\frac{n}{n+1}}{\frac{\Sigma i}{n+1}} = \frac{2}{n+1}$

Hence, option (a) is correct.

23. (b) $P\left(\frac{W}{E}\right) = \frac{2 + 4 + 6 + \cdots}{\frac{n(n+1)}{2}} = \frac{n+2}{2(n+1)}$

Hence, option (b) is correct.

24. (b) Let E be the event of getting 1 on a die.

$\Rightarrow \quad P(E) = \frac{1}{6}$ and $P(\bar{E}) = \frac{5}{6}$

$\therefore$ P (first time 1 occurs at the even throw)

$= t_2$ or t_4 or t_6 or t_8 ... and so on

$= \{P(\bar{E})P(E)\} + \{P(\bar{E})P(\bar{E})P(\bar{E})P(E)\} + \cdots \infty$

$$= \left(\frac{5}{6}\right)\left(\frac{1}{6}\right) + \left(\frac{5}{6}\right)^3\left(\frac{1}{6}\right) + \left(\frac{5}{6}\right)^5\left(\frac{1}{6}\right) + \cdots \infty$$

$$= \frac{\frac{5}{36}}{1 - \frac{25}{36}} = \frac{5}{11}$$

25. (6) Let x, y, z are probability of E_1, E_2 and E_3 respectively.

$\therefore \quad \alpha = x(1-y)(1-z)$ ...(i)

$\beta = (1-x) \cdot y(1-z)$ ...(ii)

$\gamma = (1-x)(1-y)z$ ...(iii)

$\Rightarrow \quad p = (1-x)(1-y)(1-z)$ ...(iv)

Given, $(\alpha - 2\beta)p = \alpha\beta$

and $(\beta - 3\gamma)p = 2\beta\gamma$ ...(v)

From above equations, $x = 2y$ and $y = 3z$

$\therefore \quad x = 6z \Rightarrow \frac{x}{z} = 6$

Last 3 Years' Questions
JEE Main & Advanced
(2016-2018)

JEE Main

2016

1. If all the words (with or without meaning) having five letters, formed using the letters of the word SMALL and arranged as in a dictionary, then the position of the word SMALL is **[Permutation and Combination]**

(a) 46th (b) 59th (c) 52nd (d) 58th

2. If the number of terms in the expansion of $\left(1-\frac{2}{x}+\frac{4}{x^2}\right)^n$, $x \neq 0$, is 28, then the sum of the coefficients of all the terms in this expansion, is **[Binomial Theorem]**

(a) 64 (b) 2187 (c) 243 (d) 729

3. Let two fair six-faced dice A and B be thrown simultaneously. If E_1 is the event that die A shows up four, E_2 is the event that die B shows up two and E_3 is the event that the sum of numbers on both dice is odd, then which of the following statements is not true? **[Probability]**

(a) E_1 and E_2 are independent (b) E_2 and E_3 are independent
(c) E_1 and E_3 are independent (d) E_1, E_2 and E_3 are independent

2017

4. For three events A, B and C,
if P (exactly one of A or B occurs) = P(exactly one of B or C occurs) = P (exactly one of C or A occurs) $=\frac{1}{4}$ and P (all the three events occur simultaneously) $=\frac{1}{16}$, then the probability that atleast one of the events occurs, is **[Probability]**

(a) $\frac{7}{32}$ (b) $\frac{7}{16}$ (c) $\frac{7}{64}$ (d) $\frac{3}{16}$

5. If two different numbers are taken from the set $\{0, 1, 2, 3, \ldots, 10\}$, then the probability that their sum as well as absolute difference are both multiple of 4, is **[Probability]**

(a) $\frac{6}{55}$ (b) $\frac{12}{55}$ (c) $\frac{14}{45}$ (d) $\frac{7}{55}$

6. A man X has 7 friends, 4 of them are ladies and 3 are men. His wife Y also has 7 friends, 3 of them are ladies and 4 are men. Assume X and Y have no common friends. Then, the total number of ways in which X and Y together can throw a party inviting 3 ladies and 3 men, so that 3 friends of each of X and Y are in this party, is **[Permutation and Combination]**

(a) 485 (b) 468 (c) 469 (d) 484

7. The value of $({}^{21}C_1 - {}^{10}C_1) + ({}^{21}C_2 - {}^{10}C_2) + ({}^{21}C_3 - {}^{10}C_3) + ({}^{21}C_4 - {}^{10}C_4) + \ldots + ({}^{21}C_{10} - {}^{10}C_{10})$ is **[Binomial Theorem]**

(a) $2^{21} - 2^{11}$ (b) $2^{21} - 2^{10}$ (c) $2^{20} - 2^9$ (d) $2^{20} - 2^{10}$

8. A box contains 15 green and 10 yellow balls. If 10 balls are randomly drawn one-by-one with replacement, then the variance of the number of green balls drawn is **[Probability]**

(a) $\frac{12}{5}$ (b) 6 (c) 4 (d) $\frac{6}{25}$

2018

9. From 6 different novels and 3 different dictionaries, 4 novels and 1 dictionary are to be selected and arranged in a row on a shelf, so that the dictionary is always in the middle. The number of such arrangements is **[Permutation and Combination]**

(a) atleast 1000 (b) less than 500

(c) atleast 500 but less than 750 (d) atleast 750 but less than 1000

10. The sum of the coefficients of all odd degree terms in the expansion of

$\left(x+\sqrt{x^3-1}\right)^5+\left(x-\sqrt{x^3-1}\right)^5, (x>1)$ is **[Binomial Theorem]**

(a) -1 (b) 0 (c) 1 (d) 2

11. A bag contains 4 red and 6 black balls. A ball is drawn at random from the bag, its colour is observed and this ball along with two additional balls of the same colour are returned to the bag. If now a ball is drawn at random from the bag, then the probability that this drawn ball is red, is **[Probability]**

(a) $\frac{3}{10}$ (b) $\frac{2}{5}$ (c) $\frac{1}{5}$ (d) $\frac{3}{4}$

JEE Advanced

2016

1. A debate club consists of 6 girls and 4 boys. A team of 4 members is to be selected from this club including the selection of a captain (from among these 4 members) for the team. If the team has to include atmost one boy, the number of ways of selecting the team is

[Permutation and Combination, Type 1]

(a) 380 (b) 320 (c) 260 (d) 95

2. A computer producing factory has only two plants T_1 and T_2. Plant T_1 produces 20% and plant T_2 produces 80% of the total computers produced. 7% of computers produced in the factory turn out to be defective. It is known that P (computer turns out to be defective, given that it is produced in plant T_1) $= 10P$ (computer turns out to be defective, given that it is produced in plant T_2), where $P(E)$ denotes the probability of an event E. A computer produced in the factory is randomly selected and it does not turn out to be defective. Then, the probability that it is produced in plant T_2, is

(a) $\frac{36}{73}$ (b) $\frac{47}{79}$

(c) $\frac{78}{93}$ (d) $\frac{75}{83}$

3. Let m be the smallest positive integer such that the coefficient of x^2 in the expansion of

$(1+x)^2+(1+x)^3+\ldots+(1+x)^{49}+(1+mx)^{50}$

is $(3n+1)\ {}^{51}C_3$ for some positive integer n. Then, the value of n is

[Binomial Theorem, Type 3]

Paragraph (Q. 4 and 5)

Football teams T_1 and T_2 have to play two games against each other. It is assumed that the outcomes of the two games are independent. The probabilities of T_1 winning, drawing and losing a game against T_2 are $\frac{1}{2}, \frac{1}{6}$ and $\frac{1}{3}$, respectively. Each team gets 3 points for a win, 1 point for a draw and 0 point for a loss in a game. Let X and Y denote the total points scored by teams T_1 and T_2, respectively, after two games. **[Probability]**

4. $P(X>Y)$ is

(a) $\frac{1}{4}$ (b) $\frac{5}{12}$ (c) $\frac{1}{2}$ (d) $\frac{7}{12}$

Note ***Type-1*** *Only one correct option,* ***Type-2*** *More then one correct option,* ***Type-3*** *Single digit Answer,* ***Type-4*** *Numerical value based*

5. $P(X=Y)$ is

(a) $\frac{11}{36}$
(b) $\frac{1}{3}$
(c) $\frac{13}{36}$
(d) $\frac{1}{2}$

2017

6. Let X and Y be two events such that $P(X)=\frac{1}{3}$, $P(X|Y)=\frac{1}{2}$ and $P(Y|X)=\frac{2}{5}$. Then

[Probability, Type 2]

(a) $P(Y)=\frac{4}{15}$
(b) $P(X'|Y)=\frac{1}{2}$
(c) $P(X\cup Y)=\frac{2}{5}$
(d) $P(X\cap Y)=\frac{1}{5}$

7. Words of length 10 are formed using the letters A, B, C, D, E, F, G, H, I, J. Let x be the number of such words where no letter is repeated; and let y be the number of such words where exactly one letter is repeated twice and no other letter is repeated. Then, $\frac{y}{9x}=$

[Permutation and Combination, Type 3]

8. Three randomly chosen nonnegative integers x, y and z are found to satisfy the equation $x+y+z=10$. Then the probability that z is even, is **[Probability, Type 1]**

(a) $\frac{1}{2}$ (b) $\frac{36}{55}$ (c) $\frac{6}{11}$ (d) $\frac{5}{11}$

9. Let $S=\{1,2,3,\ldots\ldots,9\}$. For $k=1,2,\ldots\ldots 5$, let N_k be the number of subsets of S, each containing five elements out of which exactly k are odd. Then $N_1+N_2+N_3+N_4+N_5=$

[Permutation and Combination, Type 1]

(a) 210 (b) 252 (c) 126 (d) 125

2018

10. The probability that, on the examination day, the student S_1 gets the previously allotted seat R_1, and NONE of the remaining students gets the seat previously allotted to him/her is

[Probability, Type 1]

(a) $\frac{3}{40}$ (b) $\frac{1}{8}$ (c) $\frac{7}{40}$ (d) $\frac{1}{5}$

11. For $i=1,2,3,4$, let T_i denote the event that the students S_i and S_{i+1} do NOT sit adjacent to each other on the day of the examination. Then, the probability of the event $T_1\cap T_2\cap T_3\cap T_4$ is

[Probability, Type 1]

(a) $\frac{1}{15}$
(b) $\frac{1}{10}$
(c) $\frac{7}{60}$
(d) $\frac{1}{5}$

12. Let X be a set with exactly 5 elements and Y be a set with exactly 7 elements. If α is the number of one-one functions from X to Y and β is the number of onto functions from Y to X, then the value of $\frac{1}{5!}(\beta-\alpha)$ is ....

[Permutation and Combination, Type 4]

13. Let $X=({}^{10}C_1)^2+2({}^{10}C_2)^2+3({}^{10}C_3)^2+\ldots+10({}^{10}C_{10})^2$, where ${}^{10}C_r$, $r\in\{1,2,\ldots,10\}$ denote binomial coefficients. Then, the value of $\frac{1}{1430}X$ is ....

[Binomial Theorem, Type 4]

14. In a high school, a committee has to be formed from a group of 6 boys $M_1, M_2, M_3, M_4, M_5, M_6$ and 5 girls G_1, G_2, G_3, G_4, G_5. **[Permutation and Combination]**

(i) Let α_1 be the total number of ways in which the committee can be formed such that the committee has 5 members, having exactly 3 boys and 2 girls.

(ii) Let α_2 be the total number of ways in which the committee can be formed such that the committee has at least 2 members, and having an equal number of boys and girls.

(iii) Let α_3 be the total number of ways in which the committee can be formed such that the committee has 5 members, at least 2 of them being girls.

(iv) Let α_4 be the total number of ways in which the committee can be formed such that the committee has 4 members, having at least 2 girls such that both M_1 and G_1 are NOT in the committee together.

	List-I		List-II
P.	The value of α_1 is	1.	136
Q.	The value of α_2 is	2.	189
R.	The value of α_3 is	3.	192
S.	The value of α_4 is	4.	200
		5.	381
		6.	461

The correct option is

(a) P $\rightarrow$ 4; Q $\rightarrow$ 6; R $\rightarrow$ 2; S $\rightarrow$ 1
(b) P $\rightarrow$ 1; Q $\rightarrow$ 4; R $\rightarrow$ 2; S $\rightarrow$ 3
(c) P $\rightarrow$ 4; Q $\rightarrow$ 6; R $\rightarrow$ 5; S $\rightarrow$ 2
(d) P $\rightarrow$ 4; Q $\rightarrow$ 2; R $\rightarrow$ 3; S $\rightarrow$ 1

Answer *with* Explanations

JEE Main

1. *(d)* Clearly, number of words start with A $= \frac{4!}{2!} = 12$

Number of words start with L $= 4! = 24$

Number of words start with M $= \frac{4!}{2!} = 12$

Number of words start with SA $= \frac{3!}{2!} = 3$

Number of words start with SL $= 3! = 6$

Note that, next word will be "SMALL".

Hence, the position of word "SMALL" is 58th.

2. *(d)* Clearly, number of terms in the expansion of

$\left(1 - \frac{2}{x} + \frac{4}{x^2}\right)^n$ is $\frac{(n+2)(n+1)}{2}$ or ${}^{n+2}C_2$.

[assuming $\frac{1}{x}$ and $\frac{1}{x^2}$ distinct]

$\therefore \quad \frac{(n+2)(n+1)}{2} = 28$

$\Rightarrow \quad (n+2)(n+1) = 56 = (6+1)(6+2)$

$\Rightarrow \quad n = 6$

Hence, sum of coefficients $= (1 - 2 + 4)^6 = 3^6 = 729$

Note *As $\frac{1}{x}$ and $\frac{1}{x^2}$ are functions of same variables, therefore number of dissimilar terms will be $2n+1$, i.e. odd, which is not possible. Hence, it contains error.*

3. *(d)* Clearly,

$E_1 = \{(4,1),(4,2),(4,3),(4,4),(4,5),(4,6)\}$

$E_2 = \{(1,2),(2,2),(3,2),(4,2),(5,2),(6,2)\}$ and

$E_3 = \{(1,2),(1,4),(1,6),(2,1),(2,3),(2,5),(3,2),(3,4),$
$(3,6),(4,1),(4,3),(4,5),(5,2),(5,4),(5,6),$
$(6,1),(6,3),(6,5)\}$

$\Rightarrow \quad P(E_1) = \frac{6}{36} = \frac{1}{6}$

$P(E_2) = \frac{6}{36} = \frac{1}{6}$ and $P(E_3) = \frac{18}{36} = \frac{1}{2}$

Now, $P(E_1 \cap E_2) = P$ (getting 4 on die A and 2 on die B)

$= \frac{1}{36} = P(E_1) \cdot P(E_2)$

$P(E_2 \cap E_3) = P$

(getting 2 on die B and sum of numbers on both dice is odd)

$= \frac{3}{36} = P(E_2) \cdot P(E_3)$

$P(E_1 \cap E_3) = P$

(getting 4 on die A and sum of numbers on both dice is odd)

$= \frac{3}{36} = P(E_1) \cdot P(E_3)$

and $P(E_1 \cap E_2 \cap E_3) = P$

[getting 4 on die A, 2 on die B and sum of numbers is odd]

$= P$ (impossible event) $= 0$

Hence, E_1, E_2 and E_3 are not independent.

4. *(b)* We have, P (exactly one of A or B occurs)

$= P(A \cup B) - P(A \cap B)$

$= P(A) + P(B) - 2P(A \cap B)$

According to the question,

$P(A) + P(B) - 2P(A \cap B) = \frac{1}{4}$ ...(i)

$P(B) + P(C) - 2P(B \cap C) = \frac{1}{4}$ ...(ii)

and $P(C) + P(A) - 2P(C \cap A) = \frac{1}{4}$ ...(iii)

On adding Eqs. (i), (ii) and (iii), we get

$$2[P(A)+P(B)+P(C)-P(A\cap B)-P(B\cap C)-P(C\cap A)]=\frac{3}{4}$$

$$\Rightarrow P(A)+P(B)+P(C)-P(A\cap B)-P(B\cap C)-P(C\cap A)=\frac{3}{8}$$

$\therefore$ P (atleast one event occurs)

$$= P(A\cup B\cup C)$$
$$= P(A)+P(B)+P(C)-P(A\cap B)-P(B\cap C)-P(C\cap A)+P(A\cap B\cap C)$$
$$=\frac{3}{8}+\frac{1}{16}=\frac{7}{16}\qquad\left[\because P(A\cap B\cap C)=\frac{1}{16}\right]$$

5. *(a)* Total number of ways of selecting 2 different numbers from $\{0, 1, 2, ..., 10\} = {}^{11}C_2 = 55$

Let two numbers selected be x and y.

Then, $x+y=4m$ ...(i)

and $x-y=4n$ ...(ii)

$\Rightarrow$ $2x=4(m+n)$ and $2y=4(m-n)$

$\Rightarrow$ $x=2(m+n)$

and $y=2(m-n)$

$\therefore x$ and y both are even numbers.

x	y
0	4, 8
2	6, 10
4	0, 8
6	2, 10
8	0, 4
10	2, 6

$\therefore$ Required probability $=\frac{6}{55}$

6. *(a)* Given, X has 7 friends, 4 of them are ladies and 3 are men while Y has 7 friends, 3 of them are ladies and 4 are men.

$\therefore$ Total number of required ways

$$= {}^3C_3\times{}^4C_0\times{}^4C_0\times{}^3C_3+{}^3C_2\times{}^4C_1\times{}^4C_1\times{}^3C_2+{}^3C_1\times{}^4C_2\times{}^4C_2\times{}^3C_1+{}^3C_0\times{}^4C_3\times{}^4C_3\times{}^3C_0$$

$$=1+144+324+16=485$$

7. *(d)* $({}^{21}C_1-{}^{10}C_1)+({}^{21}C_2-{}^{10}C_2)+({}^{21}C_3-{}^{10}C_3)+...+({}^{21}C_{10}-{}^{10}C_{10})$

$$=({}^{21}C_1+{}^{21}C_2+...+{}^{21}C_{10})-({}^{10}C_1+{}^{10}C_2+...+{}^{10}C_{10})$$
$$=\frac{1}{2}({}^{21}C_1+{}^{21}C_2+...+{}^{21}C_{20})-(2^{10}-1)$$
$$=\frac{1}{2}({}^{21}C_1+{}^{21}C_2+...+{}^{21}C_{21}-1)-(2^{10}-1)$$
$$=\frac{1}{2}(2^{21}-2)=2^{20}-1-2^{10}+1-(2^{10}-1)$$
$$=2^{20}-2^{10}$$

8. *(a)* Given box contains 15 green and 10 yellow balls.

$\therefore$ Total number of balls $=15+10=25$

$$P(\text{green balls})=\frac{15}{25}=\frac{3}{5}=p=\text{Probability of success}$$
$$P(\text{yellow balls})=\frac{10}{25}=\frac{2}{5}=q=\text{Probability of unsuccess}$$

and $n=10=$ Number of trials.

$$\therefore\quad \text{Variance}=npq=10\times\frac{3}{5}\times\frac{2}{5}=\frac{12}{5}$$

9. *(a)* Given 6 different novels and 3 different dictionaries.

Number of ways of selecting 4 novels from 6 novels is

$${}^6C_4=\frac{6!}{2!4!}=15$$

Number of ways of selecting 1 dictionary is from 3 dictionaries is ${}^3C_1=\frac{3!}{1!2!}=3$

$\therefore$ Total number of arrangement of 4 novels and 1 dictionary where dictionary is always in the middle, is

$$15\times3\times4!=45\times24=1080$$

10. *(d)* **Key idea** $=(a+b)^n+(a-b)^n$

$$=2({}^nC_0a^n+{}^nC_2a^{n-2}b^2+{}^nC_4a^{n-4}b^4...)$$

We have, $(x+\sqrt{x^3-1})^5+(x-\sqrt{x^3-1})^5, x>1$

$$=2({}^5C_0x^5+{}^5C_2x^3(\sqrt{x^3-1})^2+{}^5C_4x(\sqrt{x^3-1})^4)$$
$$=2(x^5+10x^3(x^3-1)+5x(x^3-1)^2)$$
$$=2(x^5+10x^6-10x^3+5x^7-10x^4+5x)$$

Sum of coefficients of all odd degree terms is $2(1-10+5+5)=2$

11. *(b)* Key idea *Use the theorem of total probability*

Let E_1 = Event that first ball drawn is red

E_2 = Event that first ball drawn is black

A = Event that second ball drawn is red

$$P(E_1)=\frac{4}{10},\ P\left(\frac{A}{E_1}\right)=\frac{6}{12}$$
$$\Rightarrow\quad P(E_2)=\frac{6}{10},\ P\left(\frac{A}{E_2}\right)=\frac{4}{12}$$

By law of total probability

$$P(A)=P(E_1)\times P\left(\frac{A}{E_1}\right)+P(E_2)\times P\left(\frac{A}{E_2}\right)$$
$$=\frac{4}{10}\times\frac{6}{12}+\frac{6}{10}\times\frac{4}{12}=\frac{24+24}{120}=\frac{48}{120}=\frac{2}{5}$$

JEE Advanced

1. *(a)* We have, 6 girls and 4 boys. To select 4 members (atmost one boy)

i.e. (1 boy and 3 girls) or (4 girls) $={}^6C_3\cdot{}^4C_1+{}^6C_4$ ...(i)

Now, selection of captain from 4 members $={}^4C_1$ ...(ii)

$\therefore$ Number of ways to select 4 members (including the selection of a captain, from these 4 members)

$$=({}^6C_3\cdot{}^4C_1+{}^6C_4){}^4C_1$$
$$=(20\times4+15)\times4=380$$

2. (c) Let $x = P$ (computer turns out to be defective, given that it is produced in plant T_2)

$\Rightarrow \quad x = P\left(\frac{D}{T_2}\right) \quad$ ...(i)

where, $D =$ Defective computer

$\therefore$ P (computer turns out to be defective given that is produced in plant T_1) $= 10x$

i.e. $\quad P\left(\frac{D}{T_1}\right) = 10x \quad$ ...(ii)

Also, $P(T_1) = \frac{20}{100}$ and $P(T_2) = \frac{80}{100}$

Given, P (defective computer) $= \frac{7}{100}$

i.e. $\quad P(D) = \frac{7}{100}$

Using law of total probability,

$$P(D) = 9(T_1)\cdot P\left(\frac{D}{T_1}\right) + P(T_2)\cdot P\left(\frac{D}{T_2}\right)$$

$$\therefore \quad \frac{7}{100} = \left(\frac{20}{100}\right)\cdot 10x + \left(\frac{80}{100}\right)\cdot x$$

$$\Rightarrow \quad 7 = (280)x$$

$$\Rightarrow \quad x = \frac{1}{40} \quad \text{...(iii)}$$

$$\therefore \quad P\left(\frac{D}{T_2}\right) = \frac{1}{40} \text{ and } P\left(\frac{D}{T_1}\right) = \frac{10}{40}$$

$$\Rightarrow \quad P\left(\frac{\overline{D}}{T_2}\right) = 1 - \frac{1}{40} = \frac{39}{40}$$

$$\text{and} \quad P\left(\frac{\overline{D}}{T_1}\right) = 1 - \frac{10}{40} = \frac{30}{40} \quad \text{...(iv)}$$

Using Baye's theorem,

$$P\left(\frac{T_2}{\overline{D}}\right) = \frac{P(T_2 \cap \overline{D})}{P(T_1 \cap \overline{D}) + P(T_2 \cap \overline{D})}$$

$$= \frac{P(T_2)\cdot P\left(\frac{\overline{D}}{T_2}\right)}{P(T_1)\cdot P\left(\frac{\overline{D}}{T_1}\right) + P(T_2)\cdot P\left(\frac{\overline{D}}{T_2}\right)}$$

$$= \frac{\frac{80}{100}\cdot\frac{39}{40}}{\frac{20}{100}\cdot\frac{30}{40} + \frac{80}{100}\cdot\frac{39}{40}} = \frac{78}{93}$$

3. (5) Coefficient of x^2 in the expansion of

$$\{(1+x)^2 + (1+x)^3 + \ldots + (1+x)^{49} + (1+mx)^{50}\}$$

$$\Rightarrow \quad {}^2C_2 + {}^3C_2 + {}^4C_2 + \ldots + {}^{49}C_2 + {}^{50}C_2\cdot m^2 = (3n+1)\cdot{}^{51}C_3$$

$$\Rightarrow \quad {}^{50}C_3 + {}^{50}C_2 m^2 = (3n+1)\cdot{}^{51}C_3$$

$$[\because {}^rC_r + {}^{r+1}C_r + \ldots + {}^nC_r = {}^{n+1}C_{r+1}]$$

$$\Rightarrow \quad \frac{50\times 49\times 48}{3\times 2\times 1} + \frac{50\times 49}{2}\times m^2$$

$$= (3n+1)\frac{51\times 50\times 49}{3\times 2\times 1}$$

$$\Rightarrow \quad m^2 = 51n + 1$$

$\therefore$ Minimum value of m^2 for which $(51n+1)$ is integer (perfect square) for $n = 5$.

$$\therefore \quad m^2 = 51\times 5 + 1$$

$$\Rightarrow \quad m^2 = 256$$

$$\therefore \quad m = 16$$

$$\text{and} \quad n = 5$$

Hence, the value of n is 5.

4. (b) Here, $P(X > Y) = P(T_1\text{win})\,P(T_1 \text{ win})$

$$+ P(T_1 \text{ win})\,P(\text{draw}) + P(\text{draw})\,P(T_1 \text{ win})$$

$$= \left(\frac{1}{2}\times\frac{1}{2}\right) + \left(\frac{1}{2}\times\frac{1}{6}\right) + \left(\frac{1}{6}\times\frac{1}{2}\right) = \frac{5}{12}$$

5. (c) $P[X = Y] = P(\text{draw})\cdot P(\text{draw})$

$$+ P(T_1 \text{ win})\,P(T_2 \text{ win}) + P(T_2 \text{ win})\cdot P(T_1 \text{ win})$$

$$= (1/6\times 1/6) + (1/2\times 1/3)$$

$$+ (1/3\times 1/2) = 13/36$$

6. (a,b) $P(X) = \frac{1}{3}$

$$P\left(\frac{X}{Y}\right) = \frac{P(X\cap Y)}{P(Y)} = \frac{1}{2}$$

$$P\left(\frac{Y}{X}\right) = \frac{P(X\cap Y)}{P(X)} = \frac{2}{5}$$

$$P(X\cap Y) = \frac{2}{15}$$

$$P(Y) = \frac{4}{15}$$

$$P\left(\frac{X'}{Y}\right) = \frac{P(Y) - P(X\cap Y)}{P(Y)}$$

$$= \frac{\frac{4}{15} - \frac{2}{15}}{\frac{4}{15}} = \frac{1}{2}$$

$$P(X\cup Y) = \frac{1}{3} + \frac{4}{15} - \frac{2}{15}$$

$$= \frac{7}{15} = \frac{7}{15}$$

7. (5) $x = 10!$

$$y = {}^{10}C_1 \times {}^9C_8 \times \frac{10!}{2!}$$

$$= 10\times 9\times\frac{10!}{2}$$

$$\frac{y}{9x} = \frac{10}{2} = 5$$

Solutions. (Q.Nos. 13-15)

Using standard equations

I-ii-Q

II-iv-R

III-i-P

IV-iii-S

Comparing with equations in given questions, we get

8. (c) Sample space $\rightarrow {}^{12}C_2$

Number of possibilities for z is even.

$$z = 0 \Rightarrow {}^{11}C_1$$

$$z = 2 \Rightarrow {}^9C_1$$

$$z = 4 \Rightarrow {}^7C_1$$
$$z = 6 \Rightarrow {}^5C_1$$
$$z = 8 \Rightarrow {}^3C_1$$
$$z = 10 \Rightarrow {}^1C_1$$

Total = 36

$\therefore$ Probability $= \frac{36}{66} = \frac{6}{11}$

9. (c) $N_i = {}^5C_k \times {}^4C_{5-k}$

$$N_1 = 5 \times 1$$
$$N_2 = 10 \times 4$$
$$N_3 = 10 \times 6$$
$$N_4 = 5 \times 4$$
$$N_5 = 1$$
$$N_1 + N_2 + N_3 + N_4 + N_5 = 126$$

10. (a) Here, five students S_1, S_2, S_3, S_4 and S_5 and five seats R_1, R_2, R_3, R_4 and R_5

$\therefore$ Total number of arrangement of sitting five students is $5! = 120$

Here, S_1 gets previously alloted seat R_1

$\therefore S_2, S_3, S_4$ and S_5 not get previously seats.

Total number of way S_2, S_3, S_4 and S_5 not get previously seats is

$$4!\left(1 - \frac{1}{1!} + \frac{1}{2!} - \frac{1}{3!} + \frac{1}{4!}\right) = 24\left(1 - 1 + \frac{1}{2} - \frac{1}{6} + \frac{1}{24}\right)$$
$$= 24\left(\frac{12 - 4 + 1}{24}\right) = 9$$

$\therefore$ Required probability $= \frac{9}{120} = \frac{3}{40}$

11. (c) Here, $n(T_1 \cap T_2 \cap T_3 \cap T_4) = \text{Total} - n(\overline{T_1} \cup \overline{T_2} \cup \overline{T_3} \cup \overline{T_4})$

$$\Rightarrow n(T_1 \cap T_2 \cap T_3 \cap T_4)$$
$$= 5! - [{}^4C_1\, 4!\, 2! - ({}^3C_1\, 3!\, 2! + {}^3C_1\, 3!\, 2!\, 2!) + ({}^2C_1\, 2!\, 2! + {}^4C_1\, 2 \cdot 2!) - 2]$$
$$\Rightarrow n(T_1 \cap T_2 \cap T_3 \cap T_4)$$
$$= 120 - [192 - (36 + 72) + (8 + 16) - 2]$$
$$= 120 - [192 - 108 + 24 - 2] = 14$$

$\therefore$ Required probability $= \frac{14}{120} = \frac{7}{60}$

12. (119) Given, X has exactly 5 elements and Y has exactly 7 elements.

$\therefore$ $n(X) = 5$

and $n(Y) = 7$

Now, number of one-one functions from X to Y is

$$\alpha = {}^7P_5 = {}^7C_5 \times 5!$$

Number of onto functions from Y to X is β

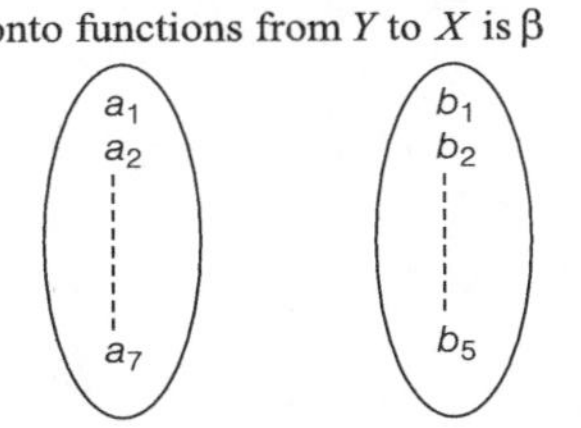

1, 1, 1, 1, 3 or 1, 1, 1, 2, 2

$$\therefore \quad \beta = \frac{7!}{3!\,4!} \times 5! + \frac{7!}{(2!)^3 3!} \times 5!$$
$$= ({}^7C_3 + 3\ {}^7C_3)\, 5! = 4 \times {}^7C_3 \times 5!$$
$$\therefore \quad \frac{\beta - \alpha}{5!} = \frac{(4 \times {}^7C_3 - {}^7C_5)\, 5!}{5!}$$
$$= 4 \times 35 - 21 = 140 - 21 = 119$$

13. (646) We have,

$$X = ({}^{10}C_1)^2 + 2({}^{10}C_2)^2 + 3({}^{10}C_3)^2 + \ldots + 10\,({}^{10}C_{10})^2$$
$$\Rightarrow \quad X = \sum_{r=1}^{10} r({}^{10}C_r)^2$$
$$\Rightarrow \quad X = \sum_{r=1}^{10} r\ {}^{10}C_r\ {}^{10}C_r$$
$$\Rightarrow \quad X = \sum_{r=1}^{10} r \times \frac{10}{r}\ {}^9C_{r-1}\ {}^{10}C_r$$
$$\left[\because {}^nC_r = \frac{n}{r}\ {}^{n-1}C_{r-1}\right]$$
$$\Rightarrow \quad X = 10 \sum_{r=1}^{10} {}^9C_{r-1}\ {}^{10}C_r$$
$$\Rightarrow \quad X = 10 \sum_{r=1}^{10} {}^9C_{r-1}\ {}^{10}C_{10-r}$$
$$[\because {}^nC_r = {}^nC_{n-r}]$$
$$\Rightarrow \quad X = 10 \times {}^{19}C_9$$
$$[\because {}^{n-1}C_{r-1}\,{}^nC_{n-r} = {}^{2n-1}C_{n-1}]$$

Now, $\frac{1}{1430} X = \frac{10 \times {}^{19}C_9}{1430} = \frac{{}^{19}C_9}{143} = \frac{{}^{19}C_9}{11 \times 13}$

$$= \frac{19 \times 17 \times 16}{8} = 19 \times 34 = 646$$

14. (c) Given 6 boys $M_1, M_2, M_3, M_4, M_5, M_6$ and 5 girls G_1, G_2, G_3, G_4, G_5

(i) $\alpha_1 \rightarrow$ Total number of ways of selecting 3 boys and 2 girls from 6 boys and 5 girls.

i..e, ${}^6C_3 \times {}^5C_2 = 20 \times 10 = 200$

$\therefore$ $\alpha_1 = 200$

(ii) $\alpha_2 \rightarrow$ Total number of ways selecting at least 2 member and having equal number of boys and girls

i.e., ${}^6C_1{}^5C_1 + {}^6C_2{}^5C_2 + {}^6C_3{}^5C_3 + {}^6C_4{}^5C_4 + {}^6C_5{}^5C_5$

$= 30 + 150 + 200 + 75 + 6 = 461$

$\Rightarrow$ $\alpha_2 = 461$

(iii) $\alpha_3 \rightarrow$ Total number of ways of selecting 5 members in which at least 2 of them girls

i.e., ${}^5C_2{}^6C_3 + {}^5C_3{}^6C_2 + {}^5C_4{}^6C_1 + {}^5C_5{}^6C_0$

$= 200 + 150 + 30 + 1 = 381$

$\alpha_3 = 381$

(iv) $\alpha_4 \rightarrow$ Total number of ways of selecting 4 members in which at least two girls such that M_1 and G_1 are not included together.

G_1 is included $\rightarrow {}^4C_1 \cdot {}^5C_2 + {}^4C_2 \cdot {}^5C_1 + {}^4C_3$

$= 40 + 30 + 4 = 74$

M_1 is included $\rightarrow$ ${}^4C_2 \cdot {}^5C_1 + {}^4C_3 = 30 + 4 = 34$

G_1 and M_1 both are not included ${}^4C_4 + {}^4C_3 \cdot {}^5C_1 + {}^4C_2 \cdot {}^5C_2$

$1 + 20 + 60 = 81$

$\therefore$ Total number $= 74 + 34 + 81 = 189$

$\alpha_4 = 189$

Now, $P \rightarrow 4$; $Q \rightarrow 6$; $R \rightarrow 5$; $S \rightarrow 2$

Hence, option (c) is correct.